Principles of

ANGIOSPERM TAXONOMY

Principles of ANGIOSPERM TAXONOMY

P. H. DAVIS, Ph.D., D.Sc.
University of Edinburgh

V. H. HEYWOOD, Ph.D., D.Sc.
University of Reading

Reprinted in India, 2022
by
Scientific Publishers (India)
5-A, New Pali Road, P.O. Box 91
JODHPUR - 342 001

E-mail: info@scientificpub.com
www.scientificpubonline.com

ISBN: 978-81-7233-725-4

Printed in India

PREFACE

In writing a textbook on the principles of taxonomy, we have aimed at providing advanced students with a comprehensive survey of the subject. The absence of such a work has, we believe, inconvenienced both student and teacher alike. Although the expanding horizons of the subject may have proved daunting enough, another major factor which must have deterred authors is that few other subjects are so open to serious controversy: on nearly every major topic there are opposing schools of thought. Until recently the principles of taxonomy have been regarded as a closed secret, only to be communicated to the initiated after a period of apprenticeship. However difficult it may be to set these principles down, and however controversial some of them may be, we are convinced that the effort needs to be made. While we do not, of course, expect that other taxonomists will agree with us on all points, we hope that our attempts at communication will shed some light and stimulate others to clarify the subject further. We have, in fact, been much encouraged in our task by the publication of G. G. Simpson's *Principles of Animal Taxonomy* which appeared after much of our book had been drafted. Although we would take issue with him on several important points, we are indebted to his important text for the clarification of several of the ideas expressed in our book. L. Benson's *Plant Taxonomy: Methods and Principles*, another major work, appeared after most of our text had gone to press.

Two significant developments in taxonomy have unfolded with almost unprecedented rapidity during the last two or three years, while our book has been in preparation. These are the fields known as numerical taxonomy and chemical taxonomy respectively. Although we have attempted to give an adequate survey of both these fields, their developments have been so spectacular that no account can be expected to be up to date for long.

In the course of this book we have drawn attention to and criticised many instances of unsatisfactory taxonomic practice. We ourselves claim no exemption from these criticisms and have probably committed many of these errors (and others not mentioned!) in our own taxonomic work. We hope our criticisms will be accepted in the constructive light in which they are offered.

We have found it necessary to draw on evidence from many and diverse fields impinging on taxonomy, fields in which we can claim no special experience. We have, however, been fortunate in receiving much help and criticism from the specialists mentioned below. If we have failed to appreciate their advice or misinterpreted it, the fault is entirely our own, and those who have helped us should not be assumed to agree necessarily with the views expressed in this book.

The literature of taxonomy is the largest in biology and extends back for over two centuries, so that our bibliography, although extensive, covers only a small fraction of the field. Wherever possible we have cited review papers or ones which give further guidance to research papers.

Acknowledgments

We owe a special debt of gratitude to Dr. W. T. Stearn (London) who has critically read Chapters 1, 2, 3, 8 and 9 in manuscript and in proof. His assistance with the sections on the history of taxonomy, typification and nomenclature has been particularly valuable. Dr. J. Cullen (Edinburgh) has read the complete book in manuscript and in proof, and has suggested many improvements and clarifications. He has also provided us with some of the original diagrams and has helped in many other ways. Mr. B. L. Burtt (Edinburgh) has also read all the galley proofs and pointed out many ways to improve the text. Professor D. H. Valentine (Durham) has read Chapters 6, 11 and 13; Dr. C. D. K. Cook (Liverpool) Chapters 6, 10 and 11 and has assisted greatly with the section on phenotypic plasticity in aquatic plants; Dr. A. F. Dyer (Edinburgh) has assisted with Chapter 6 and has provided us with information on heterochromatin; Dr. J. W. Gregor (Pentlandfield) has read the proofs of Chapters 12, 13 and 14 and offered much constructive help; Dr. J. McNeill (Liverpool) has read Chapters 8 and 9 and assisted with nomenclature and presentation of data; Professor A. R. Battersby, Dr. F. Dean and Dr. H. M. Hurst (Liverpool) and Dr. A. P. Brown (Edinburgh) have read Chapter 7; Dr. A. Manning (Edinburgh) has given much help with the pollination section in Chapter 13. We are indebted to Mr. P. McIntyre for compiling the index. Others who have helped in reading parts of the manuscript, or in other ways, include Dr. C. E. Hubbard (Kew), Mr. M. J. E. Coode (Edinburgh), Dr. P. W. Ball (Liverpool), Mr. D. M. Henderson (Edinburgh), Dr. J. A. Ratter (Edinburgh), Mr. John Lewis (London), Dr. S. M. Walters (Cambridge) and Mr. W. H. Brown (Edinburgh).

The courtesy of the following Societies, Publishers, Editors and others is acknowledged for permission to reproduce the following text figures borrowed from these published sources: The Linnean Society of London (figs. 2, 27, 29, 30), Libraire de l'Université, Genève (fig. 15), Cambridge Philosophical Society (figs. 7, 41, 42), New York Botanical Garden (fig. 8), The Systematics Association (figs. 18, 33), The Botanical Society of the British

Isles (fig. 25), Institute of Physiological Botany of the University of Uppsala (fig. 26), The Genetics Society of Canada (fig. 28), The Society for the Study of Evolution (figs. 32, 35), Carnegie Institute of Washington (figs. 31, 36), American Philosophical Society (figs. 37, 38), Long Island Biological Association (fig. 40), The Royal Horticultural Society (fig. 9), Columbia University Press (fig. 10), Professor P. Maheshwari (fig. 4), Professor A. Fernandes (fig. 16), Dr. W. T. Stearn (fig. 20), Messrs. D. C. Heath & Co. (fig. 6), University of California Press (fig. 17), The Controller of H.M. Stationery Office (fig. 21), Wilhelm Gronau Verlag (fig. 22), Messrs. Springer-Verlag (fig. 23), *The American Midland Naturalist* (fig. 14), *The Journal of General Microbiology* (figs. 11, 12, 13), Ronald Press Co. (fig. 39). Credits for figures have been scrupulously assigned in the accompanying legends, although we have not in every case been able to contact their authors for permission. To them our thanks are also due.

The manuscript has been expertly typed by Miss M. Weldhen, Mrs. P. Lawson and Mrs. R. Marshall.

Our final word of thanks is due to Messrs. Oliver & Boyd for their kindness and patience and to Mr. T. L. Jenkins for seeing the book through the press.

Edinburgh
Liverpool } *May*, 1963.

CONTENTS

ILLUSTRATIONS

INTRODUCTION

The publication of another textbook on the taxonomy of flowering plants calls for some words of explanation. The reader may wonder whether he will find here anything that has not been offered several times before. The aim of this introduction is to outline what we have attempted in this book and equally to indicate what we have not set out to do. This should make it evident in what ways our approach to the subject differs from that normally adopted.

The authors are practising taxonomists who have had to face the problems of teaching taxonomy to their students. A large selection of books is available which deals with family characters and phylogeny, with the history of classification itself, even with biosystematics and micro-evolution. But we have not found a book which deals with the *principles* of classification (as we understand them) on which the methodology and systems of classification should be based, or with the problems of using biosystematic information in practical taxonomy. Some excellent works tell us how outdated or inadequate the methods and aims of classical taxonomy are, but little attempt has been made since the nineteenth century to explain what these aims and methods should be! Indeed, much of the pedagogic literature of taxonomy either credits the reader with little intelligence or deals with an imaginary world where taxonomists devote their time to remodelling phylogenetic systems of classification or debating the various carpel theories. Such activities certainly occupy the attention of most taxonomists to some extent, but form only a very small part of current taxonomy which is more concerned with problems at the species level and below and, to a lesser extent, with the circumscription of genera.

In the herbarium, taxonomy can be (and often is) taught by imitation. Plants are classified by largely intuitive processes. But one cannot have confidence in a classification unless one has confidence in the principles on which it is based, and one certainly cannot explain taxonomy to others unless those principles are understood and convincingly expressed. In post-Darwinian times a conspiracy of silence has surrounded the principles of taxonomy. They have too often been ignored, taken for granted or misunderstood. Some scientists would even be surprised to learn that taxonomy (and taxonomists) can have any principles! It is the object of this book to seek out the principles and methodology of plant classification and to reconcile the reader to the ambivalent nature of taxonomy: a meeting ground of Science and Art.

In this book we have tried to survey a very wide field in taxonomy, from its roots to the present day, keeping always in mind the principles of classification and their practical application. This has led us to consider many disciplines besides morphology, and we hope that the extensive bibliography provided (as up to date as we can make it) will enable the reader to go more deeply into any promising field of development that interests him. We propose no new systems of classification, describe no families—there are plenty of books dealing with those topics. Taxonomy, we believe, is indivisible: we cannot accept a rigid separation of experimental taxonomy from "orthodox" herbarium taxonomy; the first is a logical and necessary extension of the second, although it may produce data of little classificatory value. A symbiotic relationship should exist between the two, but the nature of this relationship has not been adequately studied.

We have tried to find out how classification has been arrived at in the past, and how we believe it should be done today. What is applicable in other groups is not necessarily so in the Angiosperms. This has brought us in conflict with what is still the dominant school in taxonomy. Classification, many taxonomists claim, should be based on, or reflect, phylogeny. This aim, we believe, is unrealistic in a group with an extremely inadequate fossil record. As an initial procedural step, classification *must* be based on overall resemblances, and though it may then be interpreted in phylogenetic terms it cannot itself be based on phylogeny because too little evidence exists. Indeed, the whole conception of phylogenetic classification is, we maintain, a mistake except around the species level in favourable and well-studied groups; and even there phyletic relationship often conflicts with genetic relationship as expressed by phenotypic resemblances.

Classification should not be inconsistent with what evolutionary evidence is known, although there may be instances where it may be more convenient and serve more purposes if it is. Even the evolutionary principle of monophylesis of taxonomic groups may be incompatible with the purposes of general classification. We uphold the view that more than one kind of classification is possible and acceptable, and that no one can be judged best for all purposes. A classification based on overall resemblances is, however, more likely to be serviceable for more purposes than one based on known or assumed "phylogenetic" evidence. The two may frequently coincide in practice.

The opposing view, that phylogeny can and should provide criteria for classification is one with which most biologists would sympathise, but merely *willing* phylogeny will not provide the evidence on which to proceed. The thesis that the only reputable, scientific classification is that based on phylogenetic or evolutionary principles is one that cannot be accepted for the Angiosperms on either theoretical or practical grounds. The fact that many people hold this conviction very strongly does not alter the position.

The reader should not be misled into thinking that we suggest ignoring evolution or evolutionary principles. It is in fact precisely because we are impressed by such evolutionary phenomena as convergence and parallelism that we advocate the production of general classifications based on maximum number of attributes and that they be accepted as such, in their own right, and not necessarily as reflections of phylogeny.

Though the logic of this reasoning seems obvious to us, it has not seemed so to many others. It is only in the last few years that voices have been raised to question many of the generally accepted tenets of classification, particularly those concerned with phylogeny. We do not expect everyone to agree with our views, and if disagreement leads to heart-searchings and a re-investigation of the facts and their interpretation, then taxonomy should benefit by the airing. Classification has been too much taken for granted; it needs to be thought about, worried and thrashed out not only by those who talk about taxonomy but by those who practise it.

At the species level there is a similar conflict between those who wish to stress genetic and cytological criteria so as to construct taxa based on evolutionary and crossability relationship, and those who regard the morphological differentiation of groups as a prime criterion, while not ignoring other kinds of evidence. The impractability of defining species in genetic terms is now widely acknowledged, but the influence of the past few decades, when it appeared that the "biological" definition of species would generally replace the taxonomic one, still persists.

At the present time taxonomy is in a particularly "interesting condition". It is also in a rather critical state, passing through a phase of transition. An enormous amount of information (morphological, anatomical, eco-geographical, cytogenetic, chemical, etc.) is becoming available to taxonomists and requires interpretation and assimilation. The limitations of much of this evidence have yet to be assessed, and we have to guard against the danger of assuming that what is new or more difficult to obtain is necessarily a better basis for classification. How (if we can) shall we synthesise this *embarras de richesses* into classification and use it for evolutionary interpretation? This is the problem that faces us now, and it is one that is important not only to systematists but to biology in general, which taxonomy must serve as helpfully as it can. What kind of classification we shall have in the future will be greatly influenced by the principles which are adopted now.

The role of taxonomy in the teaching of biology is an important one. At a symposium on the subject held in Birmingham (England) recently, it was forcibly stated that taxonomy remains the principal agency through which the student may gain acquaintance with the diversity of organisms, the patterns of variation in the living world and of the evolutionary mechanisms which have brought these patterns about. The taxonomic approach is a focal point for students of biology and enables them to fit together into a framework of ideas a mass of otherwise unco-ordinated facts. It is unfortunate that failure to recognise these unifying aspects of taxonomy has led to the restricted role now often accorded to systematics in biological training.

Chapter 1

THE ROOTS OF TAXONOMY

"The urge to classify is a fundamental human instinct; like the predisposition to sin, it accompanies us into the world at birth and stays with us to the end."

A. TINDELL HOPWOOD, 1959

THE NEED FOR CLASSIFICATION

Classification is a natural occupation of man, recognition one of his most necessary pastimes. We have no choice but to classify and identify—they are essential processes in our daily lives. We need to know whether we are talking about the same or different organisms, and to what group they belong. "A plant's name is the key to its literature"—in other words, the key to what we know about it (van Steenis, 1957). The classification of organisms is not only meant for biologists, still less for taxonomists—it has to serve different purposes and to be used by different kinds of people, with different interests and with different capabilities and powers of perception. Classification is intended as much for other people as for ourselves, so that if it is not practical it will not be generally used. There is no better system to use today for the storage and retrieval of data than that provided by taxonomists (Rogers & Tanimoto, 1960).

Taxonomy is basic to other sciences and at the same time dependent on them. In this respect it holds a unique position. It has become a synthetic science that benefits from research in a very wide range of subjects both in the organic and the inorganic world. In this sense taxonomy is frankly a lagging science because it depends for its improvement, and indeed for its very existence, entirely on information from other fields—it has no data of its own.

At the purely practical level, biologists must know what organisms they are working with before they can pass on information about them to other people—a function of taxonomy which makes stability of nomenclature an important consideration. Certain disciplines of biology are, of course, so closely tied up with taxonomy that they cannot be practised without the use of a basic classification. These include cytology, genetics, ecology (itself the ultimate science of synthesis), phytosociology, comparative anatomy, palynology, palaeobotany and plant geography. In any of these, if the identification is dubious or wrong (within the framework of the kind of classification being used), the value of the work is greatly diminished and in many cases worthless. Experiments cannot be repeated unless the organisms used are correctly identified. Every biologist, whether he realises it or not, works with species, and his findings—even on the molecular level—may be decisively influenced by the choice of a particular species (Mayr, 1957).

In most of the disciplines mentioned above, the link with taxonomy is concerned not only with identification but also with relationships. In the applied sciences of plant-breeding, agriculture, horticulture, forestry and pharmacology the bond is particularly close and often of great economic importance. Classification must therefore provide a sound basis for the biological sciences not only in providing an accurate name (itself the function of nomenclature) but also in expressing natural relationships. The taxonomist, in fact, has serious responsibilities towards society, and it is up to him to see that he classifies plants in the most generally useful manner. A sound classification, and the information that goes with it, can suggest a host of fundamental biological problems that deserve investigation by biologists in numerous fields.

The Aims of Taxonomy

There are two main approaches in taxonomy today:

1. *An empirical approach*, based on the observed facts of character distribution. This results in a basic classification.

2. *An interpretative approach*, in which this classification is improved and interpreted in evolutionary or phylogenetic terms.

In practice these two approaches may overlap and intermingle, but an already existing classification must precede its evolutionary interpretation. New data may lead to an improvement in both approaches, but particularly in the field of evolutionary interpretation.

Among the many declared aims put forward for modern taxonomy, the following three seem widely accepted:

1. *To provide a convenient method of identification and communication.*
2. *To provide a classification which as far as possible expresses the natural relationships of organisms* (see p. 5).
3. *To detect evolution at work, discovering its processes and interpreting its results.*

Although these aims are complementary, they are not always easily reconciled. The first two mainly involve an empirical approach, while the third is concerned with phylogenetic interpretation. Taxonomists have gone a long way towards achieving the first and second aim; but the third has only been achieved in a few groups, and there mainly at the species level. At the higher levels of the hierarchy there is virtually no direct evidence for phylogeny.

When one has hacked one's way through the verbiage, one has gone considerably less far on the third road than the forest of phylogenetic trees might lead one to expect.

Our knowledge of the world's flora is extremely uneven and for this reason its classification can be considered in four overlapping phases (cf. Valentine & Löve, 1958).

(1) *The pioneer (or exploratory) phase*, primarily concerned with identification. Here the flora is known mainly from limited herbarium material, often provided with meagre field data though usually well localised. Much of the flora of the tropics falls into this category, just as the European flora did in the time of Linnaeus. Here morphology and distribution provide the data on which the taxonomist must rely. Taxonomic experience and judgment are particularly important in this phase. Limited material means that taxonomic decisions are often extremely arbitrary and can only be tested when additional material becomes available. If the modern taxonomist is better off in this phase than his eighteenth-century predecessors, it is thanks to the improved quality of herbarium material—plants are usually better collected and better labelled than they were in the time of Linnaeus. But even the pioneer stage of systematics is not within sight of completion—it is a moot point whether taxonomists will record the remaining unclassified organisms before other people succeed in destroying them (Constance, 1957; Guinea, 1960).

(2) *The consolidation phase*, in which the species have been studied both in the herbarium and field for a considerable period. Herbarium representation is rich and the regional and local variation of species rather well known. Much of the flora of southern Europe and the Near East falls into this category, though in any area there are bound to be many species that are much better known than others. Many plants, described as species in the pioneer phase on limited material, are found in the consolidation phase to be variants of other species, so that there is an increase of names reduced to synonymy. On the other hand, further exploration leads to the discovery of more new species. An increase in the number of species known may necessitate a revision of generic limits. Floras and basic monographs are produced. Increasing knowledge of the world's flora allows some evolutionary interpretation of plant relationships.

(3) *The biosystematic phase*, dealing with plants whose geographical variation is not only well known but for which cytological or biosystematic information is also available. The latter mainly concerns chromosome morphology and number, breeding behaviour and barriers, and information, derived from transplant experiments, about the constancy and genotypic control of characters. The emphasis is on micro-evolution and variation. A rapidly increasing part of the flora of North-Western and Central Europe, North America and Japan falls into this category. Its classification has special difficulties because the morphological differentiation of populations is often

not correlated with their inability to interbreed. The herbarium, which formed the basis for the pioneer and consolidation phases, becomes of diminished importance in the biosystematic phase. However, as few widespread species can be biosystematically studied throughout their range, the herbarium continues to provide the frame of reference (and the basis for accurate nomenclature) for experimental studies in limited areas.

(4) *The encyclopaedic phase* co-ordinates the other three. All available evidence is taken into consideration in order to express the taxonomic and evolutionary relationships of plants at all levels of the hierarchy. Especially at the specific and infra-specific levels, very considerable advances have been made in biosystematic investigations, and this information has been co-ordinated for the first time. Above the level of the genus, we are still only making the first halting steps in this phase of study. The physical difficulties of tracing the data to co-ordinate, and then recording them, are formidable. Encyclopaedic monographs (both of taxonomic groups and of different kinds of characters) are required before this phase can be extended to the higher groups.

The first two of these phases correspond to the "*alpha*" classification of Turrill (1938a), a preliminary classification based almost entirely on external morphology. The classifications of de Candolle and Bentham & Hooker are of this type. At the other extreme is "*omega*" classification (represented by our encyclopaedic phase), a natural classification based on all possible characters and relevant data, an ideal which may be approached but never be fully attained. Today, classification extends beyond "*alpha*" taxonomy. Biosystematics and the systematic comparison of cryptic characters provide the data that brings us a little nearer to the goal of "*omega*" classification. We cannot agree with those botanists who consider that "classical" taxonomy* and "experimental" taxonomy are entirely separate subjects with different aims and no contact between them. The distinction is one of knowledge, not of aims; it is the distinction between the "*alpha*" taxonomy of little-known groups and the taxonomy of groups which have been more thoroughly studied (A. J. Cain, 1956).

Definition and Use of Terms

Before we go on to discuss the historical development of taxonomy, it seems necessary to define some of the terms which will be frequently used in this book and to discuss their usage.

1. *Identification* implies assigning a plant to a particular taxonomic group—ultimately to the species. The plant is therefore given a name by

* "*Classical*", "*orthodox*" and "*formal*" are adjectives often used to describe classification by "normal" taxonomic methods—i.e. mainly by the intuitive assessment of morphological characters. It is often contrasted with "experimental" or "biosystematic" classification, or with "numerical" classification by mathematical means.

recognising that it belongs to a previously described taxon.* If, however, comparison with taxa of the same rank already recorded shows this plant to differ significantly from them, it may be described and named as a new taxon. Because no two plants are identical but are only treated as identical for taxonomic purposes, some botanists prefer the term *determination* to identification. Botanical *nomenclature* is the application of distinctive technical names to taxa in accordance with a set of international rules (cf. p. 282).

2. *Description* means the orderly technical recording of the characters of a group, particularly the morphological ones. A *diagnosis* is a short comparative description, stating only those characters which differentiate the taxon from its closest allies in the same rank; it usually accompanies the description of a new species. *Phytography* is the craft of plant description.

3. *Classification*, in the biological sense, is the ordering of plants into a hierarchy of classes. The product is an *arrangement* or *system* of classification designed to express inter-relationships of plants and to serve as a filing system. The term classification, however, is often used both for the process of classifying and for the system which it produces. Systems are, at least in theory, built from the bottom upwards—not suspended from the top. In practice they are usually approached from the middle!

There are basically three kinds of classification: *artificial*, *natural* and *phylogenetic*, though in practice the three overlap.

(a) *Artificial classification.* Early systems of classification were, at least in part, *artificial*—that is, they were based upon a few convenient characters for the purpose of sorting or identification, often irrespective of affinity. In Linnaeus's *Species Plantarum*, for instance, the genera of the *Labiatae* are divided between the *Diandra Monogynia* and the *Didynamia Gymnospermia* because of the emphasis given to the number (2 or 4) and relative position of the stamens in Linnaeus's Sexual System, despite the overall resemblances between the Labiate genera. A *special* classification is a particular kind of artificial classification in which one or a few attributes of the plant are used to the exclusion of others. All special classifications are designed for a particular purpose, e.g. to show breeding behaviour.

(b) *Natural classification.* Later pre-Darwinian systems have mostly attempted to be *natural* in the sense that they are based upon overall resemblances, mostly in gross morphology (i.e. macroscopic external morphology).† Such a classification is based on as many characters as possible in correlation and serves a variety of purposes. Experience shows that because resemblances or differences in a number of associated features in the plants making up a group are likely to apply to other of their features not considered in making

* A *taxon* (plural *taxa*) is any taxonomic group of any rank, e.g. *Ericales* or *Bellis perennis.*

† "Natural" was used in a different sense by Caesalpinus and Linnaeus; they used it to mean the true "nature" or "essence" of plants, a usage associated with Aristotelian principles and the belief in Special Creation.

the classification, a natural classification has a *predictive* value. Classification should therefore become more natural the more attributes we include in it (Stearn, 1961b). A classification based on morphology alone is natural with regard to the information we put into it; but it is only a step towards a wider natural classification based on the synthesis of numerous attributes. Natural classifications in use at present probably contain a considerable number of artificial groupings because of the continued use of *a priori* weighting (see p. 49).

(c) *Phylogenetic classification.* Post-Darwinian systems have mostly claimed to be *phylogenetic.* They use as a criterion the known or inferred phylogeny of taxa—in the case of Angiosperms nearly always inferred. Such systems are often expressed in the form of a phylogenetic tree showing the supposed evolution of the groups, though the Angiosperm groups themselves (especially families and genera) are usually fairly natural in the sense outlined above and delimited as they exist at the present day. A phylogenetic classification is usually based on a natural classification (i.e. a two-dimensional one) that has been modified in accordance with inferred monophyly. For such classifications, the term "evolutionary" is to be preferred to "phylogenetic". If we are classifying a sequence of fossils, it is preferable to use the term *vertical* classification (see p. 43), although it is factually phylogenetic.

The basic unit of natural classification is (usually) the *species;* these are distinguished on the basis of correlated characters—i.e. each individual species has a combination of characters peculiar to it. By assessing similarity between species, they may be grouped together into higher taxa (genera) which may themselves be grouped into higher taxa (families) until the whole range of variation is accounted for in a hierarchial system (cf. the set theory in Simpson, 1961). What taxonomists classify into species, etc., are organisms (life cycles), although it is today widely accepted that the basic unit of evolutionary change is the population of which the organisms are representatives.

A natural system does not, in theory, have an arrangement other than that dictated by overall resemblance and the construction of a hierarchy; its beginning and end are the top and bottom ranks of the hierarchy. When it comes to be expressed on paper it can be shown in a box-in-box manner, or as a *dendrogram;* these express only the relationships of the taxa to one another on the basis of overall resemblance, i.e. they are without any phylogenetic implications. The two methods of presentation are shown in Fig. 1. When a natural system is presented in a linear sequence some other arrangement must be resorted to, since taxa are seldom related to one another in a simple linear manner. A dendrogram must not be confused with a phylogenetic tree in which the vertical dimension must be *time* or *relative advancement* (cf. J. S. Huxley, 1958).

The practical value of a classification is shown by how it stands the test of time. The form a classification takes depends on the purpose for which it is intended, so that a particular classification is only "better" than another within the purpose for which it was constructed. However, a natural classification, because it contains the most information, will serve the greatest number of purposes. It will therefore be the most generally useful and be equivalent to what is often called a "general" classification. Special classifications are certainly required, but we must first have a general classification to serve as a frame of reference.

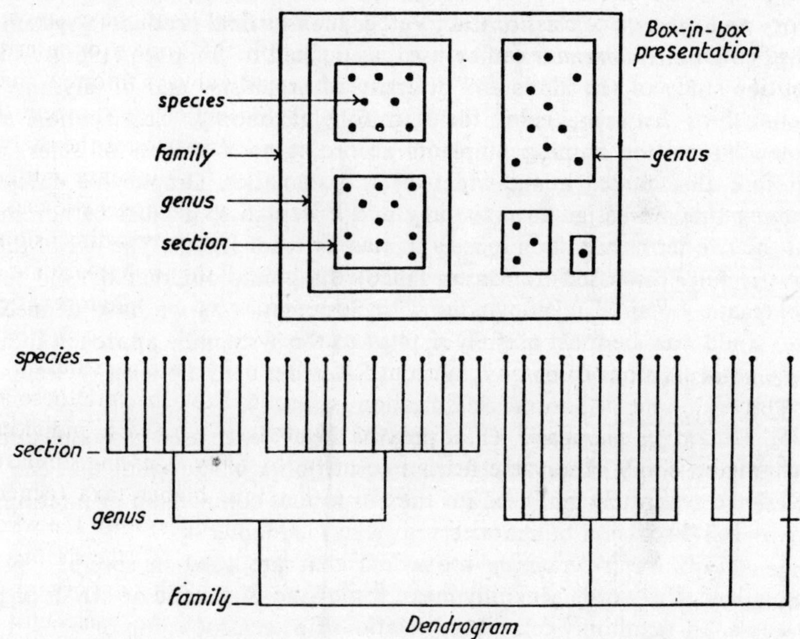

FIG. 1. Box-in-box presentation and dendrogram of relationship.

It is discontinuity of variation which enables us to have a workable classification. This is because it makes possible the perception of mutually exclusive character expressions (e.g. axile *versus* parietal placentation) and therefore the concept of character correlation (or, more precisely, the correlation of character expressions). Pre-Darwinian botanists accepted present discontinuity as a result of Special Creation. We now know it to be largely due (at least above the species level and with some allowance for the effects of neoteny in phylogeny, cf. p. 156) to extinction during the course of evolution. Evolutionary divergence has been accompanied by the extinction or survival of the ancestral types. Classification is based on character correlations and discontinuities, and cannot cope easily with continuous variation, or where variation is discontinuous but where no high correlations occur—the so-called reticulate variation found in some polymorphic species like *Biscutella laevigata*.

A *taxonomic key* is a method of analysis. Although based on characters extracted from a classification and providing a means of access to it, a key is not in itself a system of classification. It is an analytical device whereby a choice is provided between two contradictory propositions, and is particularly useful for identifying unknown organisms (cf. Chapter 9 for examples).

4. *Taxonomy and Systematics. Taxonomy* is the study of classification, including its bases, principles, procedures and rules (A. P. de Candolle, 1813; Simpson, 1961). The word (first coined by de Candolle) is often used for the theory and practice of classification, rather than its final product (a system of classification). *Systematics* can be used, as defined by Simpson (1961), as the scientific study of the kinds and diversity of organisms and of any and all relationships between them. Identification, taxonomy, classification and nomenclature (the naming of plants according to a system of rules) are therefore all included in the wider term, systematics. Despite the different meaning that we assign to taxonomy and systematics, it must be admitted that the two terms are often loosely treated as synonymous. The use of these terms is fully discussed by Mason (1950); he points out that the old term "systematic botany" (synonymous with Systematics as we have defined it here) could now be most usefully applied to the systematic approach to any of the basic divisions of botany, much as Solereder used the term "systematic anatomy".

5. *Taxonomic characters.* These provide the evidence used in classification. A character is any observable feature or attribute of the plant that can be considered separately and used for the purpose of comparison or interpretation. A full discussion of characters is given in Chapter 4.

6. *Affinity and relationship* are words that are used in several senses. Unqualified, the words generally mean some kind of genetic or evolutionary closeness. In taxonomy affinity or relationship is most commonly used in two senses:

(a) Overall similarity as it is perceived through the senses. This is what the taxonomist generally means when he says that species A is more closely related to species B than to C. Cain & Harrison (1960) have proposed the term *phenetic* relationship for this type of affinity, which thereby clearly distinguishes it from any other kind of relationship.

(b) Phylogenetic relationship, implying propinquity of descent.

In practice phenetic relationship is the most commonly used because (*a*) we seldom have evidence of phylogeny—although this has not prevented taxonomists from thinking they have, and (*b*) we assume that similarity, if enough characters are taken into account, usually reflects a phyletic connection.

Other qualifications of the term "affinity" include genetic relationship (in terms of genome similarity), cytogenetic relationship (in terms of ability to cross), chemical relationship, etc.

7. *Natural and artificial groups.* "Natural" suffers from a similar ambiguity

of usage to "affinity" or "relationship". Apart from the archaic sense in which Linnaeus employed the term (see footnote on p. 5), "natural" is still widely used in two senses: (i) for a group in which the plants are more closely related to one another in the sense of overall similarity (i.e. phenetically related) than they are to any other group, and if the connecting links between them can be seen or safely inferred, (ii) for a group whose components are known to be (or thought to be) phylogenetically related. Very often, of course, a group is natural in both senses of the term. In such cases phenetic affinity is the result of propinquity of descent. The trouble is that a group may be phenetically natural, and yet, because of convergent evolution, be unnatural in a phylogenetic sense. This is much less likely to be the case, however, if enough characters are taken into account when assessing affinity.

An *artificial* group is the opposite of a natural one, and results from the deliberate selection and weighting of characters. All special classifications are artificial, and are, of course, artificial on purpose.

An unnatural group is used in two senses: (i) for artificial groups—a condition which we usually try to avoid; (ii) for badly constructed groups, i.e. a natural group with lumps in it (e.g. *Ranunculaceae* if *Paeonia* is retained within it). The latter situation is probably still extremely common in classification, and the taxonomist usually tries to make such groups more natural by removing the black sheep. Unlike artificial groups, unnatural groups in the second sense are usually unnatural by mistake, or at least because a more natural (phenetic) group might be less convenient to use.

Above the level of the species we are usually in the dark about the facts of phylogeny, and in the absence of fossil evidence it seems essential to construct natural groups by overall similarity which may or may not reflect phylogenetic relationships of their components. In this book "natural" is used in the phenetic sense unless the context clearly implies a different usage of the term.

8. *Typology and the archetype.* During the last decade, typological and non-typological methods in taxonomy have been debated by Danser (1950), van Steenis (1957), Lam (1959), Simpson (1961) and others. Typology derives from Plato and was brought into plant classification with Aristotelian logic. It is therefore a pre-evolutionary concept and pervades all pre-Darwinian taxonomy. As the basis of Angiosperm classification was formed during that period, it is necessary that the concept of typology should be understood; its effects are still deeply entrenched in much taxonomic theory and practice today.

The basic concept of typology has recently been lucidly stated by Simpson (1961): "Every natural group of organisms, hence every natural taxon in classification, has an invariant, generalised or idealised pattern shared by all members of the group [i.e. the ground-plan]. The pattern of a lower taxon is superimposed upon that of a higher taxon to which it belongs, without essentially modifying the higher pattern." As Danser (1950) puts it, the systematic ground-plan of a natural group is an imaginary plant in which the following qualities are combined: (i) everything that the ground-plans of the component groups of the next lower category have in common (i.e. all the characters of *Primulaceae* shared by all its genera); (ii) where these ground plans diverge, all the most "primitive" conditions occurring among them. The specialised characters are ignored. Such an abstraction is called an *archetype,** *morphotype* or *ground-plan*. In conjuring up an archetype for the *Ranunculaceae*, for instance, a botanist would ignore such specialised characters as the zygomorphic perianth of *Delphinium* or *Aconitum*, the syncarpy of *Nigella*, the opposite leaves and climbing habit of *Clematis*, and define it as possessing the least specialised characters which are found scattered throughout the different genera of the family. (If this sounds like an evolutionary approach, we should remember that the pre-Darwinian taxonomists distinguished simplicity from elaboration of structure—the basis of the *Scala Naturae*. As early as 1853, T. H. Huxley treats the theory of archetypes as opposed to what we should now call evolutionary views.) Some botanists have gone so far as to equate the archetype with a common ancestor. This is a naïve and unwarranted extension of the former term. In most cases the archetype does not exist, and probably never did so.†

The preceding remarks are particularly applicable to typology as it is applied at generic and higher levels. At species level, though conforming to Simpson's basic definition, typology found its expression in a denial of variability, or at least of its importance. Variation was recognised and accepted by early typologists, but it was usually thought of as accidental or unimportant, the effect of the climate or habitat upon the ground-plan of the species created by God: an imperfection of nature. The deficiencies of eighteenth-century herbaria are largely attributable to this narrow typological concept—any specimen was thought to represent the species adequately.

(Typology or the type *concept* should not be confused with the *type method*, a purely legal device for the correct application of names by the use of type specimens, etc. The very word "type", as it is used in nomenclature, shows its historical connection with typology. To the typologists, denying the importance of variation, a type specimen really did represent the species, whereas we now recognise it as one variant of an ever-varying population; it is important only because of its permanent association with the name of the taxon. Only names have types, not species. Cf. Chapter 8.)

* Archetype was commonly employed by pre-evolutionary eighteenth-century taxonomists as a metaphysical abstraction representing the ideal or divine pattern of a taxonomic group—a concept stemming from Aristotelian principles (essence, etc.) and associated with the belief in Special Creation.

† An excellent example (most persuasively presented) is Corner's hypothetical '*Proto-Moraceae*' (1962) which is given a detailed description and figure. As he says, "taxonomy is a larder of notions awaiting thought"!

To what extent, if any, should we countenance typological methods in taxonomy today? Taxonomic groups are the product of evolution, so that any concept that has a non-evolutionary basis may at first sight seem archaic and unacceptable. At the level of the species and below there is certainly no room for a typological approach in the narrow sense outlined above. Indeed, there can be few taxonomists today who hold such extreme views and thus deny the inherent variability of species. The severe criticism directed by G. G. Simpson against typological procedures seems to be largely unnecessary; he is flogging a dead horse.

Simpson contrasts the typological concept of the nature of systematic units with the population concept. In practice this is an exaggerated antithesis. It is quite true that the species and other taxonomic categories must today be considered as made up (ultimately) of populations; these should ideally be treated in terms of frequency distribution of the different variants of each character actually present at any time. It would be idle to pretend that more than a very limited amount of taxonomy can be practised in this way. In fact, taxonomists are inevitably forced to identify material by comparison with previously collected specimens, together with figures and descriptions, since the specimens used by the taxonomist are necessarily extracted from the population and only in very exceptional cases is he provided with (or can obtain) statistically adequate samples of the population. Thus every taxonomist builds up what one might call a typological picture based on his experience of each species; this is derived from the material which he has seen in the field, herbarium, garden and laboratory and from figures and descriptions in botanical literature. He will thus come to a concept of the species which includes a greater or lesser amount of its actual variability. We might call this the *neotypological concept*, although Simpson (1961: 185) has already coined the word *hypodigm* for the use made of it in practice. It is a warranted and inevitable taxonomic procedure and a justifiable extension of typology. The hypodigm changes with our knowledge of the species.

Even in population taxonomy typological procedures are practised to some extent, since most biosystematic work is still of necessity based on inadequate samples, frequently covering a very small area of the species' total range. Extrapolation to give a picture of the total population variability has to be made. This is essentially a typological procedure. The same applies to tests of fertility-relationships of population samples, etc. In fact, it is true to say that, at least as far as widespread and variable species are concerned, analysis of a limited number of populations will give *less* idea of the range of variability found in the species than we shall get by surveying the material preserved in a great herbarium. An obvious defect of most population systematics, when it comes to drawing taxonomic conclusions, is that detailed population samples from limited areas have to be compared with a limited amount of normal herbarium material (as is readily available to the investigator) from the rest of the species' range. Frequently, due to lack of familiarity with the methods and possibilities of herbarium taxonomy, this aspect of the work receives much more cursory treatment than is necessary.

Above the level of the species we are faced with a different situation. The most important attribute of a practical Angiosperm classification should be maximum information content (based on comparative data) and as much stability as modern knowledge of the facts allows. The effects of typology are already built into our classification so firmly that we could not get rid of them without pulling down the whole structure and starting again. But how should we start again, even if we could? Natural classification, advocated in this book, rests upon correlation of characters and discontinuities of variation. These are, in fact, typological concepts which were more familiar to de Candolle than to us—indeed, the pre-Darwinian typologists seem to have understood the theoretical basis of classification better than most botanists do today. Although the units of classification are based on the populations, this is, in practice, ignored when we are classifying above the species level. In forming genera, families, etc. we are, as Simpson (1961: 48) has pointed out, using typological *procedures* even though the results may be *interpreted* in non-typological ways. In constructing a natural system there is no escape from typology. Intelligently used, it is a valid and necessary procedure.

THE DEVELOPMENT OF CLASSIFICATION PRIOR TO DARWIN

In order to survey the field of taxonomy at the present day, it is necessary to have some knowledge of the history of the subject and the development of ideas associated with it. Earlier morphological observations were seldom wasted; nearly everything became incorporated into classification in its widest sense, so that the history of systematics is an instructive ladder to climb. For convenience we can divide it into pre-evolutionary and post-Darwinian phases. The first was almost entirely concerned with gross morphology, in practice if not in theory.

1. *Classification based on Habit*

Taxonomic botany may be said to have begun with the ancient Greeks. THEOPHRASTUS (*c.* 370-287 B.C.) was a pupil of Aristotle, himself a keen observer of plants. In his *De Historia Plantarum* (Enquiry into Plants), Theophrastus appears to have gathered together the notes which he gave to his students in the Peripatetic School at Athens. It is the most complete and orderly of all ancient biological works. Theophrastus was a good morphologist, recognising the differences between polypetalous and gamopetalous corollas, superior and inferior ovaries, determinate and indeterminate inflorescences. His analytical classification (describing 480 kinds of plants) was primarily based on habit; he divided plants into trees, shrubs, undershrubs and herbs, and these into cultivated and wild kinds. Such a division had the merit of being very easy to use.

The work of Theophrastus reflects the philosophy and methods of Plato and his pupil Aristotle. It incorporates the principles of Logical Division (the division of classes), particularly the rules of Dichotomy or the "Excluded Middle"—any given object is either A or not A. Throughout the botanical work of Theophrastus, the implied questions are: What is its difference? What is its essential nature? (Hort, 1916; Turrill, 1942).

In the first century A.D. the Greek herbalist, DIOSCORIDES, wrote his *De Materia Medica*. In it he described some 600 species and gave accounts of their uses. The book (originally unillustrated) deals with perfumes, oils, spices, cereals, condiments, wines, etc., yet despite the fact that he classified plants by these properties the learned doctor showed some feeling for kinship. He even managed to recognise a few natural groups, such as a series of Labiates and another of Umbellifers. An illustrated codex (*Codex Aniciae Julianae* in Vienna) of A.D. 512, containing Dioscorides' Greek text associated with illustrations of varying quality, has proved valuable for ascertaining the original and traditional mediaeval application of his plant names, many of which remain in use today as Linnaean generic names (cf. Arber, 1938; Stearn, 1954). Dioscorides holds an important historical position in botany, since for one and a half millennia *De Materia Medica* (together with a pseudo-Dioscoridean text called *De Herbis Feminis*) formed the chief corpus of botanical knowledge throughout the Dark Ages in Europe (Blunt, 1950).

Except for a few accurate observations by Albertus Magnus (1193-1280), little botanical progress was made till the sixteenth century. The needs of mediaeval medicine led to the compilation of numerous Herbals. The earlier ones contained crude woodcuts copied from much earlier drawings which had illustrated the text of Dioscorides during the early centuries of the Christian era. These illustrations were copied and recopied by the mediaeval herbalists until they became quite unlike the original plant. Such illustrations are useless for identification, but as knowledge of medicinal plants was handed down by tradition this caused little inconvenience.

It is fortunate that several of the herbalists of the sixteenth and seventeenth centuries, however, worked in collaboration with artists who returned to the living plant. We therefore find that the herbals of Brunfels, Fuchs, Clusius and L'Obel (even when their authors have copied Dioscorides' text) often contain extremely vigorous and accurate figures. It is thanks to such fine artists as Hans Weiditz (who illustrated Brunfels' Herbal and was almost as exquisite a draughtsman as Dürer), Albert Meyer and Pierre van der Borcht that during the sixteenth century so many plants unknown to the Ancients were illustrated in woodcuts admirably portraying their habit of growth. These new plants caused some embarrassment to the herbalists who preferred to stick to the well-trodden path of Dioscorides and treated such upstarts as *Anemone pulsatilla* and *Cardamine pratensis* as "*herbae nudae*"—forlorn or destitute herbs (Arber, 1953). Throughout the Middle Ages it is the interest in the medical uses of plants that kept botany alive, and most of the new knowledge that accumulated took the form of commentaries (notably those of Mattioli, 1501-77) on the Latin translation of Dioscorides' text.

In the sixteenth century good illustrations in the Herbals, the making of collections of dried plants (however inadequate by modern standards) and the preparation of detailed descriptions by Bock and Valerius Cordus allowed great advances in classification to be made. Some aspects of these descriptions deserve comment. The emphasis of Bock was on the life history of the plant—when he describes *Ranunculus ficaria* he tells us how it behaves throughout the year. Cordus, in addition to describing the morphology of the plant, gave particular attention to its odour and taste. Taxonomic descriptions have for so long been largely static and morphological that these aspects of sixteenth-century description have something to teach us today (Arber, 1953).

CAESALPINUS (1519-1603), professor at the University of Pisa, published his *De Plantis* in 1583 (cf. Bremekamp, 1953). This is the first methodical classification of plants based on definite morphological criteria, and was moulded by those *a priori* considerations of the functional importance of certain characters that underlie nearly all pre-Darwinian systems of classification. According to Caesalpinus, the functions of most importance to the plant are nutrition and reproduction, but as all characters of the plant are involved in these functions it seems impossible to find a character which is not taken from such a part. Caesalpinus recognised the usefulness of fruits and seeds in classification. Although his first division was based on habit (probably due to an exaggerated respect for the views of Aristotle), his second was based on characters of the fruit (position, number of loculi and number of seeds). His classification stresses the number of parts, their position and their form. The result was a classification less unnatural than that of his predecessors, but as he produced it in narrative form it seems to have had little influence on his contemporaries. Caesalpinus recognised groups corresponding to the *Fagales*, *Leguminosae*, *Umbelliferae*, *Cruciferae*, *Boraginaceae* and *Compositae*, and even wrote technical descriptions of some of them. His views influenced the thinking of later botanists like Tournefort, Ray and Linnaeus (cf. Sprague, 1950). The combination of natural and unnatural groupings which we find in Caesalpinus's work results from a combination of synthetic and analytical approaches. It is a feature of all pre-Darwinian systems and persists to some extent to the present day.

The English philosopher and naturalist, JOHN RAY (1628-1705), heralded a new era in taxonomy. He formulated the principle (often neglected by later botanists) that *all parts of the plant* should be used for classification—a principle now recognised as the corner-stone of a natural system. Ray realised that variation arose both from inherent internal causes (genotypic) and from the direct influence of environment (phenotypic), but maintained that only the former was of taxonomic significance—a principle that we adopt today. Furthermore, he considered that forms should breed true within their limits of variation (*distincta propagatio ex semine*)—a view shared by Linnaeus as a criterion of species.

Ray realised the taxonomic significance of monocotyledony and dicotyledony, and named the corresponding groups although he did not use them as his primary divisions. His system of classification is presented in his *Methodus Plantarum* (1682), whose second edition (1703) is said to contain 18,000 species of plants. Although he retained the traditional primary divisions on the basis of habit, he divided these into Monocotyledons and Dicotyledons and classes based on fruit type; these were subdivided on the basis of leaf and floral characters. A remarkable feature of the book is the clear presentation of indented synoptic keys to the genera, the names of which are mostly taken from the earlier works of Bauhin. Indeed, the main purpose of the *Methodus Plantarum* was to produce a practical classification that could be used for ready reference. In many ways his classification was superior to that of the largely artificial system of Linnaeus that was to follow. As his work on the British flora shows, Ray was a better judge of species than Linnaeus, often discriminating between closely allied species where Linnaeus confused them. Ray realised that a natural system could not be achieved by a single botanist, or even in a single age.

Joseph Pitton de Tournefort (1656-1708) produced a classification that was extremely artificial; nevertheless, it was widely adopted in France, which never completely accepted the Linnaean system. Tournefort's chief importance lies in his explicit formulation of the genus concept. In his *Institutiones Rei Herbariae* (1700) he provided formal correlated descriptions for 698 genera, though many of their names had been in use since classical times or even earlier. Tournefort recognised two grades of genera—the first contained genera based on the form of the flower and fruit (e.g. *Aconitum* and *Ranunculus*), whereas the second contained genera for which vegetative differences were needed (e.g. *Pinus*, *Abies* and *Larix*, which differ in the position of their leaves). His general conclusion as regards genera of flowering plants was that one should take the flower and fruit together and add characters from other organs when necessary for their better distinction (Stearn, 1960). Most of Tournefort's genera based on floral characters were later adopted by Linnaeus, but those distinguished only by vegetative characters were not accepted. Tournefort's genus concept has proved more acceptable to later authors than that of Linnaeus; most (but not all) of his "second grade" genera are recognised today.*

2. *The Sexual System*

By the eighteenth century more and more plants had been discovered and described (some of them several times), and the difficulties of identification

* Just how far Tournefort was prepared to go in recognising "vegetative" genera is exemplified by *Quercus*. He placed *Q. robur*, *Q. suber* and *Q. ilex* in three different genera on the nature of their bark etc.!

were becoming acute. This was the position which faced LINNAEUS (1707-78), a Swede of tremendous energy and enthusiasm whose mission it was to record the works of the Creator—and who got up before dawn to achieve it.

Sexuality had been proved experimentally in plants by Rudolf Jacob Camerarius in 1694. Linnaeus became acquainted in late adolescence with the functions of stamens as male organs, and this knowledge made a profound impression on him. During his early studies he concentrated his attention on the stamens and pistils. It was their relative constancy in number which enabled him to build his basically arithmetical, so-called "Sexual System" of classification upon them (Stearn, 1960).

In his *Genera Plantarum* (1737) Linnaeus described all the genera known to and accepted by him, often basing his descriptions on one particular species with which he was familiar in the living state. In his generic descriptions only floral characters are cited, and these in a very concise and methodical manner. While Professor of Botany and Medicine at Uppsala, he published his *Species Plantarum* (1753) in which he diagnosed (or rather, keyed out) nearly 6000 species in 1000 genera. The genera (described separately in his *Genera Plantarum*) were grouped into 24 classes which were almost entirely based on the number and morphological arrangement of their stamens; these classes were subdivided into orders on the numbers of their pistils.

Above the level of the genus the Sexual System was, of course, a largely artificial arrangement (*Pinus*, for instance, appears in the same order as part of the *Euphorbiaceae*), but it was a very convenient filing system. It enabled most species to be readily identified. The *Species Plantarum* was a two-volume handbook, and Linnaeus was very economical in his presentation of the data it contained. His "descriptions" (more accurately polynomials) were limited to twelve words, the characters being carefully chosen so that the polynomials served as an unobtrusive key (Stearn, 1961b). These were followed by references to important earlier works in which the species had often been more fully described and figured; then followed a note on the plant's habitat (often very accurate) and provenance. For new or little-known species he occasionally added a descriptive note. In the margin he put the specific epithet (trivial name), e.g. *alba*, which added to the generic name gave the binomial specific name, e.g. *Betula alba*. This was regarded by Linnaeus as a convenient name for everyday use, but not as the real name for the species, which was for him the diagnostic polynomial.

Linnaeus did not invent binomial nomenclature, but first consistently applied it to the whole plant kingdom. By retaining polynomials as well he, in fact, confused the purpose of nomenclature (to provide a convenient name for international use) with that of identification—hence his dual system. The binomial provided the convenient name, while the polynomial served as a diagnostic description (Stearn, 1959). Other botanists, however, were quick to realise the advantages of the binomial system, and after the publication of the *Species Plantarum* the use of binomials, in place of the cumbersome polynomials which had been mainly used before, became standard practice. The binomial system provided such a convenient reference system that it is difficult to visualise taxonomy without it. The binomial does, in fact, simultaneously express resemblance and difference—the first by the generic name, the second by the specific epithet. Nevertheless, such is our present-day knowledge of the diversity of species that it can prove an unsatisfactory strait-jacket for certain critical groups, particularly for those which reproduce asexually.

Although Linnaeus's species concept was not always as satisfactory as Ray's, his genera were often remarkably natural and well defined. Generic limits in the *Labiatae*, for instance, are notoriously difficult, yet Linnaeus, although he disposed them in two different classes, delimited them remarkably well. There were, of course, no families recognised in the *Species Plantarum*. There is no doubt that Linnaeus succeeded in producing order out of what was rapidly becoming chaos. Whatever the failings of the Sexual System, it was a highly practical scheme which served (among other purposes) to focus attention on the sexual organs of plants instead of on their habit, which had held the field for so long.

Together with his classification went the construction of a precise Latin terminology, special botanical meanings being given to words which before had only a different (though related) classical connotation. So great was the prestige of Linnaeus that he imposed both his system and his terminology on taxonomy throughout most of the Western world.

Linnaeus' *Species Plantarum*, because of the consistent use of binary names, provides the starting-point (1753) for modern botanical nomenclature under the International Code (see Chapter 8).

Before leaving Linnaeus, we must consider his philosophy and concepts. As has been shown by Cain (1958, 1959b) and Hopwood (1959), Linnaeus was brought up in the tradition of Aristotelian philosophy. He classified plants by the use of two principles: (*a*) Logical Division (cf. p. 13), as had been done by Theophrastus; (*b*) certain *a priori* reasoning as to what were the most important (i.e. functional) organs of the plants. It was for this reason that he stressed the reproductive structures of the plant.

Stearn (1959) has given such a concise and lucid summary of the logic of classification that we cannot do better than quote his account of the principles which so strongly influenced Linnaean classification. "In terms of logic, arranging subjects into classes is classification; distinguishing divisions of classes is logical division. The same principles apply to these upward and downward processes. One is that co-ordinate classes must be mutually exclusive. The group to be divided is the *genus;* the parts into which it is divided are the *species*. The characteristics which the species possess are of several kinds, i.e. (1) those which every member of the genus possesses but which are also possessed by members of other genera; (2) those which every member and only a member of the genus possesses; (3) those which only a given member possesses. Investigating the characteristics of a group and determining to which kind they belong is a fundamental taxonomic procedure but not confined to biology. Every species is what it is and not something else because it is an expression of its essence; as expressed by Maritain 'it is by and in its *essence* that a thing possesses being or *existence*'. The definition of the particular species must provide a statement of the characteristics, arising out of its essence, that make it that and not something else. The *definition* thus covers the characteristics which the species shares with other member-species of its genus which is implied by giving it and them the same generic name, and those characteristics which distinguish it from these other members, its *differentiae*. This process is described as being *per genus et differentiam*. Simply to know this about something merely enables it to be recognised and is preliminary to other investigations. Nevertheless, the definition of some 6000 species of plants and 4000 species of animals in the *Systema Naturae* [1735 and later editions] is a monumental achievement."

As Stearn goes on to point out, there is evidence of a conflict or lack of harmony in Linnaeus's mind between the *a priori* approach of logic and the empirical approach of modern science. "The question implied is: (1) should things be put together because they conform to a definition already framed or, *vice versa*, separated because they do not do so, irrespective of other resemblances and differences, or (2) should they be linked to things with which they agree in their characteristics generally, despite lack of agreement with the definition?"

This is a basic problem in taxonomy, and Linnaeus, by the methodical cast of his mind, was biased towards the first course (*dispositio theoretica*, as he called it), and usually followed it with classes, orders and even many genera, thus achieving an artificial system. On the other hand, certain facets of his nature (or was it sublimated experience?) led him to delimit species and often genera the other way (*dispositio practica*), the empirical method of natural science leading to the delimitation of natural (instead of artificial) groups. Later in life he extended this approach to higher groups. The conflict is still with us today, and modern taxonomists still use both approaches. Whenever we determine a plant by means of a key, we use the method of Logical Division based on definition. When we determine a specimen by matching it up with others, or because we remember having seen it before, we are adopting the typological concept.

Linnaeus was first preoccupied with the genus, and adopted it as the basic, most practical unit of classification. He grouped his genera into classes and orders, and divided them into species, just as today we start with the species, group them into genera and divide them into subspecies or varieties. There were no doubt good practical reasons for Linnaeus's approach. Before defining the species he had to define the genera, and the chaos of pre-Linnaean synonymy must have made it difficult to get his species straight (in many cases he never did get them straight), and their large number made them impossible to memorise. Linnaeus took the view that botanists should be able to memorise all genera, both their characters and names—a feat which we certainly could not attempt today. Subconsciously, however, Linnaeus must have thought of genera in terms of groups of related species, even though his technical generic description was usually based on a single species. His famous dictum that it is the genus that gives the characters, not the characters which form the genus, provides the clue to the remarkably natural genera which he usually managed to recognise whenever androecial variation did not result in their being split into different classes. The majority of Linnaean genera are natural (though many have since been divided up into smaller allied genera) and must have been formed by synthesis followed by the extraction of diagnostic floral characters. How else could he have recognised natural genera in the *Labiatae*? Wherever Linnaean genera are artificial, *a priori* reasoning has usually had the upper hand.

Linnaeus did not take the *minutiae* of his Sexual System as seriously as some of his supporters or detractors. He quite often included species within a genus which did not agree with his original definition of it (or of the order or class to which it belonged) on technical floral characters; this was sometimes done accidentally, but was often a deliberate policy. Stearn (1959: 94) quotes a statement by Linnaeus's son which clearly shows that his father recognised plants by their general facies—technical characters were used by way of confirmation. This procedure no doubt accounts for the accumulation of "black sheep" that contaminated many Linnaean genera and earned the strong disapproval of Medicus (Stearn, 1961b). Such a method, however, though it resulted in a number of misalliances, was no doubt largely responsible for the naturalness of many Linnaean genera.

Species, to Linnaeus, were static entities, monotypic, genetically unitary (i.e. breeding true) and equivalent. By accepting Ray's principle that species breed true within their own limits, he was not solving the problem of delimiting them. The limits of species must be decided by taxonomists before one can say that they are genetically unitary. It is individuals that breed more or less true, and need no decision for their delimitation. Linnaeus recognised variation within species, but believed it to be due to soil or climate—i.e. phenotypic. A narrow typological view of species, together with the high cost of paper, no doubt account for the very inadequate representation of species that is characteristic of Linnaean and other eighteenth-century herbaria.

Although, to begin with, Linnaeus seems to have considered all species as created by God, in later life he modified this view and maintained that many species and genera had arisen by hybridisation (of which he had experimental evidence) between a specially created, smaller number. This neglected concept of Linnaeus was put forward in a dissertation of 1762, *Fundamenta Fructificationis submittit J. M. Gråberg* (for partial translation see Ramsbottom, 1938: 213) and summarised in the sixth edition of the *Genera*

Plantarum (1764). It is so curious, in some ways anticipating the views of J. P. Lotsy (cf. Stearn, 1957: 158; 1959: 97; see also Sirks, 1952), that his proposition merits quoting in translation:

"Suppose that the Almighty and All-knowing God in the beginning progressed from the simple to the complex; from few to many! Thus from the first principle of plant-life (*a primo Vegetabili principio*) He created only so many plants as there are *natural orders* (*ordines Naturales*). The plants of these orders He himself thereafter mixed by breeding between themselves so that there arose as many plants as at the present time exist distinct *Genera*. Nature thereafter mixed these generic plants between themselves by sexual generations (*per generationes ambigenas*), in which the structure of the flower did not change, and multiplied them into the existing *species*, excluding however from the number of species having this manner of origin the *hybrid* plants which are sterile" (transl. from *Syst. Nat.* ed. 12, **2**: 9; 1767).

There can be no doubt that this concept can only be treated as a theory of evolution (nearly a century before Darwin) in which both God and nature played their part.

Linnaeus believed that the best classification would bring together those plants that were "essentially" alike. He realised that his sexual system was artificial, and in 1738 made his first attempt at natural classification in *Fragmenta Methodus Naturalis* of his *Classes Plantarum*. In the sixth edition of the *Genera Plantarum* (1764) he proposed 58 natural orders with their included genera, many of which correspond to modern families (*Palmae*, *Gramineae*, *Umbelliferae*, *Compositae*, etc.). It was only the practical value of the Sexual System for filing and identification which led him to retain it. The lesser men who followed him, by their uncritical acceptance of the Linnaean arrangement, were responsible for delaying the development of a natural system.

The work of Linnaeus has obvious deficiencies: his Sexual System is largely unnatural above (and often at) the genus level; his species concept was inferior to that of Ray; because of his preoccupation with analysis and previous literature (which was meant to be referred to), plant description reached an all-time low that reflected the philosophical view of his day and from which it did not fully recover till the time of de Candolle; although the emphasis on floral parts drew attention to many very useful characters, neglect of vegetative characters at the genus level and above must have retarded the development of a natural system in an Adansonian sense (cf. p. 23).

But his achievement far outweighed these limitations. He put the nomenclature of the past in order; provided a practical way of identifying all plants known at the time; delimited mostly natural species and even natural genera (though his concept was broader than ours), being very accurate in his description of floral characters; firmly established the binomial system as the standard type of nomenclature; and provided taxonomy with a more exact technical terminology. Nor should one forget the success of his teaching. Linnaeus instilled great enthusiasm into his students, many of whom made valuable collections abroad and became leading botanists in their own right. It may well have been the needs of his students which led him to use the binomial system. As Stearn (1959) puts it: "His achievement was to take systems of procedure which his predecessors had used incidentally or piecemeal or on a very small scale, to analyse and evaluate them from the standpoint of their practical usefulness, and to apply them consistently, methodically and on a large scale to the whole living world as then known. . . . Whatever Linnaeus failed to do that later people with lesser burdens think he should have done, his definition and classification of these organisms [10,000 animals and plants] stands out as something no one else could do then and no one else has been able to do since." When it appeared, the Linnaean system was revolutionary, the "New Systematics" of its day.

We have gone rather fully into Linnaeus's principles and concepts because he has had a tremendous influence on taxonomy, continuing to the present. Linnaeus's ideas were largely based on Aristotelian reasoning which cannot be reconciled with evolutionary theory. A. J. Cain (1959a) claims that "all that we take to be Linnean in our present taxonomies needs complete overhaul, since most of it may now be invalid"—surely an overstatement. There is little the matter with Linnaean classification when it comes to natural groups, which include most of his genera and species. These are the bricks with which the Sexual System was made, and with which any subsequent system has been made.

Fortunately the theory which Linnaeus preached or had in mind did not always coincide with what he did in practice—a frequent phenomenon in the history of classification! Furthermore, modern taxonomy is not only riddled with typology; it is a necessary component in the building up of a natural system (see p. 5). Cain (1959b) goes so far as to suggest scrapping the binomial system, since it is in many ways inadequate to deal with the variation in nature. But therein lies its strength: if it were adequate we should not have a convenient system of nomenclature! If we dropped the binomial system, the uninomial cure he advocates would surely be worse than the disease. It is more important to see that the binomial system is used in the best possible way. We must face the fact that convenience is an integral part of a general classification, a consideration which will prevent taxonomy from ever becoming an exact science like chemistry or physics. We do not deny, however, that there is outmoded theory behind much taxonomic practice, and that we should do what we can to remedy this within the limits of a practical classification.

A most interesting paper by S. M. Walters (1961) has recently drawn attention to the early historical and philosophical background in the shaping of the Angiosperm classification of Linnaeus and his successors. The importance of these factors cannot be disputed, and there is no doubt that Angiosperm classification would look rather different if Linnaeus had lived in Singapore, for example, instead of north-western Europe. We feel, however, that Walters has overstated his case. The *names* of the higher groups would certainly be different. Generic concepts in the *Umbelliferae* might be broader without the mediaeval preoccupation with carrots, parsnips, etc. Certain genera and families might be recognised as subgenera of subfamilies of other closely allied groups; the level of the hierarchy, however, is still a matter of opinion and convenience. But by and large we very much doubt if the taxonomic groups would be very different in *content* at the present day if the foundations of taxonomy had been laid outside Europe. In this connection it is interesting to note that as early as 1789 the *Araliaceae* (which Walters suggests might have included the *Umbelliferae* as a genus (or a few genera) if Linnaeus had lived in New Zealand) was already treated by de Jussieu as a separate family closely allied to the *Umbelliferae*, although he knew only five genera in the former.

In a new interpretation of Willis's "Age and Area", Walters claims that large families and genera are large because they were recognised first, not because they are the oldest in any phylogenetic sense. These families are surely large because they are widespread, often very natural and easily definable groups, and therefore came to botanists' attention first (e.g. *Compositae*, *Umbelliferae*, *Cruciferae*). They are successful (in terms of numbers of genera and species) because they represent adaptive peaks in floral evolution (Stebbins, 1951; cf. Valentine, 1962).

Monotypic families were no doubt rare in early systems of classification because they were considered illogical and probably inconvenient. A great deal of generic splitting was done by de Candolle and his contemporaries shortly before the publication of *The Origin of Species*. This probably had two main causes: (1) the great increase in the number of species discovered resulting in genera becoming inconveniently large and polymorphic; (2) the increasing employment by systematic botanists of the mounted lens, apparently first consistently used in dissection by Kunth when preparing his monumental *Nova Genera et Species Plantarum* (1816-25).

Many of these new genera were later sunk by Bentham & Hooker. Since their time, acceptance of the theory of evolution has led to a steady increase in the number of monotypic genera and families recognised, a treatment which results from an attempt to see that accepted taxa are monophyletic. More often than not no new names need to be found for such genera—they already exist in synonymy. Walters points out that the economic importance of the *Gramineae* has been a factor in the generic division of this family, and is one which has had little effect on the *Cyperaceae* which contains far fewer genera. But the variation in spikelet structure in the *Gramineae* is considerably greater (and more discontinuous) than in the *Cyperaceae*, which makes its generic division very much easier. *Carex* and *Euphorbia* may deserve division (and certainly need monographic treatment), but if this has not been done it is not because these genera were accepted by Linnaeus but because no modern taxonomist has had the courage to tackle such very large groups; nor is there any guarantee that convenient generic distinctions could be found.

3. *Systems based on Form Relationships*

For a century after the death of Linnaeus advances were mainly in the hands of French or Swiss botanists: Adanson, de Jussieu and the de Candolle family. There was a gradual abandonment of *a priori* reasoning because the taxonomic results were so obviously unnatural. It was a period of consolidation and extensive monographic research in which the achievement of a natural system became a major preoccupation.

MICHEL ADANSON (1727-1806) published his *Familles des Plantes* (1763) only ten years after Linnaeus's *Species Plantarum*. As Cain (1959a) has pointed out, Adanson was far ahead of his time. Not only did he describe groupings more or less equivalent to modern orders and families (the latter a unit suggested earlier by Ray), but rejected all artificial systems in favour of a natural one in which all characters used were to be given equal weight. Adanson had spent four years in Senegal, where the systems proposed by Tournefort and Linnaeus proved hopelessly inadequate to deal with the rich tropical flora which surrounded him (Stearn, 1961b). He therefore tried other systems of classification, each based on a single character such as type of hair, form of inflorescence, position of ovary, etc.

In his *Familles des Plantes* Adanson produced 65 special, single-character systems, in which the position of one genus relative to another varied greatly. Nevertheless, as he found that the same genera came together in several systems, he believed that the genera which were found side by side in the greatest number of systems should be classified together because they showed the greatest number of similarities. This would result in a natural (phenetic) system of classification. As Cain emphasises, there can be no doubt that Adanson's approach was basically right in that (*a*) all organs of the plant should be taken into account when natural affinity is to be estimated, and (*b*) contemporary *a priori* principles were either superfluous or definitely wrong. Nevertheless, Adanson's attempt was a failure (cf. Stearn (1961b) on Adanson's treatment of the *Juncaceae*). Why this is so is not certain, but it seems very likely that the characters he chose were too few to produce the desired result, and probably that some of the structures compared were not homologous.

Adanson's rejection of Linnaean nomenclature, and the presentation of his account in eccentric phonetic spelling, certainly did nothing to win converts for his theory of classification. It was rejected by de Jussieu and A. P. de Candolle. The latter rejected it for two reasons: firstly, because it presupposes that we know not only all organs of plants but also all points of view from which they can be considered—a justifiable objection at that time; secondly, that it did not allow for character weighting in accordance with *a priori* reasoning—a criticism which we cannot accept. Although Adanson's mechanical or arithmetical approach did not commend itself to his contemporaries or successors, it certainly helped to establish the natural system as a goal of taxonomy, disagreement being only on how this could best be achieved. It is only in the last few years that Adansonian methods have found vociferous advocates and seem likely, in a more sophisticated form, to play a major role in the systematics of the near future. They are already being used with success in bacteriology (cf. Sokal & Rohlf, 1962).

In 1759, Bernard de Jussieu rearranged his plants in the garden at La Trianon, Versailles, moving them about until he had devised a more natural system than any that had been suggested by Linnaeus. But he never published his system, and in 1763 called in his fifteen-year old nephew, ANTOINE LAURENT DE JUSSIEU (1748-1836), to help him. After two preliminary accounts A. L. de Jussieu published his *Genera Plantarum* in the year of the French revolution: 1789. It was a remarkable work which, by taking a big step forward in establishing natural groups, is the main source of modern Angiosperm classification.

Stearn (1961b) describes de Jussieu's method as follows. He accepted long-established families with an obvious unity (like the *Gramineae* and *Cruciferae*) and looked for the characters shared by their constituent genera. These found, he then sought for similar characters in other groups held together by general resemblances. He thereby came to the conclusion that characters are to be weighted, rather than counted, and accepted a scale of relative values for plant organs: the embryo (which begins a new reproductive cycle) was of primary importance, then the stamens and pistils (whose function it is to produce the embryo), followed by the perianth and fruit which protect the reproductive organs and the embryo; vegetative organs bring up the rear. De Jussieu's principles were accepted by Lamarck and de Candolle. In so far as the flower is morphologically the most complex organ-system of the plant, it does provide the largest number of readily available characters for comparison and, quite apart from any theory of functional importance, is bound to be extremely useful in classification (Cain, 1958).

In his *Genera Plantarum*, de Jussieu divided Flowering Plants into Monocotyledons and Dicotyledons, these being subdivided into fourteen classes according to the nature of the perianth and position of the ovary. These classes (adding the "*Acotyledones*", or Cryptogams) contained exactly 100 families—a round number, Walters (1961) suggests, that de Jussieu found convenient, and into which he disposed fairly equal chunks of the chain of variation (the largest modern family, *Compositae*, being divided into three families). His families (which he called Natural Orders) were clearly differentiated, described and named, and it is thanks to him that most of the major families were firmly established. De Jussieu recognised both "definable" and "indefinable" families (as we may call *Cruciferae* and

Rosaceae respectively), though the former predominate and have of course suffered less modification by later authors.

By modern standards eighteenth-century herbarium material was very unsatisfactory, and the lack of a mounted lens must have made it even more difficult to see floral structure in such inadequate material. There can be no doubt that it was the steady increase in the number of species cultivated in botanic gardens throughout the latter part of the eighteenth century which enabled botanists to see a wide enough range of living material to build up and establish a natural system of classification.

The work of the de Candolle family (three generations) dominated botany from the time of de Jussieu until Darwin's discoveries. Augustin Pyramus de Candolle (1778-1841) was born in Geneva and trained in Paris under Desfontaines.* He was a botanist of wide interests, and in his *Théorie élémentaire de la Botanique* (1813) expounded the principles that should underly natural classification, which were essentially those followed by de Jussieu, Lamarck and the zoologist Cuvier (Cain, 1959a, 1959c). He used two methods to determine the taxonomic importance of a character:

(*a*) Thinking out *a priori* principles for what are the most essential characters, having the greatest influence on the rest of the organisms.

(*b*) Observing what characters are the most constant in a group since of necessity these must be the most essential to it.

Due to lack of physiological knowledge, the functions of organs could not be correctly assessed. Furthermore, it soon became apparent that the systematic results of *a priori* character weighting were often manifestly unnatural. As Cain (1959a) has pointed out, since constancy of characters could only be treated *after* natural groups had been constructed, the results of *a priori* reasoning were usually the same as those obtained by grouping plants together by their overall similarities, and were therefore superfluous. In practice the *a priori* approach, with its emphasis on function, was gradually abandoned in favour of empirical methods. De Candolle's system, ably discussed by Cain (1959c), is transitional between the nearly Aristotelian systems of Caesalpinus and Linnaeus and a purely empirical one.

During the later years of his life, de Candolle began to publish his monumental *Prodromus systematis naturalis regni vegetabilis* (familiarly known as the *Prodromus*), begun in 1824; it ran to 17 volumes, of which the last 10 were written by other botanists and published after his death under the editorship of his son, Alphonse de Candolle (1806-93). This *magnum opus* was, in effect, an immense monograph in which all the known families, genera and species (58,975) of Dicotyledons and Gymnosperms (not yet recognised as a group co-ordinate with the Angiosperms) were described and classified according to the de Candolle system. With regard to the delimitation and

* The most celebrated botanist in the family, A. P. de Candolle, is generally referred to simply as "de Candolle" (abbreviated to "DC." as an authority after plant names). Similarly, A. L. de Jussieu is known as "de Jussieu".

content of families and genera, this was essentially an elaboration of de Jussieu's work, 214 families (Dicotyledons and Gymnosperms) being recognised instead of 100; but he improved on the subdivision of the Dicotyledons.

The arrangement of the families in a linear sequence, however, was quite different to that of de Jussieu, who had placed the apetalous families before the petaliferous ones within both the Dicotyledons and Monocotyledons. De Candolle began with the *Ranunculaceae*, and followed a polypetalous-gamopetalous-apetalous sequence in the Dicotyledons, the Conifers coming after the "*Amentiferae*" with which they had earlier been associated by Ray (on account of the woody "cones" of *Betula* and *Alnus*). Stearn (1961a: 184) in a brief account of de Candolle's *Prodromus* writes: "His arrangement of the classes within the Dicotyledons was based on reduction and fusion of floral parts. He began with the group (*Thalamiflorae*) with the greatest number of floral organs distinct and separate from one another and placed in a descending order from this those with parts more fused and reduced in number. The consideration that it was illogical to start with little-known objects in order to arrive at those much better known doubtless made him choose for his first family *Ranunculaceae*, which had been the object of a classical essay by A. L. de Jussieu in 1773. . . ."

This lucky choice of the Ranalian group as the starting-point in the linear sequence of families (accepted as necessarily artificial by de Candolle) has done much to popularise the systems of de Candolle and Bentham & Hooker, since the arrangement often coincides with evolutionary progression. The sequence of the families in the *Prodromus* has such a remarkably evolutionary look that it is hard to remember that its construction is pre-Darwinian and that Lamarckism was not accepted by de Candolle. It should be pointed out that the sequence of co-ordinate groups (as distinct from their relationship on overall affinities) must be considered as something *extraneous* to the natural system, imposed upon it by the necessity of following a linear sequence in a book or herbarium.

The *Prodromus* was (and still is) such a very useful book that it inevitably led to the acceptance of de Candolle's system, and by 1840 it had everywhere superseded the Sexual System of Linnaeus which did not satisfy what might be described as an international urge for a natural arrangement of plants. It is a sad reflection on taxonomy today that for many of the larger genera (e.g. *Ranunculus*, *Senecio*, *Euphorbia*, *Solanum*) the *Prodromus* contains the only world-wide monographic treatment, although the number of species now known is perhaps three times as great.

Linnaeus described about 7700 species of Angiosperms in his lifetime, and thought that hardly 10,000 plants existed in the whole world. In 1813 de Candolle noted that 30,000 species of plants were known (in all groups), with another 10,000 awaiting description. Yet 58,805 species of Dicotyledons were described in the *Prodromus*, of which the last volume appeared in 1873.

Dr. W. T. Stearn tells us that, by using a new method of calculation based on the number of species accepted by monographers, he now estimates the number of known Angiosperms to be between 200,000 and 250,000 according to whether a conservative or liberal species concept is adopted.

A British contemporary of de Candolle, ROBERT BROWN (1773-1858), devised no system of classification, but his thorough research on floral morphology bore fruit in the systems of those who followed him. He recognised that the Gymnosperms were a group apart from the Angiosperms and were characterised by naked ovules—a discovery that later led to the formal recognition of the two taxa by Hofmeister. Brown was the first to understand the structure and pollination of the flower of *Asclepiadaceae*, the nature of the cyathium of the *Euphorbiaceae*, and even the morphology of the grass flower in relation to that of other Monocotyledons. It is difficult to understand how Brown could have written as he did without some intimation of evolution. What does he mean by stamens in the grass flower being "suppressed"? Could *he* have believed in the constancy of species? (cf. Wilmott, 1950a). Such philosophical problems aside, however, Brown's observations were of the greatest value to biology in general and to classification in particular. Humboldt considered him to be "*botanicorum facile princeps*".

After de Candolle, numerous systems of classification were proposed, most of them elaborations of de Candolle's system. LINDLEY'S *An Introduction to the Natural System of Botany* (1830) is remarkable for its lucid exposition of natural classification and the importance of vegetative characters. "The organs of fructification [i.e. flower and fruit] are only entitled to a superior degree of consideration when found by experience to be less liable to variation than those of vegetation"—a view expressed repeatedly in our book. Lindley divided the Dicotyledons into *Polypetalae*, *Apetalae*, *Achlamydeae* and *Monopetalae* but frankly considered these groups to be "entirely artificial". In his division of the Monocotyledons we find the precursor of Hutchinson's division of this group on the nature of the perianth. Lindley gave an excellent synoptic key to the families and a penetrating discussion of the affinities of each family and often of their morphology (see, for example, his interpretation of the perianth of *Fumariaceae* which deserves serious attention). Lindley's book is, in fact, remarkable for the amount of botany it contains both in regard to accurate observation and intelligent interpretation. It was the first comprehensive natural system to be published in English and was widely used in Great Britain and America.

Four years later the Viennese botanist, ENDLICHER, published his *Genera Plantarum* (1836-40) in which he described 6235 genera of vascular plants. This impressive compilation (published in a single volume weighing 6 lb.) recognises three primary groups in the Dicotyledons: *Apetalae*, *Gamopetalae* and *Dialypetalae* (i.e. *Polypetalae*), the first combining the *Apetalae* and *Achlamydeae* of Lindley. Endlicher's division of the Monocotyledons was radically different from Lindley's and appears to have formed the basis for

Bentham & Hooker's division of the group, although the sequence was not the same. Unlike de Candolle (but like de Jussieu), Endlicher began both the Dicotyledons and Monocotyledons with the apetalous families, thus establishing the pattern to be followed later by the Englerian school.

The pre-evolutionary era culminated in BENTHAM & HOOKER'S *Genera Plantarum* (1862-83), which, though post-Darwinian in date, was pre-Darwinian in conception and is therefore considered in this chapter. Bentham was a self-trained English botanist who had already published several Floras and monographic works (e.g. the account of the *Scrophulariaceae* and *Labiatae* in de Candolle's *Prodromus*) before he collaborated with Sir Joseph Hooker, taxonomist and phytogeographer, on the *Genera Plantarum*. This is a major landmark in botany, not so much for its system—which closely follows that of de Candolle—but for its quality. All genera of seed plants were very accurately described in Latin from living specimens or dissected herbarium material. Hooker, being Director of the Royal Botanic Gardens, Kew, had every facility at his disposal. A synopsis at the beginning of each family (of which 200 were recognised, including Gymnosperms) facilitated the identification of its constituent genera, and most of the larger genera were further divided into subgenera and sections. The geographical range of each genus was given. The careful observation of new material freed the book from the repeated mistakes of earlier works, and the fine taxonomic judgment achieved by its two authors (though Bentham wrote two-thirds of the text) set a standard for generic description and delimitation that has never been surpassed. British herbaria and those of the Commonwealth are still arranged by the system of Bentham and Hooker; most of the others follow the Englerian system.

Bentham and Hooker's system is not, of course, meant to be phylogenetic, but is certainly largely natural in the content and delimitation of its groups, particularly with regard to the families—a heritage from de Jussieu. Its arrangement, however, is in accordance with the progressive fusion and reduction of floral parts that was a feature of de Candolle's system, but the groupings above the family level differ considerably from those in the *Prodromus*. In the latter the Dicotyledons were divided into *Thalamiflorae*, *Calyciflorae*, *Corolliflorae* and *Monochlamydeae*. In de Candolle's treatment the sympetalous families were divided between two groups, those with an inferior ovary in the *Calyciflorae*, while those with a superior ovary comprised the *Corolliflorae*. Bentham & Hooker combined de Candolle's *Thalamiflorae* and the polypetalous part of the *Calyciflorae* into the *Polypetalae*, which they divided into three "series": *Thalamiflorae*, *Disciflorae* and *Calyciflorae*. The sympetalous families were all united in the *Gamopetalae* and the *Monochlamydeae* maintained as a co-ordinate group. The major groupings were, in fact, those adopted by Endlicher (1836-40). Bentham & Hooker's family concept was considerably broader than de Candolle's. They recognised 127 families of Dicotyledons in the *Genera Plantarum*, whereas 211 had been described in the *Prodromus*—a result, perhaps, of using numerous collaborators in the latter's compilation.

The Monocotyledons were not covered by the *Prodromus*, but the order in which they are listed in de Candolle's *Théorie* (1813) is largely followed by Bentham & Hooker. The "series" in which the families are grouped, however, are a modification of those adopted by Endlicher although Bentham & Hooker differ from the latter in starting both the Dicotylodons and Monocotyledons with polypetalous families in the tradition of de Candolle.

The preparation of the first volume of the *Genera Plantarum* coincided with the publication of Darwin's *Origin of Species* (1859)—indeed, the first plan was drafted the preceding autumn. Hooker, a friend of Darwin's, became an enthusiastic champion of natural selection, and wanted to reorganise their system. But Bentham was more cautious, and persuaded Hooker to carry on with the work as it had been originally planned. Darwin's views were not fully accepted by Bentham until six years after the publication of "The Origin". Turrill (1942) suggests that it was not the time factor so much as the difficulty of agreeing on a phylogenetic system and of expressing phylogenetic concepts in a linear sequence, as well as Bentham's doubt about the validity of the theory of evolution, that prevented them from publishing a phylogenetic system. It is, perhaps, just as well that they decided not to attempt it.

A combination of the systems of the de Candolles and Bentham & Hooker was adopted by Asa Gray (1810-81) during the formative period of American taxonomy.

SUMMARY OF PRE-DARWINIAN CLASSIFICATION

Before the acceptance of evolution, the development of taxonomy was largely confined to Europe. The first attempts at classification were largely based on Aristotelian principles involving Logical Division and the weighting of characters according to *a priori* reasoning. Classifications primarily based on habit were replaced by the Sexual System of Linnaeus in which floral characters were weighted above all others. The result was a largely artificial system, though Linnaeus achieved natural groups at specific and even generic level despite (and in contradiction to) his Aristotelian principles; theory and practice were out of step. Linnaeus established the binomial system, provided an eminently usable classification and exact descriptive terminology. His *Species Plantarum* (1753) provides the starting-point for modern nomenclature and lays the foundation of present-day classification.

Even Linnaeus realised that a natural system would be preferable to an artificial one, but it was left to Adanson to formulate the principles of how this might be attained—by taking all morphological characters (vegetative and floral) into account and grouping those plants together which shared the greatest number of similarities. Although reluctant to give up *a prior*, weighting, the first major step in establishing a natural system (i.e. one based upon overall resemblances) was taken by de Jussieu, who first named and described 94 Angiosperm families with their constituent genera. This formed the basis for the system adopted in de Candolle's *Prodromus*, in which, however, the families appeared in a new sequence (starting with the *Ranunculaceae*) based on increasing fusion and reduction of floral parts. *A priori* reasoning gradually gave way to empirical methods which evidently gave more natural taxonomic results. De Candolle's system was extended and improved by Bentham & Hooker in their *Genera Plantarum*, a work never surpassed in its steady judgment and in the accuracy of its descriptions. The sequence of the families (very close to that of de Candolle's) pre-adapted the Bentham & Hooker system to the evolutionary period that was to follow since it reflected evolutionary trends in floral structure. The building up of a natural system was only made possible by the steady increase in the number of plants cultivated in botanic gardens during the second half of the eighteenth century—a trend which has continued to the present day.

Typological thinking, associated with the belief in Special Creation, and the ignoring of variation (at least for taxonomic purposes) characterises the pre-Darwinian period. By the middle of the nineteenth century, however, variation was becoming impossible to ignore as a stream of new collections poured into European herbaria. Increasing attention was given to the geographical distribution of taxa, and inherited infra-specific variation was accepted and recognised in taxonomic terms. The consolidation of the Linnaean phase was followed by a period of monographic revision in which attempts at a natural system became standard practice. Although characters other than gross morphology were seldom used, detailed studies in plant morphology, particularly in anatomy and floral structure, allowed taxonomic conclusions to be more firmly based.

Chapter 2

SYSTEMATICS AFTER DARWIN: ITS MODERN BASIS

The Effects of Evolutionary Theory on Systematics

The whole course of biological thought was radically changed by the publication of Darwin's *The Origin of Species* in 1859. It influenced taxonomy by the enunciation of two important and far-reaching principles ; the effect of the first was immediate, that of the second delayed:

(1) Species are not unchanging creations but have evolved from one another in the long history of life; they form an ever-changing continuum. Out of this discovery has grown the principle that all species in any systematic group are of common ancestry—or will be if our groups are phylogenetically constructed. After Darwin, classification became avowedly evolutionary in aim.

(2) Species are not represented by types but by variable populations—the *population concept* of modern taxonomy. Darwin grasped at least the rudiments of population systematics, but the true nature of variation was hidden from him. Comprehension came with the rediscovery of Mendel's principles of heredity (cf. Wilkie, 1962) and the development of population genetics.

One might have expected that the acceptance of these two revolutionary concepts would have caused classification to undergo a profound and spectacular change. This was not the case. Post-Darwinian systems have differed little in content, though they have differed in arrangement, from those of the later pre-Darwinian taxonomists. This is because the systems of de Candolle and others were, in the content of their groups, very largely natural. As Bigelow (1958) has shown, such an approach is evolutionary in the truest sense, as distinct from being phylogenetic. Above the level of the species, the discovery and understanding of evolution, without fossil evidence, can have little effect on classification beyond provoking us to a determined effort to see that our groups are probably monophyletic, at least in so far as this is consistent with convenience of recognition. The taxonomic groups have remained much the same, though the desire for groups to be supposedly monophyletic has led to a splitting of families and genera and an increasing number of monotypic taxa.

A characteristic of post-Darwinian taxonomy has been the construction of phylogenetic trees on extremely dubious evidence. The evolution of *organs* has been mistaken for the evolution of *taxonomic groups*, and a vast amount of time has been spent in constructing hypothetical trees that might (it seems to us) have been better spent in building up a natural system based on overall resemblances. Nevertheless, these arboreal nesting activities have served one very useful purpose: they have stimulated the search for new data in the fields of morphology, anatomy, cytology, etc. which are not only of general biological interest but will come in very useful in the construction of a natural system, quite apart from phylogenetic interpretation. Of foremost importance has been the search for primitive characters, and the realisation that simple structures are often not primitively simple, but simple as a result of reduction from more complicated ones.

The effect of the population concept has, of course, been at and below the species level and will be discussed further in Chapter 11. We need only say here that it has resulted in a different approach to infra-specific classification. Interest in geographical and ecological variation led to the establishment of the *subspecies*, a taxonomic unit that, in its modern usage, has penetrated plant taxonomy from zoology and is becoming increasingly used. Another result has been the *special classifications* based on the ability of populations to interbreed under experimental conditions. Information on the distribution of *polyploidy* has steadily increased and presents the plant taxonomist with a problem in classification that is very much rarer in zoology. How cyto-genetic criteria should effect a general classification has been a hotly debated topic of the last few decades and is at the centre of the modern "species concept". Continuous variation, which cannot be adequately dealt with in the taxonomic hierarchy, has received considerable attention, and mathematical and pictorial ways devised to describe it. All these matters will be dealt with more fully in the chapters to follow.

Interest in evolutionary theory and its effect on systematics (particularly the species concept) reached its peak in the publication, mainly during and after the Second World War, of a series of books that covered both the botanical and zoological fields: Dobzhansky's *Genetics and the Origin of Species* (1937), *The New Systematics* (edited by J. S. Huxley, 1940), Huxley's *Evolution: The modern Synthesis* (1942), Mayr's *Systematics and the Origin of Species* (1942), Simpson's *Tempo and Mode in Evolution* (1944), Stebbins' *Variation and Evolution in Plants* (1950), J. Clausen's *Stages in the Evolution of Plant Species* (1951), Simpson's *The Major Features of Evolution* (1953), *Evolution as a Process* (edited by Huxley *et al.*, 1954), J. Heslop-Harrison's *New Concepts in Flowering-Plant Taxonomy* (1953), Zimmerman's *Die Phylogenie der Pflanzen* (1959), and Rensch's *Evolution above the Species Level* (1959, but originally published in German in 1954). The journal *Evolution* (which started in 1947) is largely devoted to this subject and of great value to taxonomists interested in evolutionary interpretation.

The theory of classification, which had been neglected since the time of de Candolle, has had its increasing band of exponents during the last few years, among whom should be mentioned J. S. L. Gilmour, A. J. Cain, G. G. Simpson and R. S. Bigelow. Herbarium taxonomists have often shown a reluctance to think out the basis of their methodology, so developments in this theoretical field are to be welcomed, particularly when they are accompanied by practical demonstration.

Monographic and Floristic Developments

In the practical field of "orthodox" ("classical") taxonomy, there appears to be a "cyclical succession" of monographic and floristic work (Stafleu, 1959). The first half of the nineteenth century saw great concentration in the synthetic and monographic fields (de Candolle's *Prodromus*, etc.), followed by a great expansion in floristic work during the latter half of the century, when Martius's *Flora Brasiliensis*, Boissier's *Flora Orientalis* and the British colonial Floras appeared. The first part of the twentieth century saw renewed emphasis on synthesis under the dynamic influence of Engler. Now floristic work predominates again; the enormous *Flora URSS* is nearly complete, and in the ex-colonies nationalism has replaced the earlier incentive for the production of colonial floras. The National herbaria in England are largely preoccupied with the preparation of Floras which should of course be of great practical value to biologists in the countries concerned.

Monographic work goes on painfully slowly. This is most unfortunate for two reasons: firstly, because Floras could be written much better if more monographs were available; secondly, because Systematics cannot make real progress without monographic research. Monographic work is all the more necessary today when evolutionary theory and so much non-morphological data needs to be taken into account. The recent revival of interest in the theory of classification (involving the slaying of phylogenetic monsters) requires monographic work if theories are to be tested. There is a danger that a small body of biologists will think out (but not necessarily practise) the principles of classification, while a much larger body of taxonomists will continue the necessary task of practising taxonomy by the "performing monkey" method described by Cain.* If systematics is to advance, theory and practice must be intelligently combined.

A commendable move to increase international collaboration has been made by the formation of the *International Association for Plant Taxonomy*

* "Is it not extraordinary that young taxonomists are trained like performing monkeys, almost wholly by imitation, and that in only the rarest cases are they given any instruction in taxonomic theory . . .? If we are to be scientists, we must study the fundamental principles of taxonomy, must make those alterations in our practice which follow as consequences upon them, must know, in short, what we are doing. for what purpose we are doing it, and how to do it efficiently." A. J. Cain (1959b).

(I.A.P.T.). This has published such useful works as the *International Code of Botanical Nomenclature*, the *Index Herbariorum* and the *Index of Specialists in Plant taxonomy*. It also produces a journal, *Taxon*, that has hitherto concentrated more on nomenclature and obituaries than on the principles and progress of plant systematics. Plant taxonomists interested in the basic principles of biological classification are more likely to find them in the stimulating journal, *Systematic Zoology*, which has no counterpart in Botany.

Definition and Use of Terms

Before we survey the field of evolutionary taxonomy, it seems advisable to consider the use of certain terms intimately connected with it. Several of these have been discussed by Haas & Simpson (1946).

1. *Primitive or Unspecialised, versus Advanced or Specialised*

It is difficult to define these much-abused terms (so often involved in circular argument) more clearly than in the way they are used by Simpson (1961). "The concept of primitive and specialised (advanced) are relative and meaningless unless definitely related to a particular taxon, lineage or phylogeny. Within a taxon the characteristics of the common ancestry are primitive and others are more or less specialised in proportion to their departure from the ancestral condition."

It is common knowledge that organs do not evolve at the same rate—*Delphinium*, for instance, is primitive in its free follicles, specialised in its zygomorphic perianth with a nectar spur. This out-of-step correlation of advanced and primitive characters is particularly prevalent in the lower Angiosperm families (e.g. in *Ranales sensu lato*), but when once the more specialised ones are reached (e.g. *Compositae*, *Labiatae*) the ancestral features of the Angiosperms appear to have been quite lost. The rates at which organs evolve are obviously related to the selection pressures affecting them, and to their genetic ability to vary.

Angiosperm organs tend to evolve in four more or less independent groups—those of the underground parts, those of aerial shoots, and those of flower and fruit. Within each of these four groups the organs involved often show a marked functional or ecological correlation, so that they appear to have reached a similar stage of advancement; but there is little such correlation between the evolutionary stage of organs belonging to the different groups. Obvious examples are to be found in the *Ranunculaceae* and in *Papaveraceae sensu lato* (in *Dicentra*, for instance, the species in Sect. *Dactylicapnos* retain a primitive type of rootstock while having the most specialised flowers—cf. Stern, 1961). Although this must be an overstatement of what is a more complicated and subtle inter-relationship of parts, the concept is a useful tool for the detection of convergent evolution, since similarities due to convergence are likely to be within one or other of these character groups. This concept is evidently related to the theory of "co-ordination chains" or "correlation pleiades" recognised by Sewertzoff (1931) and Terentjev (1931, 1959) in which the parts are structurally and functionally correlated and therefore show a high statistical correlation (e.g. between the different elements in xylem and phloem, or between different parts of the flower). The member of a given pleiad which shows the highest average correlation with its other members has been called the "indicator character", or simply "indicator" (Smirnov, 1924, 1938; Terentjev, 1931). The indicator of a pleiad is maximally correlated with other members of the same pleiad and therefore has the highest taxonomic significance. The set of such indicators constitutes the so-called taxonomic coefficient. (We are indebted to Professor A. L. Takhtajan for this information on Russian taxonomic theory which appears to be little known in the West. The ecological significance of correlation pleiades is discussed by Berg in *Evolution*, **14**, 171: 1960).

The different rates at which organs have evolved in the same species makes it particularly difficult to assess phyletic relationships, or even to say whether one taxonomic group is more advanced than another. For this purpose it is probably necessary to assess the average advancement of as many characters as possible. A further complication, stressed by Bigelow, is that different lineages have advanced at different rates. This means that an advanced taxon may be either more ancient or recent than another advanced group; indeed, it may even be less advanced yet younger than another (cf. Fig. 2).

Although in our view phylogeny cannot be correctly constructed (at least above the species level) in the absence of fossil evidence, we can be much more certain of the evolutionary trends that have effected *individual organs* if we understand their function and adaptive significance (cf. Cain, 1959c). We may therefore be able to estimate the general level of advancement of an organism relative to others, but that does not tell us from which taxon another has been evolved, which is what a study of phylogeny attempts to do.

Within the Angiosperms the fossil record gives us little help in establishing evolutionary trends on an absolutely unequivocal basis. Nevertheless, a great deal of circumstantial evidence—not to mention common sense—supports the majority of the dicta concerning primitive and specialised characters put forward by Bessey and Hutchinson and accepted by many taxonomists today. This evidence includes character correlation and the interpretation of natural selection in terms of "selection for what?" Although the latter must often rest upon *a priori* judgment (and in many cases may be mistaken), the fact remains that evolutionary trends (except for local cases of genetic drift) are directed by natural selection, either directly or because of linkage.

The trends in the morphology of the Angiosperms listed by Hutchinson (1959) are briefly as follows:

1. In certain groups, trees and shrubs are probably more primitive than herbs. [We should consider this the general rule.]

2. Trees and shrubs are older than climbers.

3. Perennials are older than biennials, and annuals have been derived from both.

4. Aquatic Phanerogams are as a rule more recent than related terrestrial plants.

5. Plants with collateral vascular bundles arranged in a cylinder are more primitive than those with scattered bundles.

6. The spiral arrangement of leaves on the stem and of the floral leaves (perianth parts) precedes that of opposite and whorled types.

7. As a rule simple leaves precede compound leaves. [There must be numerous exceptions here.]

8. Bisexual (hermaphrodite) flowers precede unisexual ones, and the dioecious condition is probably more advanced than the monoecious condition.

9. The solitary flower is more primitive than the inflorescence, the highest form of the latter being the umbel and the capitulum.

10. Spirally imbricate floral parts are more primitive than whorled and valvate ones.

11. Flowers with many parts (polymerous) have preceded those with few parts (oligomerous flowers), being accompanied by a progressive sterilisation of reproductive parts.

12. Petaliferous flowers precede apetalous ones, the latter being the result of reduction. (However, both primitive and derived apetaly apparently occur in the *Ranales*, *fide* Eames: 1961.)

13. Free petals (polypetaly) are more primitive than connate petals (sympetaly).

14. Actinomorphy (radial symmetry) of the flower is more primitive than zygomorphy (bilateral symmetry). [The latter condition is usually correlated with secretion of hidden nectar.]*

15. Hypogyny is the primitive condition, and from it perigyny and epigyny have been derived.

16. Free carpels (apocarpy) have given rise to connate carpels (syncarpy). (Eames, however, claims that syncarpy may also have arisen by the lateral fusion of open, involute carpels—cf. p. 180.)

17. Many carpels (polycarpy) have preceded few carpels (oligocarpy).

18. The endospermic seed with a small embryo is more primitive than the non-endospermic seed.

19. In primitive flowers there are many stamens, few in more advanced ones. This condition may, however, be reversed within the confines of a single family.

20. Separate stamens precede connate stamens.

21. Aggregate fruits (i.e. those formed from separate flowers) are more

* The evolutionary trends in the nectary of Angiosperm flowers have been the subject of papers by W. H. Brown (1938) and Fahn (1953).

recent than single fruits, and as a rule the capsule has preceded the drupe or berry.

To these we might add some further general trends:

22. Stipulate leaves are more primitive than exstipulate leaves, and free stipules probably more primitive than adnate stipules (Eames, 1961).

23. In the Dicotyledons phyllodic leaves (corresponding to petiole and leaf base) are more advanced than normal laminar leaves. (The same applies to phyllodic leaflets, as in *Lathyrus*.)

24. Spininess, at least of branches, leaves and bracts, is generally a derived condition, being an adaptation to arid environments or heavy grazing.

25. Succulence is a specialisation, and so are elaborate types of indumentum (e.g. stellate hairs, lepidote scales) though the latter may be lost so that the glabrous condition may be primitive or derived.

26. Epiphytism, saprophytism and parasitism, and consequently the characters functionally associated with them, are advanced.

27. Absence of a taproot is a derived condition. Within the Dicotyledons, rhizomes or corms (bearing adventitious roots) are more advanced than taprooted relatives (e.g. within the *Ranunculaceae*, or within *Crepis*).

28. The primitive unit of dispersal in the Angiosperms is the seed. This becomes retained in the ovary in single fruits (apocarpous or syncarpous), often accompanied by a reduction in seed number per fruit. The trend is for more and more parts of the flower and inflorescence to take part in dispersal (particularly calyx and bracts), so that finally the whole inflorescence becomes the unit of dispersal or (in tumble weeds) even the whole plant. This trend is related to special methods of dispersal.

29. Heteromorphism of floral and vegetative organs is usually derived from monomorphism. There is probably no organ of the plant which does not show heteromorphism in one genus or another. To take some obvious examples: dimorphic roots (*Ranunculus* Sect. *Ranunculastrum*); submerged dissected leaves and floating peltate leaves (*Cabomba*); dimorphic stipules (*Paliurus*); heteromorphic stamens (*Verbascum*, *Cassia*); heteromorphic fruits (*Calendula*), etc. If we compare *Magnolia* with *Salvia* or *Taraxacum*, there is no doubt the last two show considerably more heteromorphic organs than the first. This trend is obviously related to phylogenetic modification by way of neoteny and gerontomorphosis (cf. p. 156)

The trends we have listed, however, are nearly all trends in gross morphology. It would not be difficult to find even more trends in the microscopic features of the plant, many of which will be dealt with later in this book. We shall only mention here trends in pollen morphology, anatomical trends both in the plant body and in the reproductive organs (trends in wood anatomy seem particularly valuable, since they appear to be largely one-directional), trends in the elaboration of the epidermis and in the morphology and ontogeny of the stoma pore, trends in the karyotype, in the ovule and embryo sac, and in the chemical evolution of plant products, etc. It is obvious that, if we wish to assess relative advancement, suitable characters are not lacking provided we can decide which is the derived state. A survey of the evolutionary trends found in Angiosperm organs is given by Takhtajan (1958, 1959b).

Most of the trends in gross morphology have come about by three processes: *reduction*, *fusion* (cohesion and adnation) and *change of symmetry* (Stebbins, 1950). These trends may be regarded as part of a more general basic phenomenon—change in relative growth rates, or *allometry*. A fourth trend, *elaboration*, has played a less important role in evolution, appearing relatively late, although it accounts for the differentiation of vascular tissues. A biological principle of wide application can be stated here (cf. Corner, 1949a): reduction in the size and complexity of individual structures is often accompanied by an increase in the number of structures produced (e.g. follicle versus achene in the *Ranunculaceae*). Similarly, reduction in number is often balanced by an increase in size and complexity (e.g. the single seeds of *Quercus*, *Juglans*, *Myristica* and *Prunus*).

Nobody who believes in evolution can doubt that these morphological trends exist, or even that they can be seen in natural groups as they are living today. The bone of contention has been the *direction* of the trends in the absence of fossil evidence.* Apart from the difficulty of being sure of the selective factors, there is the added difficulty that most of the trends in gross morphology (and presumably most of the microscopic ones, too) can go, and very probably have gone, into *reverse* in cases where selection pressure has favoured it. A strong case has even been made out for the derivation of the perennial habit from the annual one in New World species of *Cleome* (Iltis, 1959). This conclusion is based on considerations of ecology and character association. Exceptions to reversal probably include the inferior ovary as a constantly advanced condition. Lost organs, too, are seldom regained.

Few taxonomists would doubt that the *Chenopodiaceae* have arisen from a Centrospermous stock by the loss of petals, reduction in the androecium and number of ovules (all trends present in the *Caryophyllaceae*), or that the free stamens of wind-pollinated *Compositae* like *Ambrosia* represent a derived condition within the family. But the relationship of parietal to axile placentation, or even of simple to divided leaf, is much more difficult to decide, particularly when anatomical evidence is conflicting or we can see no obvious selective advantage to indicate the direction of the trend.

Without fossil evidence such cases may remain insoluble. Sporne (1948b, 1949, 1954, 1956) has tried to break this deadlock by the use of statistical methods (cf. p. 68). However, the claim that if you can demonstrate that one character is primitive from the fossil record, then another must be primitive because it has a statistically significant correlation with it, is surely a *non*

* Trends need not be interpreted as going from one end of the sequence to the other. They may be read in either direction, from any point on the sequence.

sequitur. Not only are advanced characters sometimes closely correlated with primitive ones; the primitive feature may owe its survival to correlation with a more specialised character. Nor does the method make allowances for reversals of trend (e.g. primitive and derived apetaly) which appear to have so frequently occurred. In this case one cannot help feeling that a more subjective estimate of advanced and primitive characters (based on wide experience and common sense interpretation) might have been more fruitful than a statistical approach based on inadequate data or dubious basic assumptions. An example of the former method of calculating an advancement index is provided by a recent revision of *Cleome* (*Capparidaceae*) in America (Iltis, 1959), in which the author stresses the necessity of interpreting advancement criteria afresh for each taxonomic group. It should be realised, however, that an advancement index (however calculated) for any taxonomic group above the species is an entirely theoretical concept, analogous to that of the archetype, since it represents an imaginary average condition which may not be represented by any particular species.

Sporne (1948b), however, has clarified a point which has confused many botanists—including J. Hutchinson—in their assessment of advanced and primitive characters. It has often been assumed that an ancient group is necessarily a primitive one, and that a recent group is necessarily advanced. This is often not the case, and results from a failure to realise the implication of the term primitive. "A primitive character is one which, possessed by present-day families (or any other category) was also possessed by their ancestors." Primitive or advanced groups may be recent or ancient according to whether they became separate entities early or late in evolution (see Fig. 2).

When a character appears to be "primitive" in association with others which are obviously advanced, we must always consider whether it is (*a*) a primitive character which has persisted, or (*b*) the result of reversal of trend. In many cases it may be impossible to tell, but the clue may lie with its function in relation to natural selection (cf. Cain, 1959c), and in its relation to others in a "correlation pleiad".

The rate of evolution appears to have been frequently speeded up by aridity. Stebbins (1952) points out that there are numerous different structures which plants can evolve in adaptation to dry conditions. These include the annual habit, bulbs, corms, etc., reduction in leaf surface (reduction in size or by increasing dissection), development of succulence, elaboration of indumentum and stomata, synaptospermy,* etc. Though a few families like the *Casuarinaceae*, *Zygophyllaceae* and *Cactaceae* appear to have evolved in response to arid environments, it is more common to find trends in morphological specialisation to aridity *within* families or genera. Much of the morphological diversity in *Ranunculus*, *Alyssum*, *Dionysia* (*Primulaceae*),

* *Synaptospermy*: retention of more than one seed in the disseminule, so that the seeds are dispersed and "planted" together (e.g. *Neurada procumbens* and *Medicago* spp.). Cf. Murbeck, 1920.

Polygonum, *Silene*, *Crepis* and *Festuca* has presumably arisen in this way. Similar xeromorphic trends are found in high mountain floras where aridity, associated with other factors (such as extreme cold, strong wind and insolation), is an important selective agent (cf. Stebbins, 1954).

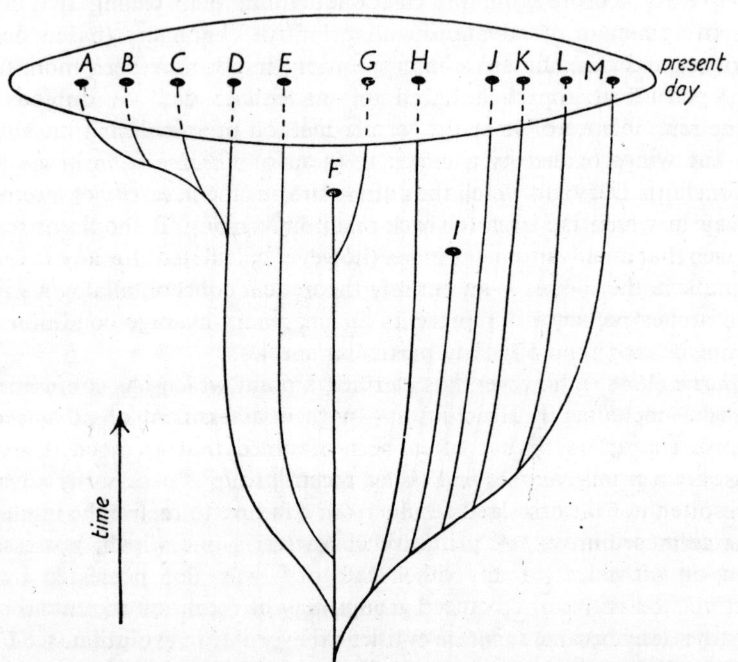

FIG. 2. Phylogenetic tree: advanced and primitive taxa. Each time a new branch arose (indicated by an angle) it is assumed that an advancement occurred; *A* and *K* are equally advanced, but *A* is much younger than *K*. *M* is more advanced, yet more ancient, than *A*, *B* or *C*. Both *E* and *G* are primitive, since they have diverged only once from the ancestral stock, yet *G* is recent and *E* ancient. *F* is the continuation of the primitive stock and is extinct. (After Sporne, 1948b.)

2. *Homology and Analogy*

Systematics is widely pervaded and influenced by the concept of homology. Both the affinity of groups and considerations of phylogeny are based upon supposed homologies. As this is the case, it is a serious flaw that the terms homology and analogy are so difficult to define, or at least their results are so difficult to distinguish.

The terms were first used in pre-Darwinian times and defined by the zoologist Owen (1848). He defined *homology* as "the same organ in different animals under every variety of form and function", *analogy* as "a part or an organ in one animal which has the same function as another part or organ in a different animal". These definitions are unsatisfactory because they leave it to us to decide whether an organ is "the same organ" or "another organ".

Darwin (1859) used the terms in both zoology and botany, giving examples from both. Homology, he writes, is "that relation between parts which results from their developing from corresponding embryonic parts, either in different animals . . . or in the same individual . . .; the latter is called *serial* homology. . . . In different plants the parts of the flower are homologous, and in general these parts are homologous with leaves." *Analogy* he defined as "the resemblance of structure which depends upon similarity of function, as in the wings of insects and birds" (i.e. resemblance in spite of lack of relationship). Darwin's definitions therefore necessitate (*a*) a knowledge of the way in which the organs are formed, (*b*) knowledge of the organism's phylogeny.

Simpson (1961) boldly defines homology as "resemblance due to inheritance from a common ancestry". (The similar characters involved are *homologous*, and the noun for them is *homologue*.) None of these definitions can be confidently applied in the Angiosperms, chiefly because there is so little fossil evidence to guide us in distinguishing homology from analogy. Furthermore, studies in morphogenesis often invalidate the frequent claim that homologous organs develop in similar positions.

Despite these serious shortcomings, the concept of homology and analogy is so widely used in systematics that it is difficult to visualise taxonomy being carried on without it today. In practice, the distinction is easy to make in closely related plants; it becomes increasingly difficult as we consider more distantly related groups, especially when they hold a very isolated position. De Beer (1951) states that the best criterion of homology in animals is comparative anatomy. The same may well be true of plants, but in any case, when in doubt one must obviously look more closely at the structure of the organs in question.

Although we have said that classification is based on probable homologies, it should be pointed out that resemblances in closely associated *analagous* organs do not necessarily mean that the plants are not closely related. This will be so in cases of transference of function (Corner, 1949b, 1958) in which the function of one organ has been taken over by another organ in the course of evolution by a shift in embryonic development (e.g. the adaptive pattern of the scarlet aril which contrasts with the black seed-coat in many tropical *Leguminosae* has been transferred to the seed-coat itself in *Adenanthera bicolor* and some species of *Erythrina*). In such cases resemblances between *analagous* organs may be a better indication of affinity than resemblance in homologous ones!

Before we continue, it seems necessary to draw attention to a general consideration which provides the background and reason for the use of these and related abstract terms. Natural classification recognises groups based on overall similarities (affinity) which are largely based on homologies.

We know, however, that this similarity does not necessarily mean common ancestry, and for an *evolutionary interpretation* it is therefore necessary to introduce the concepts of homology and analogy, parallelism and convergence, monophyly and polyphyly. These are all relative terms which can seldom be usefully employed outside a particular context.

3. *Parallelism and Convergence*

It is quite possible to define these terms, less easy to distinguish them in practice. They have also been very loosely used.

Simpson (1961), a staunch believer in phylogenetic classification, uses *parallelism* for "the independent occurrence of similar changes in groups with a common ancestry and *because* they had a common ancestry". Neither become significantly more or less similar to one another. The basis of parallelism, so defined, is therefore the common ancestral genetic basis that has enabled parallel evolution to occur—a common potential for parallelism. Simpson's definition of parallelism, although theoretically acceptable, is consequently very difficult to apply to the Angiosperms; we do not know enough about their phylogeny or genetics.

Convergence implies increasing similarity between two distinct phyletic lines, either with regard to individual organs or to the whole organism. That is, the similarities have arisen *in spite* of lack of affinity, and have probably been derived from different systems of genes—the opposite to parallelism. The occurrence of pollinia in the *Orchidaceae* and *Asclepiadaceae*, of the "switch habit" in *Equisetum*, *Ephedra*, some members of the *Leguminosae* (e.g. *Retama*) and *Polygonum* are obvious examples of convergence, although they have often been loosely referred to as "parallelism". (We may speak of parallelism in a particular *character* which itself produces convergence between the *taxa* concerned, especially when the character is an unusual one.) Convergence is generally brought about by adaptation to similar climates or habitats (mostly affecting the vegetative parts), or even to similar methods of pollination or dispersal (affecting the flower, fruit or seed).

When plants are more closely related (as assessed by overall similarities), it becomes more difficult to distinguish between parallelism and convergence. Is the remarkably similar plant-form of *Anabasis* and *Haloxylon* due to close common ancestry, or to the lack of it? In the *Pflanzenfamilien* these two genera in the *Chenopodiaceae* are placed in different subtribes on account of the different orientation of their embryos.

Before we leave the subject of convergence (discussed further on p. 47) we should define what is meant by *cladistic* and *patristic* affinity (Cain & Harrison, 1960). Similarity between two plants due to a known or inferred common ancestry, is called *patristic* similarity; in terms of phyletic lines the plants are directly related. *Closeness* of relationship in terms of phyletic lines is called *cladistic*, the word referring to the actual pathways by which a given similarity has been attained. Patristic similarity is therefore one way by which resemblance is required, the other being by convergent evolution (see Fig. 3). Patristic and cladistic similarity are both terms involving time and can only be used in a phylogenetic context. Phenetic similarity is assessed without reference to ancestors or a time dimension and is therefore essentially a horizontal concept.

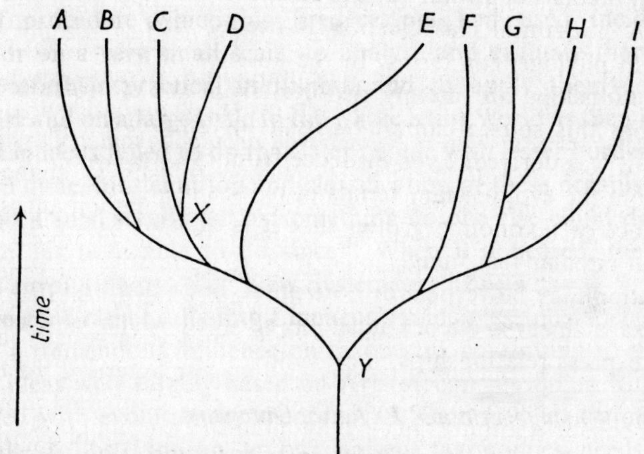

FIG. 3. Patristic and cladistic relationship. The horizontal dimension indicates morphological divergence. Species *C* and *D* show phenetic similarity, due to their patristic and close cladistic relationship, both having arisen from a close common ancestor, *X*. *E* and *F* also show phenetic resemblance, but in this case it is due to convergence. On phenetic evidence alone, *A-D* might be classified in one genus, and *E-I* in another, although *E* is cladistically more closely related to genus *A-D*. *A* to *I* are all patristically related through the stem of the tree, *Y*.

4. *Horizontal and Vertical Classification*

Two kinds of relationship exist between organisms: (1) among successive taxa in an ancestral-descendant lineage—i.e. in time; (2) among taxa as they exist at any one time. In accordance with the usual co-ordinates of tree-presentation, the first of these relationships is called *vertical*, the second *horizontal* (Simpson, 1961: 129). Vertical relationships, being three-dimensional, are very difficult to express on the printed page, and many ingenious methods have been invented for their depiction (cf. Lam, 1936: Jepson, 1944; Rodriguez, 1950; Voss, 1952; Sporne, 1956). It could be argued that one cannot classify a continuum, so that it is mainly discontinuities in the fossil record, or long periods of morphological stability, which allow one to delimit taxa vertically; but there is no difficulty in classifying horizontally at any one period of time (synchronically), as we do with modern plants.

Horizontal (synchronic) classification, though it expresses the *outcome* of phylogeny and might be described as a cross-section through it, can be constructed on a basis of overall similarity and discontinuity without any knowledge of the phylogeny of the taxa involved (i.e. a natural system). Simpson considers both types of classification to be equally "phylogenetic", but it would surely be better to say that a horizontal natural classification is evolutionary rather than phylogenetic. In the Angiosperms, to put vertical classification before horizontal natural classification is to put the cart before the horse. A horizontal classification is needed first, which may later be interpreted in phylogenetic terms. In the absence of fossil record, however, phyletic relationships are mainly deduced from the classification, so that evidence from this source cannot be used to "improve" the classification from which it was derived. Even evidence from cytological and geographical data and from the evolutionary trends of organs is so intimately linked with the acceptance of taxonomic groups that it is difficult to see how it can be used without circular reasoning.

Classification may be vertical or horizontal, but when we come to list and describe taxa a linear sequence must be adopted which of course cannot adequately express relationships.

5. *Phylogeny, Phylogenetic and Phyletic. Ontogeny*

Phylogeny is the origin and evolution of taxa; also (loosely) the study of the historical development of evolutionary lines in a group of organisms. *Phylogenetic* stresses the ancestral relationship of taxa to one another. *Phyletic* emphasises belonging to a particular line of descent (phyletic line). Both *phyletic* and *phylogenetic* refer to the course of evolution and are virtually interchangeable: which term is used is largely a matter of preference and euphony. Several cytogeneticists today employ *phyletic* in a special sense to refer to the relationships which they can determine by crossing experiments and studies of chromosome pairing. Phyletic relationship or affinity has two components, patristic and cladistic (Cain & Harrison, 1960), discussed above.

Ontogeny is the antonym of phylogeny; it accounts for the life history of the individual plant, from its development from the zygote to the production of gametes. De Beer (1951) has referred to phylogeny as a series of modified ontogenies.

6. *Monophyly and Polyphyly*

Few terms have so bedevilled taxonomic and evolutionary literature as monophyly and polyphyly. Simple definitions are inadequate.

The use of the terms monophyletic and polyphyletic is purely relative, depending on how far back in the ancestry we are prepared to go. If life arose only once on earth, all organisms are ultimately monophyletic in origin. Nevertheless, a narrower definition of monophyletic must obviously be applied if the term is to have any practical value for evolutionary interpretation. Anyone who wishes to pursue the topic should read the papers by Simpson and his critics on mammalian classification in *Evolution*, **14** (1960).

Monophyly (as recently defined by Simpson, 1961) is "*the derivation of a taxon through one or more lineages* (temporal successions of ancestral-descendant populations) *from one immediately ancestral taxon of the same or lower rank* [i.e. a family from another family or genus or species]. There are different degrees or levels of monophyly under this definition, and as a rule the level should be specified or evident in each case. The level of monophyly is specified by the category of the lowest ranking single taxon immediately

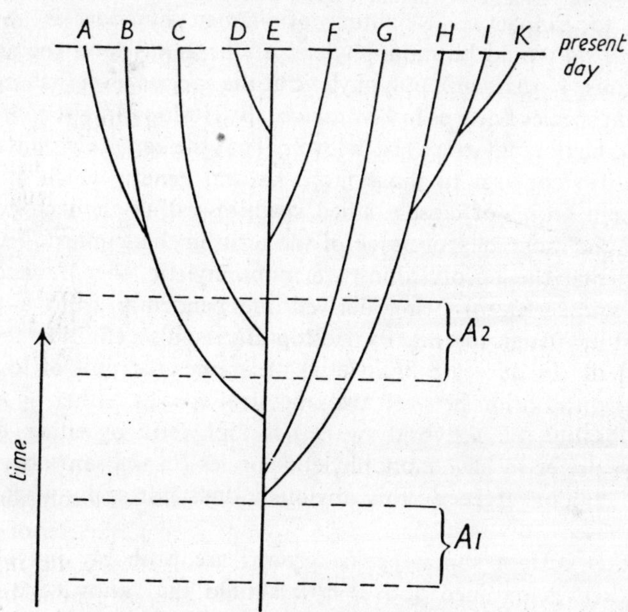

FIG. 4. Monophyletic and polyphyletic relationship. Genus *A-K* may be regarded as monophyletic if all its constituent species arose from a single ancestral population which, during interval *A1*, itself evolved characteristics that would cause a taxonomist to classify it in genus *A-K* if the ancestral population existed today. The genus *A-K* may be regarded as polyphyletic if its constituent species arose from ancestral populations which evolved, independently, during the time period *A2*, characteristics which would cause the ancestral populations to be classified in genus *A-K*. (After Heslop-Harrison, 1958a.)

ancestral to the taxon in question." Heslop-Harrison (1958a), following Wernham (1912), defines a *monophyletic group* as one *derived from an ancestral form which would be regarded as belonging to the taxon in question.*

The level of polyphyly, according to Simpson, is specified by the highest-ranking taxa, two or more of which are immediately ancestral to the taxon concerned. *Polyphyletic*, according to Heslop-Harrison, means *the derivation of a taxonomic group from two or more ancestral forms which are not regarded as belonging to the taxon in question* (see Fig. 4). By this definition, the distinction between monophyly and polyphyly rests upon a difference in time level

Let us see how these terms may be applied in practice. *Verbascum* has five fertile stamens. *Celsia* has four fertile stamens, the fifth being sterile or abortive. A similar infra-generic classification, based on seed-sculpturing and stamen dimorphism, is used in both genera, and widely dissimilar species produce vigorous (largely sterile) hybrids both within each genus and between them. There seems little doubt that *Celsia* is a "grade" genus derived from at least two different lines within *Verbascum* and owes its recognition to Linnaeus's weighting of stamen number. By Simpson's definition, *Celsia* would be monophyletic at the genus level (being derived from one genus, *Verbascum*), polyphyletic at the species level (having evolved from different species lineages in *Verbascum*). By Heslop-Harrison's definition, *Celsia* would be polyphyletic at both levels. That the genus *Celsia* is probably polyphyletic (in contrast to those more natural genera which have arisen from one or a group of closely allied species), rather than monophyletic, seems to be the most helpful usage of the term in this context. Hutchinson has claimed that the Dicotyledons are monophyletic, the *Magnoliales* and *Ranales* (*sensu herbaceo*) being derived independently from a group of hypothetical pro-Angiosperms. By Heslop-Harrison's definition they would be polyphyletic (in this case diphyletic). A species (or unit of lower rank) derived by hybridisation between two ancestral species (either by alloploidy or by stabilisation of a hybrid swarm) is diphyletic by either definition. Although we prefer to have monophyletic species (in the sense that they are not derived by convergence), this obviously does not apply when they are hybridogenous.

Given that certain supra-specific groups are probably polyphyletic in one of the senses outlined above, how should that knowledge (or likely supposition) affect our classification? It is a widely expressed principle that taxonomic groups (at least above the species level) should be monophyletic. Should they be divided up if they can be shown to be polyphyletic? The answer is one of practical convenience. To distinguish monophyletic from polyphyletic groups, in which the similarities have probably been brought about by convergence, is very difficult in the absence of fossil evidence. As Bigelow (1958) points out, the best safeguard against classification based on convergence (which will presumably be the case with polyphyletic groups) is the consideration of *as many characters as possible*. The more character expressions taxa share, the less likely they are to be convergent

Taxonomic groups at all levels of the hierarchy must be delimited by discontinuities in variation. In a probably polyphyletic group, the question the taxonomist must decide is whether or not he can detect such discontinuities so that the components of the group can be disentangled and regrouped into more natural units which can be conveniently defined. Unless this revision can be made without serious inconvenience to the user of the classification, it should not be put into effect. It would be better frankly to recognise the group as polyphyletic, as done by Hutchinson with the *Euphor-*

biaceae. The object of a classification is to classify, and if our groups do not enable us to place lower groups within them, it has failed in one of its essential purposes. Proved or suspected polyphyly may lead us to see if there is a way of reclassifying the groups into monophyletic units, but it does not mean that we should override overall similarity to achieve it. This applies with especial force to the Angiosperms, where a particular ancestry is seldom more than a likely supposition.

Another practical approach to the detection of convergence is to consider relationships within the group with particular reference to specialisation. *Minuartia*, for instance, has often been included in *Arenaria*, but if they are separated on

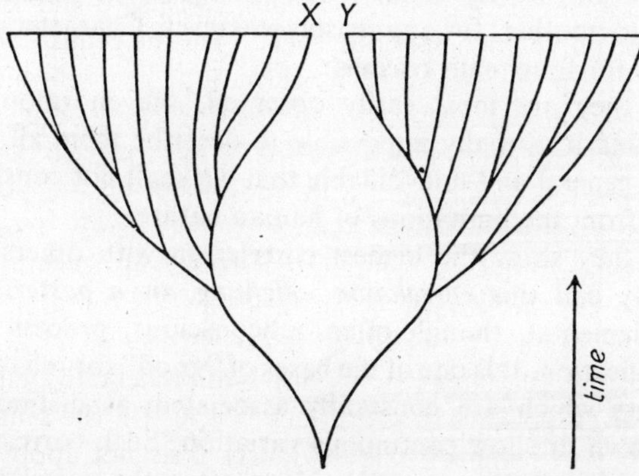

FIG. 5. Specialisation leading to convergence. The diagram shows the assumed phylogeny of two closely allied genera, each possessing a series of modern species showing trends of increasing specialisation. The horizontal dimension indicates morphological divergence. *X* and *Y* are among the most advanced species in each genus, and resemblance between them must be partly due to convergence. If *X* and *Y* were the only living species, they might well be classified in a single genus.

the marker character of the number of capsule valves (the same difference which separates *Silene*—including *Melandrium*—from *Lychnis*) they appear to form two natural coherent groups each with its own evolutionary trends and series of closely allied species. Some species in each genus show a marked superficial resemblance (and have even been collected in mixed gatherings, e.g. *Arenaria leptoclados* and *Minuartia hybrida*), but these are among the most specialised members of the two genera (see Fig. 5). If resemblance was patristic (i.e. the result of common ancestry), it would surely occur between the more primitive species in each genus, not between the most advanced; in the latter case the resemblance must be the result of convergence in relation to similar environments. In this example detection of convergence is made possible by the existence of series of related contemporary species showing trends of increasing specialisation. If the genera were much smaller, being represented by only a very few specialised species, we should have no means of detecting convergence other than by considering as many characters as possible.

In Angiosperm taxonomy, monophyletic and polyphetic origin can usually only be inferred. When applied to taxa, the terms are often loosely used in place of natural and unnatural (in the phenetic sense). The "lumps" in an unnatural taxon are often interpreted as evidence of polyphyly (cf. discussion on natural and artificial groups in Chapter 1).

7. *Character Weighting*

Weighting means, in taxonomic usage, giving greater importance to one character than to another, for any reason whatever. Characters are weighted by taxonomists for four main reasons:

1. Because they are most easily observed, the enormous number of characters making it virtually impossible to describe them all. This kind of weighting is so general and unavoidable that we shall not consider it further here. It results from the limitations of human nature.

2. Because they show the highest correlation with others in a natural group. We may call this *correlation weighting*, or *a posteriori weighting*. This is a fundamental, though often subconscious, process in making a system of classification. It is one of the bases of "good" (or reliable) characters —the characters which are constantly associated, as distinct from those which break down or show continuous variation. Such correlations may be worked out statistically (see p. 131). More often the correlations are subconsciously assessed by normal taxonomic methods. The ability to do this is probably the most characteristic and useful attribute of the good taxonomist. The human brain functions like a marvellous electronic computer, and often does its work by analysing, without our realising it, assimilated experience. Correlation weighting is the only type of weighting which *should* be used for *classification*, and probably the only method which is theoretically sound. Characters abstracted in this manner will be the ones that diagnose the group, and these "markers" will probably be the most likely characters to use for keys. In constructing the latter, however, we may for convenience need to select the characters which are most easily used or described. When new plants are discovered, it is necessary to review the group again to see if the diagnostic correlations still hold. Classification has numerous instances of what one might regard as residual correlation weighting—hangovers of correlation weighting that no longer work. *Rhododendron* and *Azalea*, for instance, may have seemed distinct natural genera when described by Linnaeus, but the discovery of hundreds of new species in the Far East has so blurred the distinctions between them that to continue separating them would be artificial.

Taxonomists are familiar with the situation in which a new species falls, on one or two technical characters, into one section of a genus and yet, on overall resemblances, appears to be more closely related to a species in another section. This situation can arise by residual correlation weighting, or because the section was originally based on characters chosen by *a priori* weighting. In the first case the breakdown of the classification has been caused by an increase in knowledge of the range of variation; in the second this need not be the case—the group was probably badly constructed initially.

3. Because the character is believed to be of particular importance for reasons other than those deduced by correlation studies—very often because of its known or inferred phylogenetic importance, or its recognition as a "conservative" character. This is *a priori* weighting (cf. Cain, 1959a). Classification is still riddled with it, particularly at generic and sectional level. We believe this to be not only unfortunate but to rest upon erroneous assumptions (cf. Chapter. 4).

4. Because they are the characters left when we have rejected others—a procedure that we may call *rejection weighting* or *residual weighting*. As a taxon is a combination of all its characters, characters are usually rejected because they are not suitable for diagnosis or the construction of keys, or for reasons of phylogenetic interpretation. These include those characters which experiment or experience shows to be more variable than others, i.e. those that are readily modified by the environment (e.g. leaf size) and those that are of sporadic occurrence and therefore without characteristic geographical distributions (cf. Wilmott, 1949). For natural classification nothing is really rejected, though variation may be interpreted in terms of known phenotypic variability and therefore ignored.

The last two methods of weighting are closely linked and are usually used for *a priori phyletic weighting*, as distinct from weighting for horizontal classification. There are two methods available which help taxonomists to decide whether to weight or reject certain characters:

(*a*) The use of fossil evidence that establishes the primitive characters of a group beyond dispute. This is an extension of natural classification to include fossil material, their relationships being assessed as though they were living (though often imperfectly known) organisms. Such a classification is phylogenetic in the fullest sense and is both vertical and horizontal. This is a practical procedure only in so far as the fossil record is partially discontinuous (thus allowing delimitation of groups), and is seldom applicable to the Angiosperms (cf. p. 55).

(*b*) By the rejection of characters that are known to be specialisations for a particular mode of life or to show great variability; all such characters (Darwin's argument goes) are likely to change readily and therefore to be unsafe guides to phyletic affinity. The weighting in the formation of families and genera of such characters as the karyotype, pollen morphology, wood anatomy, placentation and the structure and development of the stoma pore (Stebbins & Khush, 1961) rests upon the assumption that these characters are usually more "conservative" than the external morphology of vegetative and floral parts, both of which are in more direct relationship with environmental pressures and thus more readily modified. But it seems very doubtful if these assumptions are valid. As Bigelow (1958) points out, the trouble is that many of the best *taxonomic* characters are obviously adaptive (e.g. those that define the birds and the Angiosperms, the *Umbelliferae* and *Orchidaceae*, *Holcus* and *Salvia*). Adaptive characters will remain stable as long as the environmental conditions to which they are adapted remain constant; when these change, the characters will become variable unless they have lost their genetic ability to vary, as might be the case if their population size had been much reduced. Another difficulty is that selective pressures that produced a particular specialisation may be no longer operating. We have a striking example in the flower of the Bee Orchid, *Ophrys apifera*, which must have originally evolved in relation to "pseudocopulation", the method of pollination adopted by its Mediterranean allies. *O. apifera* is self-pollinated, a process made possible by an ingenious mechanism associated with the exceptionally long stalks of the pollinia which enable the pollen sacs to blow against the stigma of the same flower (Summerhayes, 1951: 310).

If species and families now tend to be natural groups (even if the rank in the hierarchy may be debatable), genera are often quaking bogs, species being transferred from one to another at the whim of the investigator. Many of these changes are undoubtedly due to the effect of size on rank (see Chapter 3), but in the highly successful "definable" families, generic changes have often resulted from the *a priori* weighting used in their construction—e.g. the details of fruit structure in the *Umbelliferae*. Where such genera are natural, the taxonomist has probably selected the diagnostic characters after subconsciously assessing the character correlations in the group. The emphasis laid by Linnaeus on floral structure is no doubt largely responsible for many glaring examples of this type of weighting by both Linnaeus and his followers—he did not always live up to his admirable dictum that the genus gives the characters. The neglect of vegetative characters has been one of the most serious errors in the history of classification, and has done much to delay the achievement of a natural system. If we accept the precept that the taxonomic value of a character rests upon its correlation value, then vegetative characters are as valid as floral ones in the delimitation of genera and families and their placing in a natural system. Hitherto they have usually been neglected above species and sectional level. And yet the most constant character separating the two great subclasses of the Angiosperms is a vegetative one—the number of cotyledons.

8. *Grades and Clades*

As J. S. Huxley (1958) has pointed out, there are three main processes at work in evolution which lead respectively to *biological improvement*, to *diversification* and to *persistence*. (Extinction is interpreted by Huxley as failure to persist, and is therefore not treated as a process.) The first two processes were fully grasped by Darwin (1859), while T. H. Huxley soon after stressed the widespread occurrence of persistence.

The process leading to improvement J. S. Huxley calls *anagenesis*; he uses this word in a more generalised sense than Rensch (1947), covering all types of biological improvement from detailed adaptation to general organisational advance. Improvement is often manifested in a succession of types, an improved type partly or wholly replacing an older and less improved type. Botanical examples of anagenesis range from gamopetalous corollas to ecotypes and polyploids.* The product and unit of anagenesis is called a *grade*. Grades are horizontal in expression, although they have been derived in time (vertically) from previous grades. Knowledge of the previous state of a grade is not essential for its definition and recognition—its characters at any moment in time define it.

Diversification covers all processes of divergence and phyletic splitting, from subspeciation and adaptive radiation up to major diversification between phyla and kingdoms. The process involved is *cladogenesis* and its unit is a *clade*. Diversification, however, is a special form of biological improvement, and it is in practice not always easy to distinguish between anagenesis and cladogenesis.

The process of persistence is called *stasigenesis* by Huxley and results in the recognition of delimitable anagenetic units (grades) at the present day. It is the persistence of grades which has often allowed them to be recognised and delimited as taxa (i.e. taxonomic grades).

There is nothing new in these concepts, but it does seem useful to have them clarified and technical terms provided for them. Although the concepts have been mainly discussed in zoology, they play a very important (and often unrecognised) part in systems of plant classification. Monophyletic groups are based upon clades. The *a priori* weighting of particular characters, however, has resulted in the recognition of taxonomic groups which are grades, not clades, and consist of taxa which have come to be classified together on account of convergence in the weighted characters. This applies particularly at the level of the genus, and supra-family groups like the *Metachlamydeae* and *Monochlamydeae*. It would seem desirable to try and reclassify grade taxa, but if that is not practicable such taxa should at least be frankly recognised for what they are—convenient grade groups.

The Data of Post-Darwinian Systematics

The post-Darwinian period has resulted in the discovery of an enormous number of new facts of biological interest in themselves and of use to the taxonomist, and in a considerable amount of speculation on the processes and course of evolution. The effects on classification have often been unfortunate.

Morphological features have continued to provide the major source of

* Anagenesis is not just another name for parallel or convergent evolution, which itself contains both anagenetic and cladogenetic components.

data used in taxonomy, but have been extended so as to interpret the phenotype in a much wider sense than previously. The study of exomorphic characters themselves has been much refined by the use of visual aids such as the binocular microscope, thus permitting more accurate perception and allowing finer details to be seen.

New sources of phenotypic data have been comparative anatomy, pollen morphology, embryology, phytochemistry, serology and immunology. Systematics has therefore become more than ever a synthetic subject, and the major advances in these newer disciplines utilised by taxonomists are outlined in the next few pages. A survey has already been published by Constance (1955). Detailed examples and a consideration of the value and limitations of the newer data of systematics are given in Chapters 5-7.

Perhaps more wide-reaching in its effects has been the application of cytological and genetical research to taxonomy, along with experimental studies on the variation of the phenotype under different conditions. These are discussed in later chapters.

Morpho-anatomical studies since the nineteen-twenties have resulted in two taxonomic conclusions that can be considered as two of the major "discoveries" of twentieth-century systematics:

1. The *Ranales* (including the *Magnoliales*) constitute the most primitive living group of Angiosperms. In no other group do we find such a concentration of primitive characters, although these are often found in different genera on account of the different rates at which the organs have evolved. Absence of vessels, scattered vascular bundles, monocolpate pollen, trimerous flowers and free, spirally arranged parts, etc. are all found in the *Ranales* (*sensu lato*). Whatever the origin of the Monocotyledons, there is no doubt that they share more features in common with the *Ranales* than with any other group of Dicotyledons.

2. The "*Amentiferae*" (a woody group with apetalous unisexual catkins, sometimes extended to include the *Urticales*) are a polyphyletic and highly specialised group. Floral anatomy shows that reduction in floral structure has affected every member of the group to a high degree: reduction or loss of perianth (*Ulmaceae*, *Salicaceae*); reduction of the inflorescence by loss of flowers or bracts, and by fusion of bracts (*Betulaceae*); reduction in the number of carpels (*Fagaceae*, *Urticaceae*); reduction of the stamens in hermaphrodite forms (*Ulmaceae*); derivation of unisexual from bisexual flowers (*Moraceae*, *Ulmaceae*). Their wood anatomy and pollen grains are also highly specialised. As the German school of Eichler and Engler have claimed that unisexual apetalous flowers are primitive, and based their classification upon this supposition, the invalidation of this claim by anatomical and other evidence is of great importance for phylogenetic interpretations.

During the last few years data on *comparative phytochemistry* have been accumulating at an increasing rate. As parallelism and convergence probably occur as frequently in plant chemistry as in morphology, the interpretation of these data presents considerable difficulties (cf. Chapter 7). Although much of the work has not been done with taxonomic ends in view, it is likely that information from this source will become of increasing importance in assessing natural relationships within both the plant and animal kingdoms, particularly where the morphologist is faced with a choice of relationships. It is applicable at all levels of the hierarchy.

Karyology—the study of nuclear phenomena, particularly of the chromosomes—is a child of the twentieth century, particularly of the last thirty years. It is of importance to taxonomy in two connections (J. Heslop-Harrison, 1953, Ch. 6): (1) The number, size and shape of the chromosomes often have taxonomic value in the same manner as other morphological characters. This can directly affect classification. (2) Chromosomes may provide a direct source of evidence relating to the nature and origin of variation in plants because the chromosomes are themselves the physical seat of hereditary material.

It has been shown that changes in chromosome number and structure can directly effect the breeding system (and thus the variation) of species. Furthermore, chromosome races are often associated with different geographical or ecological distributions. With such valuable information to be obtained, it is not surprising that Cytotaxonomy is a formidable weapon for the taxonomist interested in evolutionary studies. Its application has up till now, been mainly on the flora of North and Central Europe and North America, and until it has been extended to cover the globe, classification cannot expect to reap the full harvest from cytological research. The interpretation of its results in taxonomic terms is still a highly controversial topic.

The study of the chromosomes pairing at *meiosis* can be very illuminating. The occurrence of multivalents indicates autoploidy, segmental alloploidy or structural hybridity (cf. p. 214). In the latter case, as Heslop-Harrison (1953b: 89) points out, "we have the potentiality of a taxonomically rather disturbing situation, where races identical in gene make-up are nevertheless partly intersterile because of dissimilarity in chromosome structure".

The value of comparative karyology, both for classification and as a tool for tracing the origin and migration of species, varies from group to group.

No biological advance during the last fifty years has had such a fundamental effect on taxonomy as the discoveries about the nature of inheritance. The effects, however, have been mostly at and below the species level, and have so far had more influence on taxonomic thinking than they have had on a general (as distinct from special) classification. Mendel's research was unknown to Darwin, with the result that he was unaware of the genetic nature of variation. Since the rediscovery of Mendel's work in 1900, the study of genetics and the nature of variation, coupled with cytological investigation, has become an expanding field of research in every civilised country in the world. Emphasis on variation and particulate inheritance has led to the discrediting of a typological approach (at least for species) and its replacement by the population concept. Variation, adaptation, gene exchange and the karyotype have become the main preoccupation of a school of botanists who are known, for want of a more apt word, as biosystematists.

Biosystematics, in its broadest sense, is that part of systematics concerned with the variation and evolution of species. It is often more concerned with the *process* of evolution than it is with classification itself; several special classifications, however, based mainly on the breeding system, have been proposed. It is the relationship between these special classifications and a general taxonomic classification that has resulted in a great deal of argument. Should they (or can they?) be combined? At one time this threatened to divide taxonomists into two camps—a state of affairs that would be disastrous for Systematics in general.

Every taxonomist in the course of his studies wants to know the nature of the variation he is dealing with. What he does with that knowledge in terms of practical classification is another matter. Cytology and genetics have explained a great deal that was obscure to taxonomists before. But that knowledge may not make the taxonomist's task any easier—it can make it a great deal more difficult. Biosystematics has presented taxonomists with an intellectual challenge of the first order.

Biosystematists have proposed a series of experimental classifications based on the genetic adaptation of the plant to its environment, particularly soil and climate, and its ability to interbreed with other plants both in nature and under experimental conditions. The pioneer in this field was Turesson, who, in 1922, first formulated the *ecotype concept;* his classification was later modified and put on a more precise genetic footing by Clausen, Keck & Hiesey and by Gregor, etc. (cf. Chapter 12). Transplant experiments and breeding experiments have been the tools used in the construction of these special, experimental classifications. Failure to interbreed (in nature or the experimental plot) was recognised as of great evolutionary importance, since such populations are prevented from any further gene exchange now or at any time when they may meet in the future. They have crossed a Rubicon in evolution. There was also the belief that ability or failure to interbreed reflected the degree of genetic relationship—a supposition no longer generally tenable. The result was that some biosystematists considered (and still do) that failure to interbreed should be the chief criterion of specific rank—a view taken by Mayr and Simpson in the classification of birds. This view is not applicable to plants, for several reasons discussed in the following chapter.

Other aspects of experimental taxonomy have shown how the breeding system directly affects the variation of plants. Self-pollination and particularly apomixis (chiefly reproduction by seed without fertilisation) present the taxonomist with formidable problems, of which the worst is undoubtedly the classification of facultative apomicts (e.g. *Rubus* in northern Europe). In recent years several attempts have been made to solve the last problem, but it is doubtful if all the facts of nature can ever be satisfactorily reconciled with a practical classification.

The delimitation of taxa is based upon discontinuities in variation. The population concept, however, has drawn attention to continuities in variation which cannot be dealt with in terms of categories, and are in fact an embarrassment to the classical taxonomist. But if taxonomists are to portray the variation in nature, such situations cannot be ignored. Biosystematists have therefore devised various ways of describing continuous variation by graphical or mathematical means. Huxley (1938) proposed the term *cline* for a character gradient, i.e. any variational trend in space appearing in a population or series of populations of a species. A cline on a geographical or regional basis has been distinguished as a *topocline*, and a local one related to an ecological gradient as an *ecocline* (Gregor, 1939). Another type of variation pattern is caused by extensive hybridisation between taxa, a situation so successfully investigated by Anderson (1949) that he can sometimes predict the parents by analysing a hybrid swarm. Methods of analysis used by biosystematists supplement a formal taxonomic classification which cannot, by its very nature, give us much information about continuous variation which is nevertheless of great evolutionary importance. It is, in fact, the pattern of variation before discontinuity (making classification possible) is imposed upon it.

A great amount of unco-ordinated information has accumulated about the Angiosperms in fields other than gross morphology. Even in monographic studies, only a very limited amount of this information has been used by taxonomists in Angiosperm classification. No system constructed in the last half-century has fully capitalised on the recent accumulation of data.

Palaeobotanical Evidence

The Angiosperm fossil record suffers from several major limitations: its irregular and scanty distribution; the fact that most macrofossils belong to trees; the predominance of leaves and pollen, making the material difficult to place in reference to modern Angiosperms which have been primarily classified on floral structure (Stebbins, 1959b). The hazards of misidentification, though lessened by modern microtechniques, are accentuated by the different rates at which different organs may evolve. A Cretaceous pollen grain, for instance, though it may closely resemble that of a modern species, may in fact have been associated with other characters which, if they were known to us, would show that it belonged to a completely different group of plants.

Fossil evidence has been helpful in interpreting the evolutionary history of only a few contemporary groups, as shown, for instance, by the work of Chaney and Axelrod on the Tertiary floras of western North America (for references, see Benson, 1962). From such evidence it seems that trees have usually evolved more slowly than herbs. Although their distribution has radically altered, most modern trees for which fossil evidence is available have persisted for 5-30 million years almost in their present form.* The herbaceous flora appears to have evolved more rapidly, most of the "major" taxonomic species being, however, pre-Pleistocene in origin. Though extremely rewarding in revealing the facts of vegetational change, the abundant Pleistocene fossil record (much of it palynological) is too recent to be of much help in elucidating the phylogeny of modern species.

Evidence of rather rapid evolution in herbaceous genera is provided for *Stratiotes* (Chandler, 1923), *Sparganium* (Reid & Chandler, 1933) and the American *Gramineae-Stipeae* (cf. Stebbins 1959b: 265-267). In Britain, fossil seeds of the aquatic genus *Stratiotes* show a trend from the small, heavily sculptured types of the Eocene to progressively longer, narrower and smoother types that in the Upper Pliocene become indistinguishable from the modern *S. aloides*. In *Sparganium* there is fossil evidence for a rather irregular and interrupted sequence in the reduction of loculi from the Oligocene to the present day.

The existence of such fossil sequences, however, is the exception, not the rule in flowering plants. When it comes to the question of the origin of the Class, the missing links have not been found. In this connection it is interesting that Scott *et al.* (1960) conclude that there are no *unequivocal* Angiosperm remains prior to the early Cretaceous. Phylogenetic theory has flourished on this lack of factual data.

Although palaeobotany can be only of limited assistance in unravelling Angiosperm phylogeny, it can help to explain the present *distribution* of many taxa by the evidence which it gives on the migration of floras. This, when correlated with geological and climatic history, may provide solutions to problems of endemism and disjunction.

The following references may help the reader in a search for further information. A good review and bibliography of palaeobotanical data used in taxonomy and biogeography is given by Benson (1962, Ch. 5). Darrah's *Principles of Paleobotany* (1960) contains a useful chapter on modern techniques, and others on Cretaceous and Cenozoic floras. An excellent up-to-date work is Andrew's *Introductory Textbook on Paleobotany* (1961). Kremp & Ames's beautifully illustrated *Catalog of fossil Spores and Pollen* has been published regularly since 1957. Palaeobotanical references (covering the whole field) have appeared regularly since 1956 in the *World Report on Palaeobotany* under the auspices of the International Organization of Palaeobotany (editor, E. Boreau). Seward's very useful compilation, *Fossil Plants*, was published in four volumes from 1898 to 1919, but Hirmer's *Handbuch der*

* The closed pines of California, however, for which the fossil record is unusually good, have apparently evolved their modern species since the Pliocene (cf. Benson, 1962: 163-169).

Paläobotanik (begun in 1927) was unfortunately never completed.* The fact that there is no systematic index of macrofossils comparable to the *Index Kewensis* for living Angiosperms makes it difficult for the student of modern plants to find out if the group he is working with has a fossil history. The taxonomist must ask palaeobotanists for information and comb the various accounts of fossil floras (not forgetting those of Eastern Europe and the Soviet Union).

A symposium on palaeoclimatology (important in interpreting modern patterns of plant distribution) has been edited by Nairn (1961). Sylvester-Bradley (1956) has edited a symposium on the species concept in palaeontology—a problem which adds to the difficulties of fossil classification. Indeed it is largely the incompleteness of the fossil record which, by accentuating discontinuities, makes palaeontological classification feasible (cf. p. 49).

Classification in Post-Darwinian Times

We have briefly surveyed the post-Darwinian advances in subjects that one might expect to have had an important bearing on Angiosperm classification. Let us now take a look at those classifications which we can for convenience consider under the headings of Phylogenetic and Natural.

1. *Phylogenetic Systems*

The most striking feature of post-Darwinian systems is that they attempt not only to be evolutionary but to be phylogenetic. Such systems have proliferated with astonishing vigour, the only requirement being that no two should be alike. As books can be, and have been, written on this subject, we do not propose to pursue it in great detail here. Accounts will be found in Rendle (1930), Copeland (1940), Turrill (1942), Airy Shaw (1950), Lawrence (1951), Constance (1955), Benson (1957) and Porter (1959).

Such systems mostly fall into two groups depending on the supposed nature of the primitive Angiosperm flower. Concerning this, as stated by Sprague (1960), "two main hypotheses have been advanced: (1) that the earliest Angiosperms were wind-pollinated, and that their least altered descendants are to be found in the *Betulaceae*, *Fagaceae*, *Juglandaceae* and related families which are wind-pollinated [anemophilous] and have no petals; and also in the *Urticales*, *Tricoccae* [*Euphorbiales*] and *Piperales;* (2) that they were insect-pollinated [entomophilous] and that their nearest descendants are to be found in the *Ranunculaceae*, *Magnoliaceae* and related families which are insect-pollinated and mostly possess petals". The systems based on the first view, i.e. that the ancestral flower was unisexual and apetalous

* Although only the first volume of this work was actually printed (containing Thallophytes, Bryophytes and Pteridophytes), we are told that two more volumes (including part of the Angiosperms but excluding the *Gamopetalae*) were prepared in manuscript but never published.

(cf. Sprague, 1960), are exemplified by Engler and Prantl's and represent what we may term the "Englerian" school. The systems based on the "Ranalian theory", i.e. that the ancestral flower was hermaphrodite with numerous free parts (cf. Barnard, 1961; Eames, 1961), are exemplified by those of Bessey and Hutchinson and belong to what we may call the "Ranalian "school.

Post-Darwinian systems have, of course, their basis in pre-Darwinian systems, and like these are necessarily grounded upon an understanding of plant morphology. Among the leading morphologists of the nineteenth century was ALEXANDER BRAUN (1805-76) who made significant contributions to many fields of botany. Becoming professor of botany at Berlin in 1851, he set out to arrange the plants in the botanic garden according to a natural system of his own devising, influenced, however, by the systems of Endlicher and Adolphe Brongniart, whom he had known when studying in Paris. In 1859 Braun contributed to a local German Flora (Ascherson's *Flora der Provinz Brandenburg:* 1859-64) a synopsis of his classification of the plant kingdom which greatly influenced August Wilhelm Eichler (Braun's successor at Berlin) and Adolf Engler, and, through them, Warming and Wettstein.

Braun classified the Monocotyledons so as to show a progression from the more simple to the more complex, beginning with *Lemnaceae* and ending with *Orchidaceae*. He began the Dicotyledons with the *Apetalae*, i.e. those families with only one whorl of perianth segments or showing no differentiation into calyx and corolla,* followed by the *Sympetalae* (*Gamopetalae*) beginning with the *Lentibulariaceae*, followed by the *Eleutheropetalae* (*Polypetalae*) ending with the *Leguminosae*.

EICHLER (1839-87), who was for some years Martius's assistant at Munich and through his work on the *Flora Brasiliensis* gained an intimate knowledge of a very rich tropical flora, published in 1875-78 his *Blütendiagramme*, the result of some fifteen years' observation and study wherein he gave very careful floral diagrams for every family. In dealing with the Monocotyledons he closely followed Braun, but modified the latter's classification of the Dicotyledons by distributing the families in Braun's *Apetalae* among the families with differentiated sepals and free petals (his *Choripetalae*) which they seemed most to resemble. In his *Blütendiagramme* (1875-78) Eichler accordingly placed the gamopetalous families before the polypetalous ones (as Braun had done), but reversed this position in the Kiel *Syllabus* of 1876 and in his Berlin *Syllabus* of 1880; in the last the Dicotyledons begin with the *Cupuliferae* and end with the *Compositae*.

Eichler's system was adopted with some modifications by ADOLF ENGLER in 1886 in his guide to the Breslau botanic garden. The system was expanded in 1892 in his *Syllabus* and yet further in his two great systematic works, *Die natürlichen Pflanzenfamilien* (begun in 1887 in collaboration with KARL PRANTL) and *Das Pflanzenreich* (1900 onwards), the last being a series of

* Braun's *Apetalae* consisted of the *Podostemaceae*, *Piperaceae*, *Saururaceae*, *Chloranthaceae*, *Polygonaceae*, *Begoniaceae*, *Aristolochiaceae* and *Cytinaceae*.

detailed generic monographs (for a list, see M. T. Davis, 1957) many of them by his research students and of somewhat uneven merit. Engler himself stated in 1892 that his system, like that of Braun and of Eichler, was basically that of the French botanist Adolphe Brongniart (1770-1847) who in 1843 had already distributed the apetalous families among the petaliferous ones to which they appeared related.

Engler's *Syllabus* and *Pflanzenfamilien* covered the whole plant kingdom, a range which gave the system, for purely practical reasons, an international prestige greater than its intrinsic worth as a satisfactory system of classification. Most American and Continental herbaria are arranged by it, and numerous Floras have followed its sequence. It is the first Angiosperm system to take anatomical data into account, although in most families this decorates the text rather than influences the classification.

The Englerian system is not meant to be phylogenetic in a simple linear sense. It is a system in which the groups are built up in a step-like manner to form a generally progressional series. Engler was highly conscious of the prevalence of parallel evolution and refrained from drawing a phylogenetic tree. The Gymnosperms take their natural place before the Angiosperms, and are followed by the Monocotyledons and then the Dicotyledons. In the Dicotyledons Engler united the *Polypetalae* with the *Monochlamydeae*, as Eichler had done earlier, to form the *Archichlamydeae*—a great improvement as a natural grouping on Bentham & Hooker's treatment, though not so easy to use. But Engler maintained the *Gamopetalae* as a natural group (again following Eichler), calling it the *Metachlamydeae* (*Sympetalae*). An English summary of the system, and that of Bentham & Hooker, will be found in Willis (1948). It is surprising how much of the Englerian family sequence resembles that of de Jussieu's *Genera Plantarum*.

Engler took the flowers possessing only androecium and gynoecium as the lowest grade of organisation (cf. principle 21 of his 1892 *Syllabus*). In this first series, the flowers are naked or have a monochlamydeous bract-like perianth; a higher grade is represented by dichlamydy—the differentiation of the perianth into two series; thus two stages are evident, a lower stage in which the members of each series are bract-like in character, and a higher stage, in which there is a differentiation between the two series, the outer being represented by sepals, the inner by petals; union of petals (sympetaly) represents a more advanced stage (cf. Rendle, 1930). Engler and Prantl accordingly place the catkin-bearing families (*Casuarinaceae*, *Salicaceae*, etc.) before the *Magnoliaceae* and *Ranunculaceae*. Even the *Caryophyllaceae* precede the *Ranunculaceae* because of the former's natural association with the *Chenopodiaceae* and other apetalous families in Eichler's very natural group, the *Centrospermae* (*Chenopodiaceae*, *Amarantaceae*, *Portulacaceae*, *Caryophyllaceae*, etc.). They start the Monocotyledons with the unisexual apetalous *Pandanales*. Engler and Prantl maintain that the Monocotyledons and Dicotyledons have arisen independently from hypothetical Gymnosperms. They believe that the

Dicotyledons are polyphyletic from different Gymnosperms, and that the *Metachlamydeae* have arisen from various petal-bearing *Archichlamydeae*; most people would now agree with the latter assumption. The *Metachlamydeae*, and Bentham & Hooker's *Polypetalae* and *Monochlamydeae*, are regarded by Engler & Prantl not as natural groups but as evolutionary levels, i.e. as taxonomic grades.

Let us consider the theoretical suppositions behind Engler & Prantl's concept of the primitive flower, and those behind the very different view of it held by the Ranalian school. It is often claimed that Angiosperm flowers are derived from the strobili of Gymnosperms, although the missing links to substantiate the theory have never been found. The *Englerian school* maintains that the ancestral strobilus was unisexual, bearing either megasporophylls or microsporophylls. It gave rise to a unisexual catkin, pistillate or staminate. The bisexual flower is considered as derived from a bisexual pseudanthium*; a subtending perianth evolved later. This interpretation was strongly criticised by Arber & Parkin (1907) on three counts: (1) it presupposes that the perianth arose *de novo;* (2) Engler's so-called primitive flowers are invariably accompanied by complicated and highly evolved inflorescences; (3) the theory is phylogenetically sterile, and does not bring the Angiosperms into line with fossil plants. Sprague (1960) admits that "If Engler's hypothesis is correct, it seems strange that the bisexual Angiosperm flower should invariably have the gynoecium terminal". Anatomical studies have repeatedly shown that the "simple" flowers of present day *Amentiferae* are in fact highly specialised ones derived from hermaphrodite petaliferous ancestors with less specialised inflorescences.

Voices have, however, recently been raised in defence of the Englerian concept of the primitive flower. In Sprague's view (1960) "the conclusion is inescapable that the earliest Angiosperms were apetalous and anemogamous. Many of the groups such as the *Fagales*, *Urticales* and *Piperaceae*, may have remained in that condition, while others, like *Saururaceae* and many *Centrospermae*, may have achieved entomogamy, not by the development of petals, but by the development of colour in sepals or bracts." Hamshaw Thomas (1958) considers the appearance of bisexual flowers in the Angiosperms to be a relatively late and not a primitive feature. Of particular interest is the experimental evidence produced by Heslop-Harrison (1957a, 1958a) to suggest that all unisexual Angiosperm flowers need not have arisen by suppression of one set of organs in hermaphrodite flowers; in another group unisexuality appears to be a basic attribute, and may be a less specialised condition than bisexuality. We are not out of the wood yet!

The *Ranalian school*, originated by Bessey and greatly modified by J. Hutchinson, holds a very different view which has certainly found wider support. Its exponents believe that the Angiosperm flower arose from a

* *Pseudanthium*: a contracted inflorescence which simulates a single flower.

Gymnosperm with a bisexual strobilus bearing megasporophylls above and microsporophylls below. The lower sporophylls developed into sepals and petals by progressive sterilisation, those above into stamens and carpels. The axis of the strobilus shortened to form the receptacle. The primitive flower therefore had numerous free perianth parts, with many free stamens and carpels spirally arranged. *Magnolia* is the nearest to this primitive type of flower living today. Arber & Parkin (1907) provided the theoretical basis for his view by suggesting that the ancestor of the Angiosperms was an undiscovered Gymnosperm distantly related to that rare fossil, *Cycadeoidea dacotensis*, in the *Bennetitales*, a Mesozoic plant that bore "*anthostrobili*" in the axils of Cycad-like leaves. This ingenious speculation remains unsubstantiated and other claimants have entered the field, but there is no doubt that, on the evidence from the morphology and anatomy of living Angiosperms (including the pollen grain), there are good reasons to consider the woody *Ranales* a far less specialised group than any included in the "*Amentiferae*". The origin of the Angiosperms, however, is not the concern of this book. We are concerned with the classification of the Angiosperms as they are today.

Bessey (1915) put forward a system of Angiosperm classification that he definitely believed to be phylogenetic.* He considered the Angiosperms to be monophyletic (a claim supported by the allegedly unique "double fertilisation" characteristic of the 8-nucleate Angiosperm embryo sac followed by the formation of secondary endosperm) and produced an endearing "Cactus diagram" in which every order is shown as being directly derived from another (cf. Rodriguez, 1950). Bessey held that the *Ranales* (*sensu lato*) were primitive and that the rest of the Monocotyledons and Dicotyledons were derived from them. He put less emphasis on gamopetaly than Bentham & Hooker or Engler & Prantl, but a great deal on hypogyny *versus* epigyny. He abolished the *Polypetalae*, *Gamopetalae* and *Monochlamydeae* as natural groups, and traced two lines of development from the *Ranales:* one, hypogynous, *via* the *Caryophyllales* and ending with the *Lamiales;* the other line peri-epigynous, *via* the *Rosales*, and ending up with the *Asterales* (including *Compositae*). He derived the Monocotyledons from the herbaceous *Ranales* by way of the *Alismatales* (a group of aquatic Monocotyledons which resemble the *Ranales* in having free carpels).

In criticising Bessey, one must claim that perigyny and epigyny cannot be assumed to constitute a single evolutionary line. Character correlations most strongly suggest that these features, like apetaly, have been evolved independently in several phyletic lines. It seems an oversimplification to derive the orders (groups of families) directly from one another; the majority have probably arisen from more generalised ancestors.

There seems no reason, however, why this should apply to all families, and certainly not to all genera. The *Lobeliaceae* (if it is recognised as a

* Earlier versions had appeared in 1893 and 1897.

distinct family) could well be directly derived from the *Campanulaceae*, and the *Labiatae* from the *Verbenaceae*. Within the *Campanulaceae* such genera as *Symphyandra*, *Adenophora*, *Asyneuma* and *Zeugandra* appear to be further specialisations of the widespread genus *Campanula*, and the same must apply to numerous genera accepted in the *Cruciferae*, *Leguminosae*, *Compositae* and *Orchidaceae*. Here generic limits are often associated with different agents for pollination or seed dispersal.

J. HUTCHINSON (of the Royal Botanic Gardens, Kew) has taken up many of Bessey's ideas but superimposed upon them a primary division which is almost Aristotelian—the division of the Dicotyledons into woody and herbaceous lines. Hutchinson's system has undergone considerable modifications since its earliest (partial) presentation in the *Kew Bulletin*, 1924, but we will consider here only its latest manifestation, *The Families of Flowering Plants* (1959), in which his ideas have been pursued to their logical conclusion. He believes the Angiosperms to be "monophyletic" from an ancestral group of Gymnosperms that may have been parallel to *Cycadeoidea*. He traces a mainly woody and a mainly herbaceous line of development, starting respectively with the *Magnoliales* and *Ranales* (*sensu herbaceo*). All those Dicotyledon families which are "primitively woody" (e.g. the order *Leguminosae*), go into his division *Lignosae*, and all those which are fundamentally herbaceous into his division *Herbaceae*. Believing that the *Monochlamydeae* are polyphyletic from different polypetalous stocks, he drops this group altogether. The numerous families (now 342) are grouped into 82 orders and disposed within his two phyletic divisions (*Lignosae* and *Herbaceae*).

This treatment leads to the wide separation of some families that resemble one another rather closely on floral structure—e.g. the herbaceous *Umbelliferae* from woody *Cornaceae* and *Araliaceae*, the herbaceous *Labiatae* from woody *Verbenaceae*, herbaceous *Saxifragaceae* (*sensu stricto*) from woody segregates like *Escalloniaceae* which were included by Bentham & Hooker in the *Saxifragaceae*. Even the families of Bentham & Hooker's *Personales*, which seemed a natural group, have been widely separated by Hutchinson—the woody climbing *Bignoniaceae* from the herbaceous *Scrophulariaceae* and *Gesneriaceae*.

Hutchinson derives the Monocotyledons in the same manner as Bessey, but puts less emphasis on the position of the ovary in the Monocotyledons than he does in the Dicotyledons. The Monocotyledons are divided into three progressional groups on the nature of the perianth (*Calcyiferae*, *Corolliferae* and *Glumiflorae*), a treatment very similar to that adopted by Lindley (1830). In a recent statistical analysis of the Monocotyledons, Lowe (1961), using methods already applied to the Dicotyledons by Sporne (cf. p. 68), finds Hutchinson's classification of the Monocotyledons more natural than that of Engler. It is hard to have faith in a statistical method, however, which leads to the conclusion that epigyny in the Monocotyledons is a more primitive condition than hypogyny!

Hutchinson's phylogenetic tree is more sophisticated than Bessey's, one order not being derived from another as it is delimited today but from its ancestral stock. Its great failing seems to be its anachronistic weighting of woody *versus* herbaceous habit. Although he recognises numerous herbaceous derivatives in his *Lignosae* line, and numerous woody members within his *Herbaceae* (often within a single family), he has clung to, and indeed increased in each successive presentation of his system, his emphasis on this habit character for his primary phyletic division. Overall resemblances lead one inevitably to assume that the herbaceous habit has been derived from the woody habit again and again, and no doubt the reverse has occasionally occurred. Hutchinson has torn a likely Angiosperm phylogeny in half, and it is not surprising that the *Herbaceae* (especially when one considers floral structure) often resemble the mirror image of the *Lignosae*.* It is the most exaggerated example of *a priori* weighting to be found in post-Linnaean classification.

At a lower level, within the Monocotyledons and within his two phyletic divisions of Dicotyledons, many of Hutchinson's changes in the delimitation and relationship of families appear to rest upon an overall assessment of morphological characters—i.e. to be natural (phenetic) groupings. With his deep knowledge of tropical floras he is obviously in a strong position to assess such relationships. This being the case, it is unfortunate that so many of his changes (both phylogenetic and in the content of orders and families) rest upon intuition (which we have nothing against as a method in classification), without any consistent attempt to translate it into factual reasons to support the change. When a new system is presented, we see no reason why scientists should accept it if it cannot be substantiated. Taxonomy serves biology, and there is no point in *changing* a system of classification unless one is *improving* it—and little use in improving it if you do not persuade biologists to adopt it!

Although it is a basic tenet often repeated in this book that classification should not and cannot be based on phylogeny but only interpreted in phylogenetic or evolutionary terms, we are not against phylogenetic speculation *per se*. In itself it is a harmless pastime that makes herbarium taxonomy more interesting, and at best may lead to the discovery of additional factual information. In many phylogenetic systems the "phylogeny" has done little violence to classification, as is the case with the systems of Engler & Prantl, Cronquist and Takhtajan. The natural families have been maintained, and it is only the *arrangement* that has been altered. We have been more critical of Hutchinson's system because it has resulted in many changes which are in contradiction to overall resemblances and must therefore be considered retrograde moves. Two examples will serve to illustrate our point. Hutchinson

* We should refer the reader, however, to Deyl (1950: 260) who has at least produced a theoretical explanation to account for the evolution of separate woody and herbaceous lines within the *Archichlamydeae*.

has divided the *Ranunculaceae* into two families on the basis of whether their fruits are achenes or follicles. But there is plenty of microscopic evidence (both anatomical and cytological) to show that the achenial genera have arisen from several follicular lines within the family and therefore do not form a monophyletic group. The segregation of a herbaceous family, *Podophyllaceae*, from the woody *Berberidaceae* hardly seems an adequate way of dealing with this relict assemblage; surely *Epimedium* has more in common with *Berberis* (even their seedlings are similar) than with *Podophyllum* to which Hutchinson now relates it.

To object to Hutchinson's system, however, is not to deny that his book is a very useful one. The standard of the family description is high, and a valuable feature is the key which enables one to identify almost any Angiosperm to its family, however aberrant its characters may be. That is an intellectual feat of the greatest practical value, which makes it all the more disappointing that this painstaking work should be no more acceptable as an evolutionary system than Engler's.

The Englerian system has had few followers, but was greatly improved by WETTSTEIN in the posthumous fourth edition of his work, *Handbuch der systematischen Botanik* (1930-35), in which many of the families were rearranged in an order more acceptable to contemporary phylogenists. Here we find the *Pandanales* terminating (instead of starting) the family sequence in the Monocotyledons which he derives from the Dicotyledons in a similar manner to Bessey*. Rendle's *Classification of Flowering Plants* (1930, 1938), though the system adopted is radically that of Engler and Prantl, is invaluable for the interesting biological information that his family accounts contain.

The Ranalian school has had a host of followers, all of whom treat the Ranalian type of flower as primitive while at the same time adopting some parts of the Englerian system in various ways. We can mention only Hallier (who published numerous papers on classification, working independently from Bessey but basing his classification on the assumed strobiloid ancestry of the flower), Soó (1948, 1961), Takhtajan (1958, 1959b) Cronquist (1957—Dicotyledons only) Emberger (1960) and Firbas (1962). Takhtajan's system owes much to Hutchinson, but with the groups in the latter's "*Herbaceae*' disposed among and derived from various stocks in the "*Lignosae*"—a very plausible arrangement.

Although one is struck by the differences between these systems, it would be wrong to ignore the measure of agreement between them. It is quite possible that many of the features that have become more or less stabilised in recent phylogenetic systems do represent genuine phylogenetic relationships. By and large these agreements include: acceptance of the *Ranales* (*sensu lato*) as the most primitive group; the *Monochlamydeae* as an artificial assemblage derived from various (and often uncertain) polypetalous stocks,

* The *Stammbaum* given by Janchen (1932) graphically summarises Wettstein's view of family relationships.

the families in the "*Amentiferae*" being mostly derived from the *Rosales* by way of *Hamamelidales* (or their progenitors); the gamopetalous families as advanced groups derived from several (often uncertain) polypetalous stocks; the Monocotyledons as monophyletic and derived from herbaceous *Ranales* by way of *Alismatales* or *Butomales;* a *Liliaceous* stock as the source for both the elaborate zygomorphic *Orchidaceae* and the reduced *Cyperaceae* and *Gramineae*.

Copeland (1957), dealing with the classification of the Dicotyledons, goes so far as to claim that "the points of agreement allow us to believe that we are actually approaching the system of the future". In view of the neglect of vegetative and microscopic characters in Angiosperm classification, not to mention the surprises that may be in store for us in the homology of gynoecia (cf. Meeuse, 1961; Melville, 1962), this seems to us an over-optimistic claim. Just as most people believed in witches in the Middle Ages, so may agreement in taxonomy result from the copying of mistakes.

Taxonomists might well be more frank about families and genera of very doubtful position. By assigning them to a separate group (as done with the *Fungi Imperfecti*), attention would be drawn to our ignorance. This might encourage research into their cytology, anatomy, embryology, chemistry, etc. that could help to establish their true relationships.

Mention should be made of HAYATA (1921), who produced an original "Dynamic system" of Angiosperm classification. Hayata regarded the natural system "as a dynamic one, changing with the view of the systematizer and subject to alteration, according to the way in which it is considered. None of the species, genera or families has a fixed natural position, but has changeable positions, subject to alterations according to the criterion for comparison." The concept was based upon his "participation theory" which maintains that all plants share, or participate in, the work of certain units which he calls "genes" (a term he interprets in a different sense to that of geneticists). Hayata did not reclassify the Angiosperms, but accepted Engler's classification (and linear sequence) as a framework and put it (as he says) "into a dynamic condition". Each family is shown surrounded by several satellite families to which it is related according to the characters being considered. Thus the *Plantaginaceae* is shown as allied to the *Stylidiaceae*, *Campanulaceae*, *Boraginaceae*, *Goodeniaceae* and *Globulariaceae*. It is, as it stands, an interpretation of an accepted classification. Hayata's "Dynamic system" makes no allowance for convergence, and as he did not believe in evolution there is no reason why we should expect it to do so. Although it certainly draws attention to the reticulation of resemblances which Angiosperms show, it gets us no further. It is as though he had marshalled the facts for a reclassification but failed to translate it into classification. The "dynamic system" remains an historical curiosity.

BENSON's presentation of evolutionary relationships within the Angiosperms, published on the inside cover of his *Plant Classification* (1957), is reproduced here in abridged form (Fig. 6). This is essentially a *horizontal* classification and strikes us as being a felicitious way to present an Angiosperm system with a phylogenetic background without committing oneself to a speculative "tree". Benson accepts the woody *Ranales* as having retained more primitive features than any other group, and states that the distance on the diagram of any order from the *Ranales* and *Alismatales* is in rough proportion to the number of features of flower structure departing from those prevalent in these orders. The size of the figure representing each order is in rough proportion to the number of species composing it. The diagram represents a cross-section through a divergent phylogenetic tree without the ancestral connections being drawn in, although the shape of the slice through the orders suggests the general direction of their origin. This would seem to be as far as one can safely go in presenting on the printed page an evolutionary classification of the Angiosperms. By the use of shading, he has grouped his orders according to the nature of the perianth and position of the ovary. The informal polyphyletic groups thus formed are perfect examples of grades.

We have now surveyed some of the major "phylogenetic" classifications of the Angiosperms, and have criticised some of the points in which they differ from one another. There are, however, more fundamental objections to be raised to the construction of such Angiosperm systems, especially when they are presented in the form of a phylogenetic tree. As pointed out by Cain, Gilmour, Bigelow and others, the whole concept of a phylogenetic classification, particularly when applied to the Angiosperms, rests upon very shaky, and probably invalid foundations. Bigelow (1958) questions and finds wanting each of the three basic tenets of the phylogenetic school (printed in italics below):

(1) *Similarity corresponds to recency of common ancestry* (or, more loosely, similar organisms are more "closely related"). This assumption may be valid in many instances, but certainly not in all. Unequal rates of evolution will produce cases in which overall similarity does not correspond with recency of common ancestry. Slow evolution in two phyletic lines will tend to produce an *appearance* of recent common ancestry when actual common ancestry is remote in time. Where evolution has been rapid (i.e. where overall similarity corresponds least to recency of common ancestry), its true course is most likely to have escaped detection. Bader (1958), however, believes that those cases in which similarity does not correspond to common ancestry are not sufficiently common to negate the general principle. But as Bigelow (1959) says, we cannot claim that a principle is *true* if we cannot distinguish when it is right or wrong—however seldom it may be wrong!

(2) *Phylogeny can be constructed without fossil evidence*. Working on the very questionable assumption that similarity corresponds to recency of common origin, phylogenetic trees will show recent ancestry between slowly evolved species, whereas their common ancestry is actually remote in time;

HORIZONTAL RELATIONSHIPS WITHIN THE ANGIOSPERM

FIG. 6

DICOTYLEDONS

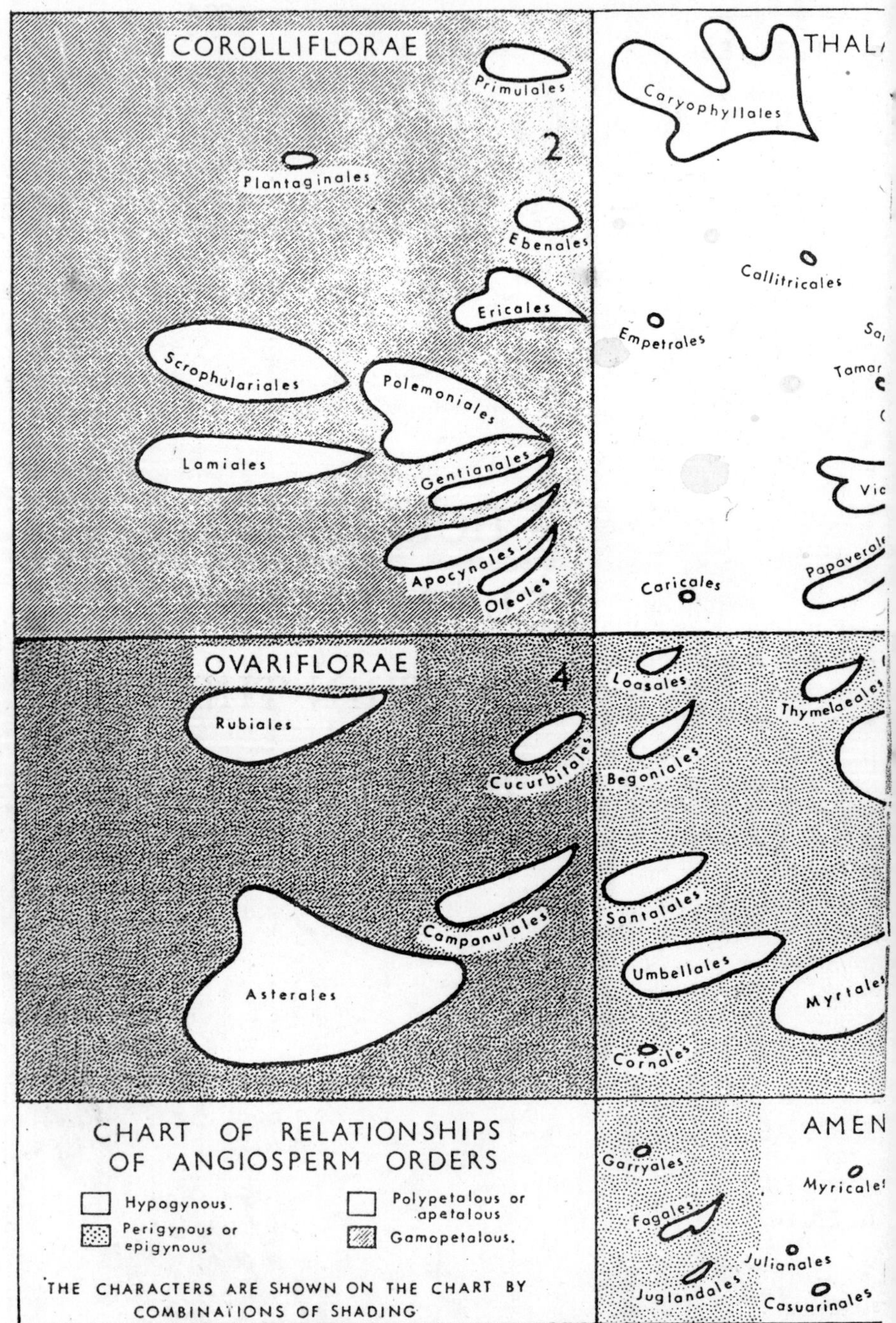

FIG. 6. Horizontal relationships within the Angiosperms, showing grades. The distance on the chart from the *Ranales* or *Alismatales* is in rough proportion to the number of features of flower structure departing from those prevalent in these orders. The size of the figure representing

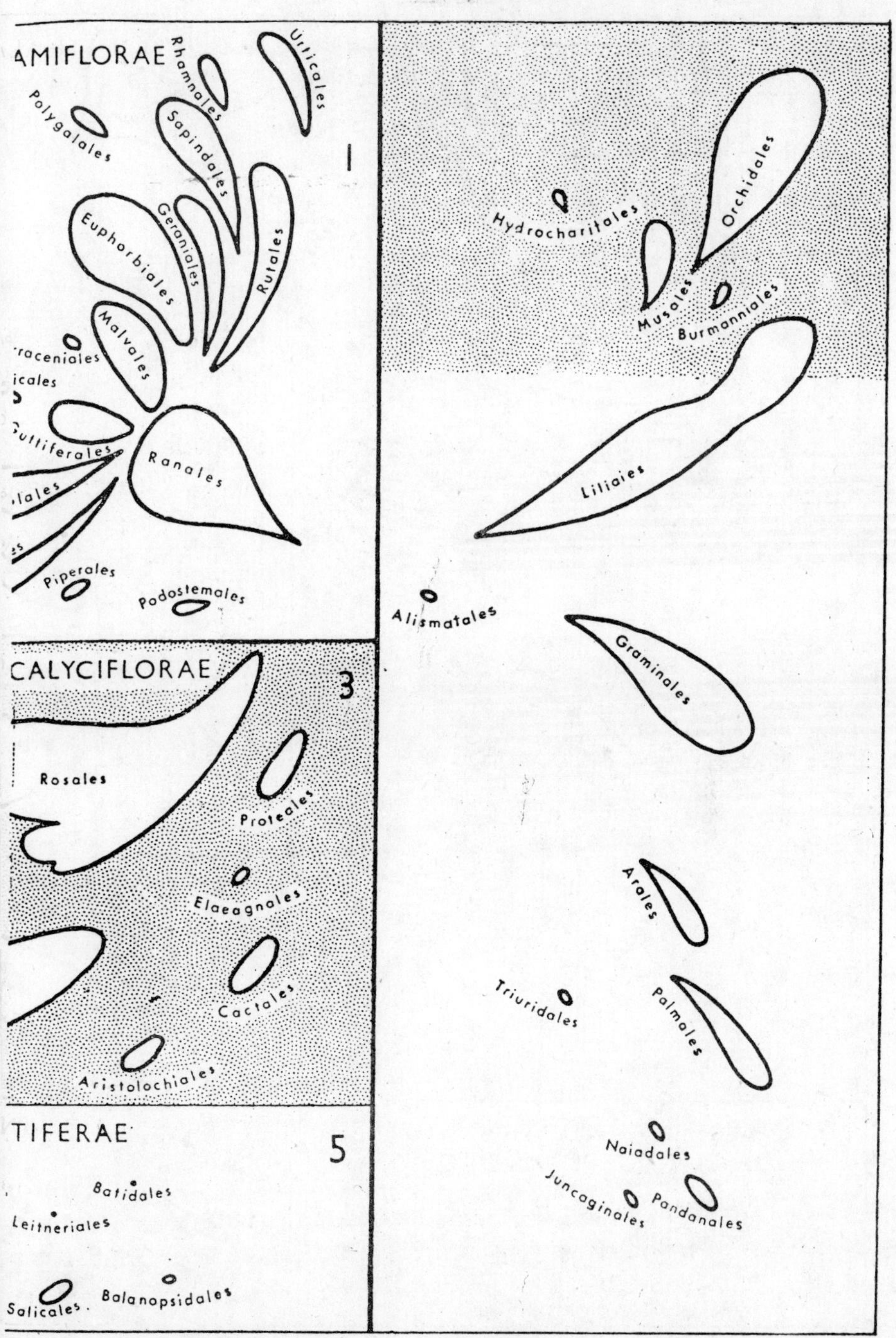

each order is in rough proportion to the number of species within it. The five informal groups (Thalamiflorae, etc.) recognised within the Dicotyledons are grade groups representing stages in floral evolution. (After Benson, 1957.)

they will show remote ancestry in groups where evolution has been rapid and common ancestry recent.

Above the species level, the claim that phylogeny can be constructed without a fossil record seems undoubtedly invalid. Phylogenists have confused the evolution of *organs* with the phylogeny of *taxa*—two completely different things.

(3) *Classification should be based on primitive, stable, conservative, non-adaptive characters and not on plastic, adaptive characters.* The fallacy in this claim, which was believed in by Darwin (1859), has already been discussed (p. 49; cf. Chapter 4). We must therefore conclude that a phylogenetic classification of the Angiosperms is impossible to achieve with any certainty without a fossil record.

The phylogenetic systems proposed in post-Darwinian times have been based on erroneous or very dubious assumptions, and we have no means of telling when they are right or wrong. It is an ironical paradox that the attempts at a natural system produced by de Candolle and by Bentham and Hooker (though the product of pre-evolutionary thought) rest, with all their shortcomings, on a firmer theoretical basis than scores of post-Darwinian "phylogenetic" systems.

Although these phylogenetic systems do not inspire us with confidence, this is not the case with some generic revisions. Although generally without fossil data, evidence from cytology, genetics and geographical distribution, combined with character correlations and morphological trends, has done much to indicate the likely course of evolution of species and groups of species. Phylogenetic classification within genera has therefore been much more successful than at family level. The publication of Babcock's *The Genus Crepis* (1947) was a landmark in taxonomy. The result of twenty-five years' work, *Crepis* was monographed on a scale that has not been achieved for any other Angiosperm genus of comparable size. Nevertheless, the classification of *Crepis* is largely based on gross morphology; information from cytogenetics, though weighted for phylogenetic interpretation, has had a very limited influence on the classification itself except in the circumscription of the genus. Where practicable, however, Babcock has tried to make the limits of his species coincide with evidence for restricted gene exchange. The first part of *The Genus Crepis*, dealing with the criteria used for classification and phylogenetic interpretation (from which he concludes that the genus arose in Central Asia), is a fascinating classic that no student of taxonomy should fail to read.

Other revisions (several on a regional basis) have been scarcely less successful in presenting plausible phylogenetic interpretations. In particular we must mention *Gossypium* (Hutchinson *et al.*, 1947), *Geum* (Gajewski, 1957) and *Nicotiana* (Goodspeed, 1954), in which attempts have been made to trace the course of evolution within the genera. In many cases, as in *Geum*, where breeding criteria have conflicted with morphological criteria, the authors have had to rely on morphology for the delimitation of sections and species. In all these studies biosystematic evidence has played a very important part in interpreting variation and in suggesting phylogenetic relationships. It must be emphasised, however, that the classification of these genera has been largely based on gross morphology and that, apart from the karyotype, very few microscopic characters have been used in their classification. Although Babcock claimed that taxonomy should rest upon the broadest possible foundation of biological disciplines, this ideal is still to be attained.

As Turrill (1942) puts it, "So much of what is termed phylogeny rests upon deductions from assumed premises which have not been subjected to adequate tests of validity. It is, indeed, a pseudophylogeny." The concept of "a true phylogeny" is one that depends on perception. In other words, phylogeny as a factual historical phenomenon exists; but perception and expression of it are an almost unsurmountable problem. There are as many "phylogenies" possible as there are types of character that one chooses to perceive (e.g. morphological or biochemical). Even in terms of external morphology, it seems very unlikely if a factual phylogeny is to be found among the forest of trees that grow up around us. It is unfortunate that phylogeny has a strong psychological appeal for herbarium taxonomists—it breaks the monotony of routine taxonomy. Families, without phylogeny, can be dull to teach and harder still to learn, so that phylogeny has become so firmly entrenched in teaching curricula that it is only now beginning to slip into the background and unfortunately taking a knowledge of families with it.

We do not suggest ignoring phylogenetic facts. It is the basing of classification on inferred phylogeny, instead of interpreting classification in phylogenetic terms, to which we are opposed.

2. *Natural Systems*

A natural classification of the Angiosperms in its fullest, neo-Adansonian sense, using all available characters (including microscopic and non-morphological ones), has not been attempted. To advocate the incorporation of microscopic and chemical data in a natural system is to put a great burden on the taxonomist. Such information is known for one group and not for another. A vast amount of investigation needs to be done, systematically, before information of this sort could be incorporated into an Angiosperm classification. Furthermore, it has to be presented in comparable and applicable form. Even with electronic computors, one must admit that the time is not yet ripe for such an attempt; we are too far from having the necessary information to compute!

Sporne (1949, 1954, 1956) and Lowe (1961) have shown how such a project might be attempted. Sporne advocates the necessity for a horizontal classification, and suggests the use of a "target-diagram" whose concentric circles correspond to relative overall advancement. Sporne calculated an "advancement index" for all families of Dicotyledons, attempting to use the fossil record and character correlations in order to distinguish between primitive and advanced characters. Each family could be placed on the circle corresponding to its advancement index, and the families then arranged next to one another according to their overall resemblances. The result would be a natural classification presented on the horizontal plane as a cross-section of phylogeny, the most primitive families being in the middle, the most advanced on the periphery—as would be the case if the general pattern of evolution is divergent (cf. Fig 7).

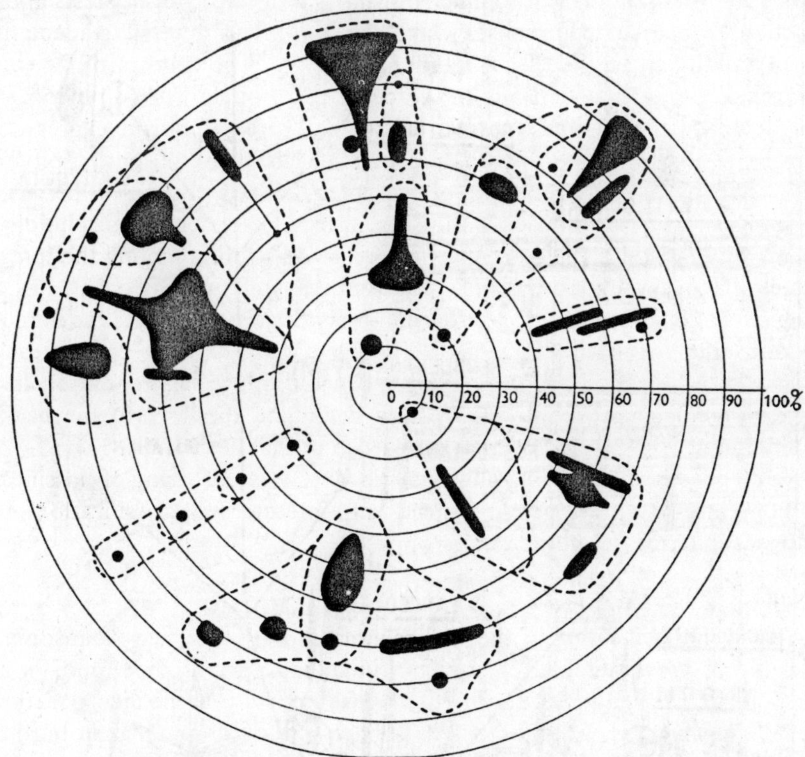

FIG. 7. Target diagram showing relationship and relative advancement of groups. The concentric circles correspond to successive grades of advancement, the black areas represent families, and the dotted lines enclose areas of affinity. The radial extent of each family represents the range of advancement within it, and the shape indicates the distribution of its constituent taxa according to their advancement. (From Sporne, 1956.)

Sporne's method of calculating the advancement index has been criticised by Turrill (1950) and Stebbins (1951), but the general method of presentation, if a satisfactory advancement index could be calculated from as many characters as possible, seems a valid and promising one. It is, in fact, the method of presentation adopted by Benson (1957) in his graphic presentation of Angiosperm relationships (Fig. 6), by Stebbins (1956) to show the relationship in the *Gramineae* and by Iltis (1959) for *Cleome* (Fig. 8). It combines

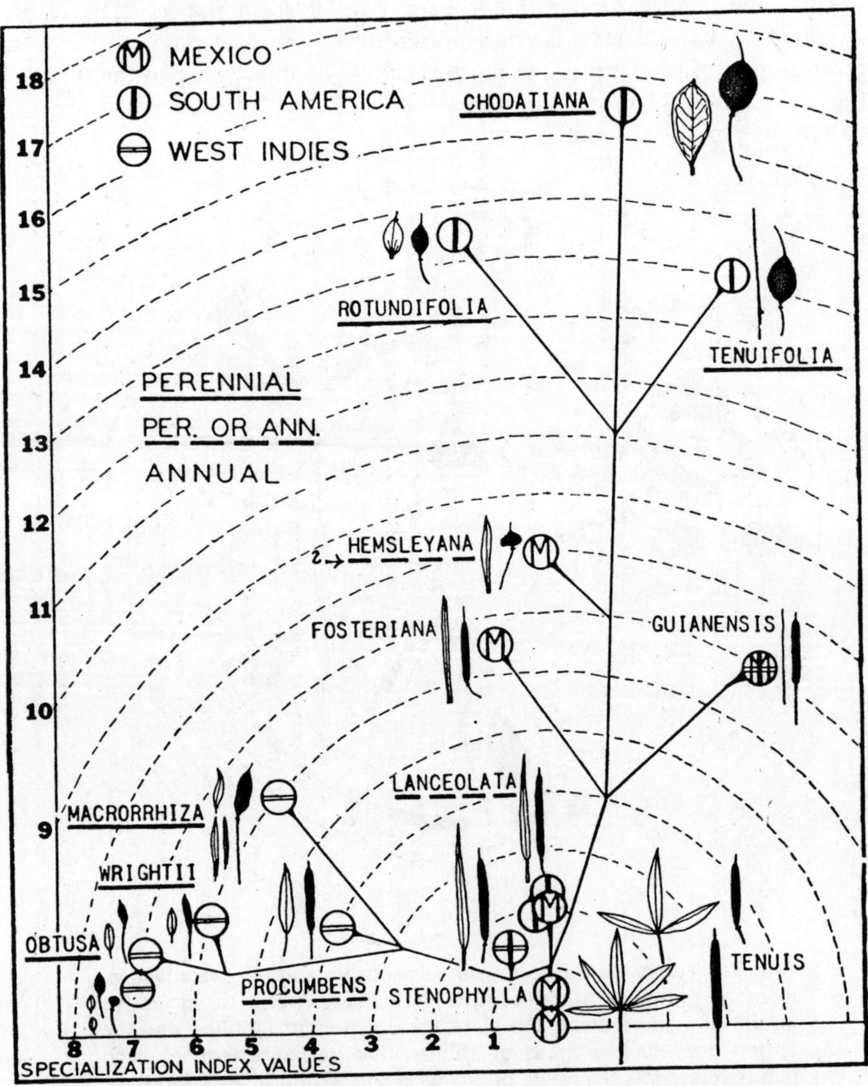

FIG. 8. Pictorialised target diagram showing relationships and relative advancement within *Cleome* Sect. *Physostemon*, largely based on specialisation index values. The ascending branch includes species with staminal apophyses, the prostrate one those with normal stamens. Note that the primitive taxa are all annual and mainly Mexican, and the most specialised taxa all perennial and either South American or West Indian. (From Iltis, 1959.)

PHENETIC RELATIONSHIPS WITHIN CAMELLIA

FIG. 9

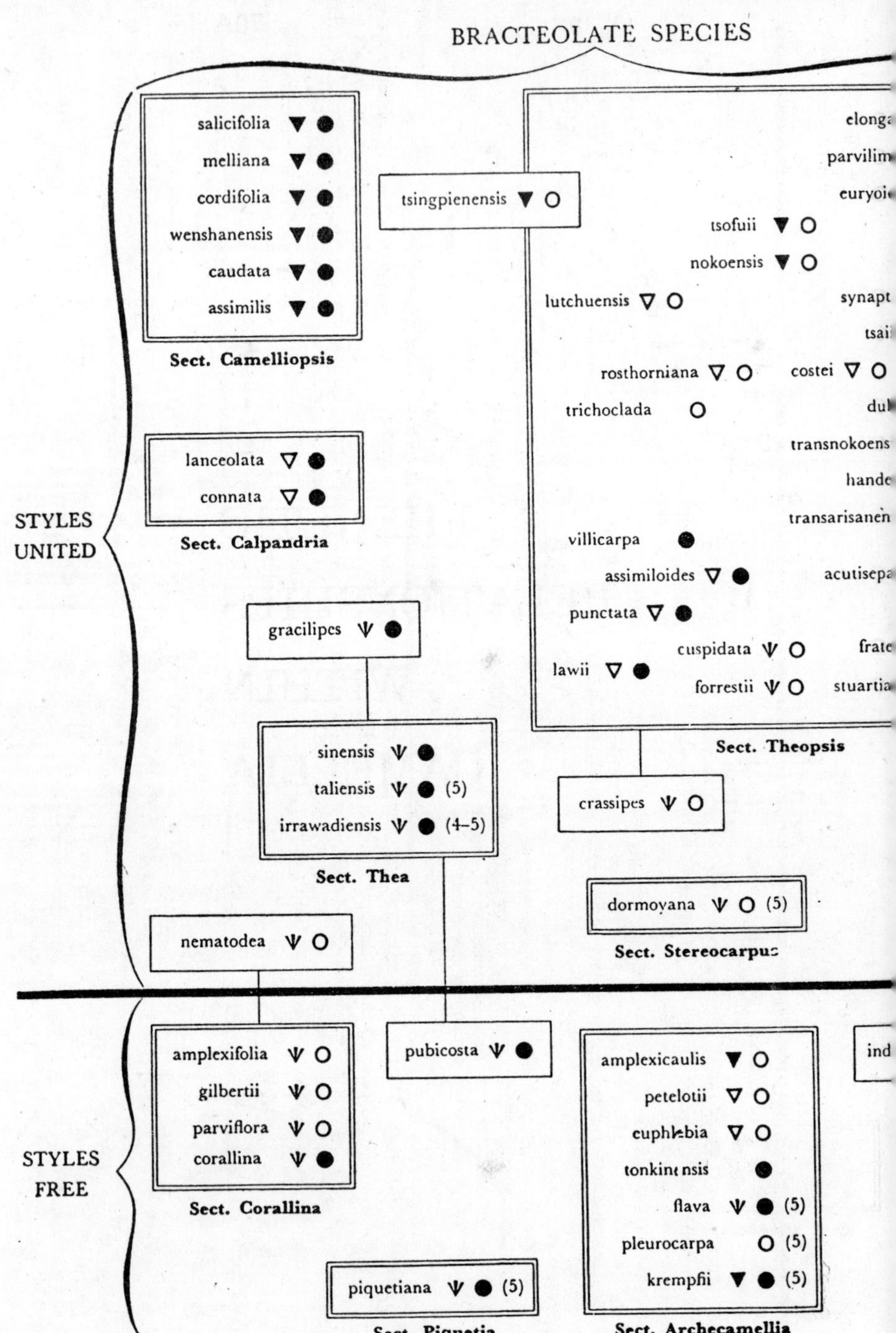

FIG. 9 Chart of phenetic relatio

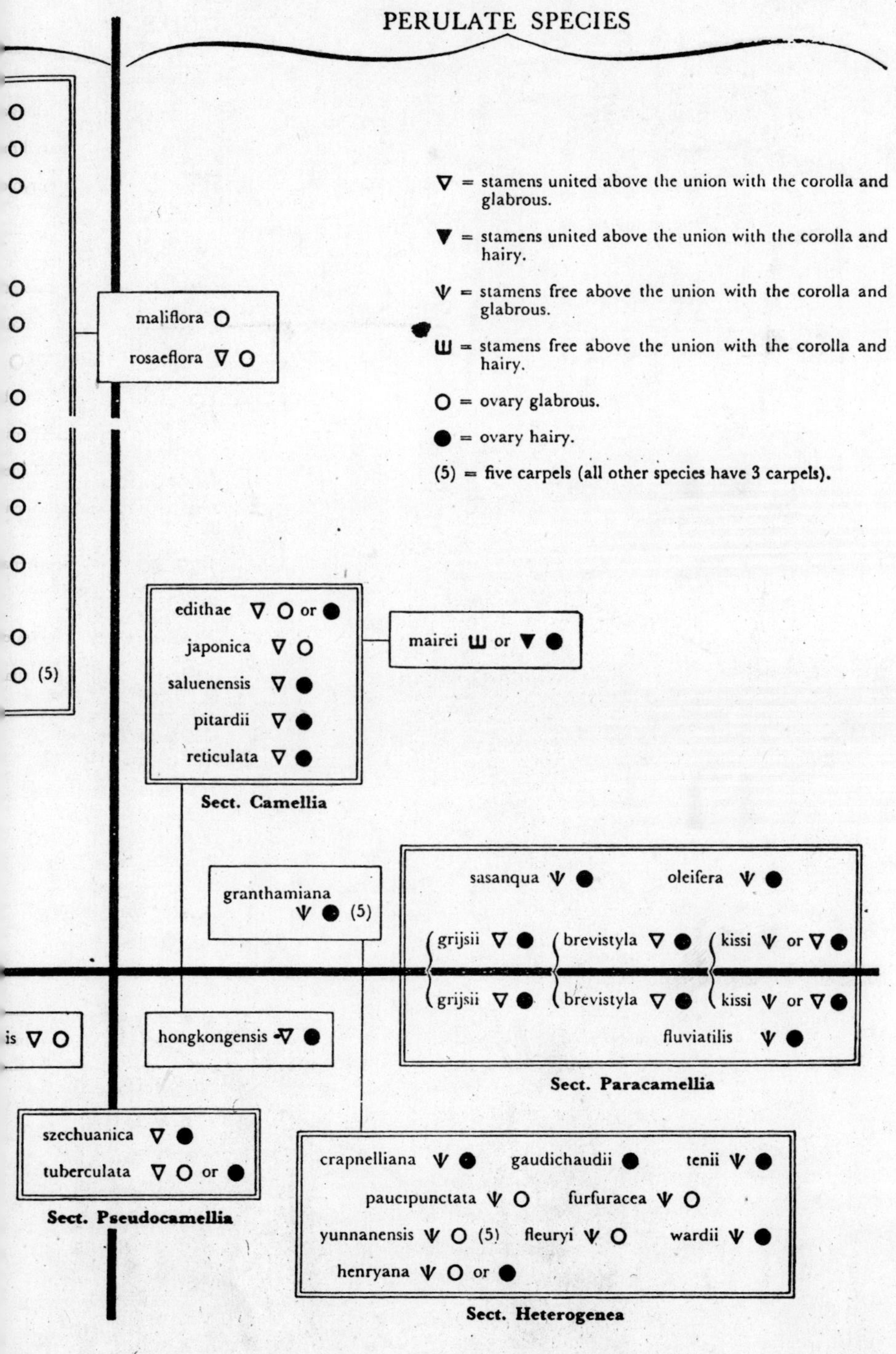

within *Camellia*. (From Sealy, 1958.)

natural classification with general level of advancement—a most sensible way of presenting and reconciling a natural classification with the facts of evolution known to us. It makes use of our knowledge of the evolution of organs, without having to guess not only from which stock a modern taxon is derived but from which relative level in time.

Systems of classification are, in practice, a mixture of natural, phylogenetic and (usually accidentally) artificial classification, though one or other of these may predominate with or without the intentions of the author. The families and genera are usually as natural as a considerable amount of *a priori* weighting has allowed. There is nothing unnatural about the *Cruciferae* as a family (whatever one may think of its genera) or about *Aconitum* as a genus (whatever qualms one may have about its species). Phylogenetic systems are to be praised not for their "phylogeny", but for the large number of natural groups they contain at almost every level of the hierarchy, though this probably tends to diminish the higher the rank. It is only in inferring the origin of families that post-Darwinian systems have gone further than the evidence has justified.

Few modern Angiosperm monographs have deliberately set out to present a natural classification and to depict it graphically, without being biased by phylogenetic assumptions. A laudable exception is J. R. Sealy's *A Revision of the Genus Camellia* (1958), whose ingenious diagram of relationship is reproduced in Fig. 9; it was designed primarily as a pictorial aid to using the text. *Camellia* is a genus for which factual information on phylogeny was almost lacking, although the evolutionary trends of certain organs (as incorporated in the diagram) seemed reasonably certain. This is the situation which faces most monographers who have primarily to base their work on herbarium material. Sealy's approach is honest, scientific and practical. Phylogenetic speculation is relegated to where it belongs—in suggesting the possible origin and evolution of the sections which appear to have been delimited as essentially phenetic groups.

Outside the Angiosperms (particularly in micro-organisms) arithmetical methods have recently been used in order to produce a natural classification with the aid of electronic computers. The methods are explained in Chapter 4. This is a rapidly expanding field of research that in the near future may come to play an important role in systematics, and is bound to widen the gulf between natural and phylogenetic classification.

Post-Darwinian Summary and Conclusions

Post-Darwinian taxonomists have had to face two evolutionary facts: (1) a phylogenetic continuum (which may be morphologically partly discontinuous if neoteny has played a major evolutionary role), one taxon changing into or giving rise to another; (2) the variability of populations and consequent acceptance of the "population concept", species being represented not by "types" but by variable populations.

The effect of these "discoveries" on classification has not been as great as might have been expected. Above the species level, recognition of evolutionary change has mainly led, not to major alterations in taxonomic groups (although there has been an increasing tendency to lower the rank in accordance with the widely held principle that taxa should be monophyletic), but to a preoccupation with their inferred phylogeny, usually on extremely dubious premises. We seem, however, to be attaining some stability in certain features of these phylogenetic systems. It is now generally accepted that the *Ranales* (*sensu lato*) represent the most primitive living group of Angiosperms and that the families in the *Amentiferae* are highly specialised, derived groups though no doubt very old ones. The fact that families and genera have not been radically changed since the time of de Candolle is because pre-evolutionary systems were largely the result of natural (phenetic) classification based on a survey of external morphology; natural classification is itself an outcome and expression of evolutionary history.

A more basic change has been at and below the species level as a result of the population concept. This has led to a series of special classifications, such as those based on the breeding system, and, on the morphological level, the recognition of geographical and ecological subspecies. Conflict between morphological and cytogenetic criteria has prevented any simple integration of cytogenetic information into general classification. The recognition of apomixis presents special problems in classification which probably cannot be adequately expressed by using binomials, although the general usefulness of the latter is a strong argument for employing them even in such groups. Continuous variation and the effects of introgressive hybridisation (processes which conflict with biological classification which is based on character discontinuities and correlations) have been studied by biosystematists; they have gone a long way in elucidating the processes of evolution at and below the species level (micro-evolution) and in describing its results.

At the level of practical "alpha" taxonomy, systematics seems to have developed a cyclical succession of monographic and floristic phases. We are now in a phase of Floras. Monographic studies are needed in order to try out theories of classification (in which there has been a sudden revival of interest) and the use of statistical methods.

An enormous amount of comparative data has accumulated from morphology, anatomy, ontogeny, phytochemistry, cytology, genetics, etc. Hitherto much of this has been weighted for phylogenetic interpretation. It has not been co-ordinated for a natural system of Angiosperm classification based upon as many characters as possible. We maintain that Angiosperm classification should express the relationships of plants as they exist at the present day, and that it should be assessed by overall resemblances. For convenience, however, the practical delimitation of taxa must be primarily based on morphological discontinuities. A natural classification will be the most generally useful, provide a sound basis for further research in numerous fields, and serve as a reference point for special classifications. Inferred phylogeny should not itself be used as the basis of classification. A natural system must come first, though it should as far as possible be made consistent with evolutionary facts and may be interpreted in evolutionary terms. Recent advances in numerical classification may have brought the attainment of a natural system, based upon a wide range of data, considerably nearer than it has been before.

Chapter 3

THE UNITS OF CLASSIFICATION

THE TAXONOMIC HIERARCHY: CATEGORIES AND GROUPS

Botanical classification involves the assembling of plants into taxonomic groups (taxa) on the basis of their relationships.* These groups are assembled into more inclusive groups, and these in turn into still more inclusive groups. The different levels of groups so produced are recognised as a series of hierarchical *categories*—what Mason (1950) calls the taxonomic structure. Categories are the levels, ranks or classes in a hierarchy to which groups are assigned. They are like a series of shelves—empty until they are filled with groups. It is important, if confusion is to be avoided, that a clear distinction should be made between categories and groups. The categories are entirely artificial and subjective and have no basis in reality—they correspond to nothing in nature since, as Blackwelder (1962) notes, they are decided upon arbitrarily before groups are assigned to them. Categories can only be defined by their position relative to other categories.

The groups placed in the categories are objective or non-arbitrary to the extent that they represent discrete sets of organisms in nature; in so far as the groups are considered as concepts of the mind, they are subjective (cf. Gilmour, 1940, 1961; Blackwelder, 1962).

Both the levels (categories) and the groups (taxa) that are assigned to them are named in the same way—species, genus, family, etc. Thus a taxon placed at the genus-level is called a genus, at the species-level, a species, and so on.

The different categories of the hierarchy are higher or lower according to whether they are occupied by more or less inclusive groups. Higher *groups* are those that are more inclusive than lower groups. It is important to note that a plant may be a member of several taxonomic groups (taxa), each of which is assigned to a taxonomic category, but is not *itself* a member of any taxonomic category—plants are not classified into categories but into groups. A taxon may belong to other taxa—a species, for example, belongs to a genus and to a family, etc.—but it can only be a member of one category (i.e., in this example, species). Categories are not made up of lower categories: the

* Relationships may be assessed in different terms according to the purpose for which we are classifying. They may be phenetic (natural), phyletic, genetic, biochemical, etc.

category family is not made up of genus-categories since there is only one genus-category! The characters shared by all members of a taxon placed in a lower category provide the characters of the taxon in the category immediately above (i.e. all the characters shared by the species of *Brassica* make up the genus *Brassica*, and all characters shared by *Brassica* and all other genera in the *Cruciferae* make up the family). Consequently, the higher a group is placed in the hierarchy the fewer will be the characters shared by all the subordinate units. Only a few of the characters possessed by a taxon serve to distinguish it from other taxa in the same rank (e.g. species) and within the same taxon of the next higher rank (e.g. genus); these are either diagnostic or differential characters. Especially at the higher levels of the hierarchy (e.g. family), no single diagnostic character may distinguish the taxa; they can only be separated by a combination of differential characters, any one of which may break down individually.

Before we go on to discuss the taxonomic hierarchy, it will be as well to list the ranks (with examples in parentheses) which are accepted in the *International Code of Botanical Nomenclature* (1961). Though all the ranks need not be used, their order may not be changed.*

Divisio (*Spermatophyta*)
Classis (*Angiospermae*)
Subclassis (*Dicotyledones*)
Ordo (*Rosales*)
Subordo
Familia (*Rosaceae*)
Subfamilia (*Rosoideae*)
Tribus (*Potentilleae*)
Subtribus (*Potentillinae*)
Genus (*Potentilla*)
Subgenus
Sectio (*Drymocallis*)†
Subsectio (*Closterostylae*)
Series
Subseries
Species (*Potentilla glandulosa*)
Subspecies (*P. glandulosa* ssp. *nevadensis*)
Varietas
Subvarietas
Forma
Subforma

* Following the practice adopted in the *Pflanzenfamilien*, supra-specific categories may be hyphened together for convenience, e.g. *Rosaceae-Rosoideae-Potentilleae.*

† A Subgenus or Section, if it contains the type-species of a genus, must bear the name of the genus unaltered and without citation of an author's name.

The Code allows for supplementary ranks to be intercalated or added if required. In few groups, however, are all the categories permitted in the Code actually filled, and still less often used in practice. That largely depends on the size of the group and its diversity—and how intensively it has been studied. The categories which are in most frequent use in the Angiosperms are probably the Subclass, Order (sometimes called Cohort), Family, Genus, Section, Species, Subspecies and Variety.

We have inherited the hierarchy. As the number of plants discovered has increased, new categories have been added. A much smaller hierarchy than we have today served the purpose of Linnaeus—more were not required at that time. Within the Plant Kingdom Linnaeus recognised only class, order, genus, species and variety. The first two categories contained taxa that were usually so artificial that they cannot be said to have had much influence on modern classification. Historically the next category to be widely used was the Family adopted (as *Ordo naturalis*) by de Jussieu.* The modern usage of the categories has obviously been greatly influenced by their history. Although in set theory (cf. Woodger, 1952; Gregg, 1954; Beckner, 1959) a hierarchical classification is built up from the bottom, in practice this is only partially true. Family, genus and species have been used for so long (the last two as the component of the binomial), that we accept these categories and, very often, the long-established taxa which represent them. When further categories are needed to deal with the quarter of a million Angiosperms that may be known today, we take these "basic" categories and divide them—e.g. genus into subgenus, section and subsection, species into subspecies and variety. If classification is natural and reflects evolutionary history, it can be assumed that a genus is older than a section, a family older than a genus. The hierarchy is therefore achieved, as it were, by jumping up the stairs three at a time, and then by coming more slowly downstairs; as our knowledge of plants increases, we need to come downstairs several times.

The *species* is now almost universally accepted as the *basic unit of formal classification.*† There is a very good reason for this. Whether defined in terms of morphological discontinuity or restriction of gene exchange, species are unique in that they are (so far as contemporary plants are concerned) *non-arbitrary as to both inclusion and exclusion.* As Simpson (1961: 115) puts it: "A group is non-arbitrary as to inclusion if all its members are continuous by an appropriate criterion, and non-arbitrary as to exclusion if it is discontinuous from any other group by the same criterion. It is arbitrary as to inclusion if it has internal discontinuities, and as to exclusion if it has external

* Earlier botanists have often used the term Order (or Natural Order) for what we now call a Family. This earlier usage of Order (adopted by Bentham & Hooker) is no longer permitted by the Code of Botanical Nomenclature.

† By this we do not mean to say that, as Blackwelder (1962) suggests, it is the smallest unit involved in the operation—that role is occuped by the individual—but that it is the smallest group that it is necessary to recognise.

continuity" (cf. Fig. 10). Taxa above the species level differ by usually containing more than one taxon and therefore show internal discontinuities. They are therefore arbitrary as to inclusion and non-arbitrary as to exclusion (i.e. their definition depends on criteria other than internal continuity). The concept, however, may be more applicable to taxa in the *Chordata* than it is to some supra-specific taxa (particularly families and orders) accepted in the

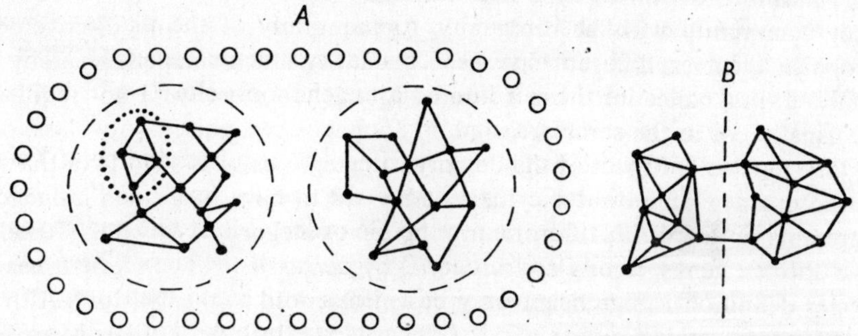

FIG. 10. Arbitrary and non-arbitrary aspects of contemporaneous taxa.

A. Infra- and supra-specific taxa.

Species are enclosed in broken lines, containing *individuals* (shown by heavy dots) which are linked to indicate continuity of variation and that they are known or assumed to be sequentially interbreeding. The species are each non-arbitrary as to inclusion and exclusion. In the left-hand species a *subspecies* is indicated by a dotted line: it is non-arbitrary as to inclusion but arbitrary as to exclusion. The two species are united in a *genus* as indicated by the line of open circles: the genus is arbitrary as to inclusion but non-arbitrary as to exclusion.

B. Diagram of a nexus of a species divided into subspecies along the dotted line. There is less continuity of variation (and known or assumed interbreeding) between the subspecies than within them (after Simpson, 1961).

Note: A group is arbitrary as to inclusion if it has internal discontinuities: it is arbitrary as to exclusion if it has an external continuity. A group is non-arbitrary as to inclusion if all members are continuous by an appropriate criterion; it is non-arbitrary as to exclusion if discontinuous from any other group by the same criterion.

Angiosperms; some of these appear, on morphological criteria, to be arbitrary as to both inclusion and exclusion (e.g. *Labiatae* and *Verbenaceae*). Simpson's concept therefore applies to an "ideal" classification rather than to the one we have today.

Species are unlike any other taxon in that they can be described and recognised without relation to any other category. In practice this is brought home to us when we sort a large collection of herbarium material. Without much difficulty (and without reference to generic limits), it is possible to sort out the material into different species (using the term in a broad uncritical sense), merely because variation within them is more or less continuous and variation between them discontinuous. We can therefore give a rough estimate of the number of species present in a collection although it would be much more difficult to estimate the number of genera—that would need a knowledge of species not present in the collection we are sorting. By the same means, a botanist in the field can draw up a life-form spectrum for a plant community without being able to name any of the plants—he can recognise them as different species. This task is made easier, it must be admitted, in a collection from a limited area; the most closely allied species are usually not in the same area.

One particular aspect of the "empty category" requires comment. This is the acceptance of monotypic taxa above the species level. *Ginkgo biloba* represents (in the present flora) a monotypic order; it is at one and the same time species, genus, family and order (*Ginkgoales*) of the class *Conopsida*. A species definition is significant only in a genus with more than one species. In taxonomic terms we can only usefully compare taxa with other taxa in the same rank and within the next higher taxon. But in the monotypic taxon we have a peculiar situation. *Ginkgo biloba* is recognised as the only living representative of an order co-ordinate with the *Pinales* and *Taxales*. It can only have been classified in this position by a process of deduction. The order *Pinales* has been built up by aggregation of genera. Yet *Ginkgo* (at least in the contemporary flora) stands alone. It is only recognised as an order because some differences which it shows from other Gymnosperms are of a similar order of magnitude to those which are used to differentiate related groups (*Pinales* and *Taxales*). A monotypic taxon can only be compared in taxonomic terms at the highest category which it represents—in this case the order *Ginkgoales* (Mason, 1950).

Principles used in assessing Relationships, Delimitation of Taxa and Attribution of Rank

Quite apart from historical usage, a large number of factors are involved in preparing a natural system of classification. The principles can for convenience be discussed under three headings: Criteria, Guides, Practical Considerations. Whether or not they are followed is largely a matter of taste although some of them, at least, are widely accepted.

1. *Criteria*

It is of course impossible to discuss criteria without reference to characters which provide the criteria (i.e. principles taken as a standard of judging). Nevertheless, as characters are fully dealt with in the following chapter, we shall not go into their detailed consideration here.

1. *Only genetically fixed variation should be used for classification.* This was a principle accepted even by John Ray, and is one which causes us some

difficulty only at and below the level of species where it is often not possible, in the absence of experimental evidence, to distinguish between environmental modifications and genetically fixed variations. In practice the taxonomist must be guided as best he can by his experience. If in doubt about the fixity of a character, it is advisable not to use it for the delimitation of taxa.

2. *Natural classification should be firmly based on correlation of characters and discontinuity of variation*, both for the assessment of relationship (in terms of overall similarity) and for the delimitation of taxa. As this has already been discussed in Chapters 1 and 2 we shall do no more than stress some points associated with it. As many characters as possible should be used in assessing relationship. We should not weight characters by *a priori* methods, but only by emphasising in our classification those characters which show the highest correlation and are *therefore* the most useful for classification. One result of this approach is likely to be the much wider use of vegetative characters in the delimitation of groups above the species and sectional level than has been used hitherto. It is not the "type" of a taxon but the range of variation and the number of intermediates which largely decide the delimitation of taxa. The assessment of relationship and the delimitation of taxa are more important (and less arbitrary) than attribution of rank.

3. *All available material should be used when classifying a group*. Characters which appeared to provide good *differentiae* when the group was first described may no longer hold when more material is available; this applies particularly to specific and generic concepts. Additional specimens frequently break down the differences used to delimit species in the pioneer phase of classification. New species described by various authors accumulate in genera originally based on far fewer species, so that a genus may become too broad and its content too heterogeneous for maintenance as a single genus; on the other hand, what were initially treated as two genera may have to be merged into one.

4. *When any taxon is revised* (be it species, genus or family), *it should be studied throughout its entire geographical range*. Above the species level this is often impossible when writing a Flora, but any taxonomic revision which entails the new delimitation of taxa and changes of rank should be done throughout the range of the group. The dangers attendant on neglecting to do so are illustrated in a paper by Tutin (1956). In discussing generic criteria, he suggests that *Lathyrus niger* (which resembles *Vicia* in its stem and leaf morphology but has the technical floral characters of *Lathyrus*) should be separated from *Lathyrus*. "Perhaps the happiest solution", he writes, "is to put it, as has been done in the past, in a separate genus, *Orobus*." This would be the most unhappy solution, because *Lathyrus* Sect. *Orobus* (as it is usually interpreted) contains a most varied assortment of species (including *Lathyrus niger*) which appear to have got there by exclusion from Sect. *Lathyrus* mainly because they lack tendrils; furthermore, the type of *Orobus* is apparently

O. tuberosus (i.e. *Lathyrus montanus*) whose leaflets have the parallel venation of typical *Lathyrus* instead of the pinnate venation of *Vicia* and *Lathyrus niger*! As accepted at present, Sect. *Orobus* is manifestly unnatural, being a heterogeneous assemblage of species, some of which in leaf form and facies show affinities with *Lathyrus* and some with *Vicia*. A thorough revision of generic and sectional limits in the *Viceae* would be a most rewarding study and is long overdue. Leaflet development and evolutionary trends in leaf morphology would make it particularly interesting.

5. *Consistency of treatment*. It is evident that no definitions of taxa or assignation of rank can be consistently employed throughout the plant kingdom. Nevertheless, there is room for more consistency than classification shows at present. Inconsistency of treatment is a very real problem and results from three main causes: lack of time for one taxonomist to study the whole group; psychological differences between taxonomists and between the different concepts they hold; revisions based on different material. Taxonomists should be as consistent as a knowledge of the facts allows, within the taxon with which they are dealing. To divide the genera in one part of a family and leave the rest of them unconsidered may sometimes be necessary but is not good taxonomy. If we recognise *Pulsatilla* and *Ceratocephalus* as genera separate from *Anemone* and *Ranunculus* respectively, there is every reason (in fact rather better reasons) for recognising *Consolida* as a genus separate from *Delphinium*. Inconsistency of treatment is very noticeable in the preparation of Floras and makes comparison difficult—a failing which the *Flora Europaea* should do much to alleviate for the European continent.

2. *Guides*

1. *Ability to cross*. The claim of many biosystematists (and of some zoological taxonomists) that *species* should not be treated as such unless their members are potentially interfertile, or, that species should be potentially intersterile with others and their components interfertile, is no longer acceptable, though in the majority of cases the factual basis of the claim may be true. It cannot be accepted as a general rule for several reasons:

(*a*) Inability to cross does not imply lack of genetic relationship.

(*b*) Morphological differences are not necessarily associated with inability to cross.

(*c*) In polyploid complexes, polyploid derivatives may be able to cross unlike the diploids from which they are derived, and this tendency increases with the level of ploidy. The derivatives are normally unable to cross with their diploid progenitor from which they may be morphologically inseparable. In such cases crossability is at variance with phyletic relationship.

In practice breeding experiments can only be applied to a limited part of the world's flora. The same applies to studies of the karyotype.

In classification the ability to cross (and the degree of fertility) can only be treated as a further taxonomic character and in a general classification must be assessed like any other character. It can be valuable when we are in doubt as to the attribution of rank on morphological characters, particularly between species and subspecies; in this connection information on crossability may serve to tip the balance.

As species of different genera are usually intersterile (with exceptions in the *Orchidaceae*, etc.), the inability to cross is of little help in separating genera. But if we shift the emphasis from the differences between genera to the resemblances between the species within them, crossability (and resemblances in chromosome structure) can help in delimiting the genus as a natural group of species (Rollins, 1953); the same would apply to sections within it.

2. *Chromosome morphology and number*. The karyotype can prove a most useful guide both in classification and phylogeny. We have already discussed the difficulties of using chromosome number for the delimitation of species, the conclusion being that it should be associated with constant morphological differences before the population can be recognised as a separate species. The karyotype, however, can be particularly useful as a guide in the delimitation of genera and in assessing their genetic relationship. In many groups virtually no help can be derived from this source, but in others it either confirms the classification based on gross morphology or suggests how a more natural classification might be achieved.

The work of Gregory (1941) on the karyotype of *Ranunculaceae* shows how the genera might be more naturally grouped on the basis of chromosome morphology and number, although this has not been translated into a revised classification. The same approach has been very helpful in the *Gramineae* in the detection of convergence. The genera in *Caryophyllaceae* Subfam. *Silenoideae* are often difficult to circumscribe, and here chromosome number has allowed species of doubtful position to be placed with confidence. In this subfamily the tribe *Lychnideae* (including *Silene*, *Lychnis*, *Agrostemma*, etc.) has the base number uniformly 12. In the tribe *Diantheae* the base numbers 14, 15 and 17 occur—for example 14 in *Saponaria*, 15 in *Vaccaria*, *Dianthus* and *Kohlrauschia*, and 17 in *Gypsophila* (cf. Heslop-Harrison, 1953b). The karyotype was used by Babcock to achieve a natural and largely monophyletic delimitation of *Crepis*; this involved the removal of several species to allied genera (e.g. *Hieracium*) and the inclusion of others in *Crepis* which had previously been placed elsewhere.

It must be stressed that chromosome morphology and number are a *guide* in classification, not an unequivocal criterion. By showing up likely phylogenetic relationships, they suggest more natural groupings. But it is up to the taxonomist to confirm these by finding correlated morphological characters so that a group can be satisfactorily defined and delimited. In many genera information on the karyotype is available for so few species that the evidence is insufficient to use as a basis either for changes in classification or for phylogenetic interpretation.

3. *Homologous variation*. The law of homologous series, worked out by

Vavilov (1922) mainly on cultivated plants, can be a very helpful taxonomic guide. Vavilov found that in closely related species and genera similar series of heritable variations occur. These are found so regularly that, knowing the series of forms within the limits of one species, we can predict the occurrence of parallel forms in other species and genera. The more closely related the taxa are, the greater will be the resemblance in their variation. Homologous variation has its basis in a series of ancestral genes. Such variation is particularly obvious in crop plants. In the native flora natural selection often eliminates certain variations, so that a rare species usually shows only part of the variation found in a commoner ally. Nevertheless, an appreciation of the law of homologous series can be of great help to the work of the taxonomist. By studying the variation in a well-represented species, he will be in a much better position to assess the variability that may be expected in an allied species of which he has very limited material. This may help to prevent him from describing extreme variants as distinct species. Although a taxonomist cannot, with certainty, make decisions on anything but the material before him, if his classification is to have stability he should try to foresee "the next move". In this connection the law of homologous series may at least indicate where caution is needed in the delimitation of species, especially new ones.

4. *Geographical and ecological distribution.* A great deal has been written, especially by German and Russian botanists, on the "geographical method" in taxonomy (see Chapter 9). All taxa (except *formae*) have a fairly well-defined geographical distribution, the outcome of evolutionary history and climate. Above the level of species, this distribution is just an attribute of taxa, but cannot be used for their definition. Nevertheless, it can be very helpful when we are trying to decide whether a group is monophyletic or not, particularly when the study of distribution is linked to the investigation of cytology and morphological trends (as in *Crepis* and *Gossypium*). Such guides, however, can only be used to *improve* a classification in so far as they are not themselves *deduced* from classification. This danger is often ignored by taxonomists and can be exceedingly subtle (Turrill, 1938a).

At the level of the species and below, geographical and ecological distribution play a rather different role in classification. Two plants growing in different areas (allopatric) may or may not be interfertile—only experimental evidence can show us that. But if they grow in the *same* area (sympatric), and particularly when they also grow in the same habitat, the situation is different. If the two plants continue distinct, it is fair to assume that there is some reproductive barrier between them. If we are in doubt whether to assign specific rank, information on their distribution and ecology may therefore be decisive. Subspecific rank is the only one in which geographical distribution can be taken as a decisive character in association with a single morphological difference. Above that level a correlation of morphological differences is required. No doubt Hedberg's recommendation (1958b) that a species should be separated by at least two morphological differences is an over-simplification of treatment in the recognition of species. (As an exception one might allow a single difference in external morphology if it were constantly associated with a high degree of intersterility—often associated with a difference in chromosome number; but in how many cases can we claim that the information (so often based on a few simple crossing experiments in the garden) is adequate?)

5. *Phylogenetic considerations.* We have already indicated (Chapter 2) that monophyly is a desirable criterion of taxa, at least above the level of hybridogenous species. Evidence for monophyly (or polyphyly) must normally be indirect, and based upon the interpretation of other evidence. This involves the study of as many characters as possible, including their relationship to one another and their functional importance, the geographical distribution of taxa, their comparative karyology and (below the genus level) ability to cross. By studying this evidence we may feel that we can distinguish between resemblance due to patristic affinity and resemblance due to convergence. In the latter case we should study the group again to see if it can be classified in a manner more consistent with its probable evolutionary history.

Phylogenetic considerations, however, are interpretative and cannot be used to construct a primary classification. They can only be used to interpret an initial classification in evolutionary terms and by revealing undesirable features allow us to improve it as far as we can without abandoning the practical aspect of classification. We must, however, be very careful as to how we have inferred monophyly or polyphyly. If these inferences have been deduced from the classification, then it is logically incorrect to use them to alter the classification. Phylogenetic inferences drawn from cytology or geographical distribution, for instance, can only be used to support or improve a classification in so far as they have not been used in its construction.

3. *Practical Considerations*

If a general classification is going to be widely used, it needs to work. We must be able to place taxa in higher taxa so that we can find them again. This means that practical considerations, which mitigate against the "objective" aspect of classification, are very important in systematics and necessitate that Science shall come to terms with Art if biological classification is to be of maximum use to science.

1. *Ease of observation.* Although we have stressed the desirability of using as many characters as possible in classification (particularly in the assessment of relationships), this does not mean that all characters are equally suitable for ready differentiation or diagnosis. Botanists have, in general, felt that humanity is best served if flowering plants are distinguished by at least some difference in gross morphology (exomorphic characters—Lawrence, 1953), though they may be associated, of course, with internal (endomorphic) differences. These "marker" characters are extracted from a much larger number considered in assessing the taxonomic position of the

group and are often the ones used in the construction of keys, since they are more constant and easily observed. Mycologists and even bryologists cannot proceed without the use of a compound microscope, but for the everyday purpose of identification it is usually felt that for the Angiosperms, which are the dominant group in the world's flora and most need to be recognised in the field, identification should be possible with good eyesight and a hand lens.

In this connection there is no doubt that the size of the flower has had a considerable influence on classification. Would *Euphorbia* be treated as a single genus if its flowers were as conspicuous as a lily's? The very narrow species concept often employed in *Paeonia* and *Pulsatilla* is no doubt a result of their having showy flowers; like pretty girls, they get a lot of attention (cf. Tutin, 1963).

2. *Size of taxonomic groups.* This is probably one of the major and most neglected factors effecting classification (cf. Walters, 1961). None of Linnaeus's genera was (by modern standards) large—in 1753 the biggest, *Carex*, contained only 29 species whereas today it contains well over 1000. The 100 families recognised by de Jussieu suggest that he considered this to be a convenient number for the division of the flowering plants, but since that time the number of plants known has enormously increased and many of the old genera and families have been divided up. J. Hutchinson recognises over 400 families, most of which are apparently based on genera which were already known to de Jussieu and were accommodated by him elsewhere. About one third of de Jussieu's genera are placed in the same families by Hutchinson, and the rest in different families. Though some scientists may find this state of affairs rather shocking, it is an inevitable effect of practical convenience which is a necessary part of classification; it is nothing for taxonomists to be ashamed of.

It is a quite commonly stated principle (cf. Mayr *et al.*, 1953) that *the morphological gap between two taxonomic groups should be inversely proportional to the size of the groups.* This is, in effect, a practical necessity, and was a principle adopted (though not in so many words) by de Candolle and by Bentham and Hooker and one which prevented them from recognising a very large number of monotypic genera and families. The *Scrophulariaceae* and *Gesneriaceae* are two large families which differ mainly in their placentation (a difference which sometimes breaks down). It is the fact that they form such large and apparently natural groups that makes them worth recognising as different families, instead of tribes of the same family. The same may be said of *Labiatae* and *Verbenaceae*, *Caprifoliaceae* and *Rubiaceae*. On the other hand, one could point to small genera in the *Berberidaceae* or *Nymphaeaceae* which might be (and indeed sometimes are) treated as separate families. From the point of view of delimitation and degree of divergence they are a good deal better as families than the pairs we have just mentioned, but because they are small there has been an understandable reluctance to raise their rank. There is no easy solution to this problem, but there is no doubt that we face a prevailing trend to subdivide old families and genera into others of the same rank. This may be partly the influence of size, but it is often due to the discovery of additional taxonomic characters (e.g. in pollen morphology, anatomy or cytology) and the increasing use of vegetative characters in the delimitation of higher taxa. We are doubtful if size, *per se*, is sufficient reason for dividing a natural group; on the other hand, it is an additional reason for doing so if other reasons exist.

Considerations of size are closely bound up with the history of taxa. The tacking of aberrant species or genera on to an otherwise homogeneous group is a rational procedure which prevents the recognition of an inconveniently large number of monotypic genera or families. Provided the taxon in question is more closely related to this group than to any other, no harm is done. In time, however, more species get discovered, and what was at first a solitary aberrant species becomes a member of a natural group. It may then become more suitable to take out this group than to leave it in the genus or family in which it was originally placed. This becomes necessary if it can be shown that the aberrant group is in fact related more closely to another taxon than to the one in which it was originally classified (e.g. *Paeonia*, originally placed in *Ranunculaceae*, is now classified in a family of its own, *Paeoniaceae* which may have more in common with *Dilleniaceae* than *Ranunculaceae*, but differs markedly from both). Here natural relationships have quite rightly got the better of any criterion of size.

We cannot leave the relation of size to convenience without considering a related phenomenon, *lumping* and *splitting*, terms which apply to a broad or narrow concept of taxa (especially species). Splitting and lumping are to some extent correlated with taxonomic purposes (Simpson, 1961). A botanist who is primarily concerned with identifying critical collections naturally has his attention focused on dissimilarities, diagnoses and keys—his main preoccupation is telling taxa apart. This is an analytical approach and he is likely to be a splitter, at least in his own group. On the other hand, a botanist who is monographing an order or family is likely to be a lumper—his approach is primarily synthetic. In addition to this there is the factor of temperament—some people are better at taking things apart than they are at putting things together!

Short of the death penalty, there will always be lumpers and splitters. Splitting tends to be accentuated with age, a characteristic of the malaise being that the sufferer seldom admits that he *is* a splitter. As Simpson points out, a taxonomist who splits in his own group frequently wishes other people were lumpers in theirs! Splitting is often the result of having a limited flora to deal with; the taxonomist may not feel "fulfilled" unless he can recognise new taxa—obviously a most undesirable state of mind. It may also result from a lack of appreciation of variability, either because the plants have not been studied in the field or because the biological nature of the group is not understood. It is not a coincidence that so many amateurs have chosen to work with apomictic groups; their revision has the psychological appeal of stamp collecting.

In little-worked areas it is most probable that an increase in the amount o material available will eventually lead to a broadening of specific limits. To describe small species in the pioneer phase can hardly be avoided, and more confusion may be caused by too wide a species concept in this stage than by too narrow a one. But to split when abundant material does *not* justify it (i.e. by ignoring continuity of variation) is inexcusable; yet it is still done, even in sexual groups (cf. *Thymus* and *Anthyllis vulneraria* agg. in *Flora URSS*). Several arguments have been put forward in favour of splitting (cf. Juzepčuk, 1958—an energetic exponent of this procedure in the U.S.S.R.) of which the following seem the most relevant:

(1) No clear line can be drawn between specific and subspecific taxa. Therefore morphologically distinguishable populations which occupy a different place in the economy of nature (rather than different areas) should be accorded similar status. Specific rank is therefore accorded to all of them—a return to the monotypic concept of species followed (in theory if not always in practice) by Linnaeus.

(2) If nature shows a vast number of such populations, it is up to the taxonomist to express the fact in his classification. He should not hide or ignore that variation of nature. The first two arguments are tenets of the Komarovian school.

(3) Since many biosystematists have emphasised intersterility as the main criterion of species, numerous "genetic" species can be described.

(4) At the level of the genus or family, small groups are more natural and easier to use than large ones.

(5) The *Index Kewensis* includes no infra-specific taxa, so only species (of the lower groups) can be readily traced.

It seems to us that none of the first three arguments is valid. They show a lack of understanding of the purposes of classification, and are unacceptable for a very simple reason. They ignore or evade continuity of morphological variation so that the "species" frequently cannot be named unless we know where its representatives came from. Such taxa are not non-arbitrary by both inclusion and exclusion and therefore do not fit the concept of species which attempts to distinguish them from all other taxa. The result is bound to be a multiplicity of "species" which even the specialist may not be able to identify. The only solution seems to be a polytypic species concept, allowing for the recognition of subspecific ranks when it is not practicable to recognise geographical (or eco-geographical) races as separate species. Many species, of course, vary very little and cannot be divided at all; they are monotypic.

The fourth argument (favouring small families and genera) is more difficult to refute. If the claim that small groups are more natural and easier to use than large ones is true, then the argument carries weight. Natural, however, is a relative term and takes the hierarchy in its stride. The fact that the sections of *Primula* are natural does not make the genus any less natural! Each case must be decided on its merits, but we very much doubt if it would be more convenient to identify a species of *Carex* or *Astragalus* if these genera were divided up. The process of sorting and identification of specimens would, in fact, take us rather longer because we should have to go through more steps to obtain the specific name. The fifth point, that no infra-specific taxa are included in the *Index Kewensis*, is an editorial limitation but not sufficient excuse for the pulverisation of species.

Taxonomy has two components: analysis and synthesis. The taxonomist must often ask himself the question: Is it more important to emphasise resemblances or differences? The answer will vary from group to group. The role of the family is one of synthesis—the expression of relationship is therefore of overriding importance. It is at specific rank that we are most concerned with discrimination. Other things being equal, binomials (species) are more convenient than trinomials (subspecies) because they are shorter. Provided some allowance is made for local hybridisation, we must be able, for communication, to refer most of our specimens to binomials. If variation is too great to allow us to do this (or the differences inconveniently small), then our species concept is too narrow to provide a workable classification.

3. *Economic importance*. Like size of flower, this must have had a considerable bearing on classification. Groups which are economically important —either for their products or for their place in agriculture or horticulture— are likely to be extensively studied, while some humble group like the *Urticaceae* wait in vain for a botanist to open the cupboard and have a good look at them, let alone study them *in vivo*. It is not that one grudges the *Orchidaceae*, *Gramineae*, *Eucalyptus*, *Nicotiana* or *Gossypium* their economic importance; it is only that one wishes *all* groups were economically as important, because the emphasis on economic groups is bound to unbalance classification. On the other hand, the extensive study of cryptic characters in economic groups has been extremely rewarding. The *Gramineae* provide a splendid example of what can be achieved in a difficult group where convergence has often obscured phyletic relationship (cf. Philipson, 1961). Not only are the *Gramineae* classified in a more natural manner than ever before; we understand them infinitely better. The same cannot be said of the *Cyperaceae* which are economically relatively unimportant.

4. *Stability*. A great deal has been made of the desirability of stable classification. Gilmour has gone so far as to consider this of overriding importance. Although we agree that stability would be desirable, we certainly do not believe that changes—and very considerable ones—should not be made. The discoveries of the present century are providing a wealth of data that should be incorporated into classification. Not to use these data to improve a natural classification is too high a price to pay for stability. By all means let changes be made cautiously, and not before the evidence is strong.

There are three different ways in which changes in a system of classification can be brought about:—

(a) *Relationship*. If it is found that a plant is overall more nearly related to another genus than to the one in which it was originally classified, then it should certainly be removed—either to another genus already accepted or to form one of its own, depending on the amount of divergence shown. Changes in this direction seem inevitable and science will not be served by ignoring them.

(b) *Rank*. We have indicated that changes of rank are very largely matters of opinion. Rank is influenced by tradition, the size of the group and the degree of divergence. Yet if we are trying to classify consistently and to produce a well-balanced classification, changes of rank cannot be avoided. In this connection it is obvious that the stability of species and genera are particularly desirable because they form the binomial which we must use for the designation of species. Yet changes in family status can be hardly less trying when we are looking up plants in a Flora or herbarium. Changes in the rank of tribes, sections or subspecific taxa result in much less inconvenience. On the whole, however, we deprecate change of rank except for motives of consistency and in the achievement of a balanced natural classification. The effect of raising the rank of a taxon is to draw more attention to it.

(c) *Nomenclature*. Changes, that are not taxonomic, in the names of plants are strictly nomenclatural and result from the correct application of the rules of the Code, particularly those concerned with priority of publication. These legalities can prove very trying and take up a disproportionately large amount of a taxonomist's time which would be better spent in studying plants. We have, however, a fairly stable set of rules, and it is to be expected that their consistent application will eventually result in a more stable nomenclature. A list of conserved (and rejected) generic names is given in the Code, but there is no list of conserved specific names, even for plants of economic importance. There has understandably been disagreement as to what principle should be followed (if not that of priority) in the choice of species names to be conserved. Though we must follow rules as they stand, we cannot help feeling that name changes for anything except good taxonomic reasons (which include ambiguity of usage) are a disservice to biology.

Usage of the Categories

We cannot do better than quote from Mason (1950): "That which is a species to one taxonomist may be a subspecies to another, and that which is a family to one may be an order to another. This difference in opinion is wholly consistent with the nature of the known facts and does no harm to the objectives of taxonomy. The empty category of the taxonomic structure has no foundation in reality and obviously cannot be objectively defined. . . . Taxonomic categories possess only relative values in so far as we are now in a position to understand them."

Nevertheless, although the categories cannot be defined, we can profitably discuss their usage, which may be a guide to the delimitation of taxa and the attribution of rank, and how the *groups* assigned to them may be defined. We shall only deal with the most commonly used categories below the level of the Subclass.

1. *Species*

The species is for most purposes widely accepted as the basic unit of taxonomy. As Darwin said, every biologist knows approximately what is meant when we talk about species, yet no other taxon has been subjected to such violent controversies as to its definition. The idea of the species as the lowest unit of biological classification has a long history: it derives from Aristotelean logic and philosophy (Crombie, 1950). Linnaeus, from whom we have inherited the concept of binomial nomenclature for species and many of the specific names still in use today, believed that species were divinely created: they were, as we have noted, discrete and true-breeding. The idea that species are made of groups of individuals (i.e. populations) goes back to Linnaeus (cf. Mayr, 1957). Many evolutionary taxonomists believe that species are formed as a result of the evolutionary processes. It is an act of faith for both Linnaean and evolutionary taxonomists that their task is to go out into nature and find these creations. The *concepts* covering these and other groups called species are, however, constructions of the human mind and cannot be defined.

Although Sirks (1952) may perhaps exaggerate in saying that on theoretical and practical grounds "the interest of nearly every right-minded biologist is still focused on the concept of species", it is nevertheless true that more than any other unit it is the daily concern not only of taxonomists but of non-systematist biologists. Species concepts have therefore to be manageable by those who have to work with them. The student of taxonomy can scarcely, then, fail to be dismayed when he is told by some that "the species is real", by others that "the species is an abstraction" and that only the individuals are real. He will soon realise that the species is, if nothing else, a problem. In fact, one of the many books and symposia devoted to this affair is entitled *The Species Problem* (Mayr, 1957).

Much of the disagreement about species concepts is artificial and obscures the large amount of agreement there is amongst various groups of taxonomists about what constitute species. The "hidden reason for so much disagreement" that Mayr (1957) feels must exist, stems largely from attempting to define what the species "really is" in all contexts: there is no universally correct definition and progress in understanding the species problem will only be reached if we concentrate on the problem of what we shall treat as species for any particular purpose. Another and probably more important source of confusion results from the failure to distinguish between the various meanings that attach to the word "species". A lucid exposition of this problem has been given by Blackwelder (1962) who notes that most of the meanings fit into either the first or second pairs of the following types:

(*a*) "Species" without any article, meaning groups of individuals (populations) recognised by any means.

(*b*) "A species", referring to one of these groups or populations.

(*c*) "The species", sometimes singular but collective, denoting the general concept covering all the groups known as species.

(*d*) "The species", "Species level", "Species category", etc., referring to the level or rank in the hierarchy at which we choose to place these groups.

Apart from these considerations it is clear that we need different species definitions for different purposes, as can be readily seen if we consider the various species definitions that are currently employed. This leads on to the main practical issue: if the species group is being subjected to two or more widely discordant definitions, for which of them should it continue to be employed?

We do not have a free hand, however, in deciding the issue. The first point that must be made is that for the purposes of nomenclatural taxonomy all organisms must be referable to species: the species is a category recognised by the International Code of Botanical Nomenclature and as such is "definable universally only in the language of formal relations and not by the property of organisms" (Mason, 1959). This is what is termed the *nominalistic* concept of the species (and other taxonomic categories). It is imposed by the acceptance of a hierarchical system of nomenclatural taxonomy since, as we explained earlier in the chapter, the categories can only be defined by reference to their relative position in the hierarchy. If we accept that the species is then an empty category, the question to be solved is with what kind of biological situation should it be filled. This has been clearly put by Heslop-Harrison (1962) when he says that if practising taxonomists wish their efforts to have legal sanction according to the International Code, they "have the unavoidable task of determining what variational units amongst all those they encounter can be fitted with least violence into [the species category]".

If we look at the "facts of nature" we may agree with Waddington (1960) that it is "an empirical fact that living organisms do not vary continuously over the whole range, but that they fall into more or less well defined groups, which are commonly called species". This would suggest that the solution could be sought by making our species coincide with the significant discontinuities that occur in nature and this is essentially what proponents of the biological species concept propose. It is, however, one of those generalisations that can be easily made in biology and is another way of saying that everyone knows what a species is in general terms, but as Waddington goes on to say, the *precise* definition of what constitutes a species is a matter of great difficulty, despite Nature's assistance. If in fact we look at the discontinuities that occur in nature, we will find that they are many and varied and of different degrees and natures. We can in fact accept (or reject) those discontinuities we choose to perceive. As Cowan (1962) has recently put it, "just as no two observers see the same rainbow, so no two biologists conceive exactly the same species". To put this in an extreme form, a traditional taxonomist will perceive and recognise as significant, morphological discontinuities; the experimental taxonomist will look for discontinuities in breeding pattern. In both instances the discontinuities are "real" in the sense in which the word is used in such contexts. Both can form the basis for species definitions and both are equally valid. Their biological significance may be different, as may be their degree of arbitrariness and practical value. We should now consider present usage in terms of these criteria and then assess the relative advantages of the different concepts currently employed (1) for general usage; (2) for special purposes.

Several classifications of species definitions have been proposed (cf. Mayr, 1957; Beaudry, 1960), so involved has the subject become, but the two main classes are:

(a) *Taxonomic* (embracing the orthodox, typological, morphological, morpho-geographical, etc.)

(b) *Biological* (including the biosystematic, genetical, cytogenetical, non-dimensional, multidimensional, etc.).

It is beyond question that the species category has been most widely employed in the first sense, i.e. for assemblages of individuals with morphological features in common and separable from other such assemblages by correlated morphological discontinuities in a number of features. The species in this sense is intended to be a collection of samples so similar that not only the describer but other users would have little hesitation in saying they were alike, recognisable and communicable. This morphological concept originated with man's original recognition of similar kinds in the world around him, such as wheat, maize, dogs, etc., and his need to communicate about them. Such kinds were in some cases quite precise and corresponded to what today we recognise as distinct species; in other cases they were species groups, sections, genera or groups of genera, etc. This historical use of the word species has recently been discussed by Cowan (1962), who quotes from John Locke's *An Essay Concerning Human Understanding* (1689): "*genera* and *species* depend on such collections of ideas as men have made, and not on the real nature of things", and again "our distinct species are nothing but distinct complex ideas, with distinct names annexed to them". Too much should not be made of man's innate ability to differentiate between "genera" and "species": it is not without significance that a pocket American dictionary defines species as "genus, family, . . ."!

The species was employed in the primarily morphological sense by Linnaeus whose *Species Plantarum* forms the starting-point of botanical species nomenclature. It was also *typological* in the sense that one individual could be taken to represent the species (see p. 10). Such crude typological usage is virtually unknown today, and as we have already indicated (p. 11) we feel that the continued use of the term typological in the archaic sense to refer to present day methods of defining species morphologically is misleading and provokes misdirected criticism.

By post-Darwinian times the morphological species concept had become much refined. It accepted the facts of variation and attempted to contain them, with greater or less success, within the limitations imposed by the species, subspecies, varietias, forma hierarchy. Geographical distribution of the populations comprising the species were taken into account and formed the basis of the morphological-geographical approach introduced by Bunge, Kerner von Marilaun, Wettstein and others, perhaps best formulated by Du Rietz (1930). Du Rietz employed the principle of discontinuity in natural variation for the definition of species: the species were to be delimited by the lines or zones of discontinuity found in nature. His actual definition of the species was "... the smallest natural populations permanently separated from each other by a distinct discontinuity in the series of biotypes* . . .". In practical terms this has come to mean that taxonomists regard their specimens as samples of natural populations, not just herbarium samples. They base their decisions about specific delimitation on the presence or absence of morphological breaks in the variation shown by these populations. The limit of the populations is judged on the basis of the distribution of the samples (and by extrapolation from other well-studied situations); the degree of distinctness of the morphological discontinuity required has to be judged on its merits for each case. Population, as employed in this sense, means an assemblage of plants with a particular distribution: its parameters are judged from morphological evidence (cf. Chapter 11). As discussed below, a consequence of the morphological-geographical approach was the recognition of the subspecies as an important taxonomic unit.

It is in this sense that most practising taxonomists employ the species concept (cf. Rothmaler, 1944, 1954; Heywood, 1958). It is intended as a generally applicable concept and takes into account all available evidence, morphological, geographical, cytogenetic, etc., but insists that the species so recognised must be delimitable by morphological characters. Such morphologically defined species "will, however, illustrate different kinds of evolutionary situation; in other words, species are only equivalent by designation, and not by virtue of the nature or extent of their evolutionary differentiation" (Heywood, 1958). Species in this sense are *phenetic*. On the other hand, Camp (1951) has proposed that the term *binom* be applied to such species, the term species itself being restricted to those units whose genetic relationships are known.

* *Biotype:* a collection of individuals which are genotypically essentially the same.

The relationship between morphological-geographical species (which are best termed taxonomic species in view of their overwhelming use by taxonomists as opposed to biosystematists, genecologists, etc.) and the type concept has already been discussed (p. 10). It is worth reiterating here that the type specimen (plate, description, etc.) may originally be regarded (Fig. 11) as the centre of a hypothetical unit (species) with subsequently discovered individuals arranged around it, their distance from the type-centre being inversely proportional to their community with the type (Cowan, 1962: 437). It is often found later, once more adequate samples have been made or when the species is being revised, that the type is in no sense typical or average for what is known of the population from which it has come; in graphic terms it is displaced from the centre of the circle. Our taxonomic judgment concerning specific status is to a large extent a reflection of our deductions on population parameters and structure, the type-specimen being merely a nomenclatural reference point.

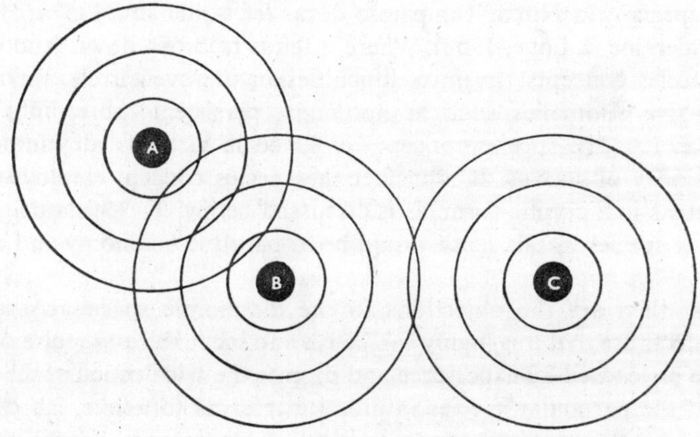

FIG. 11. Diagram illustrating the use of the type concept.
A and *C* represent the types of hypothetical species, the circles around them indicating the range of variation found in subsequently discovered individuals. Unknown plant *B* can be shown by extrapolation to be nearer *A* than *C*. (From Cowan, 1962.)

In any taxonomic investigation the earlier step in procedure is to employ a taxonomic species concept, not a biological one, since it is the only generally applicable one. Taxonomic species may then be subjected to experimental investigation and be shown to represent some particular evolutionary situation. As a result of the evidence revealed by such study, the taxonomist may reconsider the treatment of the individual case and redraw his specific boundaries, provided they are still recognisable morphologically, but he may not be unduly influenced by certain kinds of evidence (Valentine & Löve, 1958).

What then is the purpose of the taxonomic species, and how far does it effectively fulfil this purpose? From what we have said it should be evident that the purpose of the taxonomic species is to be part of a general-purpose classification based very largely on phenetic evidence to provide a "broad map of the diversity of living things" (Gilmour, 1958). There can be no doubt that such a species concept is necessary if our taxonomic system in general is to continue functioning. The taxonomic species can then be regarded as an essential first step, the key to further information. Its effectiveness has been disputed by Camp (1951) amongst others, but, as has been pointed out on several occasions, there is a very broad measure of agreement amongst practising taxonomists about the limits of taxonomic species in well-studied areas, especially in North Temperate floras (cf. Anderson, 1957a; Heywood, 1957; Valentine & Löve, 1958). Where it tends to break down (and where all other species concepts run into difficulties or are even irrelevant) is where reproductive anomalies such as apomixis, persistent inbreeding, etc. are operative. It is the species concept employed in virtually all Floras for the vast majority of species. It is in fact the species concept employed by biosystematists in a cryptic form, as is discussed below. To legislate against the use of the species in this sense would be to paralyse taxonomy and classification.

What then are the objections to the taxonomic species concept? The main defects are that it is highly subjective and incapable of precise definition: it has to be learned by experience and despite the wide concurrence as to the limits of the particular species in practice referred to above, inconsistencies of treatment do certainly occur, particularly as regards status; and unless there is consistency of treatment when faced with similar situations, our taxonomic system will become meaningless (cf. Heywood, 1960c). Apart from the phenomena of splitting and lumping already alluded to (p. 85), there are certain conventions adopted by some authors and even by national schools of taxonomy, such as the Komarov tradition discussed below.

The subjectivity or more strictly speaking the arbitrariness of the delimitation of taxonomic species can be lessened by the application of certain minimum requirements in numbers of correlated character differences for the assessment of specific and subspecific differences. Hedberg (1958b), for example, suggests for specific distinction clearly discontinuous variation in at least two independent characters. Sneath (1958) has proposed at least five correlated character differences when employing techniques of mathematical correlation for the delimitation of species by naturally occurring discontinuities in the variation pattern. The application of biometrical techniques such as population analysis and the presentation of data in the form of histograms, scatter diagrams, etc., serve to allow a more precise assessment of morphological differences, although these techniques cannot be employed without adequate material.

The taxonomic species is often criticised by biosystematists on the grounds that it is less important biologically than species defined in terms of gene pools and sterility barriers. But as Valentine & Löve (1958) comment, this objection can be taken too far, since it would be absurd to underestimate the biological significance of the morphological (and physiological) differences between populations. As they say, "The form of a plant represents its response to its environment; it has come, in the course of evolution, to have leaves of just such a shape and flowers of just such a colour and to occupy a habitat in which the soii is of a certain texture and acidity. If another plant has leaves, flowers and habitat preferences which are all different, it is arguable that these differences are of at least as great biological significance as the ability of plants to exchange genes freely."

Despite its apparent subjectivity, the morphological-geographical manner of delimiting species has been widely successful. It soon became apparent that its success was largely due to the fact that the groups it recognises very often correspond in nature to breeding populations. The notion that species are groups of interbreeding individuals, and as a corollary true-breeding, is probably as old a concept as the morphological species concept. It derives from man's observation that animals related in terms of their morphological features mated amongst themselves to produce fertile offspring bearing similar characteristics (Beaudry, 1960). This concept is well expressed in John Ray's celebrated dictum "*nulla certior occurit quam distincta propagatione ex semine*"—implying that species breed true within their own limits. In outbreeders, however, all biotypes will be somewhat different due to segregation and recombination, so that emphasis must be on the *limits*. These limits can be determined by morphological discontinuity or by breeding discontinuity, giving taxonomic and "biological" species concepts respectively.

The so-called biological concept as the theoretical basis for the species was largely developed during the last thirty to forty years along with the study of the genetical basis of variation, reproductive mechanisms and population structure. It has two components which are differently stressed in different versions of biological species definitions: (*a*) interbreeding between the component members; (*b*) reproductive isolation from other populations. A good example of such a definition is that of Grant (1957): ". . . a community of cross fertilizing individuals linked together by bonds of mating and isolated reproductively from other species by barriers to mating".

Proponents of the biological species concept rest their case on the contention that there are objective discontinuities in nature which delimit the units we should recognise as species. These discontinuities are caused by restriction of gene flow between actually or potentially interbreeding populations. As Valentine & Löve (1958) note, the species of the biosystematist (for which they suggest using the term *ecospecies*) is defined in terms of gene exchange—if two populations are capable of exchanging genes freely under either natural or artificial conditions, they are conspecific. If, on the other hand, there are *internal* barriers to gene flow (such as incompatibility or hybrid infertility), the populations are ecospecifically distinct. This kind of definition is attractive by its apparent simplicity and would seem to provide "an objective criterion of a species, something which can be determined by experiment; and it also has a biological meaning in that it marks a certain stage in the process of evolutionary divergence" (Valentine & Löve, 1958). What can be said about some biological species definitions is that the units they recognise are less *arbitrary* (though not more objective, cf. Mayr, 1957: 15) than taxonomic species in that they are based on fewer criteria to which a simple "yes-or-no" answer can be given.

All this, however, is deceptively simple, and if we look more closely at the various definitions of biological species proposed we will find that they embrace a number of elusive concepts. We will therefore consider some of the major theoretical and practical difficulties involved in applying such concepts.

(1) Reproductive isolation may be interpreted in a strict sense as meaning internal genetic or genetic-physiological mechanisms (Valentine & Löve, 1958; Clausen, 1951, etc.). Such an uncompromising concept demands that species be intersterile and the test for fertility or sterility has to be undertaken experimentally. A problem here is that the genic changes causing morphological differentiation may be greater than those causing intersterility which may in any case come at an early stage in differentiation (Gates, 1938). Genetic differentiation and sterility changes do not necessarily go hand in hand: as Stebbins has frequently observed, the visible characteristics which differentiate species (as in *Paeonia*) and the sterility barriers which partly isolate them segregate quite independently of each other. A consequence of these considerations is that specific status would be denied to markedly distinct, discontinuous groups which are generally accepted as good species, because of their actual or potential interbreeding in nature, although without detrimental effects on their persistence as distinct entities (cf. *Solidago rugosa* and *S. sempervirens*, Goodwin, 1937a, b). Huxley (1942) commented on such situations as follows: "To deny many of these forms specific rank just because they can interbreed is to force nature into a human definition, instead of adjusting [the] definition to the facts of nature." Viewed more generally, it is often found that species delimited by orthodox taxonomic methods do not coincide with units defined in terms of sterility barriers, as Gajewski (1957) found in the genus *Geum*. Many taxonomists feel that a species must be regarded as more than a group of individuals surrounded by a sterility barrier. As Grant (1958: 357) says, "Every natural species faces the pragmatic test of survival in the midst of vigorous competition. In the struggle for existence sterility barriers are sometimes dispensable, but the possession of adaptive fitness is not."

(2) A practical difficulty arises from the need to test for fertility or sterility experimentally, to which must be added the more theoretical problem of deciding what degree of intersterility is necessary (or how much interfertility is permitted) to fulfil the requirements of the test. The fertility-sterility test

is not an all-or-nothing criterion: every degree of interfertility can occur between populations, and even between different levels of polyploidy the barriers are not absolute, a limited degree of fertility being known to occur even in triploids, for example, with important consequences in terms of gene flow (cf. Stebbins & Zohary, 1959, on *Dactylis glomerata*). Apart from such considerations, the criterion is most meaningfully applicable to populations which occur in a single locality at the same time—the basis of Mayr's *non-dimensional* species concept (1957). In polytypic populations the fertility-sterility relationships may be found to vary considerably with the segment which is tested. The fact that only an infinitesimally small sample of populations can be tested is another serious drawback to which should be added the numerous technical difficulties that can arise in particular groups.

(3) In allopatric populations the fertility-sterility test becomes largely of theoretical value. If it is shown that the populations are interfertile in cultivation, this clearly tells us nothing about what status they should be given unless we are to look into the future and predict what would happen it they met. But the task of taxonomists is not to predict the future, as Lewis (1957) has reminded us, but to consider the past and the present. In a biological sense such isolated populations are species in that they are separate evolutionary lineages. Whether they will be recognised as taxonomic species will depend on our assessment of their degree of morphological differentiation. Even if allopatric populations are shown to be intersterile in cultivation, this tells us little we do not know about their biological status, as we have just indicated.

(4) The fertility-sterility test has no meaning in predominantly inbreeding or apomictic groups. This is discussed in detail in Chapter 11.

(5) If reproductive isolation is considered in a wider sense so as to include spatial, ecological, ethological and other such factors, the problems are even more intractible. As Heslop-Harrison (1955) has brought out most clearly, the barriers which operate in nature to maintain reproductive isolation are often of considerable subtlety and complexity. They evolve and function at different intensities and the question of whether populations are effectively isolated is not often easy to answer. Another point that should be added here is that two populations may be shown to be effectively isolated yet exchange genes through the intermediary of a third population.

(6) The necessary cytogenetical and experimental knowledge is only available for a minute fraction of the world's flora. This is what Simpson (1961) terms a pseudo-problem, although it is a major practical disadvantage.

It should be stressed, however, that taxonomic and biological species, although based on quite differently conceived abstractions, may coincide in practice and frequently do so in sexual, outbreeding groups. Largely because of this coincidence between taxonomic practice and biosystematic theory, the biological species concept was generalised and the biological distinctness came to be regarded as primary and morphological differentiation secondary. The role of morphology in such a viewpoint has been clearly stated by

Simpson (1961), who distinguishes between the *definition* of the species in genetical terms and its *recognition* in morphological terms. In other words, morphology provides the evidence for putting the genetical definition into practice. This would be a valid procedure if morphological evidence were not fallible but, as we have seen, it frequently is, so that morphologically distinct populations may be interfertile, and conversely morphologically uniform populations may contain more than one reproductively isolated gene pool. Even if the generalisation were true, the two situations—gene pools and morphological species—are not mutually exclusive. They are alternative descriptions, each valid in its own context (Gilmour, 1958). Which one chooses to employ would depend on the purpose of the classification.

To abandon the practical, almost universal use of the term species in a primarily morphological-geographical sense for orthodox classification in favour of a restricted usage for impractical, largely theoretical units in terms of gene pools and reproductive barriers, would surely be short sighted and unrealistic.

In summary we may regard species as morphologically definable units, made up of groups of individuals (populations), which it is assumed are usually interbreeding, the containers and expression of one or more gene pools.

a. *Subspecies*. The category subspecies was first employed by Persoon (1805-7)* but gained little acceptance until the latter half of the nineteenth century, when the patterns of (largely geographical) variation had become evident as a consequence of the extensive collections of material which flowed in to the major national Herbaria and Museums from various parts of the world. At these institutions this influx of material was studied in connection with the preparation of Floras and monographs and it was often found necessary to recognise large subdivisions of the species to accommodate certain kinds of variational unit.

On the other hand, Linnaeus did not use the category subspecies; the only infra-specific category he employed was the variety, and it has been suggested that this is one of the reasons why many botanists have never adopted the subspecies, or have regarded it as synonymous with the variety (W. F. Grant, 1960).

Towards the end of the nineteenth century the concept that species were polytypic, i.e. comprised various infra-specific units such as subspecies and varieties, became widespread. Subspecies were recognised by some authors as major morphological subdivisions of the species without particular reference to geographical distribution. For such authors as Ascherson, Rouy and Foucaud, Nyman and Engler amongst many others, it was often an artificial means of reducing the number of species, a tendency which is often deplored.

* As shown by Boivin (1962), Persoon only used the subspecies (as a category between species and variety) in three species in his *Synopsis Plantarum*; it was not employed here in any geographical sense. But cf. Fuchs (1958).

Strangely enough, the morphological-geographical approach by Kerner, Bunge, Wettstein and others discussed in Chapter 9 did not result in their using the subspecies category. Kerner (1869) and Wettstein (1898) intentionally arranged both the widely distributed species and the narrowly distributed ones (usually vicarious in distribution) at specific rank, although they considered the former to have given rise to the latter and to be better distinguished morphologically (cf. van Steenis, 1957). In Wettstein's monographs of *Gentiana* Sect. *Endotricha* and *Euphrasia*, for instance, no subspecies are recognised. As van Steenis (1957: cxc) points out, Wettstein did not pursue his theory to a logical, practical conclusion. Other botanists did, however, and the usage of subspecies for major geographical segments of species soon became quite common practice (e.g. Arcangeli, *Compendio della Flora Italiana*, 1890; Holmboe, *Studies on the Vegetation of Cyprus*, 1914), though the category has never become as widely used as it is in vertebrate taxonomy. A variation of Wettstein's treatment persists in *Flora URSS* where the minor units are frequently classified as species (no subspecies being recognised) and the major species as series (cf. p. 102).*

Modern use of the subspecies in botanical taxonomy for major geographical races was confirmed by the proposals by Du Rietz (1930). He recognised that some species showed regional differentiation (*heterofacial* species) and these races or facies were described as subspecies—populations of various biotypes forming a more or less distinct *regional* facies of a species. Subspecies so defined would show a certain degree of morphological intergradation between themselves. A similar concept was proposed by Rothmaler (1944, 1954, 1955). For him a subspecies was a race or group of plants, more or less separated by a combination of characters, but not usually isolated genetically. Subspecies should quite clearly have had a common origin—this is normally judged by having so many attributes in common that they are obviously regional representatives of the one species. They may be isolated from each other horizontally, altitudinally, ecologically or temporally, and although Rothmaler proposes that their distributions should be distinct, some degree of overlap is permissible. Some authors would regard such a zone of overlap as typical of subspecies, except of course in the case of allopatric units (cf. Chapter 13).

The subspecies has been widely accepted in a sense similar to that of Du Rietz and Rothmaler, i.e. as a considerable segment of a species with a distinct area and more or less distinct morphology, often showing some intergradation, and clearly fulfils a useful purpose (cf. Babcock, 1947a, Heywood, 1958; Valentine & Löve, 1958; Hedberg, 1958b). It is also extended to cover regional ecotypes (as in *Potentilla glandulosa*), and cases where taxa differ in chromosome number or are partly or incompletely intersterile and exhibit some correlated geographical or ecological differentiation but have an

* An account of the vagaries of the subspecies concept, including Russian usage, is given by Juzepčuk (1958).

insufficient degree of morphological differentiation to permit satisfactory treatment as separate species (Heywood, 1958; Heslop-Harrison, 1957). Clearly decisions on specific versus subspecific status will in many cases be highly arbitrary until some means of quantifying degrees of taxonomic difference are devised and generally accepted. Discussions of the problems of applying the subspecies concept are given by R. T. Clausen (1941), Weatherby (1942) and Meikle (1957).

Many subspecies (especially those with overlapping distributions) are often distinguished by several small and usually quantitative differences, but any of these characters can break down individually. In zoology the "75 per cent. rule" (cf. Mayr *et al.*, 1953) is often applied, by which at least 75 per cent. of the individuals must be morphologically different from all other individuals of the other subspecies. Many botanists would prefer a higher percentage for the useful recognition of subspecies.

There is a tendency to regard subspecies as incipient species, and no doubt in some situations this may be true. It is probable, however, that in many cases subspecific situations are the result of originally distinct species coming together and fusing—a stage in the process of secondary speciation. Cases are known where the zones or bands of intermediates between subspecies are so extensive that they might be better regarded as additional subspecies!

Some authors have found it necessary to insert a further category between the subspecies and variety to cover geographical groupings which are intermediate in scale between the two in a particular situation. The term race or *proles* is applied in such cases although it is seldom found in recent literature. The *formes* of Rouy & Foucaud's *Flore de France* (1893-1913) were such *proles* and several European authors (e.g. Sampaio, *Manual de la Flora Portuguesa*, some *Pflanzenreich* monographers) use the category (as race, raza, raça, etc.). It is sometimes translated as meaning subspecies and can lead to serious problems of interpretation and nomenclature (Heywood, 1958).

b. *Varietas*. Like the subspecies, the variety has been employed in several senses. It is widely used today for *local* facies of species (apparently comprising a few biotypes), morphologically distinct and occupying a restricted geographical area (Du Rietz, 1930; Rothmaler, 1944). Emphasis is on the small scale, more localised range of the variety, compared with the large-scale, regional basis of the subspecies (Heywood, 1959a). In this sense varieties may be geographical, ecological (ecotypes), cytological (cytodemes) or combinations of these; several varieties are often recognised within one subspecies. The variety is also used for variations whose precise nature is not understood, a treatment often necessary in the pioneer phase of taxonomy.

This distributional hierarchic construction is to some extent independent of the morphological differences involved: subspecies may differ from one another in fewer and less well-marked characters than varieties or even forms within the same species.

(c) *Forma*. The form is most often regarded as a sporadic variant distinguished by a single or few linked characters, without a distinct distribution—single-character variations which occur occasionally within populations. They are often the result of the repeated chance association of recessive genes (cf. Beaudry, 1960) and are of limited occurrence in the population. Genetically determined aberrations may be likewise treated as forms.

Although they may be very noticeable (e.g. flower colour variants), little significance is attached to forms. There may be serious difficulties in applying the definition since single features are chosen for recognition; these may refer to different organs of the same plant, with the absurd result that an individual specimen can be referred to more than one *forma*! An excellent example of this is *Papaver rhoeas* which Fedde (1909) divides into a series of leaf-form, petal-colour, indumentum and other variants (actually termed varieties, but forms in the sense we are discussing). Unless there is some special reason for doing so, such unit-feature variants should not be given formal Latin names (Burtt, 1951).

A complication is added by the fact that the same variant may occur as a sporadic form and as a local population, depending on whether the initial coloniser was heterozygous or homozygous for the differential character. In such cases it is probably advisable to use varietal rank to cover both situations.

Since the publication of Babcock's *The Genus Crepis* (1947), many taxonomists and biosystematists have come to consider the subspecies as the only formal infra-specific category worth recognising (Valentine & Löve, 1958). Taxonomists mainly working with herbarium material, however, may find one such category insufficient for monographic revision. Personally we have found the subspecies and variety adequate to deal with most of the variation which deserves recognition in formal taxonomic classification.

d. *Species aggregates*. The aggregate is a device employed to group together, for convenience, a number of species. The component species (binomials) are in taxonomic terms morphologically closely related and difficult to discriminate. The characters distinguishing them, although less pronounced and perhaps fewer in number than those that serve to distinguish between other species within the same genus, are constant and the species appear to be effectively isolated from one another. Often they are of restricted geographical distribution (e.g. mountain species of the Mediterranean flora) and frequently show little internal variation. The components of aggregate species are often referred to as microspecies—a term used for both sexual and apomictic species. The differences between microspecies, though small, are usually more constant than between subspecies.

The aggregate first gained currency in Engler's *Pflanzenreich* as the *species collectiva*. It was often a confession of ignorance (and often still is!) in that some of the binomials have been satisfactorily accounted for later as taxa of other rank; in other cases it is evident that some other disposition would be adopted if further evidence were available. Often, however, the *species collectiva* was a straightforward expression of taxonomic opinion that here was a group of small-scale, very closely related species which had more in common with each other than with other species or with other similar groups of species, and were felt to be on too small a scale to warrant recognition as fully independent species.

The aggregate device is employed in this sense by an increasing number of authors (e.g. Clapham, Tutin & Warburg, 1952, 1962). It is akin to the special terms proposed for such situations in zoological taxonomy, defined in evolutionary or biogeographical terms—the *Formenkreis* (Kleinschmidt, 1900), *Artenkreis* (Rensch, 1929), superspecies (Mayr, 1931).

There is a tendency, however, to regard the aggregate as an "experimental" category and use it as a cover for morphologically ill-defined, cytological or genetical groupings (cf. Manton, 1958, who advocates the use of the aggregate for cases where cytotaxonomic investigations of "Linnaean" species have demonstrated the biological nature and phyletic relationships of the components, as in the *Polypodium vulgare* aggregate). We do not agree with this practice of giving units Latin binary names irrespective of the degree or nature of the morphological differences and grouping them within aggregates. We feel that the aggregate should only be employed when there is a fair guarantee that the component units (segregates) can be successfully identified. It should not be used when there is little likelihood of pursuing identification beyond it (Heywood, 1963).

Before leaving the subject of species aggregates, we should discuss the use of the category *series*. This has been most frequently used for a group of obviously closely allied "Linnaean" species, i.e. as a somewhat higher grouping than the species aggregate. In large genera the series is often inserted between the section and the species. It has also been used (as in *Minuartia*) where the characters are considered too trivial, ill-defined or inconstant to justify the recognition of sections. (Its use as an informal group—e.g. Series Annuae—in synoptic keys should also be noted.)

In the Soviet Union the series has been used by the Komarovian school in a somewhat different sense (cf. Juzepčuk, 1958; Bobrov, 1959) in keeping with the general "degradation of rank" that is a feature of so many taxonomic accounts in the *Flora URSS*. Komarov's series (as he applied the category in *Caragana*) was stated to be not a morphological but a phylogenetic category; he even went so far as to choose the "basic" species from which the others were believed to have diverged! It is mostly used for groups of vicarious taxa that are believed to have evolved in geographical isolation and which (following the Komarovian concept of monotypic species) are therefore treated as species. In the *Flora URSS* the series is used very much as the species aggregate is used in Western taxonomy, or even for polytypic species whose components might be recognised as subspecies; closely allied "Linnaean" species (Linneons) often appear in different series. Bobrov advocates the use of the series for groups of sexual cross-pollinated species, *grex* (plural *greges*)* replacing it in apomictic groups.

Granted that common origin can only be inferred from phenetic resemblance, the use of the series as a category uniting closely allied species (whether vicarious or not) seems a helpful one, particularly when a rather narrow species concept is adopted for allopatric taxa.

2. *Genus*

More has been written about the concept of the genus than about any other rank above the species. This is understandable since the category is historically tied up with that of the species and cannot be separated from it in the denomination of species. The concept of the genus is as old as folk-science itself. "Wherever we look into the matter, whatever the people or the language, we find naming and classification of plants, and almost invariably a more or less well defined idea of the genus as the smallest group that almost everyone might be expected to have [a] name for in his vocabulary. It might or might not be divided up into species" (Bartlett, 1940).

We should always bear in mind that the most important "function" of the genus is to bring together species, preferably in a natural manner. Rollins (1953) writes: "It is my contention that a genus is made up of a group of closely related species. In determining the nature and even the limits of a given genus, interest should centre upon the relationship of the species. When attempting to place a species in a given genus, the primary question should be, is it related to the undoubted species of that genus?" Rollins's contention does not help us to distinguish a genus from a section, but it does focus attention on *naturalness* (in terms of overall resemblance) as the corner stone of the modern genus concept. The idea that a genus should always be readily definable on a few technical floral characters no longer holds the field as the prime generic criterion; the overriding consideration is whether it is naturally delimited.

There are really three questions to be asked when we are deciding on generic concepts:

(1) Is the group a natural one, and if not is it possible to make it so? Monophyly, in modern taxonomy often deduced by means of cytogenetic and geographical information in relation to morphology, is very helpful in deciding where the line should be drawn between two genera. Whether an unnatural group can be conveniently reclassified in a more natural manner will depend on the degree of discontinuity of variation and on how well the characters are correlated. If reclassification is a practical proposition, it should be carried out.

(2) Where should the line be drawn between closely allied genera?

* *Grex* has been used in several senses, and is recommended in the *International Code of Nomenclature for Cultivated Plants* (1961) for use with collective names of cultivars of hybrid origin. Its use in Bobrov's sense might lead to confusion.

This presents no difficulty when the genera are very distinct from one another, but is a considerable problem if the separation is not clear cut, as is so very often the case in large and natural families.

If our classification is to be natural, the modern taxonomist should be guided "not by a single arbitrary or artificial character but by the sum total of characters manifested in each group studied" (Sherff, 1940). An excellent example is found in *Bidens* which was traditionally separated from *Coreopsis* by Linnaeus and de Candolle by the retrorse barbs on its achenial bristles. This character breaks down in some North American species, so that Asa Gray went so far as to designate a specimen of the so-called *Coreopsis aristosa* as "*Coreopsis aristosa* transformed into a *Bidens*"! Several species of so-called *Coreopsis* have been found which contain forms with retrorsely barbed bristles and therefore technically referable to *Bidens*. It was left to Britton to abandon tradition and refer all the "*Coreopsis*" species showing ambiversalism in their bristle barbs to the genus *Bidens*—a decision taken on overall resemblances, not on a single technical character.

(3) Is it practicable to recognise the group as a separate genus, or would it be better included in another? The answer to this question is closely related to our second one because, if two genera are not readily separable, infrageneric rank may be preferred. This is very much a matter of opinion, but certain considerations influence our decision:

(*a*) The degree of difference between the groups. If the differences are small or rather inconstant, infra-generic rank may be preferable.

(*b*) The situation must be assessed in relation to generic concepts elsewhere in the family. Whatever the reason, genera accepted in the *Ranunculaceae* are based on more obvious morphological distinctions than those which separate most of the genera in the *Compositae*, *Umbelliferae* or *Acanthaceae*.

(*c*) The size and homogeneity of the groups and the number of intermediates between them. Two or three species intermediate between genera each of a hundred species hardly justify reduction to a single genus, whereas if these genera contain only a score of species their separation may not be justifiable on grounds of convenience. Here the principle that the size of the gap should be inversely proportional to the size of the groups furnishes a useful guide (cf. p. 84).

(*d*) Traditional usage. Changes in binary names are undoubtedly a nuisance, and the raising of sections to genera or the reduction of genera to infrageneric categories should not be undertaken unless there are strong taxonomic reasons for the change.

Whenever the traditionally accepted technical characters break down between two genera, there are only three courses to choose from: (i) we can make an arbitrary division between them, (ii) redefine them in a different way, (iii) include one genus in the other. Which course is adopted may largely depend on the considerations outlined, but personal judgment is bound to affect the results.

With regard to subjective judgment, it is encouraging that a preliminary experiment undertaken by Anderson (1957a) to test the taxonomic intuition of different taxonomists showed agreement in the concept of the genus (*Uvularia*) used in the experiment. This is an almost unexplored field of research that deserves attention.

We have already expressed the opinion that there is no reason why *vegetative characters* should not be as good as floral ones for the delimitation of taxa. Which is used should depend on character correlation and discontinuity of variation. The influence of Linnaeus has resulted in the recognition of genera primarily on floral characters, or at least in these characters being stressed in their diagnosis. Above the species level it has been traditional for vegetative characters to be used for sectional rank—as is mainly the case in *Iris*, *Saxifraga*, *Jasminum*, etc. But there is no good reason why such characters should not be used to define taxa of *any* rank. The tribe *Scilleae* of *Liliaceae* provides an example in which vegetative characters (particularly bulb structure and mode of germination) have proved most useful for the delimitation of natural genera (Chouard, 1931). The same applies to the segregates of *Cotyledon* (*Umbilicus*, *Rosularia*, etc.). Such genera have the added advantage of being very readily distinguished.

It will be apparent to many taxonomists that generic characters are often adaptive ones—the inevitable result of natural selection. So far as floral characters are concerned, they are usually adaptive in relation to a particular mode of pollination or dispersal. Indeed, a family which shows marked diversity or elaboration in *flower* structure seldom shows it in its *fruits or seeds*—and vice versa (cf. *Scrophulariaceae*, *Labiatae* and *Orchidaceae* on one hand, *Cruciferae*, *Umbelliferae* and *Gramineae* on the other). In so far as these characters reflect the adaptive evolution of the group, we may claim that classification is evolutionary. We cannot, however, accept the claim of Inger (1958) that "phylogeny . . . must be the basis of a biological classification". A study of adaptations is not (as he claims) the best means of ascertaining phylogeny and thus a stable classification. To accept such a view is to underestimate the difficulties involved in interpretation and would put Angiosperm classification on a very shaky footing. It is encouraging to find that many generic and family characters appear to be directly adaptive; if others have no obvious adaptive advantage, they are probably linked to characters which have. Our taxonomic groups must be primarily based on the discontinuity and correlation of characters, irrespective of their "biological significance", not on inferred adaptive or phylogenetic interpretations. Such considerations may, however, be taken into account when it comes to delimiting groups and particularly in the attribution of rank.

When generic limits are being revised (and this applies both to the delimitation of genera and the decision whether to accord them separate generic or infrageneric rank), it is absolutely essential that the whole group should be reviewed throughout its geographical range. Many of the "splits" recognised in Europe are unsatisfactory in a world context. We either find that several other groups in the old genus deserve similar status (as should be the case if *Chamaenerion* is separated from *Epilobium*), or that the differences claimed to delimit the genera breakdown elsewhere in their range to such an extent that generic separation is impracticable (as is apparently the case within *Saxifraga*, *Vaccinium* and *Rhododendron*).

There is much to be said for adopting fairly conservative generic limits. Genera should usually only be radically changed when they are unnatural groups or when their status is inconsistent with that adopted elsewhere in the family or tribe to which they belong. The overriding importance of synthesis should not be lost sight of when we are classifying above the species level.

When in doubt as to whether to accord generic rank to a group, there is much to be said for the *subgenus* as a suitable category; it draws attention to the group in the classification and at the same time allows people to continue to use the old binomial.

We have stressed the importance of the natural delimitation of genera and their natural subdivision. This does not mean, however, that *all* genera can be subdivided into natural *sections*. In some very natural genera it seems almost impossible to find sufficient discontinuity and correlation of characters to delimit such groups for practical use—there seem to be no natural taxa (with the possible exception of *series* for small groups of closely allied species) between the genus and the species. This appears to be the case in *Erysimum* and *Isatis* (*Cruciferae*), in *Cousinia* and *Echinops* (*Compositae*). It is not that attempts at sectional classification have not been made, but that the groups recognised are obviously artificial, many of the species being related to ones in other sections. If satisfactory natural sections cannot be found (and every attempt should be made to find them), there seems nothing for it but to rely for identification on an artificial key or on artificial (or partly artificial) groups based on a few technical characters. In a natural classification it would seem preferable not to give these any formal taxonomic rank, but to treat them as informal groups outside the taxonomic hierarchy.

3. *Family*

Many of the considerations of genera made in the last section apply with equal force to the families. Families should be natural and, where practicable, evidently monophyletic. Size, so important in the history of generic concepts, has probably played an even greater part in the concept of the family. As with the genus, vegetative characters may serve to delimit them as effectively as floral ones. The flower and fruit together, however, form the most complex and variable group of organs in the Angiosperms and show the most diversity. On these grounds alone, therefore, there are good reasons to expect floral characters to be the most generally useful in the subdivision of the Angiosperms: they provide more morphological characters for classification than do the vegetative organs. No one could deny, however, that the leaf form and stem structure of the *Palmae* provide the most striking characters of the family. Attention to vegetative characters in the Monocotyledons (e.g. the underground parts in the *Liliales*) may lead to a more natural classification than we have at present.

There are, broadly, two kinds of families recognised today—"definable" and "indefinable" (Walters, 1961). Definable families are those which are obviously very natural ones, like *Cruciferae*, *Umbelliferae*, *Acanthaceae* and *Gramineae*, and which are clearly delimited. Many of these represent "adaptive peaks" in evolution (Stebbins, 1951) and include most of the large and successful families containing a large number of genera which are often difficult to delimit. However, many such "definable" families are small (*Dipsacaceae*, *Proteaceae*, *Sarraceniaceae*). The "indefinable" families are represented by the *Ranunculaceae*, *Berberidaceae*, *Rosaceae* (the example *par excellence*) where a great diversity of structure (floral and often vegetative) is found. The genera are often very distinct, suggesting that they are probably older than most of the families in the large "definable" families. This does, however, raise an interesting question. Are these families "indefinable" because of some evolutionary reason, or is it a result of size? Is the *Berberidaceae* phylogenetically comparable (in terms of diversity) with the *Umbelliferae*, or are the *genera* of the *Berberidaceae* comparable with the family *Umbelliferae*? In an old relict group, where extinction must have caused many of the gaps in variation that we find today, the tribes or genera may be comparable, in biological terms, with the large homogeneous families. This may be so, but whether we should do something about it is another matter. It may be more consistent with the facts of evolution to divide up the "indefinable" families (as has been done with the *Magnoliaceae* and *Saxifragaceae*), but this would lead to a great increase in the number of families that might add to the burden of sorting and identification.

Arguments as to which is the more natural, species, genus or family, make little sense except within the context of the group concerned. Biosystematists have often expressed the view that the species (delimited by breeding criteria) is the only non-arbitrary taxonomic unit, but some monographers have claimed this distinction for the genus (cf. Anderson, 1940). Those who have tried to classify the species of *Rubus*, however, would probably consider the genus less arbitrary than the species whose delimitations have been obscured by facultative apomixis. One might justly claim that the genera of the *Umbelliferae* are less natural (and certainly more difficult to define) than the family. One might expect the higher taxa, since they are formed by man's aggregation of lower taxa, to become less natural as we ascend the hierarchy, but this is not necessarily the case. The Angiosperms, and even the Dicotyledons and Monocotyledons, seem rather natural divisions though it is possible that we shall not be able to retain this view in the future (cf. Meeuse, 1961).

The most unsatisfactory taxon in Angiosperm classification is the

Order ("Cohort", usually ending in *-ales*) into which families are grouped. Most of these are extremely difficult to define and the naturalness of many of them is open to question; exceptions are the *Caryophyllales* (*Centrospermae*), *Ericales* and *Personales*. In an effort to make them more natural and definable, they have been increasingly reduced in size (Hutchinson now recognises 82 within the Dicotyledons, including 342 families). Most of them are of little use as an aid in identification, and the phylogenetic assumptions which have influenced their delimitation are highly insecure. Nearly every phylogenetic system uses a different circumscription of orders (though some remain fairly stable), so that this category in the Angiosperms serves no more than to accommodate what each taxonomist believes to be a natural or convenient group of families.* The indefinable nature of most Angiosperm orders means that the family is often the first natural unit below the Subclass.† This is surely a very good reason for retaining a fairly conservative family concept, because if the families are small and ill-defined it may take considerably longer to find the genera. This is particularly the case when the family is a natural though not homogeneous one, like the *Leguminosae*, often treated as an order containing three families. The *Leguminosae* have a unique fruit and seed and no close relatives. Surely it is more useful here to emphasise resemblances than differences, so that the recognition of the *Leguminosae* as a single family instead of three is not only the most practical treatment but biologically the most meaningful.

* * *

We will close this chapter on taxonomic categories with a consideration of "dustbin" (or trash-barrel) taxa which are of rather common, if little realised, occurrence in systematics. It cannot be denied that some species are more polymorphic than others, and that polymorphy is often proportional to size of population. Nevertheless, many polymorphic species have probably been created by the taxonomist not so much on the characters they possess as on the characters they do *not* possess—i.e. by exclusion from other species.

* On the other hand, the orders recognised in the Gymnosperms are almost as well defined as the genera. This may reflect a slower rate of family evolution, the greater antiquity of the Gymnosperms combined with a larger amount of extinction, the smaller part played by hybridisation in Gymnosperm evolution (cf. Grant, 1953), or a difference in ranking to that adopted in the Angiosperms. In the Animal Kingdom, Simpson (1961) points out that the orders are well demarcated and adapted to different ecological niches. It is interesting that the genera of animals recognised by Linnaeus are now nearly all treated as families—which is far from being the case in the Angiosperms. This is no doubt related to the larger number of animals and to the greater emphasis on anatomy and microscopy in their modern classification.

† Although we follow Benson and Takhtajan in treating the Dicotyledons and Monocotyledons as Subclasses of the Class *Angiospermae*, it should be pointed out that the former are often treated as Classes and the Angiosperms as a Subdivision. The *Gymnospermae* are not co-ordinate with the *Angiospermae*, but provide a convenient group name to include several Classes of which the living representatives (according to Benson, 1957) are the *Conopsida*, *Ephedropsida*, *Gnetopsida* and *Cycadopsida*.

When the taxonomist keys out the species of a genus he is apt to take out those species which differ in certain striking or peculiar characters, with the result that he is left with one or several groups of plants that cannot be classified into any of the obvious striking species and which therefore form "dustbin" species. This situation may arise particularly easily in genera in which good taxonomic characters are hard to find (as in *Colchicum*), or in which distinctive characters are found only in the majority of species.

Dustbin species may be an unnatural agglomeration which further study may enable us to disentangle. They often contain a natural species and one or several alien elements which can be attached to other species or recognised as separate ones (e.g. *Linum mucronatum* as recognised by Boissier, part of which represents a second species, *L. scabrinerve*). In many cases they undoubtedly do represent natural species which may or may not be amenable to subspecific division on a morpho-geographical basis (e.g. *Teucrium polium* and *Alyssum murale*). What we have put in the dustbin may therefore represent an unnatural, partly natural or wholly natural group, but in all cases it has been formed by overemphasis on analytical methods, instead of synthesis, in classification. Unnatural taxa are not always, or indeed usually, formed by dustbin methods, but are also the result of *a priori* weighting.

Dustbin taxa are by no means confined to species but are found at higher levels of the hierarchy where they often represent taxonomic grades. The *Rosaceae* appears to have been formed in this way, though its tribes are more closely related to one another than to any other family. The subfamily *Caesalpinoideae* of *Leguminosae* is in part a dustbin group, containing various non-Mimosoid and non-Papilionaceous genera. *Lathyrus* Sect. *Orobus* has "grown" in a similar manner, by exclusion from the rest of *Lathyrus* and from *Vicia* (cf. p. 79). The *Saxifragaceae sensu lato* is probably a dustbin family and has been split up into several families by Hutchinson. The *Monochlamydeae* were a rag-bag group, anything being thrown into them which lacked petals. Several genera in the *Umbelliferae*, *Cruciferae* and *Compositae* appear to have been formed by exclusion from other accepted genera. *Ptilotrichum*, for instance, consists of various white-flowered Crucifers that have been separated from *Alyssum* and allied genera largely on account of their flower colour and combined to form a very artificial genus. *Carex* (according to Nelmes, 1951) probably contains some species derived from the southern hemisphere genus *Uncinia* that have lost the hook at the end of the rachilla (*C. microglochin*) or even the rachilla itself (*C. pauciflora*). Nelmes suggested referring these to *Uncinia*, but that course would not be practicable unless morphological characters could be found to delimit the resultant genera in accordance with this inferred origin.

Since dustbin methods (*methodi cloacae*) are very apt to result in unnatural groups, taxonomists should look out for them and guard against overemphasis on analysis in the delimitation of taxa. The creation of dustbin species in herbarium cupboards is referred to in Chapter 8.

Chapter 4

THE CONCEPT OF CHARACTERS

"A knowledge of the relative importance of characters can only be acquired by long study; and without a due appreciation of their value no natural group can be defined."

J. D. HOOKER, 1855

INTRODUCTION

Taxonomy, in so far as it is concerned with variation in organisms and their classification, deals with characters. Characters provide the taxonomic evidence for natural classification in the Adansonian sense and a large part of the evidence for evolutionary ("phylogenetic") classification. As Cain has recently stressed, the basis for a natural classification is entirely the characters available.* It is therefore surprising that the problems of characters *as such* should have received so little consideration by taxonomists, since the success of most taxonomic studies must depend to a very large extent on what attributes are selected and considered as characters, how they are tested and treated, and whether they are relevant for the purpose in hand. It should be pointed out straightaway that the concept of characters as recognisable, separate entities is a product of man's necessity to communicate and therefore to describe. This explains why characters are so difficult to define. This aspect is returned to later.

Although it is organisms (or life cycles of organisms) which are classified,† it is their characters which provide the evidence used in classification. Ideally the whole organism (i.e. all its attributes) should be employed, but since each individual possesses thousands of potential characters, practical limitations impose a restriction on the numbers used. As Mayr remarks, the number is limited by the patience of the investigator. This leads on to the problems of *selection*. Lam (1959) believes that it is probably not possible to make a fully relevant choice if the ultimate goal is a completely natural classification. Moreover, different selections of attributes will produce

* Throughout this chapter, natural classification is understood as meaning phenetic.

† More strictly speaking, it is our knowledge of them at any time that is classified.

different classifications or different versions of the "same" classification. The selection of features made will, in turn, depend on a number of factors such as the *purpose* of the classification (an aspect often ignored, cf. Crawshay Williams, 1961), the general convenience of the scientist, the availability of characters, and traditional usage.

Most classification hitherto has employed morphological and anatomical characters as its major source of evidence, but as is shown later in this chapter other kinds of features are now being increasingly employed, and it is by no means certain that the results will be acceptable for general purposes. It has, for example, so far been tacitly assumed that natural classifications will be morphologically expressible and comprise non-overlapping groups. Both these assumptions are open to question.

All classification, as Lam has put it, must be a compromise because of the particular selection of characters and the degree of precision and unity with which they can be defined. Human limitations are in fact the main limiting factor.

For general purposes it is widely, although by no means universally, agreed that a maximum-attribute classification is the most useful: it is the classification which allows the maximum number of generalisations to be made from it (Gilmour & Turrill, 1941). As Bremekamp (1939) has explained, it is the most serviceable because it conveys the maximum possible amount of information about the groups it contains.

Such classifications are, in theory at least, undertaken following the precepts established by Michel Adanson (p. 23) and are frequently called Adansonian. The two essential postulates are: (1) in constructing the classification each attribute selected is given equal weight; (2) taxa are based on correlations between these attributes. Further, the taxonomic groups derived by these means are of the class of concepts which have been termed *polythetic* (Sneath, 1962) "in which no single attribute is in theory sufficient and necessary for membership in the group so long as the members share a high proportion of characters" (Sneath & Sokal, 1962).

Many taxonomists find great difficulty in accepting these premises, largely because they regard them as non-evolutionary and therefore unacceptable in this day and age; but failure to accept them is to distort classification by the introduction of phyletic, logical or other sorts of weighting principles. Such principles may be justified or are indeed necessary for the construction of *special* classifications (e.g. phyletic), but are not compatible with the aims of a *general* classification. In other words, they serve a different purpose. By giving greater weight to particular evidence, a classification ceases to be a general one.

These Adansonian principles for the construction of phenetic classifications have received great support from the almost spectacular development in the past five or six years of what Sneath and Sokal have called "*Numerical Taxonomy*"—"the numerical evaluation of the affinity or similarity between taxonomic units and the ordering of these units into taxa on the basis of their affinities"—which is essentially an extension of Adansonian classification using mathematical procedures with the primary aims of repeatability and objectivity. The methods and applicability of numerical taxonomy are discussed in a later section (p. 131).

The groups recognised in natural classifications are obviously based on overall similarity. Mathematical techniques, often employing computers, are now being used for making the character correlations (see p. 135). The groups so delimited are not, however, to be interpreted as necessarily phyletic, due to a number of factors such as convergence, functional correlation and difficulties of ascertaining homologies. In considering the taxonomic value of the kinds of characters which are discussed in this chapter, a clear distinction must be made between their role in maximum-attribute (natural) classifications and special (phylogenetic) classifications (phenetic and phyletic respectively, as A. J. Cain (1959d) has recently termed them). The conversion of a phenetic classification into a phyletic one involves a further series of problems of interpreting characters and weighting them (see p. 127). It is important not to mix the two approaches, as has frequently been done in the past. The separation of phenetic and phylogenetic classifications is not generally accepted today due to the intense conviction that classification should be evolutionary and that it is the only sort of classification which should occupy the serious attention of scientific taxonomists. We must insist, however, that it is not inconsistent with evolutionary thinking to encourage this separation, and advocate great concentration on the primary task of producing general classifications based on maximum co-variation of attributes *for general purposes*. In flowering plants the possibilities of achieving a phyletic arrangement or fully evolutionary classification as it is normally understood today are limited for reasons which are discussed below. This is not to deny, however, that one of the major activities of taxonomy is the evolutionary interpretation of evidence, and part of this chapter is devoted to considering how this may be done.

Definition of Characters

From an evolutionary point of view characters should probably not be considered in isolation but in combination, since it is on character-combinations that natural selection operates. This is particularly true if one is considering the possible selective or functional value of characters, but for general taxonomic purposes it is necessary to deal with unit characters which can then be treated quantitatively. As is explained elsewhere (Chapter 2), the taxonomist's assessment of similarity is made to a large extent intuitively, and only those characters which appear to be differential between organisms or groups of them are set down on paper. The current tendency in taxonomy is to analyse these intuitive processes* and rationalise them by distinguishing the individual characters which the taxonomist assesses subconsciously together. This raises the problem of the definition of unit characters. It seems unlikely, however, that strict or objective definition can be made, due to the difficulties of deciding in many cases whether any particular attribute of a plant should be regarded as one or more characters for the purposes of making comparisons. Many of the difficulties arise from a lack of knowledge of the anatomy and morphology of the various organs and features of plants.

A general definition of a character is: *any attribute (or descriptive phrase) referring to form, structure or behaviour which the taxonomist separates from the whole organism for a particular purpose such as comparison or interpretation.* Cain & Harrison (1958) describe a single character as "anything that can be considered as a variable independent of any other thing considered at the same time".

Characters as such are strictly speaking abstract entities: it is their *expressions* or *states* that taxonomists deal with. "Sepal length" for instance may be considered as a character; but "sepal length 12mm." is an expression of that character. For purposes of comparison, a character must be divisible into two or more expressions or states (including here presence or absence in qualitative features).

For *practical* purposes a character may be defined as any feature whose expression can be measured, counted or otherwise assessed (cf. Hedberg, 1957). Examples are: seed weight, ratio of corolla to calyx length, number of teeth on a leaf, presence or absence of tubercles in leaf axils, number of transverse septa in anthers, to cite a few at random. One of the features of modern taxonomy is the wide range of characters that are being employed.

The question "what is a character?" is, however, more difficult to answer than the above definitions suggest. Many characters may be regarded as compound or multiple. For the purposes of numerical taxonomy Sneath & Sokal (1962) state that characters should be *unit characters* or if they are multiple they should be broken down into unit characters. As a working definition of a unit character they propose "a taxonomic character of two or more states, which within the study at hand cannot be further subdivided logically (except for subdivision brought about by changes in the method of coding the states)". However theoretically sound this may be, there are bound to be acute difficulties in making a decision about the logic of dividing up characters into units, a limiting factor being our own state of knowledge, skill and perception. This is one of the most serious weaknesses of any attempts to break down organisms into characters for the purposes of

* This method is essentially similar to what Matthews (1962) has recently termed *neural science*—the private, individual knowledge each man obtains for himself through his senses. It is contrasted with *formal science*—the knowledge obtained by measurement, testing experiment, and the particular kind of logic based on these evidences of reality.

objective comparisons. In other words, although the comparison of characters may be objective, the selection of what is to be regarded as a character must necessarily be largely subjective.

Let us consider an actual example: how many characters are involved in a leaf shape, for instance? The expression "obovate" may be controlled by distinct genes, some of which influence length, others of which influence width, and all of which probably interact. The concept "obovate" may be divided into components dealing with length, maximum breadth, position of maximum breadth, breadth at various intervals, length/breadth ratio, etc.—the shape being the product of differential growth rates, themselves controlled by genetically determined hormone systems. In many cases it may be sufficient to take the general outline, i.e. obovate, as a single character; in other cases it may have to be broken down and analysed biometrically. Leaf dentation and the form of the apex may be treated as different characters since they may vary independently of general outline. In compound leaves the problems are correspondingly complicated.

It follows from what we have just said that no precise general answer can be given to the question "what is a character?" This can only be considered in individual cases and what we treat as a character will depend on what we want to use it for. Even apparent absence of differential characters or expressions between individuals or groups need not indicate that they are identical: differences may well come to light after detailed study. It has been remarked that one will always find characters for separation if one tries hard enough—and, one might add, find that characters used for separation do not hold when more material is examined!

Characters may be considered from two viewpoints corresponding to two main activities of taxonomy:

(*a*) Identification, characterisation and delimitation of species.

(*b*) Grouping (classification) of these species into higher taxa (natural groups), i.e. into a system that as far as possible expresses their natural relationships.

Corresponding with these activities, characters may be regarded as analytic or synthetic: in each procedure the actual characters used are usually different in each particular instance, although there is nothing *inherently* different in them; it is rather that different kinds of character are *selected*. In identification and characterisation the taxonomist seeks to employ *diagnostic* or *key* characters, i.e. those of limited occurrence, selected so that their use alone is sufficient for reaching a correct diagnosis. In arriving at a decision about the delimitation of species numerous characters may be used, many of them of a highly specialised nature which do not lend themselves to easy description or recognition, e.g. chemical, cytological, physiological characters or those which show only a statistically significant difference. These characters are not normally given in descriptions (in Floras, revisions, etc.) which often have to be short and precise. Descriptions are really abstracts for practical convenience—they give some, at least, of the characters which are easily recognisable and comparable, not all those features which may have been used in making a decision about status. In this activity, taxonomists not only gather information (i.e. characters) together, but present them predigested. This is often not realised, especially by non-taxonomists who tend to assume that a particular classification stands or falls by the evidence quoted. It does raise the problem, however, of the amount of data that should be presented in any particular context, and clearly the purpose of the publication is a major consideration.

The purpose of a Flora is still today primarily to indicate what plants occur in a particular area, how they may be identified (by keys and descriptions) and what scientific names refer to them (correct nomenclature and synonymy). Certain subsidiary information is normally included, such as ecological and distributional data and chromosome counts (cf. discussion in Heywood, 1960a: 153-154). *Diagnostic* characters make the process of identification in such Floras easier. Discussion about them and the other characters which may or may not be given in the description is out of place and hinders rather than hastens the purposes of identification. Such information may be published elsewhere, as it often is in detailed monographic studies.

1. *Analytic and Synthetic Characters*

As noted above, characters used in identification, characterisation and delimitation may be termed *analytic*, following Just (1946). In the classification of the results of these activities, *synthetic* characters are used, i.e. characters of a constant nature which unlike diagnostic characters are of wide occurrence. The analytic or diagnostic characters which serve to recognise a group are seldom of use in synthesising the group along with others into a higher group. For this purpose one needs characters whose constancy of occurrence increases the higher the position of the group in the hierarchy. Those characters used to synthesise lower groups into higher groups serve at the same time to distinguish the higher groups from others of the same rank.

This was recognised by Darwin who noted that the classificatory value of a character which has a wide occurrence (in many forms), as opposed to one which has restricted occurrence, was in terms of hierarchical value. In other words, characters helped to determine different *ranks*. Characters which serve to separate genera in one group are of no value for such a purpose in other groups (cf. Darwin, *Origin*, Ch. XIII, discussion of Robert Brown and characters of the *Connaraceae*).

As already said, there is no inherent difference between analytic and synthetic characters: the difference is more one of the particular use made of them in a particular case. Characters which differ between taxa are usually similar in kind to those that are found varying within species populations as Simpson (1953a) has noted. He comments that "one needs only to take the diagnosis of virtually any taxonomic group to find that the same diagnostic characters are variable in other and related taxonomic groups of the same or lower hierarchic rank and that eventually they can be traced down to variation in populations" (1953a: 100).

On the other hand, certain features which characterise higher groupings in the Angiosperms, such as superior versus inferior ovary, are seldom today found to vary within populations. These are the so-called "constitutive" characters discussed on p. 122.

The distinction between analysis and synthesis in taxonomy is not always appreciated. Genera, for example, are syntheses of species, but having inherited most of our genera from previous generations of taxonomists, we find that they have, over the years, been enlarged by accretion and not by deliberate resynthesis from species. When a periodical review of genera is undertaken, this resynthesis often leads to an alteration in generic limits and consequent change in status. As an example of the use of analysis when synthesis was required, we may cite Williams's revision of *Arenaria* in which he divided the greater part of the genus into 3 subgenera based on supposed differences in structure of staminal glands, thereby disregarding the acknowledged affinities of the species as judged by overall resemblances. This is an example of analysis by *a priori* weighting of characters and contrasts with the synthetic treatment of the same group by Fenzl (1840) and McNeill (1962). It should be pointed out that staminal gland structure serves as a useful diagnostic (analytic) character to separate species (e.g. *Arenaria armeniaca* and *A. zargariana*) within the genus after it has been revised by synthetic procedures.

2. *Qualitative and Quantitative Characters*

A distinction is often made between *qualitative* (including presence or absence) characters and *quantitative* characters for descriptive purposes. Thus features assessed by size, length, etc. are quantitative while those relating to form, for example, can be considered as qualitative. Such characters as opposite/alternate leaves and placentation types are clearly qualitative. Often, however, the distinction is more apparent than real: many qualitative characters may also be expressed quantitatively (cf. leaf shape discussed above), and it is perhaps more useful to consider characters according to whether they may be assessed directly or not (Heslop-Harrison, 1952b).

Characters may be assessed directly by counting (by number), by simple measurements of linear dimensions (by size), or by angles (by shape). These are the most satisfactory from a practical point of view in that they can be easily scored, described and processed. The advantages of such characters for the purposes of biometrical or statistical comparison are obvious. (The variation of characters in populations and its causes are considered in Chapter 11).

(*a*) *Number*. Different numbers or ranges of numbers of parts (such as leaves, flower parts) are commonly employed in taxonomic descriptions The term *meristic* variation is applied to such cases where the variate can only assume discrete values. Seed number when of low value is assessed in this way too, but when the values are high, assessment by weight of numerical units (e.g. 100 seeds) is often employed. By their very nature the recording of such numerical characters seldom presents particular problems.

(*b*) *Size.* Linear dimensions, too, are commonly given in descriptions, often with a conventional indication of range and extremes, e.g. petals (3-) 5-7 (-9) mm. Overlapping values between two taxa in such characters (especially vegetative) are frequent and they are consequently not the most useful for taxonomic separation. Apparent differences may be due to inadequate sampling, especially when dealing with conventional herbarium collections. On the other hand, the differences may be on quite a small scale (such as anther length in *Centaurium* spp.) and afford a reliable criterion. Such small-scale differences are frequently neglected since they are not evident without precise measurement under strong magnification. Then there is the psychological problem, too, that small-scale characters tend to be considered unimportant, while immediately obvious ones are regarded as valuable. On the other hand, many botanists have considered characters of primary importance just because they are difficult to observe! Neither claim is valid—the importance of a character is to be judged by other criteria, as is discussed below. It is also possible that *above* certain dimensions differences between organs may be ignored, particularly when other species in the genus have relatively small organs.

In many instances differences in linear dimensions (between populations) may only be statistically significant and consequently of little value as primary recognition characters.

(*c*) *Shape.* As we have already seen, differences in shape present considerable problems although they are often more valuable for diagnostic purposes than size differences. They often contribute more to differences in facies and are instrumental in causing taxa to be recognised as distinct, but their description is correspondingly difficult. The techniques involved in the detailed specification of leaf shape by numerical indices are discussed later, but some more general problems have to be considered at this point. Simple leaf shapes, such as ovate, lanceolate, linear, etc., are bedevilled by problems of terminology and different usages. It soon becomes evident, even to relatively inexperienced taxonomists, that there is a lack of precision in the use of the various leaf-shape terms and inconsistency of usage between different Floras, etc. Nothing, of course, can be done to alter past usage although it is often possible to learn what conventions have been employed in particular countries or by particular authors. It is strongly to be recommended that today precise systems for designating conventional shapes should be followed, such as those given by Stearn (1956). There are signs that the matter is receiving international consideration. (See p. 271).

Many characters may not be susceptible to direct numerical assessment, such as colour, texture, indumentum, smell, taste, patterning, etc. They are none the less widely used in taxonomy, as is discussed in the next chapter, and clearly precise means of describing them are necessary.

Colour can be treated in several ways: the use of international standards such as the Ridgway and Royal Horticultural Society colour charts, whereby colour is divided into a number of arbitrary named classes and specimens are matched against them, is common especially in horticulture. In these cases the number of classes is higher than is normally to be advocated for other arbitrary class divisions. It may be found, however, that the range of variation in population samples is so great as not to repay the labour involved, unless colour differences appear to show some significant relationship with, say, environmental factors. Other methods of colorometric analysis may be employed, such as measurement of amounts of pigment extracted per unit area, but as Heslop-Harrison (1952b) points out, the amount of pigment present may not bear a linear relationship to the visual effect it produces. The interpretation of colour will depend on what is known about the way in which it is genetically controlled, ontogenetically and environmentally modified.

Indumentum can similarly be divided into arbitrary classes or compared against certain standards. Here again problems of terminology arise if it is divided into named classes such as pubescent, hirsute, sericeous, etc., since they are often used imprecisely and are in fact to some extent overlapping and interchangeable. Detailed analysis of indumentum often reveals differences in hair length, shape, number of cells, presence or absence of glandular trichomes, posture, etc. (see p. 154).

Sculpturing (as in seeds and fruits) and patterning (as for example the maculation in *Digitalis purpurea* corollas) are more difficult to assess. They can, given time and patience, be specified by elaborate geometrical analysis. The problems involved here are discussed by Heslop-Harrison (1952) and exemplified by his series of studies on *Dactylorchis* populations (1948, 1951, 1952a, 1953a).

Division of numerically assessable characters into arbitrary classes may also be a useful procedure. This method has the advantage that it is a great deal more rapid than very precise scoring and is probably as valuable, although care must be taken, through adequate sampling, to see that the arbitrary grades are not absurd.

Quantitative characters appear to be more frequent in the lower taxonomic categories, at specific and infra-specific levels. Thus species may be separated by leaf size, corolla length, seed weight, etc., characters which are quantitatively assessed and seldom of use at higher levels in the hierarchy. Most morphological differences between ecotypes (and subspecies), for example, have been found to be of quantitative nature and may depend on polygenic or multiple-gene inheritance; they may in fact only be appreciated after statistical analysis. On the other hand, family characters tend to be qualitative.

3. *"Good" and "Bad" Characters*

All characters are theoretically of use in classification but experience has shown that their value or use varies according to a number of factors and according to the type of taxonomic activity envisaged. Great reliance is placed on "good" characters, at least in practical taxonomy. Good characters are those that (1) are not subject to wide variation within the samples being considered; (2) do not have a high intrinsic genetic variability; (3) are not easily susceptible to environmental modification; (4) show consistency, i.e. agree with the correlations of characters existing in a natural system of classification which was constructed without their use.

We will consider this last point first. The consistency of a character will help us to make a decision between two alternative classifications: it can help us to resolve differences. In constructing a classification the taxonomist is frequently faced with conflicting evidence. In these circumstances a guiding principle would appear to be to select for the purpose of group-making that character or group of characters which agrees with the overall facies of the consequent groups. This is what we have referred to in Chapter 2 as *correlation* or *a posteriori* weighting of characters. It is essentially an intuitive or neural use of Adansonian methods of making classes according to maximum co-variation of attributes, except that facies is substituted for analysis of attributes. There are two aspects of this use of characters to be distinguished: firstly, there is the value of the character as regards making a choice between possible alternative groupings, and secondly, the use of the character as a diagnostic, i.e. for its predictive value. It is in the latter sense that it is more useful to regard a character as *reliable* since its use will allow us to make a large number of deductions about the other attributes of the groups it separates off. In both senses we are dealing with characters which have already been selected for taxonomic use and which are being then considered for a particular purpose.

It should be pointed out that there is a strong element of circular argument in this procedure of relying upon a character because it agrees with an existing classification and then saying that the classification must be a good one, since one may be doing no more than bolster up the errors of that classification. Ideally the whole classification should be broken down and the new character included along with the others for the purposes of completely estimating the correlations anew.

The other aspects of good characters, which we may loosely term their fixity, is concerned with the selection of characters for general use. Whether for diagnosis or synthesis, a good character must obviously be one that is little influenced by environmental changes. The range of expression of a character is genetically determined and the actual expression will depend on the environment-genotype reaction within that range. Good characters will have a narrow range of expression and will be easily recognizable.*

The inherent genetic variability of characters is a subject about which we often have very inadequate information. As a general rule we would agree with Rollins (1958) that "we wish to rely on those characters as indicators of relationship (or lack of it) that are so deeply seated in the genetic constitution of the species that they cannot be easily obliterated or greatly modified by the direct effects of any given, simply segregating factor or combination of segregating factors". The effects of simple gene differences on morphological features will be considered in more detail in Chapter 11. It is sufficient to mention here that many of the characters regarded by taxonomists as fundamental may be found to vary within populations and be inherited by simple genetic mechanisms. A well-known example is described in *Layia* spp. (*Compositae* subtrib. *Madiinae*) where 1- or 2-factorial inheritance governs the presence or absence of ray flowers and receptacular paleae, characters which are extensively treated by taxonomists as having tribal significance (Clausen, 1951; Clausen & Hiesey, 1958a). Other features which may be more important from a genetical and evolutionary viewpoint may have been ignored by taxonomists.

It is, however, important to distinguish between those characters which can and do vary by simple genetical mechanisms in a population and those which are shown to be simply based genetically but which are not in fact easily changed.

Thus the glabrous-fruited condition in *Dithyrea wislezenii* (*Cruciferae*), which was the main basis for describing *D. griffithsii* as a distinct species, is a simply inherited character under single gene control. The pubescent plants studied by Rollins (1958) were all homozygous and recessive for the pubescent fruit character and he concluded from his studies that presence or absence of pubescence on the fruits of *D. wislezenii* is of no taxonomic significance. Here the character in question is not only simply inherited but shows a high likelihood of change.†

It would appear to follow from this discussion that we should rely on those characters that depend on a multiplicity of genes or gene combinations for their ultimate expression, but great caution should be exercised in applying this generalisation since it is not just the genetic basis of a character that has to be assessed but its possibility of change *in practice* and the effects of such changes on the continued integrity of taxa. In other words, *a taxonomic character is only as good as its constancy, no matter what its genetic basis.*

* Plastic characters may be used to differentiate species if their ranges do not overlap or if the overlap is not too extensive.

† In other Crucifers glabrous and hairy-fruited variants can form pure populations or grow mixed together—presumably depending on whether the initial coloniser was homozygous or heterozygous for the character in question (e.g. *Isatis*, *Sisymbrium*). The same often applies to albino variants in many families.

Other aspects of genetic control of characters in assessing their value, including pleiotropism causing multiple effects, are discussed in Chapter 11.

As we have already mentioned in Chapter 1, there is a tendency to rely on certain kinds of characters, such as those pertaining to the flower, in constructing classifications. In terms of consistency and lack of plasticity floral characters probably still justify their primary role, but this is not to deny that other kinds of characters, notably vegetative ones, may not be equally important in many cases. More fundamental considerations of floral features are concerned with the concepts of constitutive and biological characters which are discussed in the next section.

Function and Taxonomic Characters

For many purposes today the plant taxonomist is seldom concerned with the functional value of the characters he deals with in delimiting, describing and classifying. In practice his concern is more with the constancy, reliability and lack of variation of characters, as we have just seen, than with their possible adaptive or functional significance (Heywood, 1959a). It was, however, a characteristic of most pre-Darwinian taxonomy that the characters chosen to construct classifications were intended to be those of high physiological or functional value (the original "natural" classifications based on the true nature or essences of organs, cf. Cain, 1958, 1959c). In fact, so little was known about physiological importance that not much more than lip service could be paid to these precepts. The gradual rejection of physiological characters as having a function in taxonomy, by de Candolle and later taxonomists, has been discussed in Chapter 1.

Some recent writers have pointed out that the role of function in taxonomy has been unduly neglected (Stebbins, 1959b; Cain, 1959c), and Huxley has introduced the concept of evolutionary grades in an amplification of earlier ideas on progressive evolution (anagenesis of Rensch) which bears on the same problem. It was Wernham (1912), however, who introduced the distinction between *biological* and *fortuitous* characters in his studies on the evolutionary origin of the sympetalous Dicotyledons. *Biological* characters are those related to some vital function or advantage; they are usually floral, such as epigyny, zygomorphy, sympetaly, etc. These biological characters were those involved in progressive evolution, marked by two fundamental tendencies—economy in production of reproductive parts and adaptation to insect visitors. *Fortuitous* characters, on the other hand, showed no such (apparent) functional relationship.

Before considering the validity of this distinction, it is worth pointing out that many Angiosperm families are recognised and described largely on the basis of such biological characters. This receives some explanation from the observation by Stebbins (1950, 1951) that the relative simplicity of plant structure makes for a large amount of parallel variation, and families are made up, as regards floral features, of combinations of the following pairs of features:

presence or absence of sepals and/or petals; polypetaly versus sympetaly; actinomorphy versus zygomorphy; numerous stamens versus few stamens; apocarpy versus syncarpy; numerous ovules versus few ovules; axile versus parietal placentation; superior versus inferior ovary. Only 86 of the possible 256 combinations of these characters were represented in the 438 families and part-families studied. Nearly half of the total families (or part-families) considered fell into one of twelve combinations indicating a strong deviation from random representation. In other words, there is little scope for diversification with such limited themes and it is precisely such features that Wernham had in mind when discussing biological characters. These selfsame characters were those distinguished by Diels as *constitutive* and by other authors as organisational characters; they are constant throughout a wide range of individuals, species, genera, etc., that is, throughout a higher taxon. Their taxonomic value is high and, as Sporne (1956) says, they are characteristic of a large area of affinity. They are contrasted with non-constitutive features which are less constant and are subject to environmental modification.

As we have noted, the number of "basic" characters in Angiosperm flowers is limited and they evolved more than once in separate lines. Epigyny, sympetaly, etc. may therefore be considered as adaptive convergences in the Angiosperms. This has serious taxonomic implications, as van der Pijl has pointed out. The occurrence of such features in flowers otherwise considered unspecialised (e.g. epigyny or perigyny in *Victoria*, *Calycanthus*; sympetaly in some *Annonaceae*) should not be taken as an indication of taxonomic affinity and used as a motive for the displacement of taxa with such features to other parts of the taxonomic system. In other words, such features should not be weighted. It seems in fact probable that many Angiosperm families are better regarded as grades as a result of parallel and convergent evolution. Whether they are, as a result, unnatural is a question requiring further research.

Although constitutive, organisational and biological characters are often regarded as synonymous (cf. Sporne, 1956) and the particular examples chosen to illustrate them in fact coincide, they differ in the functional value ascribed to them. As we have seen, Wernham's biological characters implied a functional relationship with the environment, i.e. they were adaptive; constitutive characters on the other hand are regarded (e.g. by Lam, 1959) as non-adaptive. The explanation perhaps lies in the confusion that surrounds the use of the term adaptation. An adaptation is generally defined as a feature of an organism that confers some advantage on it, while adaptation is the acquisition of such a feature by organisms. The words therefore refer both to a process and to the result of that process; as far as the organism is concerned, adaptation is active or passive (Mayr, 1942)—it is happening now or has happened in the past and is being exploited now. Function and functional are used in essentially the same sense. As Simpson (1953a) notes, it is a relationship between organism and environment that makes a character advantageous. It is difficult to define adaptation more strictly without finding that the definition hinges almost entirely on the degree to which we can recognise what may be considered an adaptive relationship. In other words, while conspicuous relationship between organism and environment (as for example in prostrate maritime races of inland species, such as *Cytisus scoparius* subsp. *maritimus*) present no problem, in the majority of cases the acceptance of the adaptive nature of particular characters of a plant depends on the ingenuity, perception, skill and indeed imagination of the botanist (cf. Barber, 1956).

For most of this century a negative and somewhat extremist attitude to function and taxonomic characters has been adopted, largely for historical reasons (cf. Straw, 1956; van der Pijl, 1960). But it would appear to stem more from ignorance than positive knowledge. Barber (1955) goes so far as to say that "the question of the adaptive value of the morphological characteristics which the taxonomist uses to separate species populations is still one of the crucial problems of evolutionary biology". He further comments, with some justification, that the museum or herbarium taxonomist often regards diagnostic characters as having no selective value, if he considers the problem at all. He is more concerned with the constancy of the features he has chosen for discrimination between groups (Heywood, 1959a). Quite apart from the problems of whether such characters may be interpreted in terms of function, some botanists doubt the value, for taxonomic separation, of characters which can be shown to be adaptive (Borrill, 1961c).* From an evolutionary viewpoint, Barber considers there are four possible explanations concerning the nature of diagnostic characters:

(1) They are the result of genetic drift causing random fixation of genes and may therefore have no adaptive value.

(2) They are the byproducts of the activity of other genes (pleiotropic gene-action). Selection for one character may be due to the pleiotropic effect of the gene controlling another.

(3) They are the result of developmental correlations such as allometric growth. If increase in size, for example, arises through selection, any feature correlated developmentally with size will also be subject to change; and if these corresponding changes are selectively neutral they will persist.

(4) They have, or have had at some time during the history of the species, some selective value and arose through natural selection acting on random mutations and recombinations.

A fifth possibility should be added. The diagnostic character may form part of a complex of integrated adaptive characters. In other words, it may belong to an adaptive complex based on an equilibrium due to the interaction of many genes, often held together by linkage.

Very careful investigation is needed to determine which of the above

* A viewpoint stemming from de Candolle's rejection of physiological characters for taxonomic comparison (cf. Cain, 1959a).

mechanisms may have been involved. Adaptation should be considered in terms of ability to survive in a particular habitat and will therefore concern such features as leaf size, internode development, etc., features which are greatly modified by the environment and therefore the least reliable. Adaptation depends, however, not only on survival of the individual but on the propagation of the race. Reproductive structures will therefore be important and the relative constancy of these structures may be promoted by natural selection.

In the grass *Melica mutica* an apparently trivial taxonomic character—a club-shaped rudiment at the apex of the spikelets—has a high selective value as an effective dispersal mechanism since it is attractive to ants. Similarly in the *Hordeeae* tribe of the *Gramineae* some of the most conspicuous of the reproductive features used in the generic and specific taxonomy of the group (shown by the glumes, lemmas and awns) consist of a series of different but efficient seed-dispersal mechanisms. In the *Compositae* similar features with similar selective values are used for taxonomic separation of genera. Stebbins (1950) quotes examples from the tribe *Cichorieae*, where the floral structure of the various genera is essentially similar and genera and species are diagnosed by differences in the involucral bracts and in the mature fruits, especially in the pappus of the cypselas ("achenes"). These are largely related to seed protection and seed dispersal mechanisms. In the *Chrysantheminae* dimorphism of the cypsela is related to different dispersal mechanisms, this feature being one of the characters separating *Chrysanthemum* from related genera such as *Leucanthemum*. Other examples are known in *Hypochaeris*, *Crepis*, etc. in the same family. (See Salisbury, 1942, and Dansereau & Lems, 1957, for a fuller consideration.)

In such groups as the *Compositae* and the *Gramineae* with reduced flowers crowded into inflorescences, the whole inflorescence often evolves as a unit adapted to the methods of pollination or dispersal, or both. Zohary (1950) has outlined trends in the evolution of the Composite capitulum in which the involucre often opens like a capsule to release the cypselas (e.g. *Senecio vulgaris*) or retains them and is itself the unit of dispersal (as in *Echinops* and *Centaurea squarrosa*). In some cases (e.g. *Crepis aspera*) the marginal cypselas are retained by the indurated involucral bracts while the others are freely dispersed—thereby achieving different degrees of dispersal. A similar result is obtained by the production of heteromorphic cypselas, the outer being epappose and the inner pappose (*Galinsoga parviflora*), or the pappus of the outer cypselas being caducous (*Senecio jacobaea*—cf. Burtt, 1961a). Loss of pappus is not to be regarded as a retrograde loss of an efficient mechanism but is often a specialisation. Although the pappus must have played an important part in achieving the vast distribution of the *Compositae*, under particular conditions it is disadvantageous (Burtt, 1961a).

One of the objections that has been raised to the hypothesis that elaborate form and colour has evolved in response to pollinating agents is that many such flowers are either selfing or apomictic. They almost certainly evolved, however, from cross-pollinating ancestors—the change in breeding system being brought about by changes in selective pressure.

Adaptive specialisation appears more clearly at the specific and particularly infra-specific level: in populations showing adaptive response to environmental conditions (Heywood, 1959a). This is discussed in detail in Chapter 12. On the other hand, Simpson is of the opinion that the adaptive significance of diagnostic characters becomes more apparent, at least in animal groups, the higher the taxonomic category. Thus higher categories should be distinguished by features of evident adaptive nature. This relationship is not, however, so obviously apparent in flowering plants, and zoologists often express surprise that floral characters which, as we have mentioned, play such a marked role in higher-plant taxonomy have been seldom studied from a functional or adaptational viewpoint.

As mentioned above, many people have denied the adaptive value of the so-called biological (constitutive, organisational) characters, such as zygomorphy, sympetaly, complex pollination mechanisms, etc. A survey of the views of such anti-selectionists as Goebel, Troll, Good, and Nelsson is given by van der Pijl (1960), who discusses their interpretation of the flower in terms of "Gestalten", automatic autogenesis or physiological necessity. During the last few decades there has been a renewal of research into floral biology, particularly in tropical countries, using the tools of ecology, genetics, zoological physiology, and it has been possible to demonstrate by experiment that the intricacies of form, colour, odour and pattern in flowers have an important role in attracting and orientating pollinating insects.

Van der Pijl points out that a major stumbling-block in considering the flower from a functional point of view was that studies were made on temperate floras where bees and butterflies were normal pollinators, and where nectaries are considered a fundamental floral attraction. In other words the climate was wrong both for understanding the flower and for interpreting the problems. If we look at the earliest known fossil Angiosperms we find that they show such features as we have previously been considering constitutive or organisational, e.g. epigyny, sympetaly, syncarpy in flowers which are otherwise "unspecialised" or simple. It is difficult to see in what way such features could be adaptive if we consider them in relation to their present day pollinators. We now know, however, from palaeontological research that in the period when the early Angiosperms were developing (Lower Cretaceous and probably Upper Jurassic), beetles (*Coleoptera*) were well differentiated and abundantly available as pollinators and that some flies (*Diptera*) were present. The *Hymenoptera* and *Lepidoptera* were not, on the other hand, sufficiently well evolved to be of much significance as the pollinators of early angiosperms. A good summary is given by Barnard (1961).

Parallelism of form in flowers known to be pollinated by similar agents may in itself be regarded as evidence of an adaptive relationship. For example.

the flowers of *Orchidaceae* and *Zingiberaceae*, *Polygalaceae* and *Papilionoideae* may be regarded as convergent adaptations in response to similar methods of pollination. Bradshaw (1958) makes the same point about parallel and repeated correlations between habitat and characters at the genecological level as suggesting probable adaptation. On the other hand, characters which are adaptive in a *broad* basic sense (e.g. the very features which define the Angiosperms or the birds) are more likely to remain conservative for longer than characters that have been evolved in relation to special conditions. For example, those floral characters related to specialised pollinators (as in *Salvia*) or to particular agents of dispersal (as in the subgenera of *Trifolium*) can be fairly safely accepted as adaptive characters that have played little part in the basic evolution of the higher groups and are associated more often with generic or specific differentiation. The cushion habit and the evolution of berries from capsules by the continued growth of the pericarp wall after fertilisation are other characters than one can reasonably accept as "superficial" specialisations that have had little to do with the basic evolution of families. Each case must be judged on its own merits.

In summary, we may agree with Cain (1959c) that the taxonomic value of characters should not be judged, for purposes of comparison, mainly in terms of their function, following the early *a priori* taxonomists such as Linnaeus and Caesalpinus; nor should we deny taxonomic value to characters which can be demonstrated as functional and rely on those which have no apparent functional value, as has been proposed by several more recent botanists such as Sachs (1906), Wilmott (1950), and Borrill (1961c).

As regards the constant characters of taxonomic groups, it is most probable that this very constancy is maintained through the forces of natural selection acting on the populations containing the characters, so that they are more likely to have a selective value than not. That we cannot comprehend in what way such characters may be functional does not mean they have (or had) no adaptive significance.

For the purposes of phenetic comparisons, knowledge of the adaptive nature of characters has little value except in so far as it stops us counting the same character more than once through the functional correlation of apparently separate characters. This question of the co-variation of characters which are necessarily correlated is discussed below. The role of function in the evolutionary interpretation of comparative data is, however, considerable.

Characters and the Definition of Groups

As mentioned above when referring to species, the characters chosen to describe a taxonomic group form an abstract for practical convenience. They do not in fact define the group. Similarly, as Simpson (1953a) says, the diagnosis of a higher taxon (i.e. above the species level) is not a description of any ancestor of the group. In other words, by abstracting from a higher taxon those characters which are common to all its members, one may get what is often referred to as a common ground-plan—some sort of general basic plan on which the group was modelled. This idealistic approach was characteristic of certain schools of archetype seekers or typologists, particularly the Germanic school. But as Cain & Harrison (1960) point out, this procedure may in fact give us less and less information about the presumed ancestor of the group, since the rank of the group from which the characters are abstracted is higher and contains a much greater diversity than any ancestral taxon. It is also quite possible that the common ancestor did not itself possess these characters in common. Features it did possess may have been lost in the course of evolution; or the common ancestor may have had the characters shown by an existing member of the group which have survived by stasigenetic persistence. It is of course possible, too, that the group may not have had a common ancestor at all if the resemblances are partly the result of convergence (cf. Sporne, 1956).

Apart from considerations of common ancestors, it is by considering the abstracted characters as in some way representing and defining the group as a whole that we fall into serious typological error. A genus or higher taxon is not defined by possessing a particular combination of floral and vegetative features: it can be differentiated or recognised by these. These characters are the ones which the component species share in common and it is useful for the practical purposes of characterisation, identification and keying that they be abstracted. (It should, however, be pointed out that since phenetically determined groups are based on overall affinity, it is possible that in some cases particular features may be modified or even lost in some of the constituent members, with the result that the group cannot be diagnosed (cf. Cain, 1954b; discussion in Simpson, 1961, pp. 93-103) but only described. A number of currently accepted genera probably come under this heading.) They are not, however, the *only* characters the group has in common, nor indeed the only diagnostic characters since others may remain unrecognised or be too difficult for practical use.

EVOLUTIONARY INTERPRETATION OF COMPARATIVE DATA

So far we have been concerned with the arrangement of comparative data derived from all possible sources so as to produce a phenetic classification based on overall similarity. The groups in this classification are recognised by their resemblances without any weighting. This is an essential first step in the production of an evolutionary or "phyletic" classification, although it will not by itself produce one. Further interpretation of the phenetic features of the groups and additional procedures are necessary if a phylogenetic classification is aimed at. It may not be possible in some groups to go beyond a phenetic arrangement, but usually it is possible to derive *some* phyletic information from the comparative data. The two most important kinds of information that can be sought are those referring to convergence and primitive characters. A further aspect to be considered is the necessary correlation of characters as already mentioned; and the genetic evaluation of characters and its bearing on the likelihood of their alteration or modification is also an increasingly fruitful source of information. Primitive characters have already been discussed in Chapter 2 in detail and are not considered further here.

1. *Convergence*

Convergent and parallel evolution has already been defined in Chapter 2. The effects of convergence on the interpretation of natural classifications have been underestimated in the past (Philipson, 1961) and, in flowering plants at least, the extent to which it has been unconsciously accepted in current classifications is still not yet appreciated. If evolutionary *divergence* only occurred in groups, then, as Cain (1959b) points out, a natural and phyletic arrangement of it would virtually coincide. However, if convergence has occurred, it might well result in forms being grouped together on the basis of their convergent features although in respect of other features they would be separated. In the absence of a fossil record it may be very difficult to tell whether a natural arrangement is also phyletic, since the extent to which it contains convergent forms cannot be determined. The assumption that forms which are most like each other will be phyletically related in the sense that they have a common ancestor is basically valid, but it does not allow for the distortion of similarity by convergence which may not even be suspected in the group in question.

The Adansonian method of classification, when thoroughly applied, is likely to overcome the effects of convergence in that the more thoroughly the comparisons are made, using as many kinds of data as possible, the less likely are convergent forms to affect the result. In other words, the large numbers of different features used will probably compensate for any convergent similarities. Groups which on the basis of their general affinities are widely separated will show resemblances in only the few convergent characters By accepting these convergent characters as indicative of phyletic relationship we would clearly be weighting them, and this is of course contrary to the principles of general Adansonian classification. Unfortunately this weighting procedure is nowadays frequently used, especially when newer kinds of evidence are employed. As a result phenetic arrangements become distorted and pseudo-phylogenetic classifications are proposed. The normal justification of this procedure is that the new line of evidence (such as biochemical, cytological, etc.) is regarded as more important, fundamental or at least more trustworthy than the traditional characters. There are two objections to this procedure: firstly, as already mentioned, in phenetic comparisons characters are not weighted, largely because we have no valid criteria for judging the relative importance of different kinds of comparative data; secondly, there is no reason to believe that these newer features are not also susceptible to convergence, deviation and even reversal, so that reliance on them alone for making classes would tend towards the production of unnatural polyphyletic taxa.

We have already commented on the dangers of reliance on a restricted range of floral features for interpreting taxonomic affinity. This is, however, an aspect of a more general problem, namely that a very large part of the higher and middle levels of classification of flowering plants was established in pre-evolutionary times and in fact much of it dates from the pre-Adansonian period. Weighting in some form or another was employed: in pre-Adansonian times weighting by taxonomists took the form of selecting "essential" characters, as previously discussed. Later, with the gradual rejection of these *a priori* principles concerning the "true" nature of characters and the introduction of Adansonian principles, weighting was still retained but in an intuitive form. In other words, there was non-phyletic weighting. It took the form of intuitively choosing some character(s) for delimiting groups largely irrespective of the co-variation in characters of the resulting groups. When the weighted diagnostic characters were also those that showed the highest co-variation, the taxon may be satisfactorily retained today as a natural group.* But when the characters selected were not those showing the highest overall correlation, they delimited an unnatural taxon or grade, the weighted characters being the grade-defining characters. In the latter case parallel evolution in different lines produced the same character which for some reason, such as its conspicuousness or a belief that it was more important, induced taxonomists to select it as a basis for delimiting a taxon. This approaches the original use of the term anagenesis by Rensch (1954), although he regarded the progressive changes involved as representing some advance in organisation or in perfection of function (cf. Simpson, 1961).

Although we are not in a position to ascertain whether the weighted features to which we have been referring can necessarily be regarded as constituting an increase in functional efficiency, it is quite clear that the recognition of parallel-lineage taxa has been widespread, and it is useful to refer to them as grades or grade-taxa, as Huxley has suggested, but without any functional implications. The example of the genera *Celsia* and *Verbascum* has already been described in Chapter 2. There, as Hylander (in Heywood, 1960a: 208) says, "in the conventional taxonomy of this assemblage the tops of different phylogenetic lines have been cut off from their base and combined in one genus *Celsia*" characterised by androecial reduction. Similarly the genera *Agropyron* and *Elymus* are often defined in a wide sense so as to include stages in evolution from different lines recognised as *Roegneria* on the one hand and *Elytrigia* on the other. Thus species of *Elytrigia* and species of *Roegneria* with *Agropyron*-like characters are included in *Agropyron*, which is consequently by our definition a polyphyletic grade; similarly with *Elymus*. Monophyletic or at least natural genera would be obtained if *Roegneria* and *Elytrigia* were recognised as separate genera, though whether they

* Cf. correlation weighting discussed in Chapter 2.

could be conveniently delimited is another matter. There must be very many numerous examples of this nature at all levels of the taxonomic hierarchy.

Convergence or parallelism resulting in close morphological similarity is more likely to occur between closely related than between distantly related forms, although it might not be possible to determine exactly what had happened in any particular case (Cain & Harrison, 1960). If we accept this general probability, it means that convergence will take place in precisely those groups where it is most difficult to detect. This is reflected in the apparently wide occurrence of grade genera in families such as the *Gramineae*, where the overall differences between genera are relatively slight due to the closely knit morphological relationships throughout whole groups of the family. Similarly in some groups of ferns, such as the *Polypodiaceae*, the difficulty of finding characters by which to discriminate genera is such that any conspicuous character tends to be seized upon irrespective of the variation in other less noticeable features.

The problem of convergence in Angiosperms, with particular reference to the *Gramineae*, has recently been discussed by Philipson (1961). We agree with him that the probabilities of its occurrence have been underestimated and the detection of many unnatural grade genera and higher groups will be the probable result of closer scrutiny of the characters on which they are currently established.

Not all taxonomists would wish to correct for convergence. For the purposes of natural classification, correction should probably only be undertaken if convergence makes a classification misleading or otherwise unsatisfactory. In extreme cases of convergence correction may be patently unwise. As Sneath & Sokal (1962) put it, if we "suppose that an absurdly extreme overall convergence had occurred so that two forms were almost indistinguishable and could readily and successfully hybridise: what is the purpose of separating them on grounds of ancestry when in all other attributes they are virtually the same?" In other words, the correction for convergence will depend on what one is trying to illustrate in one's classification.

2. *Necessary Correlates*

Characters which are regarded as distinct may be necessarily correlated, i.e. vary together, for a number of reasons. They may be *mathematically* correlated and care should be taken in such cases to avoid counting the same character more than once by assessing it in a different way (see discussion of numerical taxonomy on p. 133).

Characters may be functionally or ecologically correlated, and whether such correlations should be corrected for will depend on the taxonomic activity concerned. In assessing overall similarity, for example, such correlated characters cannot be ignored since they contribute to general resemblance, no matter what the cause. For constructing patristic groups, necessarily correlated characters cannot be regarded as separate characters and all but one must be discarded since "if one arises by convergence in a different and unrelated form but fulfilling the same function, all the others will arise as well" (Cain & Harrison, 1960: 11). It is most improbable that they will all have arisen by convergence as several unrelated characters.

There may be serious difficulties in recognising necessary correlates, especially functional ones, and they may not become evident until after taxonomic groups have been constructed.

3. *Genetics, Cytology and the Likelihood of Change*

The likelihood of a character or complex of characters to change and consequently to have evolved more than once may be adduced from a wide variety of evidence, notably cytological and genetical.

As we have already noted, simply inherited characters may be of little value for taxonomic separation (p. 120); similarly they will be poor indicators of phyletic relationship since forms in which such a character is absent may belong to the same phyletic line as others which do not possess it. On the other hand, characters which are known to have a complex genetic basis are unlikely to have evolved more than once within a general area of affinity and are therefore good indicators of monophyly. This kind of evidence is tied up with considerations of which of two characters indicated represents the derived and which represents the primitive condition.

Various examples are discussed in succeeding chapters, such as the normal derivation of inbreeding from outbreeding, derivation of polyploid chromosome numbers from lower numbers. It must be stressed, however, that common possession of the derived condition does not necessarily indicate closeness of relationship, either phenetic or phyletic. And even when the direction of change can be established beyond reasonable doubt, it may provide no more than evidence of patristic relationship and give no cladistic information, i.e. as regards the actual pathway. At the specific level it has proved possible in many cases to provide factual information about the origin of taxa, as for example by the artificial synthesis of allopolyploids.

Cytological evidence, when linked with geographical distribution, is also a valuable means of retracing the phylogeny of groups, as in cases of vicarious taxa and geographical isolates.

A general consideration of the problems of phyletic weighting of information derived from a phenetic classification is given by Cain & Harrison (1960).

Character Correlation and Discrimination: Numerical Taxonomy

Most modern classifactory procedures depend on the recognition of character combinations for the construction of taxonomic groups. The reverse side of the coin is the need to discriminate, that is to separate groups and afford means of assigning samples to groups on the basis of the individual value of particular characters. Various methods and techniques are currently employed for these tasks, ranging from the neural estimates of resemblances of traditional practice (which as Sneath (1961) points out is a subjective and imprecise statistical Q-technique) through scatter diagrams to various mathematical coefficients.

There has been a remarkable development of interest in recent years in the application of mathematical techniques to taxonomy, with the general aim of making assessments of similarity and discrimination between groups precise. These two aspects—the construction of taxonomic groups and discrimination—will be considered separately.

1. *Construction of Taxonomic Groups*

As we have already discussed in previous chapters, past practice as regards the construction of taxonomic groups has been to use some form of weighting—the *a priori* methods of pre-Adansonian taxonomists based on physiological importance of characters, post-Darwinian phylogenetic weighting, subjective opinion weighting or intuitive correlation weighting. In Angiosperms, with the absence of an adequate fossil record, phylogenetic weighting must largely be negative in the sense that characters are rejected (the negation principle discussed in the next chapter) or based on hypotheses or speculations that have as much chance of being right as wrong. As Heslop-Harrison (1962) points out, the adoption of the evolutionary hypothesis has had the paradoxical effect of impelling taxonomists to adopt a less natural and more artificial basis for the delimitation of groups, since it implies the deliberate rejection of some facts we know about the organisms we deal with. It is remarkable, too, that continued reclassifications of groups have tended to reduce the weighting given to characters so that the logical goal would appear to be equal weighting of those characters that are accepted. Much so-called phylogenetic or evolutionary weighting is what we have termed correlation weighting with evolutionary implications grafted on afterwards so as to satisfy the "snob-appeal" of evolutionary classification. While this may be an understandable human failing, it can scarcely be accepted as an honest scientific procedure. If we are not going to weight characters (other than by acceptance or rejection), we feel that it is better not to make the pretence. Such an attitude has in fact been gaining ground in recent years and is expressed in the clear separation that is urged by a growing group of taxonomists between phylogenetic and phenetic classifications.

Phenetic classification is based, as we have seen, on overall affinity as judged by intuitive or neural means, using as many characters or as much evidence as is available. It employs, in practice, correlation weighting. Because of the inability of taxonomists to translate their mental pictures of variables into a readily communicable system, mechanical methods of comparison have been developed by many workers (mostly in the fields of microbiology, where the problems are even more acute) to reduce the subjective element to a minimum in making correlations. Although such methods have been available for many years, it is only very recently that the availability of computers has enabled the actual operations involved to be mechanised and become practicable, at least in a limited number of cases. This whole development has been so rapid that it is scarcely possible to stand back at sufficient distance to assess its potential value adequately. What we shall do is to outline the various techniques involved and give the reader references which will allow him to pursue the matter further.

Methods. The best summaries to date of the methods of numerical taxonomy (or taxonometrics, as Rogers & Tanimoto (1960) have proposed) are those given by Sneath & Sokal (1962) and Sneath (1962), on which the following account is largely based. It must, however, be emphasised that a wide variety of approaches is possible and an extensive literature on techniques is accumulating. A textbook of numerical taxonomy by Sneath & Sokal is in preparation. Reference should be made to the Breslau school developed by Florek and co-workers and illustrated by Kowal & Kužniewski (1959) on *Atriplex* and *Chenopodium*. The successive steps recognised by Sneath are:

1. *Characters and operational taxonomic units.* The first step is to select the group of organisms to be studied. The lowest ranked units to be studied may be of various kinds—individuals, lines, etc., either treated as such or as representatives of named taxa (in which case care has to be taken that they are as representative of the taxa as possible); or species or genera may be employed and represented by their taxonomic descriptions, although this raises difficult problems concerning the treatment of the characters which vary in the taxon. The taxonomic entities of lowest rank used are termed *operational taxonomic units* (OTUs). Choice of them will be to some extent arbitrary and a fair latitude should be given so as to include apparently related organisms, even though they may have diagnostic characters which separate them off.

2. *Characters.* Since it is a tenet of Adansonian taxonomy that all characters are of equal importance (i.e. are given equal weighting) in constructing a classification, all kinds of character may be employed provided: (1) they are inherent in the organisms (geographical distribution is therefore of doubtful validity but not ecological characteristics, cf. Rogers & Tanimoto's use of soil types: 1960); (2) they are not susceptible to wide environmental modification; (3) they are not greatly affected by experimental or observational uncertainties. These last two requirements amount to what we have previously discussed as the goodness or reliability of characters (see p. 119). What is regarded as a good character is a matter of the scientific judgment and experience of the taxonomist: he knows what are likely to be the most relevant characters to employ although he may make errors and frequently does. Mathematically or logically correlated characters should be scored as one only so as to avoid scoring the same thing twice. Characters which are highly correlated empirically present a problem and Sneath & Sokal suggest that they should be scored as one unless we know or have

evidence to suggest that more than one factor is responsible for the two correlated features. As wide a selection of characters should be made as possible and an attempt should be made to randomise the selection deliberately: the larger the number of characters, the nearer one will approach a random selection of the characters of the genome.

A minimum of forty or fifty characters (or expressions) is needed, and if possible several hundred (Sneath, 1962). Very little reliability can be placed on a similarity coefficient S based on fewer than fifty characters.

Compound characters should be broken down into unit characters which we have defined above (p. 113), with the intention that a new item of information should be contributed by every character expression.

The characters selected then have to be coded. The simplest form is two-state coding where characters are divided into + and − which are used for simple coefficients of association. The symbol *NC*, meaning "no comparison", or O, is used when no data are available. Shared negative features are ignored. Following this system there may be instances when it is not obvious which character state should be regarded as + and which as − (e.g. lead tolerant versus lead resistant) and an arbitrary decision is necessary. Beers & Lockhart (1962) propose using the symbol O as part of a modified technique for attributes which cannot simply be regarded as + or − and which therefore may be scored twice. Multistate coding is where characters are divided into several states, 1, 2, 3, 4, etc., by breaking up a continuous variable into steps of equal weight. These can be converted into two-state values, for example by giving a separate entry for each of the states. Thus state 1 would be coded + − − −, state 2 + + − −, etc.

The data obtained by scoring the characters in the OTUs are now set up in a table in matrix form, consisting of t OTUs scored for n characters. The data matrix is called a $t \times n$ (or $i \times m$) table—the one column representing the organisms, the other representing the characters (Table 1).

Character states	Taxa (OTUs) *A*	*B*	*C*	*D*
1	+	+	−	*NC*
2	+	+	+	+
3	+	+	+	−
4	−	+	*NC*	*NC*
5	+	+	+	+
6	+	+	−	+
7	+	+	−	*NC*
8	*NC*	−	+	+
9	+	+	+	+
10	+	+	+	−
11	+	*NC*	−	*NC*
12	+	+	+	−

Table 1. Coded data table ($t \times n$ table).
See text for further explanation. (After Sneath, 1962.)

3. *Measurement of resemblance.* Three main methods have so far been devised for estimating the phenetic resemblance between the taxonomic entities:

(*a*) Coefficients of association. These are the simplest and mainly refer to characters expressed as plus or minus (two-state). Several coefficients have been proposed. A simple form used by Sneath (1957a) is a numerical index S for the similarity of each pair of organisms examined which is derived as follows:

$$S = \frac{N_s}{N_s + N_d},$$

where N_s = number of positive features shared by any two OTUs; N_d = number of features positive in one OTU and negative in the other.

(*b*) Correlation coefficients. These have been used by Michener & Sokal (1957) and others (see Sneath & Sokal, 1962).

(*c*) Measures of taxonomic distance using the convention of multi-dimensional space with one dimension for each character, the Euclidean distance between the entities being calculated by a simple formula (Sokal, 1962).

If the first method is used, the next step is the tabulation of the similarity coefficients in a $t \times t$ table with n rows and n columns for n taxa. This matrix indicates the similarity between every OTU and every other (Table 2).

Taxa (OTUs)		Taxa (OTUs)			
		A	*B*	*C*	*D*
	A	100	.	.	.
	B	90	100	.	.
	C	60	60	100	.
	D	57	50	50	100

Table 2. Similarity matrix ($t \times t$ table) based on Table 1 by comparing every taxon with every other. See text for further explanation. (After Sneath, 1962.)

4. *Cluster analysis.* Since the OTUs will probably have been arranged in a haphazard manner, they have to be rearranged so as to bring together those organisms which are most similar. This can be done approximately by shading the elements of the matrix to indicate the degree of similarity so that very similar organisms are adjacent and appear as dark triangles (Fig. 12). This allows the main phenetic groups (phenons) to be identified visually in simple cases, but more objective methods are usually preferred. These methods for identifying phenons from the similarity matrix are collectively referred to as *cluster analyses.*

The clusters can then be arranged in a hierarchic taxonomic dendrogram which summarises the main features of the cluster analysis although subject to some simplification and distortion. These dendograms are in no way to be

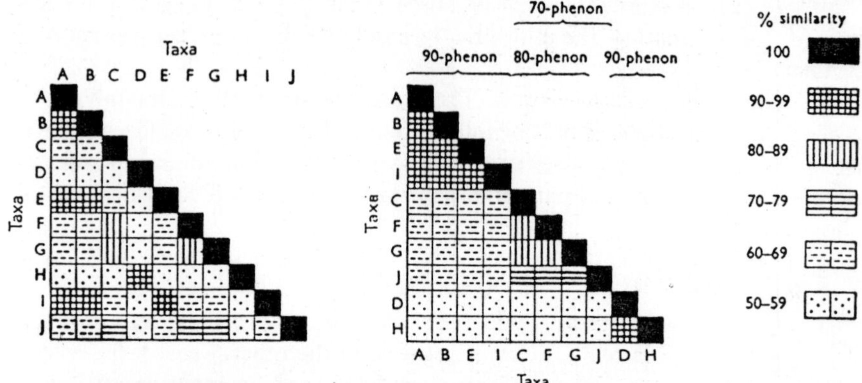

FIG. 12. *Left*: Schematic diagram showing a matrix of hypothetical similarity coefficients between pairs of taxa; the magnitude of the coefficients is shown by depth of shading.

Right: the same coefficients as in left-hand diagram, arranged by placing similar taxa adjacent to each other. Note the formation of triangles of high similarity values. (After Sneath, 1962.)

interpreted as phylogenetic trees despite their obvious resemblances. They simply represent the degree of relationship with the nearest stem (Sneath & Sokal, 1962). It should be noted that the abscissa has no special significance: it is a means of spacing out the original units used in the study. The ordinate, on the other hand, represents the similarity values, the upper end of the scale indicating maximum similarity while the lower end extends as far as is needed to unite all the stems in the particular case (Fig. 13).

5. *Phenons and rank*. Sneath & Sokal (1962) have introduced the concept of a phenon* for the definition of groups obtained by cluster analysis. They may be prefaced by a number indicating the level (percentage) of similarity at which they are formed, e.g. an 80-phenon indicates a group whose members are affiliated at no lower than 80 per cent. of the similarity scale used in the particular analysis. In practice a phenon defines groups by drawing lines horizontally across the dendrograms: these lines must be straight and parallel to the abscissa. In this way similarity coefficients may be used as criteria of rank: we can thereby delimit taxonomic groups objectively by choosing arbitrarily fixed levels of similarity. For example, in Fig. 13 by drawing a horizontal line across the dendrogram at a similarity level of 75 we have created four 75-phenons (ABEI, CFG, J, and DH). These may be regarded as subgenera, genera, etc., depending on the rank of the original taxonomic units employed in the study. Clearly such methods must be applied uniformly throughout a group under study. Phenons need not, however, be

* The term *phenon* was also one of a set of categories proposed by Camp & Gilly (1943) to describe the various kind of species situations in their pioneer paper on biosystematy (cf. Chapter 13).

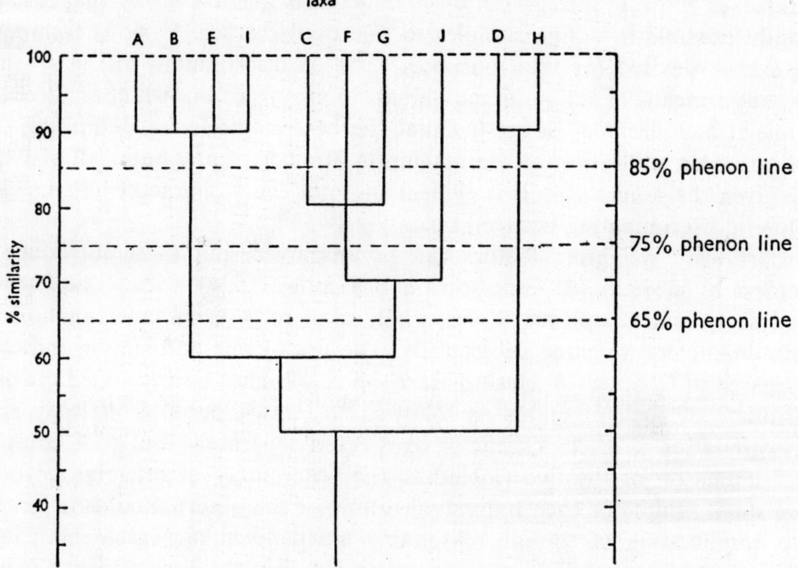

FIG. 13. Dendrogram representing the hypothetical hierarchy obtained from Fig 12 (right) drawn in the manner of Michener & Sokal (*Evol.* **11**: 130 (1957)). Ordinate indicates magnitude of similarity coefficient at which stems join to form higher ranking taxa. Horizontal lines delimit taxa of equal rank. See text for explanation of phenons. (After Sneath, 1962.)

equated with taxonomic groups unless so desired. The advantages of such methods for the delimitation of supra-specific taxa by agreed arbitrary levels of similarity are obvious in theory at least. It is premature to say whether they will prove acceptable in practice for Angiosperms.

Methods for the comparison of dendrograms derived by numerical methods with one another and with those produced by conventional means have been proposed by Sokal & Rohlf (1962). Models for the presentation of the major relationships of taxa as derived by cluster analysis have been used by Lysenko & Sneath (1959).

2. *Discrimination*

If taxonomic groups are defined in such a way as to show overlapping variation, whether by numerical methods or by other means, so that it is not evident what differences are present, some means have to be devised for discrimination. The question of keying numerical taxa has been discussed by Sneath (1962).* But as Heslop-Harrison (1962) points out, the problems of discrimination are found in their most severe form in species or population

* It may eventually be possible to include in keys a numerical estimate of the reliability of a particular attribute (Beers & Lockhart, 1962).

complexes showing intergradation in characters in such a way that it is not readily possible to assign samples to one or the other. Various techniques have been devised for such purposes, such as discriminant analysis, which serve as a means of assigning individuals to groups of known characteristics. While it has just been suggested that the best methods for delimiting taxa are based on utilisation of the maximum number of attributes all of which are given the same value, it is evident that not every character has the same value in discriminating between taxa.

Arbitrary weighting factors can be employed, but these introduce an element of subjectivity. Anderson's hybrid index (1936), widely employed to analyse hybrid complexes, is essentially akin to discriminant function but with characters weighted subjectively. The use of this and similar indices is discussed in Chapter 14. Heslop-Harrison (1952b) has clearly stated the basic postulate of such functions as follows: ". . . two populations may show intergradation in every character considered separately, but . . . complete discrimination of the two (including the potentiality of referring correctly any small sample or even individual to one or the other) should be possible if a combination of enough reasonably independent measurements can be taken into account". Precise techniques of deriving a discriminant function have been described by Fisher (1941) and have been used by Whitehead (1954, 1955) to discriminate between annual *Cerastium* species.

Although it has to be recognised that in some cases they do no more than demonstrate the obvious, such methods may be usefully employed when it is desired to separate groups showing no clear discontinuities in single morphological features but which are known from experimental samples to possess characters (such as different levels of polyploidy) which would warrant their recognition as separate units for particular purposes. However, unless the character-weighting so derived provides a means of recognition in a high proportion of cases, taxonomists will be understandably reluctant to recognise the groups as separate in orthodox treatments.

3. *Potential Value*

It would be rash to predict the reactions of Angiosperm taxonomists in general to numerical procedures, but clearly a cautious welcome should be extended. In so far as the methods are an extension of orthodox procedures, there seems no theoretical reason why they should not be employed. There are of course serious practical limitations. They do, however, imply making a clear distinction between phenetic and phyletic classification, which as we have seen is not widely accepted today. This applies at the higher levels of classification which are returned to later. At the species level, more serious objections may be validly raised. For the proponents of a biological species concept, "numerical "species are likely to be unacceptable unless some means of including the necessary genetic and crossability evidence is devised. Orthodox systematists may be no less critical, since there can be little doubt that a visually gifted taxonomist using his "natural computer" (his brain) may be more successful than a mechanical computer fed with a non-relevant selection of characters in recognising groups as species. Character selection is the weak link in this whole approach. In assessing similarity the taxonomist working neurally does not have to make consciously the abstractions we call characters: it is only when he wishes to communicate about any particular aspect, e.g. for diagnosis, that he is forced to rationalise what he recognises as a *Gestalt* of many independently varying elements, and break it down into component parts. As a consequence of this the taxonomist may produce a satisfactory division into species *despite* the characters set down in his descriptions. An interesting example of this is in the genus *Biscutella* which has been monographed by Machatschki-Laurich (1926). It is virtually impossible to employ successfully the key provided for identification of the species; nor are the descriptions of related taxa sufficiently different from one another to allow a ready separation. Yet examination of material cited from the regions concerned often shows that the species she accepted are in fact recognisable by their facies, despite her inability to convey this on paper.

It would seem therefore that statistical methods may give a less satisfactory answer than traditional methods if the characters chosen for comparison are inadequate. It is normal procedure to select those characters which *appear* to differ significantly, but as we have discussed earlier in this chapter, in many instances the definition of characters is extremely difficult. It is possible that the precise aspect of the *Gestalt* that allows taxonomic separation to be made may be missed in the breakdown into unit characters for numerical comparison. Nor is there any guarantee that the large number of characters employed in numerical taxonomy will overcome this. A serious practical difficulty is that the same character must be known for most of the members of the group being studied.

The difficulties involved in attempts to apply such numerical methods to the Angiosperms, building from the species upwards, are going to be formidable, but until they have been applied we do not know what kind of results we shall get. Some of the considerations that seem worth bearing in mind are the following.

If we attempt to use all available characters, it is possible that certain correlations will occur entirely between cryptic characters (e.g. chemical ones). This means that there may be discontinuities in variation which the taxonomist will not be able to use for the delimitation of taxa because they would be highly inconvenient. In that case we may have to ignore cryptic discontinuities for the delimitation of taxa and in the assessment of rank. It may be, of course, that chemical characters, etc. will prove to be generally in step with morphological ones—in other cases their evolutionary paths may have been different. From previous experience, however, most of those cryptic characters which *have* been studied tend to be fairly well in step with gross morphological ones. Most characters, however, whether macroscopic or microscopic, are probably prone to parallelism and to reversal of trend, so that the only safeguard against the construction of polyphyletic groups is the use of as many characters as possible. To use morphological characters alone might lead to the creation of more artificial groups than we have already—we just do not know at this stage.

It would seem that some pilot schemes should be attempted to see what the results of a natural classification, using more characters than we have ever used before, are likely to be. One might suggest three approaches:

(1) The natural classification of an obviously natural family (or genus) of moderate size, like the *Dipsacaceae* in which the present generic characters are mostly of a "qualitative" kind. Such a classification might be based:

(*a*) entirely on morphological characters;

(*b*) all available characters (including non-morphological ones).

We could then see how the result of (*a*) compares with (*b*) and how both compare with previous classifications, and become aware of the difficulties involved. There seems no point in applying the method to a difficult group until we know how it works in an easier one. We need to trust the method before we can apply it to groups of doubtful affinity. It should be mentioned here, however, that it is probably easier to recognise separate characters in a plant's morphology than in any other aspect of the plant. This is bound to put a limitation on the use of non-morphological features, since they can only be used for classification in so far as they can be separated into discrete characters.

(2) A natural classification of an animal group (using external morphology and anatomy) with an adequate fossil record so that we shall have factual knowledge of whether the groups formed by this method are monophyletic—i.e. of the part played by convergence which Cain (1959b) claims to have been so important in evolution. This factor, however, is bound to vary from group to group.

All the published examples of numerical classification that have come to our attention involve comparison of OTU with OTU. J. Cullen (1962) suggests that an attempt might be made to reverse this method by comparing character expressions and scoring the number of individuals in which pairs of expressions occur. This would result in a character-by-character table (instead of a $t \times t$ table), thereby giving a picture of individual character correlation. Whereas Sneath's method is involved with the grouping together of pre-existing taxa (i.e. OTU's), in the alternative method the lowest taxonomic groups could be extracted from the character-by-character table. In either case, however, some higher taxonomic group has to be accepted before one can analyse the plants comprising it. Similar approaches are mentioned by Williams & Lambert (1961).

Punched cards and computers can only show up the facts of character correlation. It will be left to the taxonomist to decide where, in the series of correlated discontinuities revealed, he can most conveniently insert the different ranks of the hierarchy. Convenience being an essential component of classification systems, subjective judgment can never be eliminated. Science must come to terms with Art. We might expect, however, to be in a better position to apply the rank more consistently within each superior taxon than it has often been applied before, and to delimit these taxa better, merely because we shall have more precise data on which to base our judgments. Though computers may become standard equipment in herbaria, they are not likely to replace systematists. A Court cannot function properly without a jury to assess the evidence and a judge to pronounce the sentence.

Note: Although the above discussion has been restricted to examples of Adansonian numerical taxonomy in which the characters selected are given equal value, there is no technical reason why numerical procedures in which some characters are weighted should not be employed. Some examples have been published (e.g. Proctor & Kendrick, 1963) and other unpublished ones are known to us. The problem remains of finding an acceptable basis for such weighting.

Chapter 5

TAXONOMIC EVIDENCE

I. Morphology, Anatomy, Palynology, Embryology

"Systématique moderne science de synthèse; systématicien actuel homme orchestre."

J. J.-G. LEONARD, 1959

INTRODUCTION

Taxonomic evidence (in other words, characters) can be drawn from any part and phase of development of the plant. In practice overwhelming reliance is placed on morphological characters, at least for expressing classifications, and even these have been largely restricted to the reproductive phase when dealing with Angiosperms.

The aim of this and the following two chapters is to survey the major sources of taxonomic evidence, placing special emphasis on those characters that tend to be neglected and on the "newer" kinds that are increasingly employed today. They will be considered under four headings: Morphology and Anatomy; Palynology and Embryology; Cytology (cytotaxonomy); Phytochemistry.

MORPHOLOGY AND ANATOMY

Taxonomists still rely today to a very large extent on morphological characters, both for producing classifications and for diagnostic purposes. Frequently they form, together with distributional and field data, the only evidence available; and whatever other evidence may be employed in the construction of a classification, practical considerations demand that the results be morphologically expressible. There is of course no *theoretical* reason why classifications expressed in other terms, for example chemical or genetic, should not be generally employed, provided there is sufficient evidence available, but there can be no doubt that for general use morphologically expressed classifications will be required.

There is much misunderstanding about the morphological bias of Angiosperm classification. It must be pointed out that primarily morphological classifications are in no way regarded as *ideal* by practising taxonomists: they are simply regarded as the most appropriate (and the most rapid) for the purpose of constructing a map of the diversity of Angiosperms to act as a general reference system into which other kinds of information can be fed. Morphological features have the great advantage that we can see them easily and can therefore appreciate their variability with much more facility than we can with other kinds of features. Even on herbarium specimens with which most taxonomic work has to be performed, morphological characters can be assessed with a high degree of success, as the history of taxonomy demonstrates. On the other hand, information about microscopical or endomorphic characters is often based on a single investigation and is not consistently available for the various members of a group being studied. Great progress is, however, being made in devising techniques for the apprec ation of such features as anatomy and even prediction of chromosome numbers in dried herbarium specimens, although these methods are seldom employed.

The evidence provided by morphology should not be confused with that given by "marker" characters only. In fact, the external features of a plant consist of a whole series of largely separable attributes, and even if only some of them are explicitly described, they may all be assessed at a glance, as we have explained in the previous chapter.

It should also be remembered that when fossil evidence is available for the reconstruction of phyletic sequences it necessarily takes the form of morphological or anatomical features; in Quaternary botany, too, pollen and spores and other remains are interpreted and identified to a very large extent on their morphology *faute de mieux.*

While in no way suggesting that other kinds of characters should not be sought—on the contrary, one of the aims of this and the following two chapters is to review the wide array of evidence available to the taxonomist today—it is important to remember that morphological evidence is valid on its own grounds and not merely an indirect reflection of genetics. This is well stated by Simpson (1953a), who says that, in the field of evolutionary studies, "it is the organisms, themselves, the phenotypes as agents in a changing world, and the populations of these organisms that really concern us. The implicit genetic factors are not important to us for their own sake, but only because they are among the various determinants of phenotypic evolution."

1. *Exomorphic and Endomorphic Characters*

I. W. Bailey has commented on the time and effort expended by many workers to show that certain organs or features of the plant are inherently more conservative than others and therefore should have a more dominant role in classification and phylogeny. Endomorphic features, it is argued by some anatomists, are more important than exomorphic structures. Analysis of the Angiosperms as a whole shows that such an assumption cannot be justified and that each morphological attribute is fundamentally similar. When looked at under the microscope or with the naked eye, some are found to be relatively stable in certain groups and show considerable variation in others. Bailey concludes that endomorphic characters such as anatomy are neither more nor less reliable than exomorphic features: they are of equally valid use. They give us, as he says, more strings to our bow (Bailey, 1951).

Hooker, who is quoted at the beginning of the previous chapter, had some scathing things to say about the use of trifling characters for delimiting genera and species. In this he is echoed by van Steenis (1948), who gives a list of genera which come into this category. It is, however, evident that they are judging importance of characters in terms of their constancy of occurrence in the groups they purport to delimit. There is no *a priori* reason why any inconspicuous character should be *inherently* less important than some more obvious feature. In other words, when talking about the relative importance of characters the purpose for which they are being employed should be indicated. Features may be highly important diagnostically due to their constant occurrence, but may be adjudged trivial because their smallness or inconspicuousness causes difficulties of recognition. Importance can therefore be assessed in terms of practical value.

In the case of the traditional morphological characters the current tendency is to use greater precision and more careful examination, and employ biometric and various graphic techniques for the assessment and presentation of the results. Such painstaking examination is often extended to formerly neglected characters, especially those that require the use of a hand lens or low-power binocular dissecting microscope. An often unsuspected variation is revealed in superficially similar features such as seed coats, anther filament appendages, indumentum hairs, etc. This reflects the problem, already discussed, of defining what is a character. One result has been that microspecies are now more readily accepted on the basis of small but apparently constant differences. Conversely, the separation of cytodemes has forced taxonomists to seek differences between morphologically similar populations; the marker characters are frequently features such as anther length, pollen diameter, etc. This is considered further in connection with polyploids on p. 218.

2. *Neglected Morphological Characters*

It is widely assumed today that the morphology of the Angiosperms is adequately known, but as Tomlinson (1962) has pointed out in his studies on the morphology and anatomy of the palms, this is far from true for many groups and organs. "In the present era of specialised and sophisticated botany," he comments, "the plea for more research involving simple morphological observation is likely to go unheard." Descriptive botany is not regarded as a likely source for startling new discoveries and tends to be regarded as unscientific and unprofitable. The implications of this attitude for taxonomy are alarming, since progress in our knowledge of groups depends to a large degree on the accumulation of facts which may be unexciting in themselves but which add up to significant advances. While it is understandable that scientists will prefer to devote their researches to "growing point" subjects where there is a chance of a break-through, taxonomists must also be concerned with the consolidation of information in branches of study which have long since been bypassed. Tomlinson (1962) notes that in tropical botany "there is enormous scope for research which requires the simplest apparatus: an axe, a knife and a scalpel, a clear eye, and the patience and willingness to settle down and observe elemental things". Even in temperate floras the gaps in our knowledge of simple morphological fact are often abysmal. Many morphological features are in fact simply neglected and comparative morphology is in danger of rapidly becoming a Cinderella discipline.

Neglected features may be vegetative or floral. We shall consider some of them briefly, emphasising those features which have been most neglected. By giving this survey, our intentions are to widen the scope of further study, and to draw attention to additional features that may help not only in discrimination but in assessing relationship when faced with conflicting taxonomic evidence. The reasons for the neglect of such features are varied. They may be difficult to see, such as nectaries, staminal filaments and lodicules, difficult to collect, such as palm-leaf bases, difficult to assess, such as habit, branching, shapes, surfaces, etc. Many such characters can only be studied in the field or in cultivation, and the taxonomist seldom has sufficient time or funds to allow either study to be carried out adequately. Other features, especially ontogenetic ones, are frequently overlooked.

(a) *Vegetative characters.* The majority of flowering plants can be readily identified almost as easily by their vegetative characters as by their floral structure. Vegetative characters make up the greater part of the facies of most species which allows us to recognise them visually. Ecologists are often forced to use non-flowering material for making identifications, yet it is remarkable how few keys to vegetative material have been published, and how few herbaria contain adequate material of seedlings, young plants, etc.

Some of the reasons for the comparative neglect of vegetative characters in taxonomic studies have already been discussed (Chapter 4). An important practical consideration is the greater plasticity they show compared with floral features, which makes their description more difficult. Much of this plasticity has to be assessed quantitatively, and as Heslop-Harrison (in Tutin, 1955) suggests, difficulties arise when we have to rely on size (and frequently, as a consequence, on statistical methods) for the recognition of units, while distinction is made much more feasible and acceptable when "either or" characters can be found. Runemark (1961) doubts whether taxa of any rank should be based on statistical characters derived from population analysis, but at least a minority of taxonomists would disagree with this view. Certainly quantitative or statistical differences in vegetative (or other) characters ought to be employed in deciding whether there is discontinuity of variation, but a presence or absence feature is for practical purposes desirable as a diagnostic.

It should not be thought, however, that all vegetative features are difficult to assess, and we will now consider some of the ways in which they may be employed. There is no doubt that greater attention to characters other than those pertaining to flowers and fruits would in many cases improve our taxonomy.

1. *Habit.* It is much easier to recognise habit than to describe it in many instances. Since it is often difficult, if not impossible, to assess from herbarium specimens, it is necessary to make adequate field notes, accompanied by photographs whenever possible. In taxonomic descriptions habit is seldom treated adequately, and it is interesting that it often receives greater attention from ecologists in the various life-form classifications that have been proposed. These concern such features as size, branching, spread, density, leaf outline, size and texture, rooting system, means of propagation and perennation and periodicity (Dansereau, 1957). The best known system is that of Raunkiaer (1934), which has been further refined by Braun-Blanquet (1932) and Dansereau (1945, 1957), and a survey is given by the latter author and a detailed consideration by Du Rietz (1931).* Although at first sight, life-form may appear to have little relevance to taxonomy, its significance is twofold: (1) it can be used as a means of describing and comparing the various habit features of different plants; (2) it allows a means of estimating adaptiveness and ecological adjustment to the habitat. Dansereau (1957) points out, for example, that "the phanerophyte-chamaephyte-hemicryptophyte sequence cannot be considered always a linear phylogenetic order although in some cases it is". He cites Meusel's derivation of Mediterranean rosette plants in *Echium* and in various members of the *Compositae* from chamaephytes of the candelabra form (Meusel, 1952). An interesting discussion on habit differences in *Dionysia* (*Primulaceae*) is given by Wendelbo (1961a) in a monograph of the genus.

Lems (1960) has shown how, in the Canary Islands, adaptive radiation in *Aeonium* (*Crassulaceae*) rests upon the diversification of growth form in relation to different habitats—a botanical example comparable to the evolution of Darwin's finches in the Galapagos. Though *Aeonium* provides an exceptionally striking example, adaptive radiation with respect to growth forms is no doubt widespread in the Angiosperms—the sections of many genera, for instance, are based upon such characters. However, the evolution of different life forms need not necessarily lead to speciation. Böcher (1963) draws attention to the variation in life forms and periodicity found *within* different species, and the frequent impossibility of treating this taxonomically.

* Reference should also be made to the graphical methods for comparing plant shape devised by Jentys-Szaferowa (1959), also applied by Białobrzeska & Truchanowiczowna (1960) to European *Betula* spp.

Ecotypic differentiation is often expressed in the characteristic life-form of the populations adapted to particular habitat conditions (cf. Chapter 12). Prostrate maritime ecotypes are known in many species and it is interesting that some of the most precise descriptions of habit are to be found in genecological studies (cf. Gregor 1938, on *Plantago maritima*; Grant & Hunter, 1962, on *Calluna*).

An example of the failure of the herbarium taxonomist to appreciate the significance of habit differences relates to *Cytisus reverchonii*, from south-east Spain, which was described by its collector as clearly distinct from *C. scoparius* in its compact rounded habit with all the branches curved in towards the centre, not irregularily branched outwards. On the basis of this feature, together with its unifoliolate leaves, Degen published it as a distinct subspecies but later authors dismissed it as of little or no significance. Yet in collateral cultivation at Liverpool both the habit and leaf differences are fully maintained.

Reference should be made to Corner's remarks on the shape of trees in the introduction to his *Wayside Trees of Malaya* (1952), a book full of taxonomic interest. Shape of trees depends, he says, on the shape of their crowns (round and bushy, oblong or cylindric, umbrella-shaped or flat-topped, etc.); branching is the significant factor in the recognition of trees by shape. If the form of growth of the leader shoot (monopodial or sympodial) is considered along with leaf arrangements on the twigs, a classification of tree habit types can be made.

Primary characters used by forest botanists in the tropics include presence or absence of buttresses (allowing certain families to be eliminated), and bark slashes which reveal whether the bark is soft or hard, its colour and the colour of the exudate juice.* In temperate floras bark characters (colour, texture, thickness, fissuring, etc.) are widely used for species recognition, as in *Pinus* and *Betula*.

The posture of leaves has been much neglected (cf. the role of ascending and applanate foliage in the Durian theory, Corner, 1949a), as well as that of branches, flowers and fruits, yet they are often highly characteristic of a taxon.

Much variation in growth form is, of course, environmentally controlled. The position of the resting buds can be directly influenced by temperature or moisture. The borderline, for instance, between chamaephyte and hemicryptophyte is frequently crossed within the same biotype, which only goes to emphasise what a delicate indicator of climatic conditions Raunkiaer's life forms (1934) can be. Such phenotypic plasticity is no doubt an important factor in enabling a species to become widely distributed, but should not provide a basis for taxonomic discrimination.

2. *Underground organs.* The underground parts of the plant often provide valuable characters for taxonomic discrimination, yet they frequently do not

* A survey of bark characters is given by Whitmore (1962a, b).

receive the attention they deserve due to bad collecting.* The form of the root structure is widely employed in the taxonomy of the genus *Ranunculus*. Davis (1960), for example, has divided the Turkish species of Subgen. *Ranunculus* not into the customary sections (which require revision on a world basis) but into five groupings, based largely on rootstock and habit, that are rather closely correlated with ecological preferences. When we collect a buttercup it may be in fruit or flower, but there is no reason why we should not make an effort to collect sufficient of the basal portion to allow its characteristics to be recognised. The presence or absence of a horizontal rootstock is a critical feature in the recognition of the subspecies of *Ranunculus acris* in Europe and is as important as achene morphology.

In the genus *Aristolochia* the shape of the main root—whether globose, ovoid, elongated and cylindrical, fusiform or napiform—is a constant character of major importance in delimiting species (Davis & Khan, 1961). The perennating winter buds sometimes known as *turions* are valuable in the taxonomy of some species of the genus *Potamogeton* (Hagström, 1916); bulb structure is important in *Allium*; and mention must be made of the important series of studies by Chouard (e.g. 1931) on the *Liliaceae-Scilleae* in which great emphasis was placed on the bulb and other vegetative features in the classification of the group.

Root structures are, of course, difficult to collect (see Chapter 8) and awkward to deal with in the press. But while this may explain past neglect it is no justification for continuing inadequate collecting procedures. If palm leaves can be successfully collected, surely rootstocks, corms and tubers should not defeat taxonomists. If they allow themselves to be defeated, valuable characters will be thrown away and classifications be less balanced.

Evolutionary trends in underground organs have been described in many groups, especially those referring to the perennial-biennial-annual sequence (see Chapter 2).

3. *Leaves*. No attempt will be made to survey the multifarious characters provided by leaves, but some of the more neglected aspects will be considered briefly. Leaf characters are extensively employed for the recognition of species in many genera in preference to floral features. Examples are *Alchemilla* (Walters, 1949), *Ulmus* (Melville, 1939, 1951, 1955; Richens, 1955, 1958), *Betula* (Natho, 1959; Jentys-Szaferowa, 1937, 1950a, b, 1952). In some cases complex biometrical methods are needed for dealing with leaf measurements and in consequence some of the species based on such features have failed to gain acceptance. In groups with reduced floral structures, such as the *Gramineae*, vegetative features inevitably play a great role in classification. The problem in the *Gramineae* is aggravated by the apparently narrow gamut of variation in leaf type, although a whole range of foliar characters (epidermis, stomata, histology, etc.) has been uncovered by detailed study. Some of these are considered below.

* A useful bibliography is given by Pénzes (1960).

Leaf shape in groups with divided leaves is difficult to describe and consequently difficult to visualise from the descriptions. This is especially true in the *Ranunculaceae* (e.g. *Ranunculus, Pulsatilla, Thalictrum, Delphinium,* etc.), the *Umbelliferae* (*Chaerophyllum, Caucalis, Oenanthe,* etc.) amongst others, and the position is often complicated by variation in the degree of pinnation (pinnate, bipinnate, tripinnate) which may vary from plant to plant within a population, and even from year to year in biennial or perennial species. To this should be added, of course, the notorious plasticity of the leaves of many species (see Chapter 10).

Leaf shape frequently shows a wide range of variation from the base of the plant to the apex, and especially in trees varies on different kinds of shoot. This makes it very necessary to ensure that comparable leaves are chosen for taxonomic evaluation.

Basal leaves often show peculiar features. In *Erysimum* (*Cruciferae*) there is a tendency for the basal leaves to become crowded and assume a rosulate appearance, simulating species with true rosettes, e.g. *Erysimum grandiflorum* which produces pseudo-myriophyllous forms resembling closely *E. myriophyllum* vegetatively.

Ptyxis, i.e. the rolling or folding of individual organs such as leaves, petals, etc. in the bud, was carefully observed by Lindley (1830), but nowadays receives far less attention than this interesting character deserves, although a synopsis was given by Moll (1934: 153). It is, of course, only readily seen in living material and usually only for a short period of the year, which no doubt largely accounts for its neglect by taxonomists working with herbarium material. Yet a walk round a bota nic garden in spring, observing the unfolding leaf buds, will make immediately apparent the taxonomic value of leaf ptyxis. The diversity is particularly great in the *Rosaceae*, where certain woody genera and sections in the *Pomoideae* may be distinguished by this character. The series of *Galanthus* are diagnosed by their ptyxis (Stern, 1956). It is considered of major taxonomic importance in *Primula*, as is reflected in the grouping of its sections into the "*Involutae*" and "*Revolutae*" (Smith & Forrest, 1928); involute ptyxis seems to be the primitive state in the *Primulinae* (Wendelbo, 1961a). The *Palmae* have been divided into two groups based on the way (cf. p. 150) in which the plicately-folded prototype palm leaf has segmented (Tomlinson, 1960). The term *vernation* refers to the manner in which the individual leaves are arranged within the bud before it expands. As the manner of unfolding is related to the arrangement in the bud, the terms *vernation* (Linnaeus' *foliato*) and *ptyxis* have often been treated as synonymous. However, as these characters are not necessarily functionally correlated, we prefer to recognise them here by different terms.

Ptyxis in young leaves is a difficult character to see, especially in dried material, and is sometimes misinterpreted. Wendelbo (1961a) has paid considerable attention to leaf ptyxis in his revision of the genus *Dionysia* and discussed its relationship to *Primula* in the light of such features. Sections

Dionysiopsis and *Epicamptophyllum* were described by Melchior (1943) as having involute ptyxis while in section *Anacamptophyllum* it was revolute. Wendelbo finds, however, that there is no significant difference between the species of the two sections in leaf ptyxis, and explains the involute or plain ptyxis of Sect. *Epicamptophyllum* as follows: the leaf lamina is greatly reduced in size and of a rather thick texture; consequently young leaves have little or no possibility of bending the margin in either direction, and what slight involution can be sometimes seen is a mechanical bending as a result of the close imbrication of the leaves. Dimorphism is observed in some species: in *D. revoluta* and *D. aretioides*, for example, the leaves of new shoots are flat-margined while those of the overwintering rosettes have revolute margins.

Ptyxis has been so neglected that we are providing a synoptic key to the major types drawn up for us by Dr. J. Cullen; observed examples (referring to leaves unless otherwise stated) are cited in parentheses. Intermediate states are not uncommon, as between conduplicate and supervolute; in such cases it seems preferable to hyphen the terms. In compound leaves it is usually necessary to consider the ptyxis of the individual leaflets separately from that of the lamina. In the key only the most commonly used, or least ambiguous, terms have been employed—the synonymy is extensive.

Synoptic Key to the Types of Ptyxis

A. Lamina rolled or folded parallel to the main vein, or curved
  1. Lamina with only 1 fold, or curved, never rolled
    a. Lamina folded adaxially (*Prunus laurocerasus*) . *Conduplicate**
    b. Lamina folded abaxially . . . . . *Replicative**
    c. Lamina curved into an arc (petals of *Ranunculus*) . *Curvative*
  2. Lamina with more than 1 fold, or rolled
    a. Lamina variously rolled
      a. Margins rolled adaxially (*Pyrus, Salix*) . . *Involute*
      b. Margins rolled abaxially (*Rumex, Rosmarinus*) . *Revolute*
      c. One margin rolled inside the other (*Prunus armeniaca, Musa*) . . . . . *Supervolute†*
      d. Lamina rolled within the lamina of another leaf (*Arum*) . . . . *Convolute*
    b. Lamina variously folded
      a. Two margins folded sharply inwards (tepals of *Clematis*) . . . . . *Implicative*
      b. Many longitudinal folds (*Alchemilla*) . . *Plicate*

* The segments of basically plicate palm leaves are often termed *induplicate* and *reduplicate*, instead of *conduplicate* and *replicative*, according to which way the expanding lamina splits. We could not find a simple leaf with replicative ptyxis.

† Moll (1934) treats supervolute as synonymous with convolute. Since convolute ptyxis (*sensu stricto*) often becomes supervolute at a later stage of development, there is much to be said for this view.

B. Lamina irregularly folded in all directions (petals of *Papaver*) . . . . *Corrugate*

C. Lamina folded at right angles to the main vein or rolled

1. Lamina with 1 fold only
   a. Apical part folded or curved under (*Rhodochiton*) *Reclinate*
   b. Apical part folded or curved over (petals of *Chaerophyllum*) . . . . . . *Inclinative*
2. Lamina rolled from apex to base (*Drosophyllum*) . *Circinate*

D. Lamina flat (*Galanthus nivalis*) . . . . . . *Plain*

Leaf base and associated structures have been found to be of value in some groups, such as the *Leguminosae* (e.g. in *Genista*). The occurrence of a pulvinus in the leaf of the *Papilionoideae* has been surveyed by Dormer, who finds it a primitive feature of the *Leguminosae* and of very considerable systematic and phylogenetic importance. Tomlinson (1962) has made a study of the different kinds of leaf base found in the *Palmae*, an aspect which had been previously ignored. They have been neglected in taxonomic descriptions very largely due to the fragmentary material available. Not only is living material necessary but special techniques have to be devised to deal with buds that may be up to half a metre in diameter! Two main morphological types of leaf base were recognised by Tomlinson according to whether the leaf sheath is obviously tubular or not. These were in turn subdivided into further types which are constant in each species. Their full taxonomic significance is yet to be seen, but it is extraordinary that such conspicuous characters should have been neglected for so long.

The presence or absence of persistent fibres (the remains of leaf sheaths) on the caudex of *Umbelliferae* can be a most useful character for keying out the genera. It could no doubt be accurately expressed in terms of petiolar anatomy and might prove very helpful in delimiting more natural groups.

Stipule morphology has of course provided valuable diagnostic characters for the recognition of species in numerous genera (e.g. *Trifolium* and *Viola*), but there is considerable difference of opinion about the homology and development of stipules as they occur throughout the Angiosperms (cf. Eames, 1961; Croizat, 1960).

Leaf venation has received more detailed study by palaeobotanists than by taxonomists dealing with the contemporary flora. Venation lends itself to illustration more easily than description.* Corner (1958), discussing the leaf venation of *Ficus*, shows how venation patterns can be meaningfully portrayed when their development is understood; this can often be seen "by tracing the manner in which the venation simplifies, or closes up, towards the apex of the adult leaf". This is echoed by Carlquist (1961) who observes that interpretation of the morphological nature of venation is enhanced by a know-

* An excellent illustrated summary of the main types of venation is given in Kerner & Oliver's *The Natural History of Plants*, **2**: 629-634 (1894).

ledge of its ontogeny. He also cites examples of the value of major and minor venation patterns in taxonomic and phylogenetic studies.

Epidermal characters. Palaeobotanists have perforce to rely extensively on features shown by the epidermis of their material. Taxonomists working with living material tend to neglect such characters until other lines of evidence fail, as in the case of the recognition of semicryptic polyploids. The epidermis can provide a host of features,* however, and in recent times examples of their use occur with some frequence in the literature. Walters (1953), for example, used the form and shape of the stomata and epidermal cells in his study on *Eleocharis mamillata* and allied species. In the *Gramineae* their use has reached a high level of sophistication. This is well illustrated by the figures in Prat's recent review (1960). Prat demonstrated in 1932 that one can distinguish a *Festucoid* type of epidermis characterised by its simplicity (siliceous cells of simple shape; no bicellular hairs) and a *Panicoid* type characterised by complicated siliceous cells, and bicellular hairs; this second type is further subdivided. In addition, there are special elements which occur only in certain taxonomic groups below the rank of subfamily: and there are those features which are present in all groups.

The presence or absence of these features allows *structure* of the epidermis to be characterised and permits a division into major groups. To pursue the analysis it is necessary to consider the shape of each kind of element which can vary from genus to genus, and their distribution, which can be expressed in a schematised way as a *dermogram*. The dermograms may be indicated for each species under consideration as a *dermotype* which Prat suggests may be used in a way comparable to karyotypes. This idea might well be explored for other groups of plants.

There are numerous examples of the value of particular epidermal features in the taxonomy of grass genera. Sörensen employed the structure and size of the epidermal cells with considerable success in his revision of the Greenland species of *Puccinellia* (1953); Tutin (1955) extended the examination to the British species and also found that they allowed the separation of otherwise difficult taxa. The genus *Puccinellia* can moreover be distinguished from *Glyceria* by the "ripple" (instead of smooth) walls of the intercostal long-cells (Church, 1949).

Borrill (1961a) has investigated the taxonomic value of epidermal papillae which are found in one of the Mediterranean races of *Dactylis glomerata* "subsp. *hispanica*". Populations allied to the race in question, var. *hackelii*, were found in nineteen localities, and Borrill concluded that those having this feature in common formed a natural group which he distinguished as a new species, *D. marina*. Borrill regarded it reasonable to consider *D. marina* as having arisen from a single ancestral group of tetraploids sharing a pool of genes which was marked in the phenotype by epidermal papillae. As a putative diploid ancestor he proposed *D. smithii* from the Canary Islands

* The review of Linsbauer (1930) should be consulted.

which resembles *D. marina* in a number of features and has a proportion of epidermal cells with papillae. Another possible ancestral type is *D. ibizensis* from the Balearic Islands, although this does not have epidermal papillae. He comments, however, that Stebbins had informed him that there was evidence to suggest that the F_1 hybrids between *D. ibizensis* and *D. smithii* tend to produce more papillae than *D. smithii* alone. In view of the unnecessarily limited amount of material examined by Borrill, we feel that his arguments and conclusions must be regarded as decidedly tentative.

Borrill (1961b) also investigated epidermal characteristics in the diploid subspecies of *Dactylis glomerata.* He found that certain features of the cells were of value in grouping populations and the groupings agreed with the geographical distribution of the component subspecies.

In his posthumously published studies in *Agrostis* and related genera, Björkman (1960) paid considerable attention to the presence of an epidermal network in the lemmas of many species (the so-called *Trichodium* net). This was found to have considerable taxonomic value in delimiting groups in this very difficult complex of grasses. All *Agrostis* species with a minute palea possess the network. It was reported in *Calamagrostis tenella* by Paunero (1953), but in the rest of the genus the network is absent; Björkman used this as part of the evidence for transferring this species to *Agrostis* (as *A. schraderiana*).

Stomata and associated epidermal cells are providing an increasingly important source of taxonomic characters. The structural features of the individual stomata may be important, but equally valuable may be their patterns of distribution. Presence or absence of stomata on the upper leaf surface is often a good diagnostic feature: it is a constant character separating, for example, *Alnus subcordata* and *A. orientalis* which are otherwise difficult to distinguish. Watson (1962) has surveyed the taxonomic significance of stomatal distribution and morphology in the *Epacridaceae* and found that members of the tribe *Styphelieae* typically possess only adaxial stomata (on the sepals) while in the *Epacrideae* they are usually abaxial. This agrees with Bentham's primary division of the family based on ovary and fruit characters.

The evidence which is accumulating concerning the value of the morphology and arrangement of the epidermal cells adjacent to the guard cells as a taxonomic guide in the Monocotyledons has been pointed out by Metcalfe (1961). In many, if not most, Monocotyledons the stomata are of the *paracytic* type with a subsidiary cell on either side lying parallel to the guard cells. In other Monocotyledons the two epidermal cells at the narrow end of the stomata differ in appearance from the other cells of the epidermis, so that each stoma of this type is accompanied by four subsidiary cells—two lateral and two polar. The term *tetracytic* was proposed for these. In other cases the stomata were *anomocytic*, i.e. not accompanied by recognisable subsidiary cells. These have a more restricted taxonomic distribution. A fuller review of the variation in these subsidiary cell complexes in the leaves of Monocotyledons was undertaken by Stebbins & Khush (1961). They found that the pattern of development was characteristic for each of the cell complexes and was as important taxonomically as their final form. The type with many subsidiary cells was considered by them to be probably the most primitive from which other types had been derived, and they drew a number of phylogenetic conclusions concerning the relationships and taxonomy of the current systems of Monocotyledon classification. Within the limitations of the restricted amount of material examined, the type of subsidiary cell complex appeared to be constant within a genus and usually within a tribe or family. A much fuller investigation is required, however, since some variation in these subsidiary cell types is to be expected and has in fact been found by us in several groups examined.

Epidermal and stomatal characters are widely employed to separate members of polyploid pairs (cf. Chapter 6), but precautions are needed since the *size* of stomatal cells, often used as an indicator of the degree of polyploidy (Sax, 1938), varies according to the age of the plant and the part chosen (Schwanitz, 1951). With this in mind, Brandt (1961) in a cytotaxonomic study of *Veronica prostrata* made stomatal measurements on leaves collected at the same time of day from plants at the same stage of development. The leaves were situated on the same part of the stem and all the stomata were measured in a median sector of the lower surface of the leaves. When comparing stomatal size in herbarium material Brandt found that further factors have to be taken into account such as the age of the plants, time of collection, state of preservation, quality of drying, etc. The size of epidermal cell surfaces of the corolla proved more valuable in the Veronicas as an indicator of the degree of polyploidy in dried material, clearly separating subsp. *prostrata* ($n=8$) and subsp. *scheereri* ($n=16$). Reference should be made to Brandt's paper for details of the very careful techniques he employed.

4. *Indumentum*. Great diversity is shown in the indumentum of many groups which has often great taxonomic significance. While the degree of hairiness is usually much affected by environmental factors (Chapter 10), the nature of the hairs, trichomes or scales is found to be more or less constant in many species.* In the *Cruciferae* the type of hair—simple, branched or stellate—was used by O. E. Schulz (1936) as a major criterion, along with nectary disposition and shape, in the subdivision of the family into tribes, genera, etc. It is, for example, one of the features separating the closely related genera *Hutchinsia* and *Hornungia*.

The extent of variation in such characters in the *Cruciferae* is not yet known in sufficient detail, and in some species the presence of particular hair types varies in different parts of the plant. In *Alyssoides utriculata*, for example, the distribution of stellate and bifurcate hairs in the indumentum

* A very detailed account of the morphology, development and function of trichomes has recently been published by Uphof (1962).

of the leaves appears to vary according to a geographical pattern and is not closely correlated with other features. Plants from the Alps have stellate hairs on the rosette leaves, while plants from Italy, Romania and the Balkan Peninsula have bifurcate hairs as well and are often distinguished as a separate species, *A. graeca*. In Bulgaria the bifurcate hairs occur also on the stem leaves.

In the genus *Digitalis* some species have only glandular hairs (e.g. *D. grandiflora*, *D. nervosa*, *D. lutea*), while others have a mixture of glandular and eglandular hairs (e.g. *D. purpurea*, *D. lanata*). Presence or absence of glandular hairs has long been used to characterise the various components of the *D. purpurea* group (Martínez, 1931) and detailed studies of the distribution of the different kinds of hairs on the sepal margin has lead to some interesting results. Not only may the hairs bear glands or not, but there are modes at different numbers of cells per hair; and there are statistically significant differences between the distributions of sepal hairs of northern and southern populations of *D. purpurea* in the Iberian Peninsula, and between populations from north and south of the Pyrenees. These hair distributions are correlated in some cases with other features, such as seed colour, and are likely to provide a basis for taxonomic distinction (Evenson, Heywood & Spilsbury, unpubl.).

The evidence derived from a study of the trichomes in *Rhododendron* (Cowan, 1950) conflicts with the views put forward by Hutchinson (1946) as to the primitive species. He considered that the non-lepidote species with 7-10 petal lobes and numerous stamens (Series *Falconeri*) were most primitive, but these have the most elaborate trichomes. In view of pentamery being the usual condition throughout the *Ericaceae*, it is difficult to believe that numerous floral parts is not here a derived character. The trichomes do, however, afford useful characters for taxonomic separation at infrageneric and specific levels. A later study on trichome form in the genus has been made by Seithe (1960) and has considerable taxonomic implications.

The complex indumentum found in the *Labiatae* has long been recognised as a most valuable character in the discrimination of critical species (as in *Stachys*, *Salvia* and *Scutellaria*). In some groups, however, it needs to be used with caution. In *Geranium*, for example, glandular versus non-glandular hairs on the pedicels apparently provide the only diagnostic character separating two geographical subspecies of *G. ibericum* in the Near East; yet in the same area *G. collinum* produces glandular and eglandular pedicels on the same individual!

To be of use for critical discrimination, indumentum must, of course, be described in detail, each type of hair present (and its disposition) being separately recorded. The distinctions can be very subtle but are often remarkably constant. Townsend (1962) finds that *Galeopsis ladanum* and *G. angustifolia* (*Labiatae*) are best distinguished by the nature of their calyx indumentum; the hairs of the former are smooth and glistening while those of the latter are dull due to their walls being covered with minute papillae.

(b) *Ontogenetic characters*. We stressed at the beginning of this chapter that all stages of development of the plant should be examined for possible characters. Ontogenetic features have not received much attention from taxonomists and a brief introductory paragraph concerning the principles of ontogeny follows.

The relationship between phylogeny and ontogeny (the development of the individual) is a subtle and complicated one which has been less studied in plants than in animals, where developmental stages are more clear-cut. De Beer (1951) rejects Haeckel's so-called "biogenetic law" which accepts ontogeny as a brief recapitulation of phylogeny, considering the claim to be misleading. On the other hand, it is certainly true that embryos and young plants of related forms usually resemble one another more closely than the adult stages.* We will cite here some of the conclusions so clearly stated by de Beer (1951):

> Phylogeny is due to modified ontogeny.
>
> New characters may appear at all stages of ontogeny, and by *heterochrony* may be retarded or accelerated, so as to appear later or earlier in subsequent ontogenies.
>
> Characters in early stages of ontogeny have (provided they are not too specialised) played an important part in evolution by *paedomorphosis* (i.e. by neoteny—discussed below—and embryonic deviation), resulting in large structural changes without loss of plasticity.
>
> Characters present in the late stages of ontogeny have played an important part in evolution by *gerontomorphosis* (i.e. phylogenetic change as a result of modification of characters which were already adult, by means of adult variation, hypermorphosis and acceleration) resulting in relatively small structural changes with loss of plasticity.
>
> Paedomorphosis and gerontomorphosis may act alternately in the phylogeny of a race, the former producing racial "juvenescence", the latter racial "senescence".
>
> *Recapitulation*, i.e. the pressing back of adult ancestral stages into early stages of development of descendants, does not occur.
>
> Paedomorphosis explains gaps in the fossil record.

Medawar (1955) believes that neoteny, so far from being a curiosity, is an entirely fundamental tactic of the evolutionary process. He notes, too, that "most of the affinities that were at one time considered good evidence of recapitulation can now be treated as paedomorphic transformation: Haeckel is still the hero, though his portrait hangs upside down". Although written in a zoological context, this may also be true in plants.

In the seedlings of many vascular plants, the leaves first formed by the apical meristem are often of more primitive structure than those formed later. Well-known examples are the scale-leaved Junipers which produce long seedling leaves resembling the mature foliage of other Conifers, and the

* Carlquist (1961: 33) illustrates the way in which "ontogeny recapitulates phylogeny" in the glandular trichomes of *Blepharizonia plumosa* (*Copositae*).

phyllodic Acacias which bear bipinnate leaves in the seedling. These plants mostly grow in arid climates with marked seasonal distribution of rainfall: the seedling stage, with less specialised leaves, grows under more mesophytic conditions than the adult. In *Cytisus reverchonii* the first seedling leaves are trifoliolate, but by the 3- or 4-leaf stage all the leaves are unifoliolate as in the adult plant; in the latter there is occasional reversion to trifoliolate leaves on juvenile shoots. In many plants with divided leaves, however, the seedlings leaves are probably simple or less divided than those of adult plants and grade into them by a series of connecting forms. In herbaceous perennials a rather similar sequence is repeated annually by rejuvenescence. Though a genetic character of the species, the leaf series is easily modified by the environment, especially by light intensity and water supply. Shading tends to prolong the juvenile phase and dry conditions to accelerate the oncome of adult foliage and the reproductive phase (very obvious in annuals). Although there may be a general tendency for more "primitive" leaf types to be found in the seedling than persist in the adult, there is no doubt that both stages need to be in harmony with the environment in which the species grows.

The probable basis of embryonic and seedling similarity in plants is summarised by Stebbins (1950) as follows: "First, mutations which act relatively late in ontogeny are less likely to disorganise the whole process of development and so have a general deleterious effect than those which alter the earlier stages. The mutations established will therefore usually be those which effect development at the latest possible stage for the modification of the mature structure in the direction of a selectively advantageous shape. That part of ontogeny which is least modified by the sum total of these late-acting genes will show embryonic similarity. Second, most mutations which alter the form of the mature organ change the rates of growth processes. Therefore, the longer the genetically different organs grow, the greater is the difference between them at maturity: and, conversely, the younger they are the more alike they will be. Finally, the fact that seedlings often live under different environments from adult plants, and usually make different demands on the environment, may cause a particular type of structure to retain its selective value in the seedling after it has already been lost in the adult plant."

A most interesting account of the relation between ontogeny and phylogeny has been given by Takhtajan (1959a), who recognises neoteny as an important phenomenon in the evolution of vascular plants. *Neoteny* (literally, protraction of youth) leads to the premature consummation of ontogeny and may be defined as the attainment of the reproductive phase at a juvenile stage.* Unlike animals, plants grow by an open system of growth. Fixation of

* A distinction is sometimes made between *neoteny* (i.e. retention of some juvenile characteristics by the adult) and *paedogenesis* (i.e. sexual maturity in the juvenile phase). Although this distinction may be drawn (at least in theory) in animals with a metamorphic life cycle, it would be extremely difficult to apply in plants. We follow de Beer and Takhtajan in using the term neoteny in a broad sense.

an early stage of development in the higher plants usually implies the fixation of its lower tiers of appendages and the dropping out from ontogeny of tiers that would have arisen later. Whole parts of a plant may therefore be eliminated if it flowers and fruits at an early stage of development. A striking example is found in some species of *Streptocarpus*, one of whose cotyledons becomes immensely enlarged and persistent, giving rise to the inflorescence on the base of the midrib. Many examples are known of trees blooming at the seedling stage, and it has often been claimed that the herbaceous habit in the Angiosperms has been derived from the woody habit by neoteny: the transition may be either gradual (via the shrub) or abrupt. In either case neoteny results in the arresting of the herbaceous stage of stem structure.

Takhtajan suggests that neoteny has been of the greatest importance in plant evolution because it provides a way in which highly specialised groups may retain their evolutionary plasticity and give rise to a totally new line of development. Paedomorphosis could account for some of the discontinuities in Angiosperm variation. Absence of transitional forms at various levels of the hierarchy need not always be the result of extinction or the imperfections of the fossil record. It may result from neoteny and more or less sharp embryonic divergences leading to a discontinuity in the sequence of adult organisms. Phylogenetic sequences may therefore be partly discontinuous, with major breaks which correspond to the stages of paedomorphosis in the evolution of groups. Takhtajan suggests that the female gametophyte (embryo sac) of the Angiosperms has probably arisen by neoteny from that of the lower Gymnosperms. Also the Monocotyledons have probably been descended from dicotyledonous ancestors by the substitution of one cotyledon for two at the very earliest stage of development, thus giving rise to a whole Subclass. A close relationship has been recognised for some time between the *Araceae* and *Lemnaceae*. The adult structure of *Lemna* is so similar to the seedling *Pistia* (*Araceae*) that the evolution of *Lemna* by paedomorphosis from the aquatic monotypic genus *Pistia* seems a likely possibility (see Takhtajan 1959a: fig. 10).

It can hardly be doubted that mutations resulting in the retention of juvenile foliage into the adult stage can explain some of the taxonomic differences which separate species of Angiosperms. Seedling leaves are lobed or divided in all species of pears (probably a primitive condition in the *Pomoideae*), but whereas in *Pyrus communis* seedling leaves of this type are quickly replaced by entire leaves, in other species (e.g. *P. pashia* and *P. regelii*) lobed or pinnate juvenile foliage persists in the adult or reappears annually on juvenile shoots (Rubtsov quoted in Takhtajan, 1959a). Vassilzenko (1961) illustrates examples between allied species within *Sophora* and *Glycyrrhiza* (*Leguminosae*) in which simple juvenile leaves are retained later in development. Although botanists usually consider the bracteate raceme a more advanced condition than a series of flowers borne singly in the leaf axils, the latter could also be a derived state. In those specialised annual species of *Veronica* Sect. *Alsinebe* (e.g. *V. persica*) with solitary axillary flowers, the condition might well have arisen by neoteny from ancestral types with bracteate racemes, flower-formation taking place at an earlier stage of development in association with the evolution of the annual habit in arid climates.

A neotenic relationship apparently exists between some annual species of *Ranunculus* in which the progression from simple to dissected leaves has been arrested at an increasingly early stage in the series *R. scandicinus—R. cornutus—R. marginatus*. A somewhat similar kind of relationship, involving annual rejuvenescence, exists between allied perennial species. The outer rosette leaves (in the nature of juvenile foliage) are usually simple or much less divided than those formed later in the rosette or in the lower parts of the flowering stem. In some species this serial progression is either telescoped, so that the change-over to dissected leaves is rapid, or protracted so that the rejuvenescent leaf-form persists into the later stages of development. Tomlinson (1960) has described similar relationships in the palms. In such cases foliar differences are presumably controlled by genes which alter the timing at which particular leaf-forms are produced, thus extending or telescoping the inherited sequence (Davis, 1960). Such trends can presumably go in either direction, depending on selection pressure. The pinnate leaves in *Fraxinus* and *Pistacia* appear to show a similar sort of relationship. Allied species and varieties differ from one another in the number of leaflets, but also show a serial progression in the same character according to the leaf's position on the shoot.

Angiosperm classification is traditionally based on the mature plant in its reproductive phase. Yet genetic differences and resemblances can find expression at any stage in the life cycle, and may be of vital selective value in the seedling stage. Melville (1953) has drawn attention to the serial changes in leaf-form that distinguish one species from another (as in *Gossypium* and *Epacris*) and calls it the *leaf spectrum* (cf. p. 333). It would be a welcome development in taxonomy if it became standard practice to describe not only the reproductive phase of plants, or even to illustrate or graph the leaf spectrum, but to describe the stages of their ontogenetical development, preferably under controlled conditions. This would show up additional differences and resemblances, thus providing more characters for a natural classification, and could not fail to be biologically interesting. The contention of Orton (1955), though made in a zoological context, surely applies to biological classification in general: "the full characterisation of a species comprises all of its genetically regulated attributes throughout its ontogeny and thus, ideally, judgment in taxonomy and evolution should be based on a unified summary of the data available from the entire developmental pattern rather than from a single stage".

Embryological studies are beginning to have important taxonomic effects, but seedling development has not had the systematic attention it deserves. We shall no doubt, for convenience, continue to use mainly morphological characters from the adult stage for the practical delimitation of taxa, but ontogenetic characters should be used in the assessment of relationship.

Identification of seedlings is of great importance to plant breeders (Haskell, 1961), ecologists and others. Seedling morphology often provides characters of taxonomic value.* Seedlings are described and figured in all Biological Flora accounts of the British Ecological Society (published in *Journal of Ecology*), an example to be encouraged, but the drawings often stop when development is becoming ecologically and taxonomically interesting (cf. Gilbert & Leonard, 1954). Seedling characteristics played an important role in a study on the geographical variation in *Pinus contorta* (Critchfield, 1957): seedlings from various geographical regions were grown under relatively uniform conditions in a greenhouse and supplemented by field and herbarium studies, an important technique in tree groups where the possibilities of growing plants to maturity are limited.

(c) *Floral characters*. The inflorescence, flowers and associated structures form the traditional basis for much taxonomic and descriptive work and no attempt will be made here to survey the field. A good account is given by Lawrence (1951: 59-85). Despite the apparently privileged role in taxonomy of floral morphology and structure, characters derived from the inflorescence and flowers may in many instances be overlooked, or they may be difficult to appreciate and their significance inadequately assessed. The problems of establishing homologies between organs or features in different groups are especially acute.

A welcome feature of many recent revisions and monographs is the detailed scrutiny to which the parts of the flowers are subjected, often with surprising results. In many groups only inadequate morphological studies of the flower have been made and the reliability of the characters derived from them is uncertain.

An outstanding example of floral morphological studies carried out in connection with a taxonomic revision is that of Sharsmith (1961) on the genus *Hesperolinon* (*Linaceae*). Distinctive features studied were the adaxial petal appendages of considerable complexity, and the attachment of the petals to the rim of the cup, which merit the recognition of the genus as distinct from *Linum*. Sharsmith notes that "from 1877 to 1955 a series of both accurate and false observations have been made on the flower morphology of *Hesperolinon* and the related *Linum*, with both reasonable and unreasonable interpretations of these observations and at the same time an almost total disregard of both by North American botanists". This work is cited an example of the large

* A basic systematic survey of Angiosperm seedling morphology is given by Lubbock (1892) in two illustrated volumes. Surveys of seedling anatomy are given by Gravis (1943) and Tronchet (1930).

scope for original comparative work that remains to be accomplished, often employing little more than the simplest techniques and apparatus, but careful and painstaking observation.

Even such obvious organs as the calyx may provide macroscopic characters of considerable value which have been overlooked. For example, calyx type separates two main groups in the *Anthyllis vulneraria* complex, a feature long neglected in many countries yet easily observed when pointed out (Cullen, 1962). Similar considerations apply to bracts, as in *Calystegia* (*Convolvulaceae*) where the degree of inflation of the bracts and their shape clearly distinguishes species in the *C. sepium—silvatica* complex although past writers have largely ignored them or not appreciated their diagnostic value.

It is most impressive to note how accurately nineteenth-century botanists observed and illustrated detailed floral features, despite the primitive instruments of magnification at their disposal. The binocular dissecting microscope today greatly facilitates the careful observation of the less obvious details of floral structure. For example, Zohary (1956) in a revision of the genus *Tamarix* in Israel made a detailed study of the configuration of the floral disc. This had been investigated previously by Arendt, who concluded that it had no diagnostic value. Zohary found, however, that it was highly constant in certain species and seldom found any marked variability or change in the disc after flowering, as claimed by Arendt.

The *staminal disc* and associated *nectaries* (cf. Fahn, 1953; Frei, 1955) are widely employed in the taxonomy of many groups. In the *Cruciferae* a very wide range of nectary types is found and these are widely employed, together with other features, for generic and tribal classification (Schulz, 1936). A serious problem about this kind of character is the difficulty of observing it and a consequent lack of information about its variability.

Stamens are often neglected (apart from meristic variation) yet they may help to orientate a taxonomic study where variation in other floral features is limited. An example is the genus *Alyssum*, where staminal appendages provide good specific diagnostic characters. Anthers and their appendages are traditionally employed in some groups (e.g. *Ericaceae*, cf. Palser, 1961). Antheral classifications have been important in the genus *Eucalyptus* (Bentham, 1866; Blakely, 1934; cf. Penfold & Willis, 1961). *Staminodes* are valuable in some genera, notably *Scrophularia*, where they often provide the best means of separating otherwise similar species. Again, the extent of variation is seldom adequately known. Filament hairs provide key characters in the genus *Verbascum* (Murbeck, 1933).

Many of these floral features are part of, or connected with, pollination mechanisms. The exact nature of many of these has yet to be elucidated, even in such well-known families as the *Orchidaceae* (cf. the studies of Vermeulen on the *Ophrydeae*, 1959). The problem of homologies is emphasised by Carolin (1960), who has investigated the structures associated with the style within or on which pollen is presented to insects in the *Campanulaceae*, *Goodeniaceae* and *Stylidiaceae*. The similarity in positioning of these structures has been largely the basis for the uniting of these families in the *Campanulales* by some authors, along with the *Compositae* and *Calyceraceae*. The results of his investigations provided evidence for questioning the naturalness of the order: he found that the pollen presentation structures of the three families were not strictly homologous, the *Stylidiaceae* standing quite apart in this respect. There has apparently been convergent evolution in different lines in this group of families in relation to the morphological structures associated with pollen presentation and protection; taxonomic relationships have been partly obscured by these.

Even the *numbers of parts* of the flower are not always accurately observed. When the numbers are high, they are often assessed only in general terms (stamens numerous, carpels many, etc.) and no attempt made at accurate counting, yet they may be of great value in circumscribing groups. Burtt (1961a) in an interesting paper on the functional evolution of the *Compositae* notes how seldom generic descriptions give the *upper* limit of the number of flowers in a capitulum while the *lower* limit is frequently recorded. As he says, there is very little actual numerical data about flower heads containing more than thirty flowers, and the same is true of number of fruits. Similarly, in *Ranunculus* there is often significant variation in the number of petals, stamens and achenes yet few Floras indicate the facts with precision. Races of *Ranunculus ficaria*, for example, are statistically separable on such features (cf. Heywood & Walker, 1960).

A revealing study by Haskell (1949) showed that the number of stamens in *Stellaria media* was quite different from that reported (and copied) in many Floras. The species has normally been described as possessing 3, 5 or 10 stamens, whereas scorings made of populations samples from various parts of England gave means of 3 and 4. We have confirmed the accuracy of Haskell's observations by studying this feature in northern populations of *S. media* (*sensu stricto*) as a class exercise. Part of the confusion stems from the fact that older Floras included *S. neglecta* as a variety, thus accounting for the higher stamen number. Often an inaccurate observation is made through inclusion of more than one taxon within a species, and when in time the other taxon is recognised and removed, the description of the original species may only be correlated to the extent of removing the apparent diagnostic characters of the new species but still leaving in the residue of its attributes. Resulting inaccuracies are frequently perpetuated in Flora after Flora.

Flower colour poses a number of problems, some of which have already been mentioned (p. 118). The chemical basis of pigmentation is discussed in a later section. Colour changes during ontogeny (e.g. in *Boraginaceae*), colour variants in populations, and changes after anthesis (as in *Cruciferae*: *Brassica*, *Rhynchosinapis*, etc.) and on drying for the herbarium (e.g. *Alyssum*,

Lotus) are all phenomena that have to be borne in mind. A useful discussion of variation in flower colour is given by Moreau *et al.* (1960).

The gynoecium and associated features likewise contribute their quota of characters—ranging from the pattern of papillae on the stylar branches in Composites to the degree of pubescence of young ovaries.

Inflorescences. The inflorescence provides useful characters in many genera such as *Linum*, *Bellevalia*, *Alyssum*, *Silene*, etc. Apart from the general type of inflorescence, these characters include differences in posture of the pedicels during and after anthesis, the angle and degree of branching (often showing subtle distinctions), and the manner of opening and development.

Inflorescence types have been classified by Rickett (1944, 1955) and are discussed by Eames (1961), who also considers the phylogeny of the inflorescence. Problems arise in practice through the difficulties of applying the various descriptive terms, especially in herbarium material, and in assessing the degree of variation shown by the inflorescence, especially in respect of degree of branching which may be greatly affected by ecological factors including grazing.

This is well illustrated by the genus *Silene*, where great reliance is placed on the type of inflorescence for making groupings and for the construction of keys to the species. Terminology here is difficult. In the perennial sections paniculate inflorescences are found with more or less opposite branches bearing groups of 3-flowered dichasia. It is possible to derive many of the other inflorescence types in the genus from this paniculate type and Chowduri's arrangement (1957) of the genus follows this logical pattern. It is of significance that no annual species possess this type of inflorescence. Reduction of the terminal panicle produces the narrow spike-like kind of inflorescence in which the opposite branches may be very short-stalked (the inflorescence then being pseudo-verticillate), or they may be reduced to single flowers. Further reductions may suppress some or all of the branches, and this series of more or less irregular alternately branched panicles is terminated by the solitary terminal flower.

Individual variations due to environmental conditions affect the development of the inflorescence in *Silene*, with the result that no key which relies on the form of the inflorescence can be expected to work with more than moderate success, and small or depauperate specimens, whether of annual or perennial species, may key out to the wrong species (Walters).

Species of *Nepeta* have been based largely on the degree of branching of the inflorescence by Pojarkova (*Fl. URSS*, **20** (1954)), but are regarded by Hedge (1961a) as variants which occur sporadically throughout the area of the species complex to which they belong.

(d) *Seeds and small fruits.* Comparative seed morphology has been largely neglected since Gaertner's classical studies *De Fructibus et Seminibus Plantarum* (1788-1805). This has aggravated the difficulties of identifying unknown seeds and much recent work on seed morphology and anatomy has been directed to the production of keys and descriptions for ease of identification. Seeds frequently receive summary treatment in Floras, and even in detailed taxonomic studies their potential importance in providing characters of value for comparative or interpretative purposes is seldom adequately explored. This is all the more surprising since it appears that seeds are relatively stable in their outer morphology and more particularly in their internal organisation. Differences between them are often constant and are valuable diagnostically. A survey of seed structure is given by Harz (1885).

Variations in the external characters of seeds are expressed in shape, size, colour, surface and markings, and appendages such as arils, caruncles and elaiosomes. There is wide diversity in the definition of these terms, which are often applied conventionally to certain groups. Although too numerous to provide a satisfactory basis for a general classification of seed types, they are of great value in separating closely related genera and species, as the following examples show. In the genus *Chenopodium* the markings of the testa provide taxonomically useful characters to separate species although they require the use of a low-power compound microscope if they are to be correctly observed (cf. Cole, 1961). Similarly, such features are invaluable within the genera *Euphorbia*, *Hypericum*, *Montia*, amongst others. A recent example may be taken from the genus *Drymaria* (*Caryophyllaceae*), where Duke (1961) finds that the seed provides "a rather critical indication" of the systematic position of species. The features of the seed that vary between species are shape, sculpturing and colour.

Too much reliance should not be placed on such features, however, without adequate assessment, since it is known that they may be susceptible to environmental modification. Thus New (1959) found that variation in the number of papillae on the seeds in *Spergula* species could be induced by the environment. The constancy of many seed characters suggest they have a high selective value. This is not surprising since adaptations to varying dispersal mechanisms of the seeds would be expected. Detailed studies can often reveal the way in which such adaptations work. H. E. Green (personal communication) has related many such differences to the various methods for natural sowing: considering the need for some degree of soil covering by cultivated seeds, he examined various species to see how this could be brought about in nature. Apart from such obvious adaptations as the hygroscopic awns of *Erodium* species, he found that in small seeds of species growing on exposed ground there was a striking production of mucilage which could have the effect of anchoring the seeds to the ground and attracting soil particles. Such mucilage may be produced in streams from transparent tubes, as in *Hornungia petraea* (*Cruciferae*) when moistened. Similar phenomena occur in *Draba* and other genera. Presence or absence of mucilage delimits natural groups of species in *Alyssum*. All sections have constant mucilage with the exception of Sect. *Odontarrhena* and Sect. *Aurinia*, in which it is almost absent.

Mucilaginous seeds may in fact be used as a character to separate genera of the *Cruciferae* such as *Sisymbrella* and *Nasturtium*.

Another example of the use of seed differences as generic characters is in *Kohlrauschia* and *Gypsophila:* some species of the former genus are remarkably *Gypsophila*-like in appearance but can be clearly separated by their seeds. The configuration of the testa has recently been used by Ball & Heywood (1962) to separate diploid and tetraploid cytodemes of *Kohlrauschia prolifera*. They have shown not only that the tetraploid *K. nanteuillii* can be recognised by its seeds but that it is intermediate in its testa pattern between diploid *K. prolifera* and another diploid species, *K. velutina*, suggesting that it may be of allopolyploid origin. Intermediacy in other features supports this suggestion, but it has not yet been confirmed cytologically.

In some genera such as *Atriplex* (*Chenopodiaceae*), certain species may produce two kinds of seed which differ morphologically and in their germination capacities. Hard and soft seeds are cited by Huxley (1955) as a reproductive morphism. Germination is delayed in hard seeds, while soft seeds germinate without a resting period (Crocker & Barton, 1952). This phenomenon is advantageous in that a single plant, e.g. in *Tuberaria guttata* (*Cistaceae*), may produce seeds which germinate in a few days and others which do not germinate till the following spring and therefore have a selective value as a means of overwintering in a hard season.*

In agriculture the structure of seeds and the form and dimensions of the hypocotyl and other structures are widely employed for the recognition of races of economic plants such as *Brassica* species (Madueño Box & Palau Medina, 1941; Berggren, 1962). Weed seeds are similarly of economic importance and several seed atlases have been prepared to allow the easy detection of contaminant seeds in crop material.

Internal anatomical features of the seed, although more difficult to observe, are even more valuable in taxonomy than external characteristics. The embryo particularly shows a high level of constancy within taxa and is now increasingly used for characterising groups. A summary of seed anatomy of Angiosperm families was published by Netolitzky (1926), and recent important papers include the survey by Martin (1947) of the internal morphology of 1287 genera (mainly from the United States), Isely (1947) who gives the seed characters of seventeen families, and McClure (1957) who added descriptions of a further twenty families.

Apart from the well-illustrated roster of families and genera given in Martin's work, the importance of his study lies in the attempts he makes to introduce more precision in the description of seeds and their various features. He points out, for example, the inconsistency of usage found in the term endosperm, some authors using it to cover any and all non-embryonic

* The adaptations of seeds and fruits are discussed in detail in a long series of publications by Salisbury (e.g. 1942, 1961) concerned with the reproductive capacities of plants. See also Dansereau & Lems (1957).

storage tissue within the seed, regardless of whether or not it is partly or wholly perisperm. Correction of this loose usage would be difficult. On the other hand he gives five size designations to express the relative proportions of embryo to endosperm which are usually loosely implied by such terms as minute, small, or large embryo, scant or copious endosperm. His valuable system is based on the use of quarter units of volumetric proportion of embryo to endosperm (Fig. 14). He also classified seeds into a number of basic embryo types and suggested a series of evolutionary trends.

An elegant investigation of seed structure was made by Corner on the *Leguminosae* with a view to indicating the importance of the seed in the arrangement of the subgroups of the family and their position relative to the *Rosaceae*. Subdivision of the *Leguminosae* into subfamilies (or into separate families) has been the subject of much discussion and the structure of the seed has brought additional characters to bear. Corner (1951b) showed that

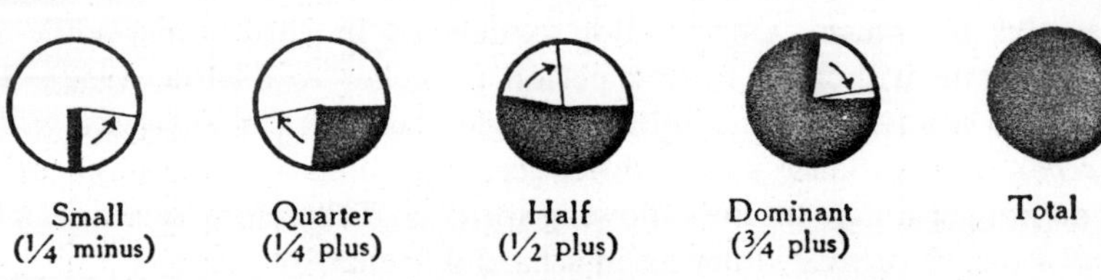

FIG. 14. Diagram of embryo: endosperm proportions. (From Martin, 1947.)

they could be used (*a*) to separate the Leguminous seed from the Rosaceous seed; (*b*) link the Mimosoid and Caesalpinoid seeds; (*c*) distinguish the Papilionaceous seed as a specialization of the Mimosoid; (*d*) recognise the *Swartzieae* as a fourth group (subfamily) linking the other three as was proposed by de Candolle in 1825!

Corner concludes that seed structure was probably of great importance, perhaps even down to generic level in the *Papilionoideae* where features of the vascular supply are used. The Rosaceous seed has none of the characteristics of the Leguminous seed and the alliance of the *Mimosoideae* with the *Rosaceae* is not justified. A key to the various groups is given.

In the *Cruciferae* great reliance is placed on seed characters, together with features of the fruit, nectaries and type of hairs in the indumentum, for generic and tribal divisions. They are also employed to some extent at specific level, but the constancy of such features has not been adequately investigated. Variation within species of such characters as embryo orientation (widely used as a generic feature) is known and there are cases of species being described on the basis of seed characters later found to lack constancy (cf. Heywood, 1961, for example in the genus *Ionopsidium*). As noted by Tutin (1956), the taxonomic changes in the family made by Schulz, based largely on such comparatively trivial features, are largely in agreement with the cytological evidence regarding basic numbers shown by Manton (1932). On the other hand, tribes based on such features may not be natural and sometimes separate what in other characters appear to be closely allied genera. The strictures of Hooker that "notwithstanding all these nice distinctions, the genera of Crucifers as at present defined are often as artificial as they are difficult" are still largely true today. While seed characters may in this family be used as diagnostic features, they should not be weighted to form groups without reference to other characters.

Fruits, particularly small ones, have frequently been treated by taxonomists in a like, cursory manner. In the *Compositae* the cypsela offers many valuable characters as has been already mentioned in connection with adaptation (p. 124). The tendency to dismiss them as trivial and of no great taxonomic value is largely, we feel, a result of lack of detailed study combined with difficulties of observation. In the *Umbelliferae* details of fruit structure have been given much greater weight (cf. Baumann, 1946; Klan, 1947), and Briquet (1916) suggested that a critical study of the carpology of certain Composite groups, completed by anatomy, must become the basis of their systematics, the role of fruit structure becoming analogous with its position in the *Umbelliferae*. In the *Chrysantheminae* fruit characters combined with embryological features have provided a satisfactory basis for the subdivision of the group (Heywood, 1954a, 1959b).

3. *Anatomical Characters*

(a) *Introduction*. Anatomical characters play an increasingly important role in the formulation of natural or phenetic groups. They are also widely used in other aspects of taxonomy and have been largely applied to the elucidation of "phylogenetic" relationships. Wood anatomy has been an especially fruitful source of data and has often revealed possible affinities or has thrown new light on existing taxonomic arrangements, sometimes indicating their weaknesses and possible faults.

The application of anatomical data to the solution of taxonomic problems is not new but as Metcalfe (1954) observes it is only within recent years that sufficient data have accumulated to allow us to draw firm conclusions.*

Bailey (1951) has written a wise paper on the use and abuse of anatomical data in classification and phylogeny. As already discussed, he rejects the tendency to regard such endomorphic characters as being inherently more important than exomorphic ones. The use of these characters depends, he says, on the objectives in mind. In other words, we must distinguish the *purpose* of our study. Anatomical data can be employed either for practical purposes such as identification, group-making, etc., or for studying phylogenetic trends and degrees of relationship. The two should not be confused although

* A valuable source book for the role of anatomical characters is Carlquist's *Comparative Plant Anatomy*, 1961—"A guide to taxonomic and evolutionary application of anatomical data in Angiosperms".

they have become almost inextricably linked, very largely through failure to keep the different purposes separate in the past.

The goal of phylogeny (what is sometimes termed "a truly natural classification") is not, as Bailey says, undesirable, but controversies arise largely due to "excessively circumscribed points of view and to dependence on unreliable premises". These are chiefly the tacit and unwarranted assumptions that plants can be arranged in valid phylogenetic sequences or their relationships can be established on the basis of data from single lines of evidence or from single organs without corroborative evidence. This has been put very clearly by Swamy & Bailey (1949), who say that "before attempting to arrange surviving Angiosperms in phylogenetic series, it is essential to obtain reliable evidence regarding salient trends of evolutionary specialisation in the various organs and internal structures of these plants. Such evidence can be acquired only by comprehensive and time-consuming investigations of the dicotyledons and monocotyledons as a *whole*."

It is against this background that we should examine the contribution of anatomical studies to classification and phylogeny. We should be sceptical of those who propose classifications or phylogenetic trees based primarily on such evidence, especially when they are considered as offering possibly better alternatives to existing *general* schemes.

(b) *Anatomy and the delimitation or identification of taxa.* Although we can trace the use of anatomical characters in taxonomic descriptions back to Caesalpinus, it was not until the mid-nineteenth century that anatomists became concerned with the systematic importance of internal structures of plants. One of the pioneers was Auguste Mathieu, who gave the distinctive features of the wood of each genus and species included in his *Flore forestière* (1858). He was probably the first to propose the systematic value of wood structure. In the work cited he stated that a careful examination of the wood of nearly all indigenous species had shown that their structure was constant throughout a genus and even within a family when it was fairly natural. This viewpoint was taken up later by Solereder (1885) in a work on the systematic value of anatomical structure in the Dicotyledons and in 1899 in his classic *Systematische Anatomie der Dikotyledonen*, later to be translated into English (1908) and succeeded in 1950 by Metcalfe & Chalk's *Anatomy of the Dicotyledons*, which was an attempt to summarise all that was known at that time of the descriptive anatomy of Dicotyledons. A guiding scheme for the description of general and special histology is given by Moll (1934).

Other important treatises include many works on tropical woods such as Lecomte (1918) on the Woods of Indo-China and (1922) the woods of Madagascar, Janssonius on the woods of Java, Perrot on the wood of Madagascar, the Gabon and the Ivory Coast, Wagemann on the woods of Chile, etc. Most of these accounts are well illustrated and contain data of great value for taxonomy. The texts of Esau (1953, 1960), Boureau (1954-57),

Eames (1961) and Bailey's collected papers (1954) should also be consulted.

Much of the work on anatomy has been concerned with the structure of wood, and as Metcalfe (1954) points out this is a weakness since it treats of this as though it were a subject apart from the rest of anatomy. Economic considerations have made this concentration on wood anatomy inevitable and it is not surprising that some of the most important taxonomic results have come from this branch. Before considering these in detail we shall mention some recent studies on the anatomy of other features.

Metcalfe & Chalk suggested in their *Anatomy of the Dicotyledons* that the structure of the petiole might provide characters of taxonomic value and R. A. Howard (1959) has recently published the results of such an investigation. He found in a survey of three-dimensional vascular patterns of the petiole in woody plants features of systematic value both for identification and for assessing relationships. He devised a system of petiole types based on such features as leaf position, presence or absence of stipules and their vascularisation, nodal structure, number of traces, etc. These allowed identification of fertile or sterile material to family, generic or even in some cases specific level.

Leaf anatomy has been widely employed in taxonomically difficult groups such as the *Coniferae* and the *Gramineae*. In *Pinus nigra* it has been known for very many years that the component races (Austrian, Corsican, Dalmatian pines, etc.) differed in their leaf structure and many attempts have been made to define these differences more precisely. One of the latest and most detailed investigations (Vidaković, 1957) uses epidermal and hypodermal structure, breadth and height of the needles, number of layers of the transfusion parenchyma as well as the more traditional feature, number of resin canals. The differences were computed statistically and expressed in μ and a key provided. Not surprisingly difficulties are still experienced in determining the races within this complex group! In *Pinus halepensis* differences in leaf anatomy do not appear to be at all correlated with other characters. Other examples in the *Coniferae* will be found in the works of Gaussen (1942-60).

In the *Gramineae* the reduced floral structure has caused taxonomists to seek other characters and the great economic importance of the group has added a further stimulus. Anatomical features are prominent among these additional characters. Tissue arrangements are very diverse in the major taxonomic groups. The tissues that show the most differences are the endodermal sheath (whether small or large, with or without chloroplasts), the parenchyma sheath (present or absent) and the mesophyll, particularly the arrangement of the cells just outside the parenchyma sheath. This was clearly demonstrated in W. V. Brown's recent survey (1958) based on a study of 72 genera. The importance of grass leaf anatomy in systematics was realised by Duval-Jouve (1875) and has been much studied since then. Details can be obtained from Brown's paper, where he recognises six main types on the basis of their tissue arrangement. They cannot, however, be segregated into the two traditional subfamilies—*Festucoideae* and *Panicoideae*.

Brown's results are on the other hand consistent with certain alterations in grass taxonomy which have been proposed on other grounds, but it is their phylogenetic significance that he stresses. He has constructed a diagram in which the relationships of the six groups and their constituent tribes are indicated, together with their relative evolutionary specialisation worked out on the basis of all known characters, not just leaf anatomy.

At a more practical level, leaf anatomy in the grasses may have a specific diagnostic value. Paunero (1952) found for example that a study of leaf sections provided some good specific differential characters in *Alopecurus*. In *Agropyron* Jansen (1951) also found that species and hybrids could be distinguished by transverse sections, In general it appears that such features often corroborate conclusions obtained after a study of floral, morphological and sometimes cytological data.

A notable study on the anatomy and histology of the leaves of 587 species of *Rhododendron* was carried out by Cowan and his collaborators at Edinburgh (Cowan, 1950; Hayes *et al.*, 1951), already referred to in connection with dermal appendages (p. 155). The systematic value of the results of this study were not as helpful as had perhaps been hoped. Thus a primary division of the genus into groupings proportioned by leaf anatomy followed "a new and independent line". For secondary divisions they sometimes agreed with other characters, sometimes not. At species level the distinctions are seldom diagnostic.

A wide range of anatomical variation may be shown by a single individual, a factor that is sometimes overlooked. In the *Marantaceae* Tomlinson (1961) found that anatomical variation seems to be more related to the ecology and habit of the species than to their taxonomic position. He concluded that such anatomical features are of little value to the taxonomist for the delimitation of taxa of any rank within the family. It is often difficult to assess the extent to which anatomical characters may be modified by environmental factors (or indeed how much they vary at all, whatever the reason). Thus Pellegrin's studies on the anatomy of *Genista* led him to certain sectional rearrangements, and recently, influenced by this work, Vicioso (1953) delimited a new section *Cytisospartum*, but it has proved difficult in a later investigation to confirm some of Pellegrin's and Vicioso's results, partly because of the lack of comparability of descriptions, and partly due to the variation found within the species studied (Gibbs, unpubl.).

An example of the successful use of comparative anatomy for the revision of a genus can be taken from the genus *Mouriri* (*Melastomataceae*), which was divided into three sections by Cogniaux on the basis of the calyx lobes (free or fused) and on pollen grain shape. This treatment has not proved satisfactory and Morley (1953) in a sectional revision using a wider range of characters, which were necessarily morphological and anatomical because of the woody nature of the genus, was able to segregate the genus into three subgenera and eighteen sections; of these one section corresponded to one of those recognised by Cogniaux while the rest have little or no relation to his classification.

Some general principles of the role of wood anatomy in taxonomy have recently been enunciated by Leroy (1960):

1. Recourse may be made to wood anatomy at all taxonomic levels. At the family level it has been successfully used along with other lines of evidence to establish the systematic position of the "primitive" vessel-less Angiosperm families *Winteraceae*, *Trochodendraceae*, *Tetracentraceae* and *Amborellaceae* (Bailey & Nast, 1945a, b; Bailey & Swamy, 1948). Anatomical evidence also supported the case for considering the remarkable genera *Paeonia* and *Austrobaileya* the types of independent families (Bailey & Swamy, 1949; Lemesle, 1948).

At the generic level one of the most useful contributions of wood anatomy is to provide evidence for the positioning of taxa of uncertain affinities. Carlquist (1959) warns that to "be able to use and interpret such anatomical data the taxonomist must clearly understand the significance of the structural features described by the anatomist and their range and pattern of variation. Anatomists can co-operate by utilising existing taxonomic systems for their choice of material and should have an understanding of the general morphology and taxonomic relations of their chosen plants." He cites as examples of the successful application of such principles the outstanding work of Bailey and his collaborators in such groups as the *Degeneriaceae* and other groups mentioned above (see Bailey, 1954).

Identification down to the generic level or to species group level is often possible, but there are some cases where species can be recognised by their anatomical features (although it may be necessary to determine the genus by exomorphic criteria first). Simple spot characters can be illustrated by the genus *Juglans* where *J. cinerea* can be distinguished from *J. nigra* by the shape of the ray cells in section: in *J. cinerea* they are elliptic, in *J. nigra* they are round; oxalate crystals are also present in the woody parenchyma of the latter species but absent in the former. A number of recent revisions of North African genera include anatomical keys as well as the traditional morphological ones (Sébastian, 1956, on *Phillyrea*; Nègre, 1957, on *Frankenia*; Marmey, 1958, on *Marrubium*; Panellati, 1959, on *Bupleurum*; Veuillet, 1959, on *Elaeoselinum*; Vindt, 1960, on *Euphorbia*). Other examples of anatomical keys and descriptions will be found in the journal *Tropical Woods*. The most ambitious attempt at specific identification by wood anatomy is that of Greguss (1955), which gives keys for all species of contemporaneous Gymnosperms based on wood anatomical characters. Although in theory sound, it breaks down in practice and needs considerable revision.

2. Anatomical discontinuities, in general, correspond to taxonomic

divisions. This has to be considered along with Leroy's third proposition.

3. In general, secondary wood behaves as a very conservative feature in evolution. It might be predicted that certain features of the anatomy of the wood have evolved at a considerably slower rate than external features. The relative stasigenesis of these features would lead to an apparent conflict between morphology and anatomy in some cases. That there are such disagreements is now well known, although agreements tend to be stressed in the literature. The resolution of such conflicts is the task of the taxonomist and the general principle involved is well enough known—all evidence from all organs and sources has to be harmonised to arrive at a natural delimitation of taxa.

This line of argument is no more than an extension of the general principle so well stated by Bailey on several occasions (e.g. 1957) that each organ or part may be relatively uniform in certain taxa while highly variable in others. The rates of evolutionary specialisation in different organs, different structures, internal and external, anatomical and morphological, are frequently not closely synchronised. This can have important consequences for the delimitation of taxa and is fundamental in interpreting evolutionary relationships on the basis of phylogenetic trends.

Not only may anatomical features evolve at different rates from other characters but in different directions, in parallel, or convergently. The illuminating work of Jansonnius (1950a) mentions such examples.

An interesting fact pointed out by Jansonnius (1950b) is the wide *infra*-specific variability in wood anatomy in large genera which was greater than the *inter*-specific differences in small genera. This is not necessarily reflected in their respective morphology. In a study of 25 species and varieties of Javanese *Quercus* he found that the differences in wood structure were often very small between species and the infra-specific variation between samples very large. Anatomical keys of small genera are in consequence easier to construct than for large genera because of the great variation obscuring clear identification in the latter.

The evolutionary significance of this variation needs further study. It has been suggested that larger genera are more recent and the intermediate forms still extant, while the smaller are more ancient and the intermediates have been eliminated in the course of evolution. Such a hypothesis needs to be treated with considerable reserve and supporting evidence from other sources should be sought. It does, however, point to the fact that we need extensive further investigations of the wood anatomy of individual taxa in most groups to determine the range of variability of these characters in the same way as morphological characters are tested.

(c) *Lines of evolutionary specialisation in stelar structure*. During the last four decades plant anatomists, notably Bailey and his associates, have established a number of lines of evolutionary specialisation in the structure of the stele. Excellent summaries of this work are given by Bailey (1957), Tippo (in Gundersen, 1950), Carlquist (1961) and Eames (1961).

These series have been arrived at by both intensive study of the variation in trachearystructure in different parts of the same plant and in different samples of the same species growing under various environmental conditions, and by extensive sampling of randomly selected Dicotyledons and Gymnosperms.

The results were tabulated and analysed statistically and as a result an impressive series of evolutionary changes in size, structure and pitting of tracheary cells in the Dicotylodons as a whole were found. Similar series of changes have been found in the Monocotyledons.

Bailey asserts that these trends of specialisation in the Dicotyledons and Monocotyledons are amongst the best established, most complete and reliable which are known in either plants or animals. Not only are they reliable, but there can be little doubt about the direction followed by these trends: they are largely one-directional and irreversible. Vessels evolved by the modification of tracheids and the most primitive vessels have members closely resembling tracheids. Tracheids cannot be derived from vessels.

The following table summarises the features of the most primitive type of vessel compared with the most specialised types.

PRIMITIVE	SPECIALISED
Long	*Short*
Slender in relation to length	Comparatively broad in relation to length
Gradually tapered, extensively overlapping ends (in tang. L.S.)	Truncated ends
Thin-walled	Thick-walled
Angular in T.S.	Oval or rounded in T.S.
Multiperforate in overlapping ends	Single perforation at end of each cell
Intervascular pitting scalariform	Intervascular pitting alternating-multiseriate

The primitive type of Angiosperm vessel differs from the tracheid only in the dissolution of the pit membranes in a number of the pairs of bordered pits at maturity. Such multiperforate or scalariform perforations preceded those with simple perforations (one large pore in each end). Among those vessels with scalariform perforation plates those with many bars and narrow openings between them preceded those with fewer bars and wider openings. Thus there was a gradual reduction in number of the perforations, and a fusion of the bars between them to produce a wide single opening. It should be noted, however, that modification in the perforation of the vessel walls is ontogenetic.

There has been a similar sequence of specialisation of pitting in the side walls (intertracheary walls) from scalariform bordered to transitional, opposite, and finally alternate.

Vessels with scalariform pitting are known in about 110 families (Eames): they may be the only type present as in the woody *Eupomatiaceae*, *Himantandraceae* and *Annonaceae*, or they may occur to varying degrees. They are most frequently found in families which are regarded as relatively unspecialised in other features; scalariform pitting is not found in the most advanced families.

Not only have linear series of tracheids become modified phylogenetically to produce vessels, but other tracheids have become increasingly fibre-like in form, and at the same time the elements have been reduced in diameter and thickness of wall; the pitting, too, has changed from scalariform to circular through reduction in number and size.

In the wood parenchyma the diffuse arrangement of the cells is regarded as more primitive than the various aggregate distributions. Bailey notes, however, that although the available statistical data demonstrate that the various aggregates of parenchyma are derived from a primitive diffuse distribution, they do not form a reliable basis for arranging the various patterns of diversification into a single linear sequence of increasing structural specialisation (Bailey, 1957). Nearly 90 per cent. of woods examined having primitive vessels have diffuse parenchyma. The amount of wood parenchyma may be fairly constant or may vary from a small amount to a large amount, but its pattern of distribution tends to be constant except in those families which show a wide range of floral structure (Eames, 1961).

All these features are found in the vertical system of the wood. In the transverse system there is much variation in the form and cell arrangement of the parenchymatous cells which constitute the xylem rays. Several trends of specialisation occur, but they cannot be arranged in a single series. Thus heterogeneous rays are regarded as less specialised than homogeneous rays, which are apparently a highly specialised type. Rayless xylem appears to be the most specialised condition but it can be acquired either ontogenetically or phylogenetically.

All these sequences of evolutionary specialisation have been arrived at independently of any system of classification, so that circular arguments based on taxonomic assumptions of what is thought to be primitive are avoided. The general degree of assumed primitiveness or advancement of the taxa in which the various types of vessels and anatomical features occurred is not taken into account.

It must be stressed that although it is possible to arrange the Angiosperms as a whole according to a system of increasing specialisation in wood structure, a general system of classification so based would be as artificial as any other using restricted evidence. While it is true that certain of the different aspects of tracheary modification are as a whole closely synchronised—in other words the various types of wood cells evolved in specialisation together—there are numerous exceptions. As Bailey says, "any particular evolutionary trend, however reliably established, applies to a single part of the plant *only* and does not *by itself* provide a reliable basis for classifying the plants from which the evidence was obtained in a similar phylogenetic sequence, since evidence from other parts of the plant may negate such an arrangement" (Bailey, 1957). This may be termed the *negation principle*.

Parallel or convergent evolution may in fact have occurred frequently in the evolution of the xylem. Vessels have evolved independently in most of the major groups of tracheophytes; they have evolved independently in the Dicotyledons and in the Monocotyledons (Bailey, 1944; Cheadle, 1953). Primitively vessel-less Dicotyledons occur in systematically diverse families and the group *Homoxyleae* constructed for them is unnatural when morphological and floral characters are taken into account. This suggests that vessels have evolved independently in several unrelated stocks of primitive Dicotyledons, not just once.

Vessel-less Dicotyledons also occur in several other groups due to secondary loss. They are absent in aquatic genera in families such as the *Nymphaeaceae*, *Ceratophyllaceae*, *Podostemaceae*, and in heterotrophic groups such as *Monotropa*, and in succulents. This is interpreted as failure of potential xylem elements to undergo typical differentiation and maturation (Bailey, 1953; Cheadle, 1953; Eames, 1961). Takhtajan (1958), however, considers the *Nymphaeales* to be a primitively vessel-less group.

When these various evolutionary trends are considered by themselves they are of use primarily in *negating* affinities. For example, the apparently recent origin of vessels in the Monocotyledons does not permit them to have been derived from the *Ranunculaceae* or *Piperaceae* as has been suggested in some theories concerning the origin of the group. Plants with high levels of tracheary specialisation cannot have been the precursors of groups with relatively simple anatomical features. The *Amentiferae* likewise may be discarded as putative ancestors of the Dicotyledons in view of their specialised xylem. In this sense the anatomical evidence is comparable with cytological evidence and often equally conclusive (see Chapter 6). And like cytological data it can only serve as positive evidence of relationship or evolutionary derivation when corroborated by other lines of evidence. In other words, similar combinations of anatomical characters may occur in widely separated families or genera as regards their overall systematic position. Primitive features may persist in diverse groups between which their is no close relationship as judged in terms of other characters.

Phloem. Although less studied than xylem, trends of specialisation in Angiosperm phloem have been proposed as a result of the work of Hemenway (1913), Esau *et al.* (1953) and Zahur (1959). To a large extent the trends are parallel to those found in the xylem. Thus the sieve cell becomes specialised by progressive shortening, the end walls become decreasingly oblique; and there are corresponding trends of specialisation in the sieve plates.

Specialised phloem generally occurs with specialised xylem and may be better correlated than floral with vascular structure.

Internal phloem occurs in a wide range of what are considered fairly advanced families such as *Myrtaceae*, *Convolvulaceae*, *Campanulaceae*, *Cucurbitaceae*, *Onagraceae*, *Solanaceae*, and *Apocynaceae*. It is absent in the so-called primitive woody families and is unlikely to have been lost secondarily. Those families which do possess this feature are not closely related on other grounds.

Although the various deviations and localised evolutionary trends just mentioned detract from the value of wood anatomy for the purposes of deducing relationship in higher taxa, they do assume an increasing importance the lower down the taxonomic hierarchy one goes. These very deviations may in fact serve to distinguish between otherwise similar taxa, for example at generic level. It is at this finer level of analysis that much further research in wood anatomy is required, for little is known, as Bailey points out, of the variation in anatomical characters in different parts of plants, between individual plants grown under different habitat conditions, and even within the same population.

If we compare the results of wood anatomy to existing systems of Angiosperm classification some general conclusions can be drawn. Metcalfe (1954) finds that in his experience nearly all the taxa above the level of the family (with a few exceptions such as the *Centrospermae*) in all the well-known Angiosperm systems are heterogeneous and therefore presumably artificial. This has been shown by tabulation of the diagnostic anatomical characters and the heterogeneity becomes immediately evident no matter whether the systematic arrangement of the families follows that of Engler & Prantl, Bentham & Hooker or Hutchinson. He notes a tendency, however, for some of Hutchinson's orders to be more homogeneous than those of the other two arrangements (perhaps because they are smaller).

The general effect has been to show that there has been parallel evolution of stocks of woody plants around the *Magnoliaceae*. Wood anatomy does support in general the evidence from floral anatomy concerning the primitive condition of the Ranalian complex. Conversely it has emphasised the derived condition of the *Urticales* and amentiferous families: the *Urticaceae*, *Moraceae*, *Fagaceae*, *Betulaceae* (Tippo, 1938).

Although there are several cases where wood anatomy supports Hutchinson's arrangement, in other cases his division of the Dicotyledons into herbaceous and arboreal lines is not borne out. The herbaceous habit would appear to have arisen repeatedly in primitive woody groups, not just once. In the Monocotyledons, wood anatomical studies have cast doubt on the primitive nature of contemporaneous *Butomaceae* and *Alismataceae* in view of the presence of highly specialised vessels in their roots; it confirms, however, the specialised nature of the *Gramineae*.

A valuable survey of the bearing of anatomy together with other evidence on the evolutionary modifications and probable relationships of a selection of primitive Dicotyledons and of the *Helobiales*, *Liliales* and *Palmae* is given by Eames (1961) in chapter 11 of his recent book. Readers are advised to refer to this for details and further bibliographies.

Summary of the role of comparative anatomy in taxonomy

The salient points of the above discussion may be listed as follows:

1. Anatomical features may have a value as diagnostics or for determining relationships and evolutionary trends of specialisation.

2. As diagnostic characters they may be useful at all taxonomic levels but tend to be less reliable at specific level and above the rank of family within the Angiosperms, although for very different reasons.

3. Identification to the genus by traditional exomorphology is normally needed before anatomical identification can be pursued to specific level.

4. Above family level in the Angiosperms, heterogeneity in anatomical structure suggests polyphyletic origin (with few exceptions).

5. Endomorphic criteria are not of equal value in all taxa. They are relatively constant in some groups, highly variable in others.

6. Environmental factors may cause variation in anatomical features. This is often overlooked and our knowledge of such variation is highly incomplete.

7. Systematic anatomy has no special merits of its own to place it on a higher plane than any other approach to taxonomy. It is the use of endomorphology to supplement exomorphology (Metcalfe).

8. Resemblances in anatomical characters can arise by parallel and convergent evolution. Similarity may be due to the repeated origin of the features in question in different taxa. Even similarities in structural specialisation may be due to similar independently occurring trends

9. Trends of different anatomical features are not always closely synchronised.

10. As a consequence of the above points, anatomical evidence by itself cannot form the basis of a general classification or phylogeny. It must be interpreted along with other data from morphology, floral anatomy and morphology, palynology, cytology, etc.

11. When anatomical features are highly correlated with other features they may be used as diagnostic or marker characters, but not until the correlations have been established.

12. The phylogenetic value of anatomical characters is mostly in providing negations of derivations rather than affording positive evidence for lines of descent.

13. The overall trends of specialisation in the structure of the xylem are firmly and reliably established and provide almost incontrovertible evidence as to nature of primitive woody plants.

(d) *Floral anatomy*. Much of the major classification of flowering plants has been based on characters provided by the floral parts. The reasons for this, partly historical and partly deliberate as a result of the view that reproductive organs show great conservatism, have been discussed in Chapter 1. Such classifications appear to have been reasonably successful as judged by their almost universal use today. There is reason to believe that these floral characters have been relied upon extensively because other equally stable and conservative characters are not easily visible.

It is not our intention to survey the field of floral anatomy but simply to point out that there is still very considerable debate as to the nature of the flower (according to Eames a term too loosely used) and its homologies with other parts of the plant.

Since floral morphological characters are slow to change, it is natural to suppose that their vascular supply is likewise conservative. The vascular system of the flower has been extensively studied and the results are regarded of great significance in understanding the nature and trends of evolution of the floral parts and in deducing taxonomic relationships.

These studies have involved the nature of the perianth, androecium and carpels, and the inferior ovary. Interpretations are based on the fact that, after reduction or fusion of organs has taken place, traces of the ancestral condition can still be found in the vascular skeleton—residual supplies going to organs no longer there, or fusion or suppression of traces where organs have fused. Traces to lost petals are common, for example, in some apetalous flowers such as *Salix*, *Quercus*; to lost stamens in the *Scrophulariaceae*, *Labiatae*; to lost carpels in the *Ericaceae*, *Caprifoliaceae*.

The vascular anatomy of the flower has therefore been used to give some idea as to probable lines of evolutionary descent. It has been especially valuable in considerations of the nature of the "simple" flower—whether simplicity is due to reduction and specialisation or to true primitive nature.

There is a danger, however, that nutritional demands of the flower may obscure any evolutionary trends. The work of morphogeneticists such as Wardlaw must cause some doubts as to the interpretation of some aspects of vascular anatomy. Observations such as the excision of the leaf primordium having the effect of stopping the development of the leaf trace are significant. One of the major critics has been McLean Thompson, who put forward the theory, in a series of papers from 1931 onwards, that the vascular anatomy of the flower has a significance only in relation to the nutritional demands of the floral parts and had nothing to do with phylogeny. He considered that the real nature of the flower could only be determined by developmental morphology based on ontogenetic studies and nothing else. Although his and other theories of acarpy* gained little currency, it would be unwise to dismiss them as containing nothing of value. While most taxonomists today would agree with Parkin that McLean Thompson's views would seem to lead us nowhere in the understanding of the phylogeny of the flower and may rejoice the hearts of those who seem inclined to hark back to the formal morphology of pre-Darwinian days, a more correct position probably lies between the

* E.g. Grégoire (1938); Plantefol (1949).

two extremes. Eames (1961) endorses the comment that theories of acarpy "hardly come within the province of morphology". The evidence from morphogenetic studies on the control of marked floral differences as a result of minute changes in hormone balance during the early developmental stages warns us to proceed with great caution.

The field of floral anatomy has been reviewed by Leroy (1955), Tepfer (1953), Douglas (1957), Puri (1951, 1952a, b), Eames (1953) and others and is surveyed in the recent book (*Morphology of the Angiosperms*, 1961) by Eames. The flower is conceived by the majority of botanists as a determinate stem tip (receptacle) bearing sporophylls and commonly other appendages that are sterile. A new theory of the Angiosperm flower has been proposed by Melville (1960b). He started with the assumption that the carpel theory (stemming from Goethe and de Candolle) might be wrong and reinvestigated the vascularisation of a range of Angiosperm families by means of bleaching and clearing techniques. As a result he proposed that the "basic structure of the Angiosperm ovary was a leaf bearing on its midrib or petiole a dichotomous fertile branch". The various kinds of gynoecia found in modern flowering plants could be derived from such a structure. The male organs were similar in basic structure. To avoid confusion this basic structure was termed a "gonophyll" and the theory the "gonophyll theory". The male, female and bisexual analogues can be referred to as "androphylls", "gynophylls" and "androgynophylls" respectively. The flower and inflorescence are both regarded as condensation products of the primitive gonophyll or gonophylls. Only a brief exposition of the gonophyll theory has so far been published by Melville*: the evidence so far put forward is plausible, but there are many hypothetical steps involved.

Because of their bearing on the concept of the carpel and the nature of the flower, reference should be made here to the proposals of Lam (1947, 1948, 1950, 1955) that the Angiosperms were derived from two stocks—*phyllosporous* with the ovules borne on leaves, and *stachysporous* with the ovules borne on the stem—a viewpoint refuted by Eames (1951, 1961).

Eames is primarily interested in the nature of primitive Angiosperms, and summarises the evolutionary trends found within the Angiosperm flower as follows:

1. From many parts, indefinite in number, to few, definite in number.
2. From three or four sets of appendages—perianth, androecium and gynoecium—to one.
3. From spiral to whorled arrangement of appendages.
4. From freedom of floral parts to fusion—connation and adnation.
5. From radial symmetry (actinomorphy) to bilateral symmetry (zygomorphy).

* A fuller account has now been published by Melville (1962)—note added in proof.

In the androecium of the *Ranales* Eames finds strong evidence of the leaf-like nature of the sporophyll, both in morphology and vascular supply. He refutes the claim that the stamen arose from a complex branch system bearing terminal sporangia—an extension of the telome theory (Zimmerman, 1930; Wilson, 1953) by which the organs of vascular plants are derived from dichotomies of the original branch system (a theory which derives its strongest support from the *Psilophytales*). As Eames says, "Consideration of the primitive plant body as made up of basic units, telomes, is doubtless of value for the understanding of the more primitive taxa, but its value to the interpretation of the higher taxa, where axis and appendages have become established as morphological units, is doubtful".

At the level of Angiosperm morphology, the telome theory seems no longer relevant or applicable. It is in the *Ranales* (*sensu lato*) that Eames finds convincing evidence for the conduplicate nature of the carpel (Bailey & Swamy, 1951) as shown by *Drimys* and *Degeneria;* in these genera the free carpel is still unsealed, with the ovules submarginal. Closure of the carpel has apparently come about in two ways: (1) by the bringing together of the abaxial surfaces (conduplicate closure); (2) by the inrolling of the carpel walls (involute closure). Syncarpy has also probably arisen by: (1) adnation of closed conduplicate carpels; (2) adnation of still open, involute carpels. In the former case, closure of the separate carpel must precede syncarpy, but not in the latter case (cf. Eames, 1961; figs. 74, 75 and 84). With two ways of deriving a syncarpous ovary, it follows that trends in placentation are difficult to interpret, such end-types as single basal or suspended ovules being derived in more than one way (Eames, 1961: fig. 90).

Placentation is one of the basic characters used in family delimitation and plays an important role in deciding taxonomic relationships. It is true to say, however, that placentation may in many instances be regarded as a neglected character. Taxonomists may be misled by superficial resemblance in placentation between groups in their search for affinities. Thus the follicular carpels of the *Butomaceae* have been adduced as evidence for relationship with the *Ranunculaceae*, yet the seed arrangement and details of carpel structure are quite different in the two groups. This recalls the crucial problem of what constitutes a character.

A study of the vascularisation of the carpel will greatly assist in understanding placentation. Two main techniques are available for the study of floral vascular systems: (*a*) serial sections; (*b*) clearing and bleaching techniques. Sporne (1958) puts in a plea for the more extensive use of clearing techniques since they have the dual advantage of being rapid and permitting the observation of the vascular system in three dimensions. He stresses the need to make comparisons between flowers of comparable age (cf. Tepfer, 1953) and notes that comparisons should be made between the primary system of the flower and the primary system of the stem to discover evidence of intrinsic similarity between floral organs and leaves. He does not propose that such clearing techniques (Sporne, 1948a) should replace the classical method of staining serial sections.

The need for both approaches is well illustrated by two recent papers on the nature of the placental column in members of the *Caryophyllaceae*. Bocquet (1959a) observed a central vascular strand in the placental column of a species of *Silene* (*Melandrium*) sect. *Gastrolychnis*. By clearing techniques he found it was not a prolongation of the floral axis but a portion of the placental vascular system. He concluded that placentation in this species was only free-central in appearance but that the ovules really arose on carpellary tissue. On the other hand, Moeliono (1959), using sectioning and staining, came to opposite conclusions in a study on *Stellaria media*. Bocquet & Bersier (1960b) were able to resolve the apparent contradiction in a further study by demonstrating that the vascularisation of the ovary shows the "primitive" organisation, while serial sections show the morphogenesis of the outer parts of the organs which are more advanced than the vascular system. Both techniques should therefore be employed in such cases. Contradictory results may be expected in some cases since the actual evolutionary tendencies of a group will not be the same as its primitive structure as revealed by the vascularisation.

This phenomenon of "direct development" is discussed by Zimmermann (1959), who terms it "neoteny", in relation to the development of the ovule in the *Ranunculaceae:* in *Ranunculus*, *Anemone* and *Clematis* the ovule is not formed on the carpel margin but opposite the carpellary bud which the carpel later envelops. The vascular system is arranged, however, as in the primitive type of ovary structure as found in *Helleborus*. In the *Caryophyllaceae* placentation is often an example of direct development where placentation which is axillary in origin is substituted by free-central placentation: the vascular system shows the primitive structure with the ovule attached to the carpel; sections show the recent evolutionary tendency for ovules to arise on the axis. There is no real contradiction here, but two ways of looking at the same facts (see also Barnard, 1957, on the placentae in the *Gramineae* and Eckardt, 1957, for a contrary view).

The partitions in ovaries showing axile placentation may disappear in the course of phylogeny or ontogeny (Hartl, 1956). In the *Caryophyllaceae*, there is no structural difference between the entirely unilocular ovary of a *Melandrium* and a partly septate ovary of a *Silene*. In all species of these two groups the base of the ovary is plurilocular at least in early stages of development, and the degree and rate of breakdown of the septa varies according to the species. Unilocular versus plurilocular ovaries are of no value in separating *Melandrium* and *Silene*, a fact which supports the recent tendencies to unite them in one genus, *Silene* (Chowduri, 1957; Bocquet & Baehni, 1961).

Detailed discussions on placentation are given by Eames (1951, 1961), and Lawrence (1951) gives a useful summary.

In epigynous flowers the interpretation of placentation and the origin of placentae is very complex (cf. review by Douglas, 1957). Anatomical and ontogenetical evidence supports de Candolle's claim that the inferior ovary is in most cases appendicular—that is, formed by the fusion of appendages and their traces to the ovary wall. The only examples of axial invagination are, on anatomical evidence, *Rosa*, *Calycanthus*, *Cactaceae* and *Santalaceae*. Puri (1952a), however, is doubtful if anatomical evidence is valid for distinguishing between these two types of ovary. Meeuse's recent paper on the *Pandanaceae* (1961) provokes much controversy on the question of carpel homologies.

Burtt (1961b) suggests that Eames is too rigid in his interpretation of the perianth, regarding petals as sterilised stamens, or differentiated from sepals, according to the number of their traces. But this, comments Burtt, is a point of resemblance, not a proof of origin: transitional organs (as in *Nymphaea*) which intergrade between petals and stamens would be better accepted for what they appear to be—transitional organs. The interpretation of perianth homologies needs very careful handling, but is obviously of great importance when it comes to considering the relationships of "apetalous" families.

The application of floral anatomy to taxonomy is clearly more limited than that of other branches of anatomy in view of the technical and interpretative difficulties involved. Floral anatomical characters can scarcely be regarded as practical characters, although the new rapid clearing techniques make their study even in herbarium material more feasible.

Most results of such studies have been applied to the fields of phylogenetic systematics but they can be valuable in suggesting taxonomic relationships and delimitation. According to Eames (1953), the floral anatomical characters of families and genera are generally well marked and even specific characters may be quite clear in some cases, as in the genus *Vaccinium*. The detailed studies of Palser on the floral morphology and anatomy of the *Ericales* should be cited (e.g. Palser, 1951, 1954, 1961). Lindsey (1938) has shown that floral anatomy supports the separation of *Menyanthes* from the *Gentianaceae*.

A profitable field of study is the correlation of floral anatomy with secondary wood anatomy. This has been successfully employed in the *Amentiferae* at the family and generic level.

In the *Aceraceae* floral anatomy has contributed data supporting the inclusion of *Acer negundo* in the genus *Acer* and not as a separate genus. *A. negundo* is a highly specialised species, dioecious and anemophilous, but anatomically the flowers retain the characteristics of unspecialised species such as *A. pseudoplatanus* and *A. spicatum*. Hall (1954) has shown that dioecism is not completely established in *A. negundo*—the species has genes that under certain conditions lead to the development of stamens in pistillate flowers. He points to the dangers of inadequate sampling of a species as a basis for phylogenetic conclusions. The evidence from phytochemistry, on the other hand, supports the separation of *Negundo* as a distinct genus

(Towers and Gibbs, 1953)* and the cytological evidence has to be considered, too. This is a good example of the need for a considered synthesis of evidence from various fields before drawing taxonomic conclusions.

Comparative anatomy in botany (unlike zoology) threatens to become a neglected discipline. This is a poor look-out for systematics, because a satisfactory classification depends upon the interpretation of as many facts as possible. Nor can we make much use of anatomical (or any other) information if it is not available for the whole group we are studying. One can only hope that the rapid clearing techniques advocated by Sporne and others will lead to a revival of interest in this important factual field, since it allows floral anatomy to be studied far more quickly (and often with more certainty) than ever before.

Certain questions can, of course, only be solved by developmental studies. But before we can find out how tissues develop, we must know what those adult tissues are and where they are to be found. That is the "pioneer phase" of systematic anatomy, and it is still far from complete.

Palynology and Embryology

1. *Pollen Morphology and Taxonomy*

(a) *Introduction.* Taxonomists have been prone to overlook pollen-morphological characters although they were fully appreciated by early authors such as Lindley, von Mohl and Fritschze (1830-40). Rarely, as Faegri (1956) says, does the taxonomist take the pollen grains into account when describing or revising a group of plants, and when descriptions are given of the pollen grains they tend to be superficial. As a result of the lack of data, palynological† characters scarcely enter into current taxonomic thinking. However, pollen morphology is highly significant for the taxonomist and there is little doubt that it would be more extensively employed if the relevant data were more easily accessible.

It was only in 1935 that the first comprehensive book on palynology was published by Wodehouse—*Pollen Grains*, now a classic work. Erdtman published a treatise on pollen analysis in 1943, and in 1952 the first part of his manual *Pollen Morphology and Plant Taxonomy*, dealing with the Angiosperms, appeared. This latter placed at the disposal of taxonomists a summary of pollen morphological characteristics of all Angiosperm families. Wodehouse in particular has considered the evolution of pollen grains (1936a) and their value in classification (1936b). A review is given by Erdtman (1963).

Largely as a result of the stimulus of these authors, there has recently been

* Studies by Momotani (1961, 1962a, b) on the taxonomy and protein affinity of the *Aceraceae* support retaining *Negundo* within *Acer* (as a subgenus) although *A. negundo* is isolated in respect of its proteins from other species.

† The term palynology was proposed by Hyde & Williams to cover all studies of spores and pollen grains.

a great increase in comparative analytical studies of pollen morphology and wall structure. Results of great systematic and phylogenetic value have been obtained and these have often been in agreement with conclusions drawn from other fields of study, particularly anatomy.

By 1956 the collections of preparations of pollen and spores at the Palynologic laboratory at Bromma, Stockholm, had reached 20,000, the majority from non-Scandinavian plants. On this basis the occurrence of individual pollen or spore types has been followed throughout the Angiosperms as well as the Gymnosperms, ferns and moss groups. Clearly the more material available, the sounder a basis these studies will have.

(b) *Characters of the pollen and their role in taxonomy*. The main characters of taxonomic value in pollen grains are: number and position of furrows, number, position and complexity of the apertures, and the form of sculpturing of the exine. Variation is also shown in size and general shape, but this is of less diagnostic value; in some cases, however, spectacular differences have been found, as mentioned below.

Two basic types of grain occur in the Angiosperms—monocolpate and tricolpate:

Monocolpate: these are provided with a single furrow which develops on one side of the grain remote from the point of contact in the tetrad. This is characteristic of most Gymnosperms, the Monocotyledons and certain Dicotyledonous families, especially the woody members of the *Ranales* and members of other families, e.g. *Piperaceae*, *Chloranthaceae*.

Tricolpate: these have three meridionally placed furrows. They are characteristic of most of the Dicotyledons where they are the basic type from which other types have been derived.

Lines of specialisation starting from both the monocolpate and tricolpate appear to have culminated in the acolpate (without a furrow) and pancolpate (many furrowed) grains. Thus a great diversity of derived types is found in the Dicotyledons, including both an increase in the number of furrows and pores and (especially in wind-pollinated plants such as the *Gramineae*) reduction or loss of pores. Care must be taken, however, in interpreting such apparent trends as indicative of taxonomic relationship without reference to other lines of evidence.

Elaboration of the exine is characteristically found in insect and bird-distributed pollen, while in derived anemophilous groups it has become smooth and thin. Other apparent adaptations of the exine have been described, such as spines in the *Compositae*, crests in *Trapa natans* (*Trapaceae*), etc.

The taxonomic and evolutionary importance of pollen morphology may be at specific, generic or higher levels. In many cases the type of pollen of a taxon is characteristic and constant (the taxon is then termed stenopalynous) and may be exclusive to that group, e.g. the thick-walled grains of the *Gyrostemonaceae* of Australia. In other cases, and even within otherwise natural groups, the types of pollen may vary considerably in size, aperture, stratification of the exine, etc. In such taxa, which are termed eurypalynous, the pollen may be of great value to the taxonomist for subdivision into lower taxa. Stenopalynous families are, as a rule, very natural such as the *Asclepiadaceae*, *Cruciferae*, *Gramineae*, *Labiatae*. Eurypalynous families are mor frequent—some of them appear to be natural, others are clearly heterogeneous. It is interesting to note that the *Verbenaceae*, which is generally considered closely allied to the *Labiatae*, is unlike that family eurypalynous.

Pollen morphology has proved of value in the classification of the *Saxifragaceae* which tends to be a dumping ground for several rare plants of uncertain affinity due to the imperfectly known material available. The palynologist can often assist here, since fragmentary material may possess pollen grains, examination of which can suggest taxonomic position. Thus *Berenice arguta*, which was placed in this family, was shown to have Campanuiaceous grains and further investigation of its other features showed that it would indeed be better accommodated in the *Campanulaceae*. Another interesting case is that of *Trisyngyne*, a New Caledonian tree, male specimens of which were placed in the *Euphorbiaceae* by Baillon. Female specimens were later found and *Trisyngyne* transferred to the *Fagaceae*. Examination of the pollen grains of the male specimen would have shown, however, that they are exactly the same type as those of *Nothofagus* and moreover precisely of a special kind of which the oldest geological finds were made in the Cretaceous. An excellent survey is given by Erdtman (1954) of examples such as these of the value of palynology in taxonomic studies.

Like families, genera may be eurypalynous or stenopalynous. The pollen of related genera may, however, be quite different, as in *Salix* which has long and narrow 3-furrowed grains compared with *Populus* which has spherical grains without distinct apertures. The form of the pollen grain has proved invaluable in classifying the genera of the *Acanthaceae* and it seems likely to be of great help in achieving more natural groupings in the *Compositae*. The pollen of the *Compositae* has many features of interest: the sculpturing of the exine is, as a rule, highly elaborate, but there are tendencies to simplification and loss of sculpturing in anemophilous genera. Anemophily is regarded in this family as a recently derived condition and the reduction in sculpturing of the exine appears to be closely related to different pollination methods. Recent studies by Stix (1960) into the stratification of the pollen wall in the *Compositae* had added data of great systematic value.

(c) *Ultrafine structure of pollen grains*. Pollen wall structure has in recent years become the subject of very detailed and critical study, employing phase contrast and ultraviolet microscopy as well as more conventional light microscopy (Erdtman, 1954, 1956, 1959). Ultrafine sections of grains have also been studied by the electron microscope, a field pioneered by Fernandez-Moran & Dahl (1952) and Muhlethaler (1953). With the improvement of techniques (Afzelius, 1954, 1956) it has been possible to demonstrate fine details of structure of the walls (sporoderms) and their stratification. Electron microscopy has shown up new details in the surface of pollen grains, even in Monocotyledons where none had been found before (Bradley, 1958).

Van Campo (1959), well known for her studies of the pollen of the Conifers (recently used by Gaussen in his monograph of the genus *Pinus*, 1960), has discussed the importance of the endexine* in systematics and phylogeny. She points to the dangers of interpreting pollen only in terms of general appearance and especially the characters of the ektexine.* The ektexine and endexine ought to be studied separately: there is for example a tendency for the aperture in the endexine to remain constant while the ektexine proves to be variable, as in members of the *Umbelliferae*.

Although there is evidence to suggest that the fine structure of the pollen wall may have a potentially high value (cf. Larson & Lewis, 1961), both for taxonomy and phylogeny (fossilised pollen retains its structural details, as has been shown by electron microscopy, Ehrlich & Hall, 1959), too few species have been studied and techniques are not yet sufficiently standardised for firm conclusions to be drawn. Interpretation of the different layers of the pollen wall which have been recognised is still open to discussion, but it seems fair to say that pollen morphology is a great deal more complicated and potentially more useful than was originally envisaged (Erdtman, 1960). Taxonomists must ruefully look forward to interpreting such vertiginous statements as the following from Erdtman just referred to: "A baculation may also be found within sexinous tegilla (some tegilla actually seem to consist of densely packed, radial, baculoid elements)"!

(d) *Other pollen characters.* The systematic value of the nuclear condition of the pollen grains is considered by Dufou (1961). At the time of anthesis Angiosperm pollen is 2-nucleate or 3-nucleate. In a survey it was found that the Monocotyledons (*Liliaceae*) studied were 2-nucleate, the polypetalous Dicotyledons 2-nucleate, gamopetalous Dicotyledons 3-nucleate, and apetalous members often 2-nucleate, There were, however, exceptions to these—the *Centrospermae* for example were uniformly 3-nucleate. The three-nucleate condition is regarded as more highly evolved than the 2-nucleate. Although the presence of three or two nuclei in the pollen was generally constant within a family, some families were heterogeneous. It would be premature to draw conclusions.

Pollen morphology has played an important role in the splitting of genera. An excellent example is *Polygonum sensu lato* which contains many different types. It has been found that most of the groups into which the genus has been divided in the past are distinct in their pollen type and Hedberg (1946) used this together with general morphology and chromosome numbers to suggest a division into seven genera: *Koenigia, Persicaria, Polygonum, Pleuroptero-pyrum, Bistorta, Tiniaria* and *Fagopyrum*. This arrangement is accepted by a

* The exine consists of two layers—the ektexine and the endexine. Some palynologists recognize four layers in the wall: ektexine, endexine, mesine and intine.

number of recent workers. Pollen characters have been extensively employed by Wendelbo (1961b) for generic separation in his studies on the *Primulaceae*. Pollen may be successfully used for identification of species of annual *Ranunculus* in the absence of fruits (Davis, 1960).

A phenomenon which provided useful data for understanding taxonomic genetic and evolutionary problems is pollen dimorphism associated with heterostyly (see p. 372). In some cases studied the dimorphism of stigma and pollen is spectacular (Baker, 1956). In the genus *Rudgea* the amount of difference between the two kinds of pollen is as great as that between genera and higher taxa in other cases.

New classifications of pollen types, each more complex than the last, appear in rapid succession (cf. G. Erdtman, 1963). That of Erdtman and Straka (1961) is based on one feature only—the apertures (tremata). It is termed the "NC-system" since its classes are established according to the number, position and character of the apertures. Since there are 7 N-classes, 7 P-classes and 7 C-classes, a total of 343 pigeonholes is available to accommodate the grains. The field of descriptive pollen morphology lends itself to the creation of extensive terminologies, a subject which Erdtman has pursued with great vigour and erudition. The results are impressive but, we feel, self-defeating due to their very complexity, as reference to the Erdtman and Straka paper will readily show.

Pollen studies have contributed to the elucidation of phylogenetic relationships. They have been brought to bear, along with other lines of evidence, on the problems of the primitive Ranalian Dicotyledons and their relationship to the Monocotyledons (see discussion in Eames, 1961: chapter 11). In agreement with other evidence the facts of palynology do not support the existence of a sharp line of demarcation between the Monocotyledons and the Dicotyledons. Monocotyledonoid characters occur in the *Annonaceae*, *Magnoliaceae*, *Nymphaeaceae*, *Piperaceae*, etc.; on the other hand some Monocotyledons exhibit certain Dicotyledonoid characters. Pollen grains with composite apertures (where the inner part of the aperture has not the same outline as that of the outer part) have only been found in typical Dicotyledons.

Kuprianova has studied the pollen characteristics of the whole of the Monocotyledons (1948) and some of her conclusions are very suggestive. She considers that the *Helobiae* are not related to the other Monocotyledons but ought to be considered specialised *Polycarpicae* with Ranalian affinities. Her conclusions were, however, based very largely on palynological characters and other features were not sufficiently taken into account in working out her schemes.

Summary

Taxonomy has been shown to benefit from macroscopic, microscopic and ultramicroscopic pollen characters. Although the results are frequently in agreement with established classifications, they must be treated with caution and not interpreted without reference to other lines of evidence. The interpretation of certain structures, the extent of variation and the elucidation of trends are still causing concern and it must be remembered that pollen characteristics are subject to parallelism, convergence and probably reversal, as are other characters employed by the taxonomist.

Palynological evidence may be used to place taxa of uncertain affinities, to suggest rearrangements, withdrawals and separations, as well as corroborating other lines of evidence. That it is now possible to study even fine structural details in fossil pollen opens up possibilities of new sources of phylogenetic evidence. We can confidently expect pollen morphology to play an increasing role in both comparative phenetic taxonomy and in evolutionary interpretations.

2. *Comparative Embryology*

(a) *Introduction.* Since the discovery of double fertilisation, the morphology and development of the embryo has been the subject of much study. Excellent detailed treatments have been published by Johansen (1950) and Maheshwari (1950) to which the reader is referred. Maheshwari interprets embryology in a broader sense than Johansen so as to include those events that take place in the ovule prior to fertilisation as well as afterwards (i.e. including embryo and endosperm); he also includes the development of the male gametophyte and fertilisation. Cave (1959) adopts a similar broad view in outlining those embryological features that are of taxonomic value. Such characters are to be found in three generations: (*a*) the stamens and carpels of the diploid generation and micro- and macro-sporogenesis; (*b*) the haploid micro- and mega-gametophytes; (*c*) the new sporophyte generation including both embryo and endosperm. The features of taxonomic value she lists are: (1) anther tapetum; (2) delimitation of microspores; (3) mature micro-gametophyte; (4) coverings of nucellus: number of integuments, aril, obturator; (5) size of nucellus; (6) persistence of nucellus; (7) nature of archesporium; (8) cover cell; (9) hypostase; (10) embryo sac type; (11) mature megagametophyte; (12) endosperm; (13) haustoria; (14) embryo; (15) mature seed.

Impressive though this list may be, the data contributed by embryology to taxonomy are rather scarce and are generally not well known, and although individual workers such as Cave herself have made important contributions to the solution of taxonomic problems there is much in the subject that is difficult to interpret and technical problems add to the complications.

Corner (1951a), in a review of Maheshwari's *An Introduction to the Embryology of the Angiosperms*, comments on the different classifications for various kinds of embryo sac, endosperm and embryogenesis. He says that the different types seem to have little or no taxonomic significance and that they are either examples of polyphyletic convergence from diverse families or the classifications are not critical enough and apparent similarities have been confused.

(*b*) *Descriptive terminology of ovules.* The primitive ovule in the Angiosperms appears to have been anatropous (but see below), with two vasculated integuments, massive parietal tissue in the nucellus and (according to Sporne, 1954b) nuclear endosperm. Bocquet (1958, 1959b) has recently proposed a new terminology for the description of ovules (see also Bocquet & Bersier, 1960a). Classic division has been into orthotropous, anatropous, campylotropous, hemitropous and amphitropous. The new classification is based on the following two hypotheses: (*a*) That there are only two fundamental types of ovule—the orthotropous and anatropous. Other types have been derived

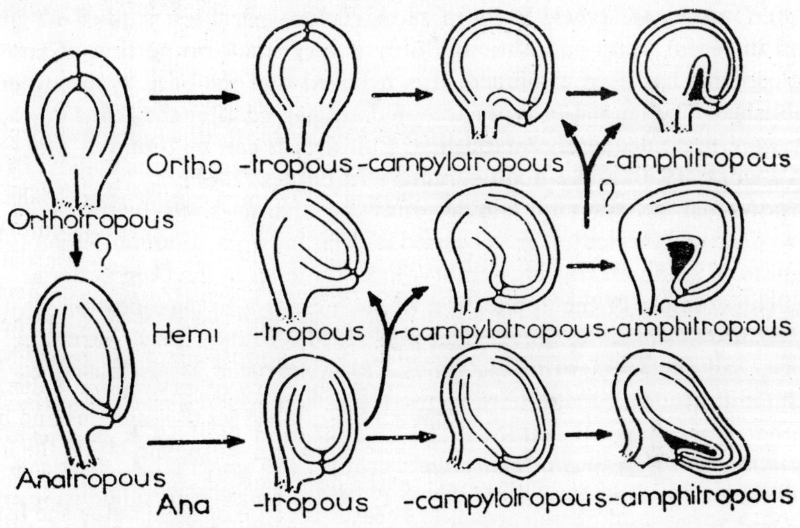

FIG. 15. Illustration of ovule terminology, and the probable evolutionary relationship of the nine types. (After Bocquet & Bersier, 1960a.)

from these so that two series may be proposed—the orthotropous and the anatropous. A third series—the hemitropous—has to be added. It is derived from the anatropous type and develops parallel to it. (*b*) Campylotropy and amphitropy are superimposed on these other three basic types. They appear successively in the course of ontogenetic development.

It is difficult to decide whether the anatropous or the orthotropous ovule is more primitive, but Bocquet & Bersier (1960a) believe that palaeobotanical and ontogenetic evidence suggest that orthotropy is the earlier condition.*

As a result of these two hypotheses, nine terms are proposed for the description of ovules, as shown in Fig. 15.

* Corner (1962) maintains that in the *Moraceae* orthotropy is derived from campylotropy.

Systematic and morphogenetic evidence is presented by these authors to confirm the validity of their views. Although (as in all such cases) these evolutionary series or trends may be susceptible to a reverse interpretation, they present a useful working hypothesis and have allowed certain apparent embryological contradictions to be resolved. For example, the reniform ovule of *Phaseolus* seems nearer the reniform ovule of the *Caryophyllaceae* than it does the spherical ovule of neighbouring genera such as *Pisum*. However, the ovules of *Phaseolus* are hemicampylotropous and those of *Pisum* hemi-amphitropous and both belong to the anatropous series and are therefore related. The Caryophyllaceous ovule (orthocampylotropous) belongs to the orthotropous series and is not related to the Leguminous types.

(c) *Embryo sac types.* Embryo sac structure has been studied in great detail in recent years and although only a very small proportion of known Angiosperms has been examined, it is believed that the basic types are now established. The ten major types accepted are named after the genus in which they were first adequately established. Full details can be found in Maheshwari's work (1950) and a good summary in Eames (1961).

Individual genera and families may be characterised by the type of embryo sac and this feature may be used in deciding on taxonomic delimitation and natural affinities. Harling's embryological studies in the *Compositae* are an excellent example of the application of such criteria to classification. In the *Anthemideae-Chrysantheminae*, Harling (1951) found great variation in embryo sac development in the genus *Chrysanthemum* as circumscribed by Hoffmann. Mono-, bi- and tetra-sporic types were found as well as variation in other embryological features. The classification of the *Chrysanthemum-Leucanthemum-Pyrethrum-Tanacetum* group of genera is particularly troublesome (see summary in Heywood, 1954a) and a new subdivision was proposed by Briquet as a result of his studies on the structure and anatomy of the cypselas. This gained limited acceptance, but it was found by Harling that the embryological evidence lent considerable support to Briquet's arrangement. An assessment of the embryological data together with the anatomical and morphological characters persuaded Harling to revise the classification of the group in a way which largely agreed with Briquet's views, although he preferred to recognise as subgenera some of the groups such as *Leucanthemum*, *Argyranthemum* and *Tanacetum* which Briquet regarded as genera. Embryological features also support the recognition of *Chrysanthemum balsamita* as a separate genus, as was proposed by Desfontaines (1792).

Other generic arrangements in the *Chrysantheminae*, *Anthemideae* and *Astereae* are supported by embryology (Harling, 1950, 1951, 1960). It is interesting to note that in these cases embryology has served largely to confirm the validity of genera previously proposed during the last century; these failed to gain acceptance in more recent times, largely as a consequence of the broad circumscriptions of genera accepted by Bentham & Hooker

(*Genera Plantarum*) and Hoffmann (in Engler & Prantl, *Die Natürlichen Pflanzenfamilien*) which have been followed by most authors. What embryology *cannot* do is indicate how different groups must be before they should be recognised as genera, subgenera or sections, any more than other lines of evidence can do. What is important, however, is that the taxonomic units recognised should, as Harling says, be homogeneous and as natural as possible.

In the *Gramineae* the systematic value of the embryo and endosperm was first recognised by van Tieghem and since then has been extensively studied together with many other seed characters. Surveys of recent progress are given by Yakovlev (1950), Jacques-Félix (1958) and Reeder (1957, 1959). Reeder notes that in the majority of cases there is a very high correlation between embryological features and other characters such as leaf anatomy, chromosome number, etc. A survey of embryo types in grasses has been published by Kinges (1961).

(d) *Other features*. Embryological studies are seldom so helpful for taxonomic purposes as those described. Cave's studies on generic delimitation in the *Liliaceae* (e.g. Cave, 1948, 1953) are, however, outstanding, and other examples may be found discussed in Maheshwari's recent book (1950).

The genus *Paeonia* has attracted much attention from morphologists, anatomists and others with a view to determining its systematic position (see summary in Eames, 1961). An interesting piece of evidence is reported by Yakovlev & Yoffe (1959), who have shown that there is a coenocytic phase in the embryogenesis of *Paeonia* unlike other Angiosperms (although found in the Gymnosperms).

Developmental differences have also been studied in grass embryos and it has been found, for example, that specific variation in the number of layers of tunica is of taxonomic significance (Brown *et al.*, 1957).

The need for caution in the interpretation of embryological data is pointed out by Mohan Ram (1959) and Harling (1954). The latter author reinvestigated the embryo sac development of *Vittadinia triloba* (*Compositae*) reported by Palm (1922). He concluded that at least part of the embryo sac in Palm's material had been unreduced and that he had dealt with a species which was totally or partly agamospermous. It was known that Palm sometimes utilised botanic garden material without checking its taxonomic identity, and after some detective work Harling decided that Palm had been working with *Vittadinia triloba* hort. which is correctly *Erigeron karvinskianus* DC., an agamospermous species.

Clearly, a strict check of the taxonomic identification of material used for embryological studies is vital, as indeed it is for any other systematic studies. It is exceedingly difficult to assess to what extent early results may be suspect due to such reasons, in the same way that many of the earlier chromosome counts have had to be rejected.

The different types of early embryos (proembryos) have been designated by reference to the name of the family in which they occur, thus we have the Asterad, Onagrad, Chenopodiad, Solanad, Caryophyliad and Piperad types. Much of the discussion about the phylogenetic relationships of these different types must be dismissed as speculation (see details in Johansen, 1950).

Chapter 6

TAXONOMIC EVIDENCE

II. Cytology

Introduction

The two basic sciences of cytology and genetics are interwoven in modern approaches to taxonomy ("experimental taxonomy", "biosystematics") and have, as Keck (1957) says, been combined in the vigorous hybrid, cytogenetics. The contributions made by cytology to taxonomy form an outstanding feature of the past few decades. Cytological data are derived mostly from the nucleus and pertain to chromosome number and morphology, and chromosome behaviour at meiosis. They may be used for purposes of comparison or for evolutionary interpretation.

Cytotaxonomy is the discipline which seeks to study variation and explain variational discontinuities and relationships in terms of cytology. It studies chromosomes in all their aspects, both in natural populations and in experimental cultivation. A review of the foundations of the subject is given by Löve (1954a) and a more detailed treatment in Darlington's *Chromosome Botany* (1956). The literature of cytology in relation to taxonomy and classification is already vast and only some of the major features can be outlined in this chapter. Most of the examples we have chosen can be duplicated many times over from other sources.

A special role is claimed for cytological data in taxonomy since the chromosomes are the seat of the hereditary material. From an interpretative viewpoint this claim is justified, but chromosome number and morphology as comparative data have to be considered on a par with other kinds of taxonomic evidence. This is discussed further below.

The term *karyotype* is applied to the appearance of the somatic chromosomes at mitotic metaphase. Various features of the karyotype are used in taxonomy for comparative purposes: these are mainly chromosome number and morphology which form the common and basic phase of cytotaxonomy.

Chromosome Number

The importance of chromosome number as a taxonomic character is that it is probably one of the most constant single features employed: all individuals within a species usually have the same chromosome number, although there are exceptions. There may be somatic instability in polyploids and hybrids whereby numbers of chromosomes within individual plants are not always constant (Vaarama, 1949; Sachs, 1952). In *Hymenocallis calathinum* (*Amaryllidaceae*) Snoad (1955) found from 23 to 83 chromosomes in root-tip cells in four plants from a clone; there was also a range of numbers, although smaller, at meiosis. Jones (1954, 1955) has pointed out that a distinction must be made between the cytology of artificially raised seedlings and that of field populations. In the former case the chromosome numbers will show the potential variability of the population which may be quite different from the actual variation permitted by natural selection. Thus in *Holcus mollis* he found that counts in seedlings resulting from open pollination showed up to seven numbers which had not been found in plants collected in the field. As he says, these "experimental" counts represent potentialities as far as natural populations are concerned; whether they would survive under natural conditions is not known. The general point to be stressed is that cytological analysis in natural populations is the most efficient means of determining the situation in a species: less intensive sampling may result in the inherent cytological variation, which occurs in many species, being overlooked.

Of the estimated quarter of a million species of flowering plants about 20,000 have been analysed cytologically, or less than 10 per cent. (cf. Larsen, 1960a; Löve & Löve, 1961). But even this figure is highly misleading since thousands of these species, probably the majority, have been counted only once, often from a single plant raised from seed obtained through botanic garden exchange arrangements. This seed may not be of known wild origin. Another source of error is misidentification of the material counted: it has been suggested that anything from a third to a half of published counts are for one reason or another erroneous. In the case of older counts faulty technique may have been responsible for errors, but it must not be thought that modern cytologists are infallible. Publication of a chromosome number is in some respects like the publication of a new species in that the actual number lends a certain credence to its validity without any further investigation, just as new species must be accepted until there is evidence to the contrary.

Löve & Löve (1961) go so far as to say that "the fact remains that most chromosome numbers are still determined either by cytologists lacking taxonomical training or by taxonomists lacking cytological skill and experience". This is, of course, especially acute in critical groups where species and infra-specific groupings may be determined with difficulty even by a specialist. Unless the botanist responsible for chromosome counts in such groups is one of these specialists, the counts should be accepted for the taxa with considerable caution.

Apart from problems of identification, few species of wide distribution have been adequately sampled from more than limited parts of their area, so that little information is available about the possible cytological variation within them. As Larsen (1960a) says, "one may compare this situation with a herbarium in which each species is represented by one single sheet only". It is in fact not so satisfactory since the one herbarium sheet is available for examination but the cytological preparation is seldom available for inspection.

The importance of preserving voucher specimens of plants whose chromosome number has been counted cannot be too highly stressed. It is easy enough to make a fixation of a cultivated plant, count its chromosomes and later find that the plant has died or is no longer suitable for preservation. A further complication is that the specimen may be preserved at a stage which does not show all the features necessary for accurate identification. For example, many cytological voucher specimens of *Ranunculus acris* cannot be identified beyond the species level since the numerous subspecies into which it is divided require the basal portion of the plant and mature achenes for positive determination.

Well-preserved voucher specimens permit later studies to ascertain if morphological features are associated with cytological features. In *Kohlrauschia prolifera*, for example, diploid and tetraploid plants with distinct geographical distributions occur (see p. 165), but morphological separation of the cytological races proved difficult for several years. Ball & Heywood (1962) were able to find constant differences in seed morphology between the diploids and tetraploids and were able to confirm this by re-examining the cytological voucher specimens made by Böcher & Larsen (1955b).

We have prefaced this consideration of chromosome numbers and taxonomy by these admonitory remarks, since so much weight is placed on cytology in making taxonomic decisions and evolutionary interpretations that it is as well to realise what technical pitfalls stand in the way. Literally millions of chromosome countings will have to be made before our knowledge of this aspect of flowering plants approaches what we know of them taxonomically in terms of morphology and distribution. Premature conclusions should not be drawn.

The major source works for cytological references are: Tischler (1950), Darlington & Wylie (1955), Löve & Löve (1961), Cave *et al.* (1958). In addition, there are numerous lists covering individual countries or regions such as Maude (1939) for the British Isles, Löve & Löve (1948) for N.W. Europe, Tarnavschi (1948) for Rumania, although they quickly become outdated. Many Floras today include the chromosome numbers of the species, sometimes referring to counts made on material growing within the area in question, sometimes to counts recorded for the species from any area. References are seldom given. Various review publications refer to newly published counts, e.g. *Biological Abstracts*, *Excerpta Botanica*, *Plant Breeding Abstracts*, *Resumptio Genetica*.

Chromosome numbers recorded for Spermatophyta vary from $n=2$ in *Haplopappus gracilis* (*Compositae*) to $n=154$ in *Morus nigra* (*Moraceae*), probably around n=250 in *Kalanchoe* species and n=263-5 in *Poa litorosa* (Hair & Beuzenberg, 1961).

Three broad classes of chromosome number relationship within taxonomic groups may be recognised (Warburg, 1949):

(1) Throughout the whole group (such as a genus) no deviation in chromosome number is found. For example, all species of *Pinus* and some related genera have n=12; in *Quercus* likewise, and most other members of the *Fagaceae* examined, all species have the same number, n=12. In these and several other examples chromosome number is clearly of no value for taxonomic discrimination *within* the group, although the constant number may be a useful character to distinguish the group from others. In some cases the same number occurs throughout a group except for one species.

(2) The basic number of chromosomes (conventionally indicated as x) is found to be exactly multiplied in various members of a group giving a polyploid series. The common, lowest diploid number ($2x$) by doubling may produce tetraploids ($4x$) and octoploids ($8x$). Triploids ($3x$) and hexaploids ($6x$) may be formed through hybridisation between the different levels. Individuals of this kind, whose somatic chromosome number is an exact multiple of the basic number, are termed *euploid.* An example of a polyploid series is found in *Taraxacum* with $2n$=16, 24, 32, 40, 48. A remarkable range of polyploid numbers is found in the family *Malvaceae* (Skovsted, 1941):—

	x	$2x$	$3x$	$4x$	$5x$	$6x$	$7x$	$8x$	$9x$	$12x$	$15x$
x=5	5	10	15	20	25	30	35	40			
x=6		12				36			54	72	90
x=7	7	14	21	28	35	42		56		84	
x=11	11	22	33			?66					
x=13	13	26	39		65						

A very frequent phenomenon is the polyploid pair, such as *Nasturtium officinale* a diploid ($2n$=32) and *N. microphyllum* ($2n$=64) a tetraploid, *Cardamine hirsuta* ($2n$=16) and *C. flexuosa* ($2n$=32), and *Saxifraga hyperborea* ($2n$=26) and *S. rivularis* ($2n$=52). Such apparently simple multiple relationships are regarded as of great value to the taxonomist: the different chromosome numbers serve as good markers of other differences or pointers to where other discontinuities may be found, and they often indicate the presence of a reproductive barrier, although this must not be automatically assumed. Many polyploid pairs turn out upon further investigation to be parts of much more complex situations, as is discussed later.

Some cytologists still maintain that the presence of two chromosome numbers in related forms is, with certain exceptions, an indication that they belong to different species (Löve, 1951). "It is but logical to regard all taxa differing in chromosome number as distinct species" (Löve & Löve, 1961). A difficulty arises from the fact that morphological differences associated with polyploidy may be well marked or scarcely perceptible, and all intermediate states are known. Further, the differences may not be constantly associated and may even in some cases be reversed. The interpretation of the significance of polyploidy in relation to taxonomy is still highly controversial and is discussed in detail below (p. 212 seq.).

The somatic number is occasionally doubled but with one or two chromosomes added or dropped. When this occurs in an allopolyploid it may be referred to as a *secondary polyploid.* An example is *Dahlia merckii* which has 18 haploid chromosomes, whereas the other species of the genus have a base number of 8. It is assumed that two of the eight chromosomes are present three times; cytological confirmation of this assumption (and in other cases too) comes from the occurrence of secondary pairing (see Darlington, 1937: 239-243). Darlington suggests that the uniform basic number of the *Rosaceae-Pomoideae* ($x=17$) is derived from the primary basic number of 7 of the *Rosa* group (cf. Darlington & Mather, 1949) and he believes that the addition of a chromosome pair to a presumed tetraploid number may have occurred in several genera (1956: 85 fig. 7). A very different but taxonomically more plausible interpretation of the origin of the *Pomoideae* is that they are diploidised polyploids which arose from ancient hybridisation between the *Spiraeoideae* ($n=9$) and the *Prunoideae* ($n=8$) or their ancestral stocks* (see Stebbins (1950: 359-61); and Gajewski (1957: 370-72) for further discussion and references).

Polyploids, it seems, are more able to withstand such alterations in basic number than are diploids (see p. 200).

Dibasic polyploidy occurs where two different basic numbers are combined in a single polyploid. The combined dibasic number can be referred to as x_2 as in secondary polyploids (tribasic, x_3 when three sets are added together). An example can be taken from *Brassica*, where the hybrid between *B. oleracea* $2x=18$ and *B. rapa* $2x=20$ produced on doubling the allotetraploid *B. napus*, the Swede turnip $4x(=2x_2)=38$. Different inferred dibasic numbers may characterise subgenera, as in *Veronica* ($x=7, 8, 9$; $x_2=15, 17$; $x_3=26$).

Darlington considers that "The dibasic polyploid has bearings beyond the convenience of dividing genera or tracing the phylogenies of species . . .". ". . . . every dibasic polyploid must be derived from the crossing or hybridisation of two diploid races, or species, or genera, which differed in basic number. Where these ancestral types still survive there can be no doubt that the origin is polyphyletic. And if it can be polyphyletic where the base numbers of the ancestors differed, it can also be polyphyletic where they were the same."

In instances like the Brassicas conclusions about dibasic polyploidy can

* Taxonomists worried by the difference between the gynoecium of the *Pomoideae* (inferior) and that of its postulated ancestors (apocarpous), may find some consolation in the "atavistic" mutation in *Malus* (*Pomoideae*) which produces a superior, apocarpous gynoecium (Masters: 1869).

be safely drawn, but in larger groups it is easy to fall into errors in drawing inferences about the origin of apparent dibasic numbers, and the temptation to play chromosome arithmetic and accept the results must be resisted unless there is strong supporting evidence from other sources.

A word should perhaps be added here about the difficulties involved in determining basic numbers. The basic (or base) number is defined as the primitive or original number from which polyploid numbers have been derived (Swanson, 1958). In practice the basic numbers of a genus are usually calculated from the lowest diploid numbers known in the group (Löve & Löve, 1961). In some cases the diploids may no longer exist or are not recognised as such and only apparently polyploid numbers are reported within the genus concerned. A theoretical basic number has to be inferred in such cases. There may be internal evidence of polyploidy which helps to confirm inferences, or there may be no evidence, as in *Zea mays* ($n=10$) where the haploid number of 5 in the related genera *Coix* and *Sorghum* suggest strongly that 5 is the ancestral basic number (cf. Darlington, 1956).

A point that must be made is that it is impossible to talk about the basic number of a group until it has been decided how the group is to be circumscribed. This is well illustrated by the *Compositae-Helenieae*, where the commonest basic numbers found are $n=8$, $n=7$ and $n=6$, but the group is so obviously polyphyletic that it is not possible to speak of an original basic number for the tribe (Raven & Kyhos, 1961). In other words, the taxonomy has to be sorted out first.

(3) The numbers found within a group bear no simple numerical relationship to each other (as of the kinds just discussed). Here there are changes in number without multiplication of complete sets of chromosomes and they are most frequent at the diploid levels. In some of the best known cases the changes in the haploid basic number seems to imply a gain or loss of chromosomes one by one. These are known as *aneuploid* series (in contrast with euploid series mentioned above), although the term *aneuploidy* can be used to cover two distinct kinds of phenomena. In the first case there is an increase or decrease in number whereby the same genetic material becomes distributed in a different number of chromosomes, leading to a change in the basic number. In the second case the basic number is unaltered but the genetic material present is changed through the addition or loss of chromosomes—an unstable condition which is less important in long-term evolution than the first. The term *dysploid* is sometimes applied when the aneuploid sequence does not apparently form a particular series. An example is the genus *Carex* which contains a wide range of numbers from $n=6$ to $n=112$ with multiples of 5, 6, 7 and 8 (Heilborn, 1939). It is, as Swanson (1958) remarks, "a classic example of cytotaxonomic confusion", and in the evolution of many of the species agmatoploidy (see p. 226) and polyploidy has probably been involved (Löve, Löve & Raymond, 1957).

There are various mechanisms by which increases or decreases in chromosome number may be brought about, such as irregularities during division of the cell or changes in the chromosomes themselves (fragmentation, misdivision of the centromere, etc.).

It has been suggested that increase in basic number can come about without any change in the *content* of the chromosome set, as by misdivision of the centromere whereby unpaired chromosomes at the first meiotic division break across into two halves which go to opposite poles, each arm forming a new independent chromosome with a "terminal" centromere (Darlington, 1939; Darlington & La Cour, 1950; Darlington, 1956), but this has so far only been observed under experimental conditions. A more frequent method of increase appears to be through polyploidy followed by loss of chromosomes.

Although increases in chromosome number are more easily assimilated by the plant than decreases, which may lead to nutritional and other disturbances, ways are known, however, by which decrease in number may arise. A frequent method of chromosome decrease is through unequal interchange between chromosomes. This has been documented in *Crepis*, where the main trend in chromosome number is a progressive decrease in basic number from 6 to 3, step by step, one chromosome at a time. A decrease from $n=4$ in *Crepis neglecta* or some related species to $n=3$ in *C. fuliginosa* appears to have arisen through an unequal reciprocal translocation, so that one chromosome of *C. neglecta* (or a related ancestor) lost nearly all of its long arm. The chromatin attached to the remaining centromere and the centromere itself is not found in *C. fuliginosa* and has presumably been lost (Togby, 1943; Babcock, 1947a). A similar derivation of *C. kotschyana* ($n=4$) from *C. foetida* ($n=5$) has been described by Sherman (1946), and other examples in the *Cichorieae* are given by Stebbins *et al.* (1953).

The basic set may, on the other hand, remain the same though supernumerary chromosomes are added to it. There are two mechanisms:

(*a*) *Polysomics and Monosomics*. The first case is the type of *aneuploidy* where irregularities at meiosis result in cells which are deficient for a particular chromosome or have it duplicated: $n-1$ or $n+1$. From these complementary cells, individuals with $2n-1$ and $2n+1$ will arise on fertilisation by normal gametes. Individuals with $2n+1$ are termed *trisomics*, those with $2n-1$ are termed *monosomics*. Trisomics are "polyploid" for a particular chromosome.

Trisomics are known from several plant species and have been studied in detail in *Nicotiana* (Smith, 1943), *Lycopersicum* (Rick & Barton, 1954) and *Zea* (Rhoades & McClintock, 1935) as well as in economically less important groups such as *Antirrhinum* (Rudorf-Lauritzen, 1958). The classic case is, however, found in *Datura* (*Solanaceae*), which has $n=12$ and thereby twelve possible trisomics. These were all produced experimentally and identified by Blakeslee (1922, 1930, 1934), who also showed that each had multiple phenotypic characteristics which allowed them to be recognised. A certain amount of unbalance is produced by these extra chromosomes, but the capacity of the plants to vary is not greatly affected. The addition of two extra chromosomes (tetrasomes) does, however, seriously weaken their viability and hence their powers of competition in nature. In tetraploids the disequilibrium is less marked.

The taxonomic and evolutionary significance of these trisomics is limited by their genetic unbalance. However, although their phenotypic expressions and difference are comparable to those found separating species, they cannot be considered as deserving taxonomic rank: they do not breed true from seed and can only be maintained vegetatively (Blakeslee, 1934).

The loss of a whole chromosome in a diploid has much more serious consequences than the gain of one. Diploid monosomics are normally inviable and are seldom found. Aneuploids that do occur are most likely to be polyploid where the loss of a chromosome can be more easily assimilated than by a diploid as pointed out above. Examples are known in Tobacco (*Nicotiana*), Wheat (*Triticum aestivum*), etc. They may be phenotypically distinguishable or not.

Monosomics may arise in obligate apomicts such as *Taraxacum* spp. where they have been studied in detail by Sörensen & Guðjónnson (1946). Here the plants are triploid ($3x = 24$) and derived from probably sexual diploid ancestors and may give rise to seed with 23 chromosomes by the loss of one chromosome of one of the sets. Eight disomic triploids ($3n-1$) are possible according to which chromosome has been lost. All these strains may be recognised by their leaf shape and by their chromosome morphology, but are best not given formal names.

(*b*) *Accessory Chromosomes* (B-chromosomes). The second kind of supernumerary chromosomes found are, as Darlington (1956) says, something altogether new. They have been defined as chromosomes which "occur in variable numbers in some members of a population and (which) may be absent in other individuals without noticeable effects. They are usually small and heterochromatic and do not pair with the A-chromosomes at meiosis" (Müntzing, 1959).

The B-chromosomes pair with each other. Since they can be dispensed with altogether in some plants in populations in which they do occur, they are sometimes referred to as "ghost" or second-class chromosomes! Their effect on the plant is not specific as with A-chromosomes.

Reviews of the nature, origin and significance of accessory chromosomes are given by Darlington (1956: 19-27), Müntzing (1959) and Larsen (1960a: 233-237).

Accessory chromosomes were observed 50 years ago by Lutz in the insect genus *Diabrotica* and were first detected in plants in *Zea mays* (Longley, 1927; Randolph, 1928, 1941). Since then there have been numerous reports and according to Darlington hundreds of species possess them; they may be discovered in a fairly high proportion of plants. So far, they appear to be restricted to certain families.

The effects of accessory chromosomes are various. Their genetic effect may be little or none; or it may be deleterious by upsetting division in pollen grain or embryo sac, as in *Secale cereale* (Müntzing, 1959) and *Centaurea* (Fröst, 1956), thereby reducing fertility. In small numbers they may be advantageous. In *Festuca* Bosemark (1957) found that the presence of one or two B-chromosomes increased the vigour of the plants; similar results were obtained by Fröst (1958b) in *Centaurea*.

Adaptive significance: In several cases studied the frequency of accessory chromosomes differs notably from population to population as, for example, in *Phleum phleoides* (Bosemark, 1956b), *Secale cereale* (Müntzing, 1957, 1959). Darlington (1956) states that in any large population, either wild or cultivated, containing B-chromosomes, plants with the latter exist side by side in competition with those that do not possess them. Often it is possible to show some correlation between the different frequencies of B-chromosomes and ecological or climatic factors. In *Festuca pratensis*, for example, Bosemark (1956a) found that there was a positive correlation between the number of accessory chromosomes in a population and the clay content of the soil, although the correlation was not a simple one. In *Centaurea scabiosa* Fröst (1958a) found that in Scandinavia and Finland there was a marked east-west difference in frequency of accessory chromosomes. The geographical differences were largely correlated with differences in humidity, B-chromosomes being generally high in areas of low humidity. Lewis (1953) found a similar correlation in *Clarkia* between the presence of accessory chromosomes and adaptation to arid conditions.

Accessory chromosomes are more frequently found in diploid species than in tetraploids (Darlington, 1956). They have not so far been found in long-lived woody plants. It is suggested that the B-Chromosomes may increase variability and thereby adaptability, and that this stimulus is less necessary in polyploids than it is in diploids.

Origin and Nature: In plants accessory chromosomes may be of recent derivation or they may have arisen originally from A-chromosomes. They may represent centric fragments of telocentric arms of A-chromosomes formed as a result of misdivision (Darlington, 1956). This has been suggested for *Oenothera* (Cleland, 1951) and *Caltha* (Reese, 1954). In *Clarkia* the differences between A- and B-chromosomes are somewhat arbitrary. The B-chromosomes here represent duplication of part of the normal genome complement (Lewis, 1953).

The variations in number and presence of B-chromosomes introduce difficulties when it comes to interpreting chromosome numbers. As Darlington (1956) says, "they break almost all the rules of genetics and physiology which rest on the uniform behaviour of A-chromosomes". He regards the great amount of unspecified variation in chromosome numbers recorded in the literature as being largely due to B-chromosomes; they would also explain discrepancies between counts recorded by different authors for the same species in many cases. This goes to emphasise even more acutely the enormous amount of research into chromosome number that is still needed (cf. p. 194).

Chromosome Morphology

In addition to changes in chromosome number, the basic haploid complement is subject to variation in the form, size and volume of the chromosomes, including number and size of the satellites, and the amount and distribution of heterochromatin. Comparison of karyotypes is possible in certain groups, where the chromosomes are large enough to be studied in detail, and provides data of taxonomic value. In other groups the laborious and lengthy techniques necessary are not profitable.

In general, an alteration in the karyotype indicates that more profound changes have taken place at the submicroscopic level, although chromosomal changes and rearrangements can take place which are not reflected in the morphology of the karyotype, as Swanson (1958) points out. He comments that in the comparative cytological approach "criteria from all levels—morphological, cytogenetic, and cytochemical—must be integrated, for any one set of data can be misleading". Goodspeed (1954) writing on *Nicotiana* says that "because of presence of basic genom similarities relating members of a given section and because of specific dissimilarities, which in some cases identify individual species within sections, chromosome morphology . . . has phyletic significance as a commentary on validity of taxonomic conclusions based on evidence from external morphology, distribution and cytogenetics".

1. *Chromosome Size*

The absolute size of the chromosomes of a karyotype is a fairly constant feature of species and is genotypically controlled (Thomas, 1936). The reasons for the differences in size observed are not, however, evident. Size is no indication of genic content, as is seen by comparing closely related genera or species. The classic example is in the *Droseraceae*, where the total volume ratio of the chromosomes of *Drosera: Drosophyllum* is 1:1000. In the *Liliaceae*, *Trillium* has chromosomes 100 times larger than those of the related genus *Medeola;* even species within a genus may vary in the same ratio, as in *Euphorbia* (Darlington, 1956).

In general, the Monocotyledons have larger chromosomes than the Dicotyledons, although there are exceptions. Large chromosomes are found in the genus *Paeonia* which was formerly included in the *Ranunculaceae*, and this adds further support to its placing in a separate family which has been proposed on other grounds. Examples of studies on chromosome size are: Anderson & Sax (1936) on *Tradescantia* (*Commelinaceae*), Huziwara (1959) on *Aster* (*Compositae*), Kurita (1958, 1961) on *Ranunculaceae*. The latter author (1958b) reports that the total chromosome volume in the 14 chromosome species of *Anemone* studied by him was much larger than that found in the 16-chromosome species; in *Ranunculus*, on the other hand, the 14-chromosome species showed no difference in total volume from the 16-chromosome species. The difference probably reflects the different origins of the 14 chromosomes from the 16—by chromosome arm fusion and centromere loss in *Anemone*, associated with enlargement of the chromosomes, and loss of the entire chromosomes in *Ranunculus* but without such enlargement of those which remain.

In an investigation of the cytology of species of six genera of the *Liliaceae-Polygonateae*, Therman (1956) found two trends—in *Streptopus* species and North American representatives of *Disporum* studied there were short chromosomes (2-5μ), while the other groups including East Asiatic species of *Disporum* possess larger chromosomes ($4-14\mu$). There were also other cytological differences—in chiasmata and resting nuclei.

Many evolutionary trends in chromosome size have been described within families and lower taxa. In polyploid series, where the evolutionary sequence of species is normally beyond question, it has been found, as in *Dianthus* (*Caryophyllaceae*) and *Chrysanthemum* (*Compositae*), that the polyploids have smaller chromosomes than diploids, suggesting that reduction in size is an adaptation to increase in number (Darlington, 1956). Likewise annual species may have smaller chromosomes than perennials, as in *Crepis* (Babcock *et al.*, 1942). There is in fact a loose correlation between large chromosomes and lack of evolutionary specialisation although numerous exceptions occur. Increases in size may be associated with evolutionary advancement, as in the *Gramineae*, where genera adjudged to be the most advanced have larger chromosomes than those considered less evolved (Stebbins, 1956). Likewise in *Nicotiana* there is apparently no correlation between decrease in chromosome size and increase in specialisation of the plant. *N. anomala*, which has the largest chromosomes of the 12-paired species, is regarded as highly specialised in terms of morphology and physiology.

Smaller chromosomes may be found in hard-wood plants than in their herbaceous relatives—this is probably due to the small dimensions of the cambial cells which restrict mitosis and require small chromosomes for mechanical reasons (cf. Stebbins, 1950).

It is quite evident that although broad generalisations about some evolutionary trends may be advanced, the exceptions and reversals of trend may be so common that each group has to be treated on its own merits. Misuse of what is an extremely valuable phyletic tool is a danger to be guarded against.

2. *Chromosome Morphology and Relative Size*

We have so far considered differences in absolute size of entire chromosome complements, but some of the most obvious characteristics of karyotypes are the size and shape of the chromosomes relative to each other within the same complement. These differences are often represented in a diagrammatic form known as an idiogram or karyogram. Pioneers of the study of the karyotype in systematics include Levitsky (1931a, b); Levitsky & Tron (1930) and Navashin (1925, 1926). Recent illustrations of karyotype differences between species are given by Babcock *et al.* (1942), Babcock (1947a) on *Crepis*; Goodspeed (1954) on *Nicotiana*; Fernandes (1951, 1953, 1959a, b) on *Narcissus*; Maugini (1953, 1956,) on *Scilla*; Battaglia (1957) on *Allium*; Langlet (1932), Kurita (1958), Barros Neves (1944) on *Ranunculus*; Sato (1942) on *Liliaceae*; Sato (1960) on *Zingiberales;* Ono (1960) on *Heterotropa* (*Aristolochiaceae*); Schotsmann (1961) on *Callitriche*, to cite just a few representative examples.

Viewed at mitotic metaphase, the individual chromosomes of some groups of plants show marked differences in size and shape, in other cases the chromosomes may be quite uniform; or the differences may be gradual between the various chromosomes. The constant position of the centromere also allows chromosomes to be characterised by the relative length of the arms. Further differences may be shown by the presence and position of secondary constrictions marking off bodies known as satellites when they are distal.

Markedly different forms of chromosomes occur. They may be known as V-types (symmetrical), with two equal arms and a median centromere; and asymmetrical types—J-types with the long arm much longer than the short one, and I-types or rod-shaped with a subterminal centromere. All intergradations between these types are found.

It has proved possible to work out evolutionary trends in these features in some groups. In *Aconitum* (*Ranunculaceae*), for example, the karyotype varies from species to species in respect of chromosome length and position of spindle attachment; the annual species which are regarded as the most recently evolved have the most "specialised" karyotypes. Similar trends are found in the tribe *Helleboreae* as a whole—the most asymmetrical karyotypes are found in the most advanced genera *Delphinium* and *Aconitum* (Levitsky (1931b). Further discussion of the karyotype trends in the *Ranunculaceae* is given by Kurita (1958a). He considers that the karyotypes of *Delphinium*, *Lycoctonum*, *Aconitum*, *Trollius* and *Caltha* with many J-type chromosomes or with the largest V-types with a satellite to be more advanced than those of *Actaea*, *Cimicifuga*, *Anemonopsis*, *Clematis*, *Eranthis*, *Anemone*, *Pulsatilla* and *Ranunculus*. *Actaea* is regarded as the most primitive in possessing more V-type of chromosomes of large size than any other group.

In general, there appears to be a correlation between asymmetry in the karyotype and evolutionary specialisation of the plants in habit and morphology. This is illustrated by the genus *Crepis* (Babcock, 1957a) and is further discussed by Stebbins (1950). There are, however, exceptions to this general rule, symmetrical karyotypes being found in groups regarded as highly evolved or specialised morphologically.

An interesting situation occurs in the *Agavaceae*, as discussed by Darlington (1956). There the chromosomes are of two types, long and short, the long being of the J-type and the very short (microchromosomes) appearing as though they had been derived from the short arm of the longs (McKelvey & Sax, 1933). The genera *Agave* and *Yucca* possess 5 longs and 25 shorts. They were formerly placed in separate families, the *Amaryllidaceae* and *Liliaceae* respectively, but Hutchinson united them in the *Agavaceae*—a decision which is not only justified on general taxonomic evidence but which is supported by their karyotype similarity.

In general, these changes in symmetry and shape are due to various structural alterations such as translocations, inversions, duplications and deletions. When it is possible to trace the ways by which they have arisen in a particular karyotype, this gives valuable information on the origin of the group concerned. On the other hand, structural changes may occur which are not reflected in the morphology or number of the chromosomes: the order of genes in the arms may be shifted by (paracentric) inversions, for example, a feature of considerable evolutionary importance since it may lead to intersterility between plants without any accompanying phenotypic change.

3. *Heterochromatin*

Heterochromatin is a term applied to all those chromosome regions which show a difference in their appearance, compared with the euchromatin, at some or all stages of division. This wide definition includes nuclear materials with widely different properties and, no doubt, functions. It includes such commonly found types as the sex chromosomes, or parts of them, in animals and some plants; the precociously condensed regions of small plant chromosomes; and some accessory, or B, chromosomes. In the first two cases at least, only differences in the degree of coiling seem to be involved, and there is no reason to think that such regions are genetically inert. On the other hand, there are other types (as in the autosomes of *Drosophila* and some mosses) where the structure of heterochromatic regions, which may be reduced or added to with apparently little effect, may differ more fundamentally from that of euchromatin. In most cases the distribution of heterochromatin can add nothing to any conclusions which may be reached concerning the relationships of species and individuals based on chromosome morphology.

There is one type of heterochromatin, however, which does reveal relationships which cannot be deduced from chromosome morphology. This type, found very sporadically in plants and animals with large chromosomes (Dyer, 1963), appears as condensed chromocentres at interphase, but as underspiralised segments during mitosis at low temperatures only.* The position of these segments varies from species to species and individual to

* Dyer (1963) records the certain occurrence of cold-sensitive heterochromatin in the following plant genera: *Vicia* (*Leguminosae*), *Cestrum* (*Solanaceae*), *Adoxa* (*Adoxaceae*), *Trillium*, *Paris* and *Kinagusa* (*Trilliaceae*), *Fritillaria* and *Tulbaghia* (*Liliaceae* sensu Bentham), *Secale* and *Hordeum* (*Gramineae*). In animals it is found in *Mecostethus* (a grasshopper), *Triturus* (newts), *Bufo* (toads) and *Rana* (frogs).

individual, forming a spectrum of "H-patterns" characteristic of the species. Similar or overlapping H-patterns are found in related species, as illustrated particularly clearly by *Trillium* (Bailey, 1952, 1954; Darlington & Shaw, 1958; Kurabayshi, 1958) and *Fritillaria* (Dyer, 1963). Parallel polymorphism of H-patterns seems to reflect a recent common ancestry. More distantly related species may show little or no similarity.

The analysis has been carried even further in *Trillium kamtschaticum* in northern Japan (Fukuda & Kozuka, 1958). Here each population is characterised by its spectrum of H-patterns and the relative frequencies of each pattern. Comparison with similar data from other populations throughout the species' range provided information on the history and structure of each population and its present relationships with its neighbours. Though valuable as an interpretive tool, such a refined analysis is below the level of taxonomic classification.

Most of the plants with the type of heterochromatin exemplified above flower early in the season under cold conditions. Dyer (1963) concludes that the occurrence of heterochromatin "appears to be correlated with the ability of species with large chromosomes to undergo cell division at low temperatures". Cold-sensitive heterochromatin appears to have arisen in several unrelated Angiosperm groups and to have enabled the genera concerned to spread into colder areas. In this connection it is interesting to compare the distribution of *Trillium* species with their H-content (Dyer, *in litt.*). Among the American species those with least heterochromatin have the most restricted, and southerly, distributions, whereas those which extend further north have, in general, the most heterochromatin. A similar relationship exists between the Japanese species of *Trillium*. Within the Japanese species *T. kamtschaticum*, the only one in which detailed studies of heterochromatin pattern have been made throughout the species' range, it can be shown that those chromosome types restricted to the north of the range have a much higher heterochromatin content than those restricted to the south; those occurring throughout the range occupy an intermediate position. In both continents the genus appears to be spreading from south to north, and the amount of heterochromatin to be increasing.

4. *Chromosome Behaviour at Meiosis*

Valuable information can be obtained from a study of the way the chromosomes behave at meiosis. It may permit a distinction to be made between taxa, as has been demonstrated by Garber (1958) in the genus *Collinsia* (*Scrophulariaceae*). He employed chiasma frequencies as a means of grouping species within the genus: those with "sessile" flowers possessed a low mean number of chiasmata while those with pedicellate flowers had a relatively high number of chiasmata. A classification of organisms according to the behaviour of chiasmata at metaphase is given by Darlington (1937: 110-111); such behaviour may give some indication of affinity. Chiasma frequencies and distribution are known, however, to vary widely between sexes, between closely related species, with environmental conditions, with genes, etc., so that their taxonomic significance varies and their occurrence may in fact cut across established classifications.

More frequent use is made of meiotic behaviour as a means of indicating relationships through the kind and degree of pairing that takes place, since it may show that hybridisation has occurred, indicate structural differences in the parental chromosomes such as reduplications, or explain the causes of sterility.

Great weight is often placed on the degree of chromosome homology in hybrids as an estimate of the degree to which the parental species are related. The chromosomes derived from the two parents are often not homologous at meiosis and do not therefore pair; if the numbers of the parents are different, pairing will of course be incomplete.*

The Taxonomic Value of Cytological Data

1. *General*

As we noted earlier in this chapter, cytological characters are often regarded as of predominant importance in taxonomy since the chromosomes are more closely connected with the mechanism of heredity than other features employed in taxonomy (Darlington, 1956; Stebbins, 1959; Löve, 1960). As Warburg (1938) comments, this is true to some extent, since chromosome number and morphology, being directly inherited, often give direct evidence regarding origins of forms, as in polyploids. But those cytological characteristics which are visible are not necessarily those most directly concerned in the hereditary mechanisms. A distinction has to be drawn between the value of cytological information as providing comparative data, and its value for making evolutionary and interpretative deductions.

The problem is acutely posed by Smith-White (1954), who writes, "If phyletic relationship is accepted as a basis of taxonomy, cytological characteristics must have a significance exceeding that of ordinary morphological characters." Even more pungent is the statement of Löve (1960a) that "we must remember that the chromosomes are *not* just another character comparable to the superficial morphological characters taxonomists are forced to use for the identification of herbarium material. The chromosomes determine the characters, whereas the characters do not determine the chromosomes. It is evident . . . that to describe the chromosomes as only another useful character, comparable to the hairiness of the stem or the colour and form of the flower, is therefore hardly a slight but a serious misunderstanding of the principles of evolutionary biology." Taken literally, such overemphasis on chromosome characters suggests a misunderstanding of the principles of classification!

* However, it has been shown in *Triticum* that chromosome pairing can be under relatively simple genetic control (Riley, 1960).

Flower colour may be used as a marker of discontinuous groups in a classificatory sense; so may chromosome numbers. Chromosome number, for example, cannot by itself be relied upon for taxonomic division unless it correlates with other characters, as was pointed out by Smith (1933) in one of the pioneer papers on the role of cytology in taxonomy. More than one character has to be used to make a group. As Smith-White (1954: 302) notes in a discussion of the chromosome numbers in the *Rutaceae-Boronieae*, a taxonomic grouping based purely on chromosome numbers would be unsound. Reliance on *any* one character without reference to other evidence is, as we have pointed out repeatedly, incompatible with natural classification. What is often meant by the greater significance of chromosome numbers is, presumably, that given a choice of groupings, those marked off by chromosome number are more important than those indicated by a morphological criterion. In other words, chromosomes should be *weighted* as characters.

Weighting, as we have seen, can be of several kinds (Chapter 2). As for any other line of evidence, *a priori* weighting of chromosome number must be rejected. On the other hand, chromosome number may be given *correlation weighting*, in which case it is no different from any other kind of character. The situation where there is a choice of groupings is seldom so simple in practice, for unless the groups delimited by chromosome number are natural (in the sense of showing maximum covariation of attributes) they would not be better for general classificatory purposes; if they *are* natural, then morphological features (such as flower colour) may well be available as markers. It helps little to say in the latter case that the morphological markers are a reflection of the underlying chromosome differences: both are reflections of a classificatory phenomenon. Even if the aim of a classification is avowedly phyletic, chromosome number or other cytological evidence still cannot by itself provide a satisfactory basis.

The value of chromosome numbers is that since they are known so often to show a high degree of stability, and since they so often correlate with natural groupings, they may be extremely valuable in suggesting *where* to look for discontinuities in the variation pattern, i.e. they may tell us where a division may be made. But the division into taxa cannot be made unless the other evidence available confirms its validity. Classification must make sense!

Likewise chromosome numbers often serve as a pointer for taxonomic reappraisal, as in the major classification of the *Gramineae* (Avdulov, 1931), but themselves cannot form a basis for taxonomy: they must be related to other phenetic evidence in a meaningful way, as Babcock (1947a) stressed in his monograph on *Crepis*.

A few examples will illustrate this. Smith (1933) showed that within the genus *Primula* cytological characters may vary widely: in some sections the chromosome numbers agreed with morphological groupings, in others they did not. Likewise in some sections large cytological differences correspond to small morphological changes and vice versa. Chromosome characters were, he found, subject to the same kind of variation as other characters, and parallelism may be the explanation of the discord between morphology and cytology in some cases. This is also true in *Crepis* (Babcock, 1947a), and in the *Rutaceae-Boronieae* (Smith-White, 1954) where the series 9-8-7 has apparently occurred twice within *Boronia* and other groups. Very many other examples are known of such parallel alterations in chromosome number and other cytological trends.

For comparative taxonomic purposes we see no clear reason why cytological characters need necessarily be considered as of special importance, except in so far as they are more often good indicators of where discontinuity may be found than are many other kinds of character. We must insist, however, that there are no grounds for considering them as *a priori* more important for group-making.

On the other hand, when it comes to interpreting a classification and establishing various kinds of relationship, cytological evidence is often of the greatest importance. It may show the *direction* of evolutionary change, indicate which groups are derived from others or at least which derivations are not possible, following the negation principle discussed in the last chapter. For example, a group with a diploid chromosome number cannot normally be derived from one with a tetraploid number. This kind of evidence combined with a knowledge of geography and other data will often tell us a great deal about the history of a group. Cytology will often give evidence as to the mechanism whereby the evolution of a group has come about. Different chromosome numbers will often indicate the presence of strong reproductive isolation between groups, and evidence from chromosome pairing at meiosis will give an indication of phyletic relationship.

2. *Family Level and Above*

Darlington & Mather (1949) have constructed a diagram showing how the changes in basic number of woody flowering plants are related to phylogenetic schemes such as that of Hutchinson. But as Darlington (1956) notes, it has to be understood that the names of the families in such a scheme "are taken to represent directly not their external structures, but their chromosome numbers; and the relation of the two is just the problem to which we cannot give a uniform solution". In other words, one has to presuppose a classification with which one may not agree and circular arguments are difficult to avoid. Such diagrams are certainly suggestive of phylogenetic derivations and add an important dimension to such speculations.

For classificatory purposes chromosome numbers have little value at the higher levels of classification where there is seldom much dispute about the delimitation, as opposed to the relationships or rank of most families. The presence of an anomalous chromosome number in a group may, however, be an indication that a change of classification is needed.

Few studies of the cytological relationships of families have been published.

An example is that of Warburg (1938) on the taxonomy and relationships of the *Geraniales* in the light of their cytology. Within families chromosome number may suggest improved arrangements of tribes and genera. Good examples are afforded by the *Ranunculaceae*, *Cruciferae* and *Gramineae*.

The cytology of the *Ranunculaceae* has been studied by Langlet (1932), Gregory (1941) and Kurita (1958a). The chromosome sets of the family fall into two distinct types—the large R-type as found in *Ranunculus* and the small T-type as found in *Thalictrum*. When taken in conjunction with chromosome number, the karyological data provide a more satisfactory basis for subdivision than that provided by fruit type in that a more natural arrangement ensues. In the previous arrangements of the family, genera with large and small type chromosomes and different basic numbers ($x=7$, 8, 9, 10) were grouped together in the same tribes. Rearrangement according to cytological features has removed certain taxonomic anomalies, although further work is needed.

Kurita (1958a), for example, considers that *Nigella* ($x=6$) and *Adonis* ($x=6$) are far removed from other genera in the family in respect of their karyotype; *Paeonia* ($x=5$) is also distinct karyotypically and has been removed on various other lines of evidence to a separate family.

The studies of Manton (1932) on the cytology of the *Cruciferae* were remarkable in the way that the different basic chromosome numbers fitted in with the subdivision of the family by O. E. Schulz on the basis of apparently trivial characters such as indumentum type, nectary arrangement, etc.

We have mentioned in Chapter 5 the way in which Avdulov's cytological studies on the grasses (1931) suggested major rearrangements within the *Gramineae* which agreed with those which leaf anatomical studies indicated. The major subdivisions of the family accepted today are in fact characterised by chromosome number and size amongst other features (cf. Prat, 1960; Stebbins, 1956).

In the *Epacridaceae* the tribe *Epacrideae* has a probable basic number of $x=6$; but the usual haploid numbers are 13 and 12 and only one case of polyploidy is reported. The tribe *Styphelieae*, on the other hand, has basic numbers of $x=4$ and $x=6$ from which are built up by polyploidy and aneuploidy haploid numbers of 4, 6, 7, 8, 9, 10, 11, 12 and 14. The tribes are also different in their pollen types—the *Epacrideae* with tetrads while in the *Styphelieae* they are usually monads (S-type) with variants in the pattern of pollen development associated with aneuploidy (Smith-White, 1948, 1955a, 1959).

Chromosome numbers in the different tribes of the *Rutaceae* show a similar diversity: the basic number is $x=9$, which is characteristic of all tribes except the *Boronieae*, where secondary numbers are more frequent, forming a more or less continuous series from 7 to 19 (10 and 15 excepted). Different basic numbers are regarded as characterising genera and constitute a first level of change in the family. Current classification largely agrees with this criterion of generic status for groups with different base numbers (Smith-White, 1954, 1959).

The difficulties of determining basic numbers in the *Compositae-Helenieae* have already been mentioned. Chromosome studies have also been made on various other tribes such as the *Cichorieae* (Stebbins *et al.*, 1953). In the latter tribe it has been found that karyotype evolution does not always go in step with morphological specialisation. Exceptions are found to Levitsky's hypothesis that primitive species have karyotypes with chromosomes similar in size and possessing median or submedian centromeres, while more advanced species tend to have chromosomes which are relatively unequal in size within the same complement. These exceptions are found in the subtribe *Microseridinae* and in *Hieracium*, *Sonchus* and *Taraxacum*. Phylogenetic decrease in basic number occurs repeatedly as a series of separate trends in various genera.

3. *Generic Level*

As mentioned in connection with the *Rutaceae-Boronieae*, different basic numbers are characteristic of genera. There are very many other examples where this is found. In the *Cistaceae*, for example, *Cistus* and *Halimium* have a basic number of $x=9$ while *Helianthemum* has $x=8$. This lends support to the recent tendency to place *Halimium* near *Cistus*, and not include it in *Helianthemum* as done previously.

This reliance on basic chromosome number as a criterion of generic status cannot, however, be accepted as a general rule, since on the one hand a basic number may be constant throughout a tribe, subfamily, family or even higher taxon, and on the other hand parallel development of more than one number may occur within different genera. In other cases more than one basic number may occur within a genus and not be correlated with taxonomic division, as, for example, in *Cochlearia* (*Cruciferae*), where "species" with different basic numbers and different levels of polyploidy on these basic numbers may be virtually indistinguishable morphologically. Thus *C. pyrenaica* ($2n=12$, $x=6$) cannot be satisfactorily separated from *C. alpina* ($2n=28$, $x=7$).

The validity of different numbers in characterising infrageneric groups is not supported in this instance, since it appears that chromosomal evolution has been independent of morphological evolution. There may likewise be a lack of synchronisation of other aspects of karyotype evolution and other genetic or morphological evolution, as in the genus *Lathyrus*.

Recent examples of chromosome numbers supporting generic status include the following. Favarger (1960b) has found that the two species *Cicendia filiformis* and *Microcala pusilla* (*Gentianaceae*), which at one time were placed in the same genus *Cicendia*, have different base numbers, $n=13$ and $n=10$ respectively, which supports their generic separation on the basis of ovarian morphology, stigma and calyx. He points out, however, that we may remain sceptical of the value of basic number as a general generic criterion in the *Gentianaceae* if we consider that within a single section of *Gentiana* one may find several numbers ($x=7$ and $x=11$ in sect. *Cyclostigma*; $x=11$, $x=13$ in sect. *Crossopetalum*, etc.). However, the sections themselves may not be natural! Again, all species of the genus *Jovibarba* (*Crassulaceae*) possess $n=19$, while in *Sempervivum* the basic numbers are $x=16$, 17, 18, 20 and 21; this has been taken to support the re-establishment of the former genus (Zésiger, 1961) which differs from *Sempervivum* in its hexamerous flowers.

4. *Specific Level and Below*

Most cytotaxonomic literature is mainly concerned with problems at the species and infra-specific level. Surveys of the chromosome numbers of the species occurring in particular areas are often published. In addition to the major lists already noted on p. 195, examples include Löve & Löve's cytotaxonomical conspectus of the Icelandic flora (1956), and Jørgensen, Sørensen & Westgaard's taxonomic and cytological survey of the flowering plants of Greenland (1958). Series of studies on the cytotaxonomy of the flora of various countries have been published: Baksay (1956, 1957, 1958) on Hungarian Pontic-Mediterranean plants; Favarger (1949, 1953, 1959, 1960a, 1962) on Alpine plants; Larsen (1955, 1956a) on Mediterranean plants; Larsen (1960b) on flowering plants of the Canary Islands; Skalińska (1950), Skalińska *et al.* (1957, 1959, 1961) on Polish plants; Hara (1954, 1956, 1962) on Eurasiatic plants.

An outstanding series of studies on the cytotaxonomy of the genus *Narcissus* has been published by Fernandes. In a paper (1951) summarising the phylogenetic relationships of the species as evidenced by their cytology, he has proposed a new classification based on karyological studies, morphology, ecology and geographical distribution. This is illustrated in Fig. 16. In the evolution of the species of the genus most kinds of chromosomal change have been involved—autopolyploidy, allopolyploidy, polysomy, euploidy and aneuploidy and B-chromosomes.

Cytological differences between species are frequently in terms of aneuploidy and/or polyploidy. Similarly a great deal of infra-specific cytological variation due to these factors occurs and will be considered in more detail in the next section.

Polyploidy and Taxonomy

In the early days of cytotaxonomy there was a tendency to regard polyploidy as a phenomenon requiring special treatment. Today we realise that the incidence of polyploidy is so high that it is almost exceptional if a group under consideration does not show it. In the taxonomic treatment of groups showing polyploidy we have been to some extent handicapped by those botanists who seek to apply rigid rules concerning particular sorts of evidence

CYTOGENETIC RELATIONSHIPS OF SPECIES IN RELATION TO THE CLASSIFICATION OF THE GENUS NARCISSUS

FIG. 16

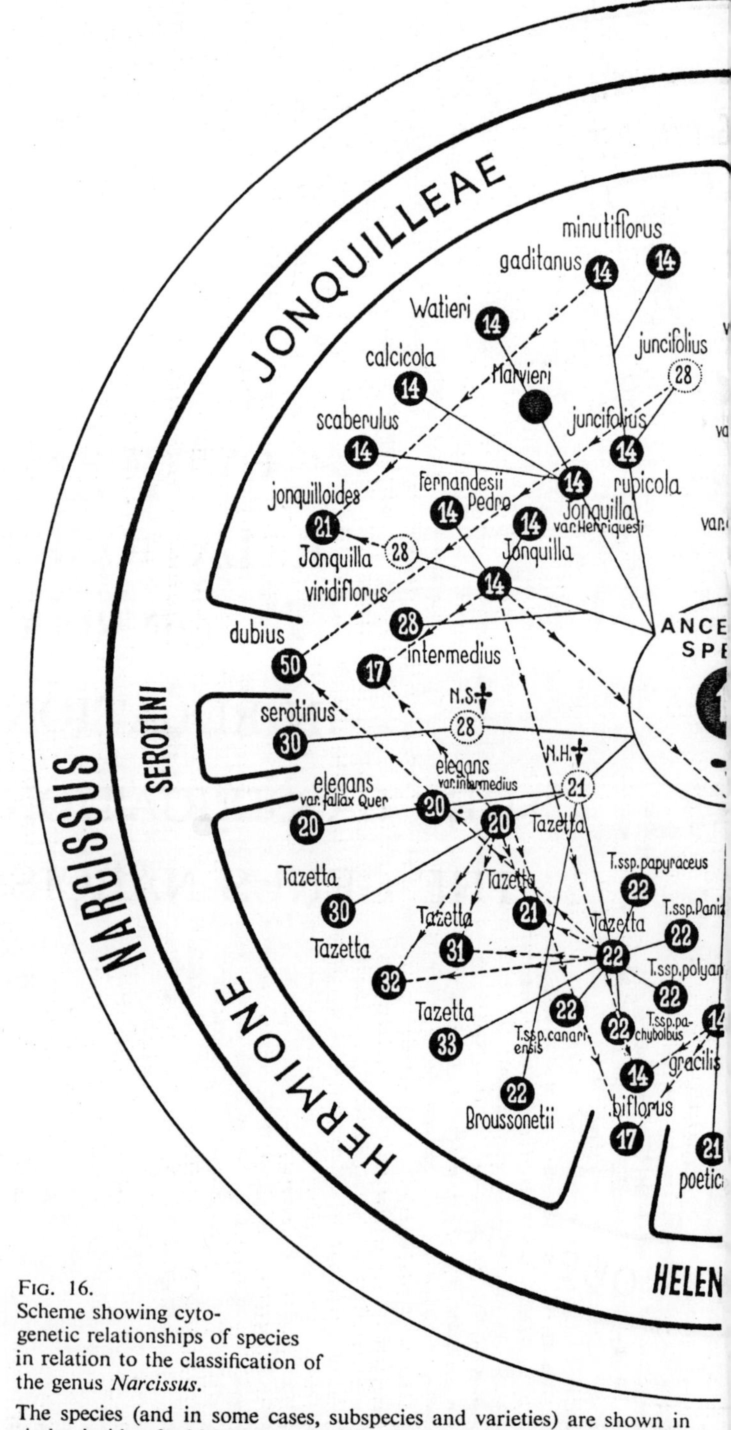

FIG. 16.
Scheme showing cyto-
genetic relationships of species
in relation to the classification of
the genus *Narcissus*.

The species (and in some cases, subspecies and varieties) are shown in circles inside of which are given their respective chromosome numbers. The dotted circles correspond to polyploids which have not yet been found in nature but which probably occur. The daggers indicate forms which no

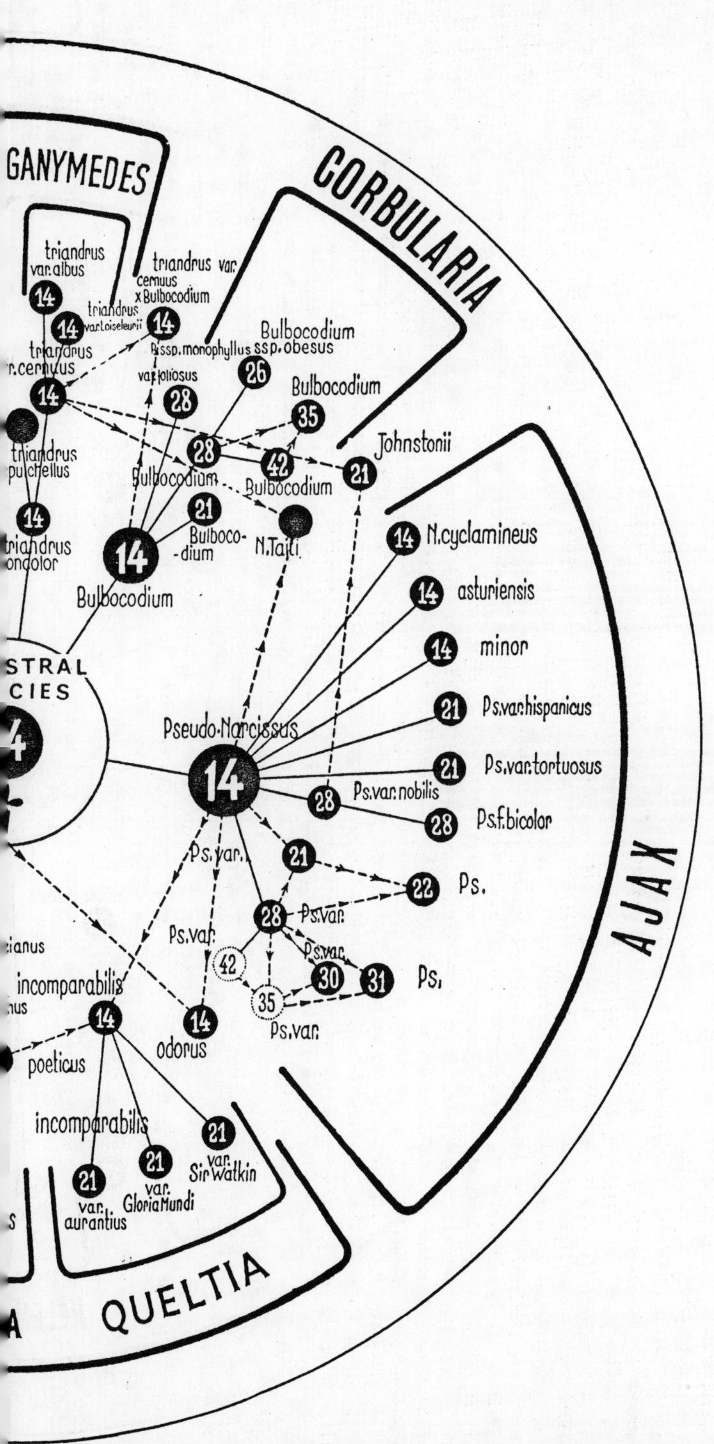

longer exist. N.S. means the ancestor of Sect. *Serotini* and N.H. the ancestor of Sect. *Hermione*. The crosses from which other forms have been derived are indicated by broken lines and arrows which unite the parents with the hybrids. (After Fernandes, 1951.)

used in classification. Thus, there is a school of cytotaxonomists which insists that diploids and polyploids must always be kept apart as separate species, no matter what the apparent degree of morphological differentiation involved. This attitude is based mainly on the assumptions that chromosomal evidence is more important than morphological evidence, as discussed above (p. 207); and that diploids are separated from tetraploids by genetic barriers through the formation of sterile triploids. Polyploidy in fact illustrates most acutely some of the major conflicts to be found in taxonomy today.

1. *Frequency of Polyploidy*

Published estimates suggest that polyploidy has been involved in the formation of anything from a third to a half of known Angiosperms (Stebbins, 1947, 1950; Löve & Löve, 1949). These estimates have been queried by De Wolf (1957), who calculated that on the basis of the counts reported in Darlington and Wylie's *Chromosome Atlas* (1955) only about 6 per cent. of Angiosperms and Gymnosperms were known cytologically and a third of these, about 5000 species, gave evidence of a polyploid origin. "Thus, to date about 2 per cent. of the total number of species have been proved to have polyploid representatives."

Most of the available data on chromosome numbers are from temperate floras; only very limited and sporadic information is available for the incidence of polyploidy in tropical and subtropical floras. When more is known, total estimates may well have to be revised.

Various attempts have been made to explain the different incidences of polyploidy in different geographical areas with factors such as the increased frost resistance of polyploids compared with diploids (Löve & Löve, 1943; Löve, 1951); and there appears to be a certain positive correlation between the percentage of polyploidy in a flora and its latitude. There is some evidence to suggest that polyploidy is frequently associated with extreme conditions in general. Detailed discussions are given by Manton (1950), Mesquita Rodrigues (1953), Löve & Löve (1957) and Reese (1961). Such apparent correlations may, however, be biased because of other correlations, as between polyploidy and herbaceous as opposed to woody habit (Stebbins, 1938), so that a high or low incidence of polyploidy may be a reflection of the composition of the flora concerned in respect of its habit. Temperate floras would therefore be expected to possess higher polyploidy because of the greater percentage of herbaceous species than a tropical flora with a higher percentage of woody species. As Swanson (1958) says, any conclusions drawn concerning the extent and geographical distribution of polyploidy must be regarded as tentative in the light of knowledge so far available.

Apart from such considerations, in Temperate floras not only are polyploid counts being published with increasing frequency, but apparently simple diploid-tetraploid pairs are frequently shown on further investigation to be part of much more complex polyploid situations.

2. *Types of Polyploidy*

In simple terms, a polyploid is an organism whose somatic complement is a multiple of the basic number of chromosomes of the group to which it belongs. A customary division is made between autopolyploidy (autoploidy) and allopolyploidy (alloploidy) according to whether there is replication of the chromosome complement of a non-hybrid diploid, i.e. multiplication of essentially similar genomes, or of the chromosome complement of a hybrid between two diploids, i.e. multiplication of dissimilar genomes. If similar genomes are indicated each by a different letter, A, B, C, etc., a diploid may be represented as AA, an autotetraploid as AAAA, while an allotetraploid would be indicated as AABB derived from a doubling of the hybrid AB formed from two diploid species AA and BB.

In practice it is not always easy to make this distinction since there are many cases where both kinds of polyploidy are involved, especially at higher levels where situations known as autoallopolyploidy occur (Stebbins, 1947, 1950).

Verification of the kind of polyploidy may be made by a study of meiotic pairing. In autopolyploids (autoploids) the chromosomes come together and may form univalents and multivalents at meiotic metaphase—in autotetraploids the four homologous chromosomes may pair, forming quadrivalents. This is taken as evidence of the similarity of the genomes which are interpreted as having come from the same species. In allopolyploids (alloploids) the chromosomes tend to pair normally as in diploids, since the chromosomes from each parent pair amongst themselves.

In cases where diploid hybrids between species show a high degree of pairing and have high fertility,* polyploids formed from them will show mainly bivalents but also some quadrivalents and trivalents showing that some segments of the chromosomes are homologous in the two genomes. These are known as *segmental allopolyploids* (Stebbins 1947, 1950)—see p. 221.

The occasional fallibility of these distinctions between auto- and allotetraploids are pointed out by Nordenskiöld (1945)†, and the position is neatly summed up by Swanson (1958): "allopolyploidy and autoploidy, in a sharply defined cytological sense, are only the extremes of a spectrum of meiotic homologies ranging from complete homology between genomes (autotetraploidy) to a complete lack of homology (allotetraploidy)." Segmental allopolyploidy occupies an intermediate position.

Despite these difficulties of definition, the taxonomic, evolutionary and

* This is not necessarily true: pairing may be good but the hybrid sterile, e.g. *Primula kewensis* ($2n$).

† See also discussion by Roy (1962) who considers that genetic segregation ratios provide a safer criterion to judge the type of polyploidy than the kind of pairing.

physiological differences between clear-cut cases of autopolyploidy and allopolyploidy are such that it is convenient to consider them separately. The variation shown by the different types of polyploid may be quite diverse, and it is important to remember that this variation may be the result of differences in the constituent chromosomes which may be of a genic or structural nature (Keck, 1957).

In autopolyploids formed by the doubling of intra-specific crosses (i.e. between different races or ecotypes of the same species), the resultant chromosome sets may show differences due to these factors. The taxonomic consequences may be quite considerable, since there may be segregation towards one or other of the constituent genomes as a result of the differences in the chromosomes.

In very many cases, probably the majority, evidence as to the kind of polyploidy operative is indirect since cytological confirmation has not been attempted.

3. *Autopolyploidy*

Occasional autoploid individuals are known to occur as aberrants in many species. Examples of such accidental tetraploids have been found in *Lotus uliginosus*, Larsen (1956b), *Trifolium arvense*, Böcher *et al.* (1955a), *Onobrychis caput-galli*, Larsen (1955, 1956a), and others are mentioned by Böcher (1961a:35).

The occurrence of sporadic tetraploids (as well as triploids) may prove to be widespread as cytological analyses of species are extended: Darlington (1956), for example, suggests that most diploid species contain some tetraploid individuals. Böcher (1954) found accidental polyploids scattered in populations of *Pulsatilla pratensis* (*Ranunculaceae*) in N. Denmark and treated them as var. *duplex* since they were distinguishable by gigas features, but later (1960b) he considered they would be best treated as f. *duplex*.

It has been suggested by Heywood (1960b) that there is much to be said for regarding this kind of autopolyploidy as a *grade* character (cf. Chapter 2). Autoploid individuals may be thrown up independently and repeatedly by the diploid pool of the population. These individual autopolyploids depend for their persistence on their ability to reproduce and find the appropriate ecological niche for survival. In some cases they may come to be regarded collectively as a distinct taxon, as in *Deschampsia alpina* which Hedberg (1958a) has suggested consists of a polyphyletic assemblage of clones that have arisen by polyploidisation of various potentially viviparous races of *D. caespitosa* (cf. Nygren, 1949). In Scottish populations there are such continuous transitions in morphological features between diploid plants referred to *D. caespitosa* and tetraploids referred to *D. alpina* that a specific separation between them is not considered possible. The individual autoploid derivatives would be satisfactorily treated as *formae*, as Böcher suggested for the case of *Pulsatilla pratensis* mentioned above.

Stebbins (1959b and *in verb.*) is of the opinion that very few successful autopolyploids have arisen from a single diploid population. He knows of no example of a recently evolved natural polyploid (as opposed to artificially produced) for which it is the most likely origin. In fact, autopolyploids have probably been mainly derived from hybridisation between differently adapted diploids, such as completely interfertile ecotypes of the same species. Despite their hybrid origin they are cytogenetically to be regarded as autopolyploids—so-called interracial or intra-specific polyploidy. A generally accepted example of a genuine natural autotetraploid is *Galax aphylla* (*Diapensiaceae*) of eastern North America (Baldwin, 1941), where the diploids and tetraploids are virtually inseparable morphologically, occupy the same habitats and have the same distribution. A critical re-examination of this case is desirable.

It is generally considered that autopolyploidy has little taxonomic significance since it is essentially an intra-specific phenomenon (Keck, 1957), but there are certain evolutionary consequences which may be reflected in their taxonomy, as discussed below.

Differences between diploids and their artificial autotetraploid derivatives are frequently of the following kind: (1) gigantism as a result of increase in cell size in various parts of the plant, (2) increased physiological vigour, (3) sometimes distinct ecological preferences and different geographical distributions. Other differences are discussed by Stebbins (1950: 301-5). An example illustrating these differences is *Cuthbertia graminea* (*Commelinaceae*) described by Giles (1942).

Taxonomically usable differences may not be associated with autopolyploidy. In fact, on the basis of the nature and degree of morphological differentiation involved, autopolyploids (and to some extent allopolyploids, too) may be divided into cryptic, semicryptic and distinct, and the consequent taxonomic status afforded to them ranges from specific rank to no formal recognition at all.

As we have already noted, however, some cytotaxonomists believe that every polyploid level should be treated as a distinct taxonomic species. Intra-specific polyploidy would not be recognised by them except as a sporadic low-frequency phenomenon in populations. That this viewpoint is not too widely held is evident from recent literature. Many cases have been described where a recognition of polypoid races as separate species has not been considered appropriate or indeed possible. This is, in fact, one of the most controversial aspects of modern taxonomy.

The theoretical background of polyploids at specific rank is that polyploidy causes the immediate separation of an effectively isolated unit or gene-pool, an aspect stressed by Valentine's useful terms abrupt and gradual speciation (Valentine, 1949). But to regard polyploidy as an automatic indication of specific status is an oversimplification, for the autoploid individual, athough reproductively isolated from its diploid, may or may not survive and propagate. As Böcher (1961a) says, "It is not the doubling up of a chromosome complement which is crucial but rather the success of the tetraploids. If they are able to find a separate ecological niche, isolation mechanisms will soon remove them from their diploid ancestors and, as evolution proceeds, they will reach one day a stage which makes it natural to regard them as species." Autopolyploidy, therefore, provides an isolation mechanism, but isolation by itself is not a specific criterion, for species must exist as independent differentiated populations. Initially there may be no genetic differentiation and no morphological distinguishing features which can be satisfactorily appreciated. On the basis of the single character of genetic discontinuity, two gene-pools may be defined and two "genetic species" may be deemed to have arisen, but the separate evolutionary divergence of the two units, although no doubt likely in many cases (cf. *Narcissus bulbocodium*, discussed below), is a matter for prediction. At best it is always a matter of opinion, as Böcher comments, and from a practical point of view the task of the taxonomist is to describe relationships as he sees them, not as they might be in the future (cf. Lewis, 1957).

The major taxonomic problems posed by autopolyploidy are concerned with assessing the degree and constancy of the morphological features of markers associated with it. The ascertainment of chromosome number is not practicable in most circumstances: it cannot, for example, be determined from herbarium material (although frequently seeds may be available in recently collected material for germination for root-tip counts). Even if the services of a cytologist are available when the taxonomist is not sufficiently skilled cytologically, there may be many technical reasons why the chromosome count cannot be established, and delays are inevitable.

Consequently taxonomists may have to rely on polyploid indicators or markers (cf. Bell, 1954). This raises a whole series of problems which have seldom been adequately considered. We will now consider some examples illustrating these problems.

4. *Cryptic Polyploidy*

The term *cryptic polyploidy* may be applied to these cases where morphological markers are absent or virtually so. It is applicable to both autoploids and alloploids or to situations where the exact nature of the polyploidy is not known. An interesting case is described by Fagerlind (1937) in *Galium verum* and *G. mollugo* (*Rubiaceae*), in both of which diploid and tetraploid races (cytodemes) occur which cannot be distinguished morphologically from each other. The situation is further complicated by the fact that extensive hybridisation takes place at the tetraploid level. Normal taxonomic treatment here is to regard the diploid and tetraploid cytodemes together in each case as one species. A biosystematic classification which adhered strictly to the delimitation of breeding populations would regard the two tetraploids between which gene flow occurs as one species (hologamodeme), thus separating them from the diploids, their immediate progenitors, from which they differ only in chromosome complement. This classic example is given here since it illustrates so well the different approaches of the taxonomist and cytogeneticist to such problems. Which solution is adopted depends on what one is trying to show.

In *Lathyrus pratensis* cryptic polyploids have been described by Larsen (1957). Diploids have been found in Scandinavia and Central Europe while tetraploids have been found to be restricted to South and West Europe. Flowers size differences were shown to be unreliable over a wide range of material—tetraploids normally having the largest flowers, but three British samples were found to be tetraploid and small-flowered; the morphological variation and degree of hairiness within the two races was greater than that between them; no difference in pollen size could be appreciated.

Cryptic polyploidy occurs in several species of *Narcissus*. In *N. bulbocodium* forms with $2n=14$, 21, 28, 35 and 42 occur. The tetraploids grow mainly at the edge of the distribution area of the species, although sometimes mixed populations of diploids and tetraploids occur, or they may grow in neighbouring localities but under different ecological conditions. Morphologically the polyploids cannot be satisfactorily separated from the diploids; they are not given taxonomic recognition, although they may represent incipient species (Fernandes, 1951).

Many other morphologically indistinguishable autopolyploids have been described and examples are given by Böcher (1961a: 35), Darlington (1937), Larsen (1957), Jones (1954, 1961).

Frequently autopolyploids are distinguished from their respective diploids statistically in a number of features such as epidermal cell size, number and size of stomata, pollen diameter, seed size, etc. We have already mentioned in Chapter 5 some of the pitfalls of employing such features as a consequence of their variation in response to a number of factors. The fallibility of pollen size as a polyploidy indicator has been described by Böcher (1960b) in *Campanula rotundifolia*. He found that in the Greenland diploids the pollen may attain the same dimensions as in European tetraploids. There may be internal factors which influence pollen size so that some plants may have large grains and some small grains irrespective of their level of polyploidy. An overall clear-cut separation between diploids and tetraploids is not possible.

Perhaps the commonest situation is where morphological differentiation between different polyplotypes is not clear-cut or easily perceptible. These may be termed semicryptic polyploids. Taxonomic treatment should reflect these situations. Examples include *Arenaria ciliata*, which has recently been studied cytotaxonomically by Favarger (1960a). He found as a result of examining material from 24 provenances that subsp. *moehringioides* from the Western Alps and from the Northern and Insubrian calcareous Alps and Prealps was tetraploid ($n=20$) while subsp. *tenella* was octoploid ($n=40$) in the Valais and Grisons, but duodecaploid ($n=$c. 60) in the Swiss National

Park. It has not been found possible to separate the octoploids and duodecaploids morphologically, but plants from Gantrisch showing a certain amount of gigantism, especially in the flowers, were found to have $n=$c. 120. The chromosome counts for subspp. *moehringioides* and *tenella* supported previous taxonomic treatment based on morphology, ecology and geographical distribution; the nature and constancy of the differences between them did not warrant specific separation.

Similarly subspecific treatment was advocated by Brandt (1961) for two cytological races in *Veronica prostrata* where he found that subsp. *prostrata* from Central and Eastern Europe with small pale-blue flowers had $n=8$, while subsp. *scheereri* from sub-Atlantic Western Europe with larger dark-blue flowers had $n=16$. A means of distinguishing the two subspecies in herbarium material, based on corolla tube epidermal cell surfaces and number, was devised which appears to be valid in most cases, although "ultimate recourse has to be made to the caryological study".

Many other examples of this nature have been described and a useful discussion of the recognition and taxonomic treatment of chromosome races has been given by Favarger (1956).

A few examples are known of autopolyploids (using the term loosely) which appear to have diverged clearly and consistently from their diploid progenitors. Provided that the morphological characters separating them can be unambiguously appreciated, recognition at specific rank is appropriate. Thus *Saxifraga hyperborea* ($2n=26$) and *S. rivularis* ($2n=52$), a diploid-tetraploid pair, are maintained as separate species, the most conspicuous feature distinguishing them, apart from differences in pollen diameter, being the presence of runners in the tetraploid. This association between features associated with vegetative reproduction and polyploidy has been noted in several other cases. It is found, for example, in *Biscutella laevigata* where the tetraploids are again generally characterised by runners (Manton, 1934).

An elegant demonstration of a polyploid series has been given by Löve & Löve (1958a) in the *Triglochin maritimum* complex where the numbers range from $2n=12$ to $2n=144$. The North European plant (*T. maritimum* sensu stricto) is octoploid, while a West American plant previously regarded as a variety of *T. maritimum* has been found to be 16-ploid and raised to specific rank: it is distinct in its growth habit, size and ecology. Other specific units are apparently recognisable in the complex and found to correspond to certain ploidy levels. A similar example has been described by Böcher (1959) in *Trisetum spicatum* agg. (*Gramineae*). The widespread arctic plant is tetraploid whereas the low-arctic Temperate North American plants are hexaploid and had been previously recognised as *T. molle*. The tetraploids and hexaploids are distinct morphologically as well as in distribution and ecology.

While such clearly distinguishable polyploids may be expected to occur, caution should be taken before drawing conclusions on limited samples. Even apparently well-established cases may break down upon further investigation. A few examples will now be considered. A detailed study has been made by I. Hedberg (1961) on the morphological differences purporting to separate diploid and tetraploid *Anthoxanthum odoratum* (*Gramineae*). The species in the strict sense is tetraploid ($2n=20$), but diploid plants have been recorded from several countries (Norway, Sweden, Switzerland, Poland) and were distinguished by Löve & Löve as a new species, *A. alpinum*, the epithet referring to its occurrence in alpine and sub-alpine localities. The various morphologicai features given by Löve & Löve and other authors to separate the diploids and tetraploids were thoroughly examined by Hedberg, who reports that she could not find any reliable distinction between them on herbarium specimens. Despite a very careful analysis it appears impossible in many cases to determine whether a herbarium specimen belongs to the diploid or tetraploid race without access to the chromosome number. An interesting point that comes out of Hedberg's study is that the characters chosen to separate the cytodemes in different countries are not the same, or the ranges are different and not applicable from one country to another. K. Jones (*Chromosoma* 15: 248-274. 1964) suggests that *A. odoratum* is of hybrid origin.

A complex situation is found in *Valeriana officinalis*, a widespread European species which has been studied taxonomically and cytologically by many workers. It contains diploid, tetraploid and octoploid cytodemes, but although there is an apparent correlation between polyploid level and morphology in some parts of its distribution area, this does not hold true for British or West European material. In Britain, plants that have the morphological features normally associated with diploids cannot be separated from known octoploids. It has not yet proved possible to produce a satisfactory treatment which acknowledges the polyploidy situation (cf. Skalińska, 1947, 1951; Sprague, 1952).

In *Ranunculus ficaria* which contains diploid, triploid, tetraploid, pentaploid and hexaploid cytodemes in Europe, with different although somewhat overlapping distributions, one of the tetraploid cytodemes has been found to possess tubercles developed in the leaf axils (Marsden-Jones, 1935). This cytodeme has been recognised in Britain and elsewhere as a separate variety, subspecies or species. Apart from the presence or absence of axillary tubercles, various floral and fruiting features have been employed to separate the diploid and tetraploid cytodemes, such as mean petal number, stamen number and carpel number and type of petal. In an investigation of populations in the north of England, Heywood & Walker (1961) have shown that these morphological features are not reliable cytological markers. Thus it was shown that although the mean petal number has been found to differ between diploid and tetraploid populations, the position of the mean is not correlated overall with the level of polyploidy. Again, although two types of petal are commonly recognised—broad and overlapping in the diploid, and narrow, not overlapping in the tetraploid—an intermediate type was found to occur so commonly that it was recognised as a third type. It was found in both diploid and tetraploid plants! In general, there is no difficulty in distinguishing between diploid and tetraploid populations in any one locality; the difficulty is to find features, other than chromosome number, which serve as overall *differentiae*. There is unpublished evidence to suggest that not even the tubercle character is constantly associated with polyploidy.

In many instances it is being found after extensive investigations that correlations between polyploidy and taxonomic differentiation break down in certain areas and no satisfactory overall correlation is possible. There may even be a "reversal" of characters in some parts of the distribution range, so that plants with the features normally associated with a level of polyploidy are found to be diploid. This has been noted in *Ranunculus ficaria*, *Valeriana officinalis* and in *Orchis maculata* and *O. fuchsii*, to cite a few examples. The explanation appears to involve two main factors: (1) the polyploids have arisen polytopically and are consequently different morphologically amongst themselves although they are sufficiently similar to be placed in the same taxon. These differences will be more pronounced the more differentiated the diploid species, especially in the case of interracial autoploids. (2) The variation due to chromosomal differences associated with polyploidy are largely independent from those due to genetic or environmental factors.

It is evident from the above considerations that conclusions about polyploid indicators must be delayed until a wide range of material has been examined both morphologically and cytologically. It is becoming increasingly apparent that in certain cases the value of apparent markers are highly suspect and break down upon wider investigation. In practice, this means that a chromosome count has to be made on samples from a wide range of material to check the constancy of polyploid markers before they can be safely utilised in lieu of chromosome counts. Unless this is done, conclusions may be misleading. Quite clearly such painstaking verification is seldom put into practice and often chromosomal evidence is used to propose or support taxonomic separation on the basis of a few counts which are apparently correlated with morphological characters. It is to be expected that several such decisions will have to be revised in the light of later evidence.

Another problem of morphological markers of polyploidy arises when they are extremely difficult to appreciate and can only be used by a specialist with accuracy. It is questionable whether any useful purpose is served in such cases by giving the cytodemes specific status, knowing that the non-specialist has little chance of making a successful identification. And how is one to draw the line between usable and non-usable characters? Decisions will evidently be very subjective.

5. *Segmental Allopolyploidy*

As we have already noted, segmental allopolyploids occupy an intermediate position between strict autoploids and strict allopolyploids. On account of some multivalent pairing at meiosis, they usually have a lower fertility than strict allopolyploids.

One of the best known examples of a segmental allopolypioid is *Primula kewensis* (derived from *P. floribunda* and *P. verticillata*) which arose in cultivation. Natural segmental allopolyploids include *Zauschneria californica* (derived from *Z. cana* and *Z. septentrionalis*), *Delphinium gypsophilum* (from *D. hesperium* and *D. recurvatum*) and probably *Lotus corniculatus* (Tomé & Johnson, 1945). Segmental allopolyploids may be more common in nature than strict allopolyploids (amphiploids) and autopolyploids, but as they simulate autopolyploids they have often been confused with them. Stebbins (1950) postulates that segregation and selection over a long period of time should lead to the evolution of fertile, genetically stable types from segmental allopolyploids.

Segmental allopolyploids have two distinct properties. (1) Their ability for genetic segregation, both in respect of the morphological differences between the parental species and the chromosomal differences that produced the sterility barrier between them. Selection pressure will effect the segregation of these chromosomal differences in two ways—either in the direction of typical autopolyploidy or typical allopolyploidy, according to the constitution of the individual chromosomes. (2) They can be expected to form partly fertile hybrids through backcrossing with autopolyploid derivatives of either parental species. Hybridisation may obscure or obliterate the morphological differences and the genetic barrier which existed between the parental autopolyploids. This phenomenon accounts for the treatment (*fide* Stebbins, 1950) often accorded segmental allopolyploids in taxonomic literature. They may often have been placed in the same taxonomic species as one of their diploid ancestors; the existence of a segmental allopolyploid may even result in the two ancestral diploids being combined in one species together with the allopolyploid. Though such a treatment would not reflect the inferred phylogeny, it would be a justifiable procedure in a general classification if morphological intergradation necessitated it. In other words, although many cases of segmental allopolyploidy no doubt remain to be recognised, knowledge of their cytological status will not necessarily alter the taxonomic treatment accorded to them.

6. *Allopolyploidy*

Hybridisation followed by polyploidy has long been recognised as one of the principal ways by which species originate. Allopolyploids may in fact be considered fixed hybrids which have overcome their sterility by chromosome doubling. Ample confirmation of this theory has been received from the artificial synthesis of many allopolyploids, some of which are known to occur naturally. Classic examples include the intergeneric cross, *Raphanobrassica* derived from *Raphanus* and *Brassica* (Karpechenko, 1928) and *Galeopsis tetrahit* ($2n=32$) synthesised from *G. speciosa* ($2n=16$) and *G. pubescens* ($2n=16$) by a series of hybridisations and backcrosses (Müntzing, 1932).

As Keck (1957) notes, the evidence for the allopolyploid origin of a species must be to some extent circumstantial so long as its actual formation was not observed by a botanist. And in practice the evidence used to suggest allopolyploidy in many cases involves morphological and to a lesser extent ecological intermediacy, coupled with the fact that the chromosome complement is the sum of those of the two putative parents. On such grounds the authenticity of many presumed allopolyploids seems fairly convincing. In a few cases detailed analysis of meiotic pairing of the chromosomes has added further confirmation. An example is *Saxifraga nathorstii*, which is believed to have originated by the doubling of the hybrid between *S. aizoides* and *S. oppositifolia* (Böcher, 1941). *S. nathorstii* is endemic to north-east Greenland and is morphologically intermediate between the other two species. It is completely fertile. It has 26 pairs of chromosomes while the assumed parents have 13 pairs each, and there is other cytological evidence to support its alloploid nature.

Other examples of naturally occurring allopolyploids which have been artificially synthesised or for which there is strong cytological and morphological evidence include the following:

Phleum nodosum ($2n=14$) × *P. alpinum* ($2n=28$)→*P. pratense* ($2n=42$) (Gregor & Sansome, 1930; Nordenskiöld, 1937).

Poa supina ($2n=14$) × *P. infirma* ($2n=14$)→*P. annua* ($2n=28$) (Tutin, 1957).

Nicotiana silvestris ($2n=24$) × *N. tomentosiformis* ($2n=24$)→*N. tabacum* ($2n=48$) (Kostoff, 1938).

Spartina maritima ($2n=60$) × *S. alterniflora* ($2n=62$)→*S. townsendii* ($2n=124$) (Huskins, 1931; see also Marchant, *Nature* 199: 929 (1963)).

Brassica rapa ($2n=20$) × *B. oleracea* ($2n=18$)→*B. napus* ($2n=38$) (*U*, 1935; cf. Sinskaja, 1960b).

Gilia millefoliata ($2n=18$) × *G. achilleifolia* ($2n=18$)→*G. clivorum* ($2n=36$) (Grant, 1954).

Many further examples are discussed by Clausen *et al.* (1945) and by Stebbins (1950).

An interesting situation is found in *Nasturtium microphyllum* (*Rorippa m.*) with $2n=64$ which was originally considered to be an autotetraploid of *N. officinale* ($2n=32$), largely on the basis of its morphology and chromosome number. It was later found that artificially produced autotetraploids were quite different in morphology from *N. microphyllum* and even form sterile hybrids with it (Howard & Manton, 1940). It was further found that the hybrid between the spontaneous tetraploid and *N. officinale* possessed 16 bivalents and 16 univalents at meiosis, indicating that one of the two genomes in the tetraploid is completely non-homologous with the genome of *N. officinale*. The tetraploid was later described as a new species (*N. uniseriatum*) although it had already been previously detected on taxonomic grounds and called *N. microphyllum*. It is now speculated that it is an allotetraploid derived from a hybrid between *N. officinale* and a species of *Cardamine*!

In autopolyploidy, taxonomic problems are largely concerned with assessing the degree and constancy of the morphological distinguishing features of the polyploids. As we have seen, they may be amenable to treatment as varieties, subspecies or even, in rare cases, species. The taxonomic problems posed by allopolyploidy, on the other hand, are somewhat different. Although allopolyploids are sometimes more distinct taxonomically than autopolyploids, they have in several cases been overlooked by taxonomists because of their close morphological similarity to their parents and have not been recognised until their status was known. It is equally true that in some instances they have been described by taxonomists before their cytological status was known but subsequently considered as insignificant minor variants until they have been shown to be alloploid (cf. *Kohlrauschia nanteuilii*, Ball & Heywood, 1962).

This is understandable if we consider that an allopolyploid tends to fill the morphological gap between its parents as well as combining their genomes (as expressed in their vigour, physiology, ecological tolerance, etc.). Also there may be backcrossing to one or other parent with continued selection and segregation of genes to that parent. As a result, there may be no evident morphological discontinuity between the allopolyploid and the parent species. To the taxonomist AA and part of AABB may be so similar as to be regarded as one unit, while BB and part of AABB may be regarded as another. Or the whole complex may be so closely connected as to appear to form one variable species and the polyploidy may go unrecognised (cf. Grant, 1957). It is even possible for the different allopolyploids (AABB, BBCC, etc.) formed from morphologically similar diploid species (AA, BB, CC, etc.) to be themselves so similar that they are united in the one species. This is more likely to occur when the distributions of the diploids and tetraploids are essentially sympatric (see p. 423).

The taxonomic treatment of allopolyploidy may be as difficult as that of autopolyploidy, although for different reasons. For one thing, a minimum of three units (two diploid and one polyploid) has usually been involved at any one level, not just a diploid-polyploid pair.

When the differences are clear cut, recognition at specific level is to be recommended on both taxonomic and biosystematic grounds. The differential features may be quite small as occurs with autopolyploids: in *Nasturtium microphyllum* details of the seed coat are the most reliable. If morphological discontinuities are not evident, taxonomic treatment may come into conflict with biosystematic considerations to some extent. In so far as the allopolyploid combines the genomes of two other species, it is clearly eligible for consideration as a further species; in so far as it is a hybrid, it may be treated taxonomically as such. When a considerable amount of backcrossing has occurred, linking up the diploids and allopolyploid inextricably, no satisfactory taxonomic division may be possible. There is indeed no valid reason why all plants can be expected to fit into neat categories once cytological and genetical information is available, any more than can be expected on morphological grounds. This is even more evident in the case of complex polyploid situations as discussed in the next section.

The solution to the kind of polyploid situation where there is morphological intergradation due to continued hybridisation between the components may be to recognise the various units as subspecies. The use of the species aggregate is an alternative treatment which has been advocated recently (cf. Manton, 1958; Böcher, 1961a), although the use of aggregates as a cover for unsatisfactorily delimited groups is not, in our view, to be recommended (see Heywood, 1963).

7. *Other Situations*

In several cases both autopolyploidy and allopolyploidy may have been involved in the formation of a taxon. In *Holcus mollis* a range of numbers has been found with the pentaploid ($2n=35$) the commonest, at least in Wales. The tetraploids are considered to be autoploid in origin but there is evidence to suggest that other polyploid populations found in the group are of alloploid origin (Jones, 1954). Other examples from the *Gramineae* include *Calamagrostis canescens* (Nygren, 1946) and *Phleum pratense* which we have already cited as an alloploid (p. 223) though plants resembling it have also been synthesised from *P. nodosum* by autopolyploidy (Nordenskiöld, 1949). Possibly *P. pratense* is an autalloploid, i.e. autoploid in respect of one genome, and alloploid in that it contains another genome also (AAAABB).

Keck (1957) cites another, highly complicated case, in *Pentstemon* (*Scrophulariaceae*) where *P. attenuatus*, a hexaploid species probably of alloploid origin, contains four confluent and evidently interfertile subspecies. It is thought that *P. albertinus*, a diploid, may be ancestral to all four subspecies, another, tetraploid species, *P. confertus*, may have been involved in the origin of subsp. *attenuatus*, a further tetraploid, *P. globosus*, probably concerned in the origin of subsp. *militaris*; and two other species, containing both diploid and tetraploid races, appear to be the probable ancestors of the other two subspecies. Yet Keck concludes that "despite their diverse origins, it is the consensus of taxonomic opinion today that these several elements should be classified within a single variable species".

In many cases polyploidy may be only one aspect of the variation of a group in which other factors such as aneuploidy, apomixis and vegetative reproduction are important. An excellent example is the *Cardamine pratensis* group which has been thoroughly studied by a number of workers recently, notably Lövkvist (1956). Here chromosome numbers of $2n=16$, 24, 28, 30, 32, 38, 40, 42-55, 56, 60, 64, 73, 56-80, 84, 88, 90, c. 96 have been found, this oscillation of numbers depending on two basic numbers (7 and 8), aneuploidy and the existence of hybrid swarms, as between two octoploids (56 and 64) and two decaploids (72 and 76). Aberrants also occur. Polyploidy is a continuous process in the complex and together with hybridisation does not permit the subdivision of variable groups within it into single-chromosome number species. There is some correlation between chromosome number and water content of the soil—the higher the water content, the higher the chromosome number in the *C. pratensis—palustris* plants. Another factor involved is vegetative reproduction. In such a group there is no clear connection between chromosome number and polymorphism in certain parts of the complex, and taxonomic treatment must be somewhat of a compromise. Some form of provisional aggregation of units and liberal use of observations are necessary in Flora accounts; segregates should not be named and described formally when they are only genetically but not morphologically distinct.

Apomixis may be associated with polyploidy and several well-documented cases have been described in the *Gramineae*, for example. *Poa alpina* contains a bewildering array of variation with a range of chromosome numbers from 14 to 56, pseudovivipary, pseudogamy, and chromosomal aberrants. In a detailed study of the complex, Müntzing (1954) considers that although certain biotype groups in some parts of the distribution area could form the basis of taxonomic units, in other regions the variation is so great that a taxonomic treatment is not feasible. In this and similar situations, polyploidy is just one facet of the variation and strict adherence to it alone as a framework around which to build a taxonomic arrangement would yield results of little taxonomic value (Heywood, 1960b: 188).

A special complication is found in a few genera where the chromosomes possess diffuse (or polycentric) centromeres, permitting the survival of fragmented chromosomes. The phenomenon known as *agmatoploidy* or *endonuclear polyploidy* has been described in *Luzula* (Nordenskiöld, 1951, 1956, 1957, 1961), *Carex* (Löve, Löve & Raymond, 1957), *Juncus* (Löve & Löve, 1944), *Scirpus* and some related genera (Sharma & Bal, 1956). A detailed discussion and extensive bibliography is given by Löve, Löve & Raymond (1957), who note that partial agmatoploidy is widespread in the *Cyperaceae* while complete agmatoploidy is typical of the related family *Juncaceae*.

The situation in the genus *Luzula* has recently been summarised by Nordenskiöld (1961). Cytological variation recorded is of the following kinds: (1) in some taxa all material has the same diploid chromosome number ($2n=12$); (2) normal polyploidy; (3) endonuclear polyploidy (agmatoploidy); (4) aneuploidy. The last two phenomena are a result of changes in the size and number of the chromosomes within a certain level of ploidy, and all four are found in the *L. campestris—multiflora* group.

Endonuclear polyploidy and aneuploidy are found in taxa which have chromosomes with diffuse centromeres. Studies on the morphology of the mitotic chromosomes indicate that there is a successive replacement of each of the 12 long chromosomes of the genuine diploid by two chromosomes, which are about half as long as the original long ones. When all twelve long chromosomes of a diploid are replaced by two small chromosomes, a polyploid number is obtained despite its diploid genetic constitution, hence the term endonuclear polyploidy. If some only of the 12 chromosomes are replaced by fragmented chromosomes, aneuploid numbers result (incomplete agmatoploidy).

Morphological distinctions in the group are not clear cut, although it is easier to separate the diploids morphologically than the polyploids. The diploids are normally accorded specific or varietal rank while the polyploids are often grouped within the collective or aggregate *L. multiflora.* Nordenskiöld concludes that the morphological features used by taxonomists to characterise species has little or nothing to do with the cytological and genetical variation they show.

In *Carex* Sect. *Capillares,* where auto- and allo-polyploidy appears to have been associated with agmatoploidy in species formation, Löve, Löve & Raymond (1957) were able to assign the different chromosome numbers to separate species.

8. *Polyploid Pillar Complexes*

In the kind of group commonly referred to as the polyploid complex (Babcock & Stebbins, 1938), where a series of diploid pillars support a superstructure of polyploids, the taxonomic problems are like those found in allopolyploids but in a much more acute degree. Difficulties arise because of the presence of similar or identical genomes in some of the species at the different levels, and from the hybridisation which occurs at the various polyploid levels and even between levels to some extent. The consequences are that morphologically similar units occur at the different levels in the complex because of their possession of common genomes, and the hybridisation has the effect of filling in the morphological gaps between the units, especially at the polyploid levels (D. Smith, 1961).

In the *Sanicula crassicaulis* polyploid complex (Bell, 1954) it is clear that polyploidisation has taken place at the hexaploid and octoploid level many times (sometimes by backcrossing with the diploid level and by other hybridisation). At these high levels the polyploids represent different combinations of the same genic material that is found at lower levels, and selection as well as random fixation has given rise to parallel series of variants at the different levels. Satisfactory taxonomic division within the group is not possible because of this parallel morphological variation and because of the independence of much of it from variation due to polyploidy itself.

Such polyploid pillar situations, where a number of closely related diploid species bear a tetraploid or higher polyploid superstructure whose members cannot often be individually or collectively separated morphologically from the diploid members, are probably not infrequent and it is likely that many will be detected by wider investigation of groups in which polyploidy is known to occur. Excellent examples are found in: *Achillea* (Clausen *et al.*, 1940, 1948: Ehrendorfer, 1952, 1953, 1959; Schneider, 1958); *Aegilops* (Kihara, 1954); *Artemisia* (Ward, 1953, 1959); *Bromus* (Stebbins & Togby, 1944; Stebbins, 1956); *Campanula* (Böcher, 1960b); *Casuarina* (Barlow, 1959); *Dactylis* (Stebbins & Zohary, 1959; Böcher, 1961b; Borrill, 1961a, b, c); *Danthonia* (Brock & Brown, 1961); *Galium* (Ehrendorfer, 1961); *Helianthus* (Heiser & Smith, 1954; D. Smith, 1961); *Knautia* (Ehrendorfer, 1962); *Phacelia* (Heckard, 1954, 1960).

Variable factors include the presence of apomixis or structural changes, the degree of distinctness of the polyploid units and of the diploid pillars, the extent of gene-flow between the levels. A few of these cases will now be considered in more detail.

Dactylis glomerata: Although studies are by no means complete, it is evident that this group contains ten or more diploid species or subspecies, each with a fairly circumscribed ecogeographical distribution and more or less clearly separable morphologically, and showing a narrow range of variation. Superimposed is a large tetraploid, widespread superstructure showing a more or less continuous range of variation which combines most of the diploid characteristics. The tetraploids represent about 95 per cent. of the genus and some of them are so similar to the diploids that they cannot be morphologically separated from them, especially where diploid and tetraploid populations are in contact. They show typical autoploid behaviour, but it is not known whether the parallelism between diploids and tetraploids is due to the latter having arisen *in situ* or whether there has been an adaptive radiation of diploids and tetraploids in the same environment (Borrill, 1961b). The diploid subspecies are interfertile and capable of gene-exchange. There is, moreover, one-way gene-flow from diploid to tetraploid level through the intermediary of triploids.

As regards taxonomic treatment, if primary weighting is given to polyploidy as a species criterion, following some current biosystematic practice, diploid and tetraploid species, each with more or less parallel ranges of variation, could be proposed. Such species would not be distinguished from each other by any character or character-combination, since they are polyphyletic assemblages held together primarily by their common diploidy or common attainment of polyploidy. The impracticability of such an arrangement hardly needs stressing. It is, however, a tenable grade system of classification and it is worth emphasising that here a biosystematic classification and a grade one can coincide (it is not, however, the only biosystematic treatment possible). Such a treatment has in fact been proposed in the case of *Biscutella laevigata* (Manton, 1934), although it has not so far proved feasible to put it into practice on taxonomic grounds (Guinea & Heywood, unpublished). An alternative treatment is to recognise that genetically and morphologically the whole complex is so closely knit that a diploid-tetraploid separation must be arbitrary, and to regard the whole complex as a single species with parallel series of subspecies, one at the diploid level, one at the tetraploid level. Grades are here accommodated as an intra-specific phenomenon. Such a solution has in fact been proposed by Stebbins & Zohary.

Phacelia magellanica complex. Heckard has shown that six basic diploid species may be recognised, each morphologically distinct and reproductively isolated, either by geographical or genetic barriers, or both. Each diploid has a morphologically similar counterpart at the tetraploid level which also contains a series of intergrading forms which combine features of two or more diploids or even features not found in them.

The morphological intergradation between the tetraploids nearly always occurs where their ranges approach or are juxtaposed. The source of this intergradation is hybridisation and consequent gene-flow between the plateaus of morphological uniformity; introgression has increased the variability of the tetraploids. As a result the whole complex is an assemblage of interrelated plants within which lines must be drawn somewhat arbitrarily in order to delimit taxa. Some hexaploids also occur. The relationships of the group are shown in Fig. 17.

In attempting a taxonomic treatment, the diploid units are distinct and may be recognised as species (together with some subspecies). As regards the tetraploids, a logical treatment would be to regard the intergrading assemblage as one species with various subspecies corresponding to the morphologically uniform plateaus. However, the extremes of the tetraploid assemblage are so far apart morphologically that Heckard has proposed recognising each tetraploid plateau as a species despite the connections between them. Furthermore, the tetraploids resemble their diploid counterpart more closely than they do other tetraploids, so that it is necessary to place the diploid and tetraploid counterparts in the same taxon. "The result is a series of poorly defined groups which lack the degree of discontinuity usually desired in species but which are best considered species lest the entire complex be reduced to one huge unmanageable species with an unwieldy number of infra-specific taxa."

From the above examples it is quite clear that the taxonomic treatment of polyploid complexes often presents virtually insoluble problems. Certainly no doctrinaire solution can be successfully applied. If the polyploids intergrade and yet are given specific recognition, the treatment is unsatisfactory and regarded, as D. Smith (1961) says in the case of *Helianthus*, with mental reservations. If the whole complex is united under one species, it may be regarded as too cumbersome, as Heckard comments for *Phacelia*. The problem of the recognition of polyploids which closely parallel diploids to such an extent that they cannot be morphologically separated from them remains unsolved. Here there is an evident conflict between biosystematic theory which would regard the diploid and tetraploid units as separate

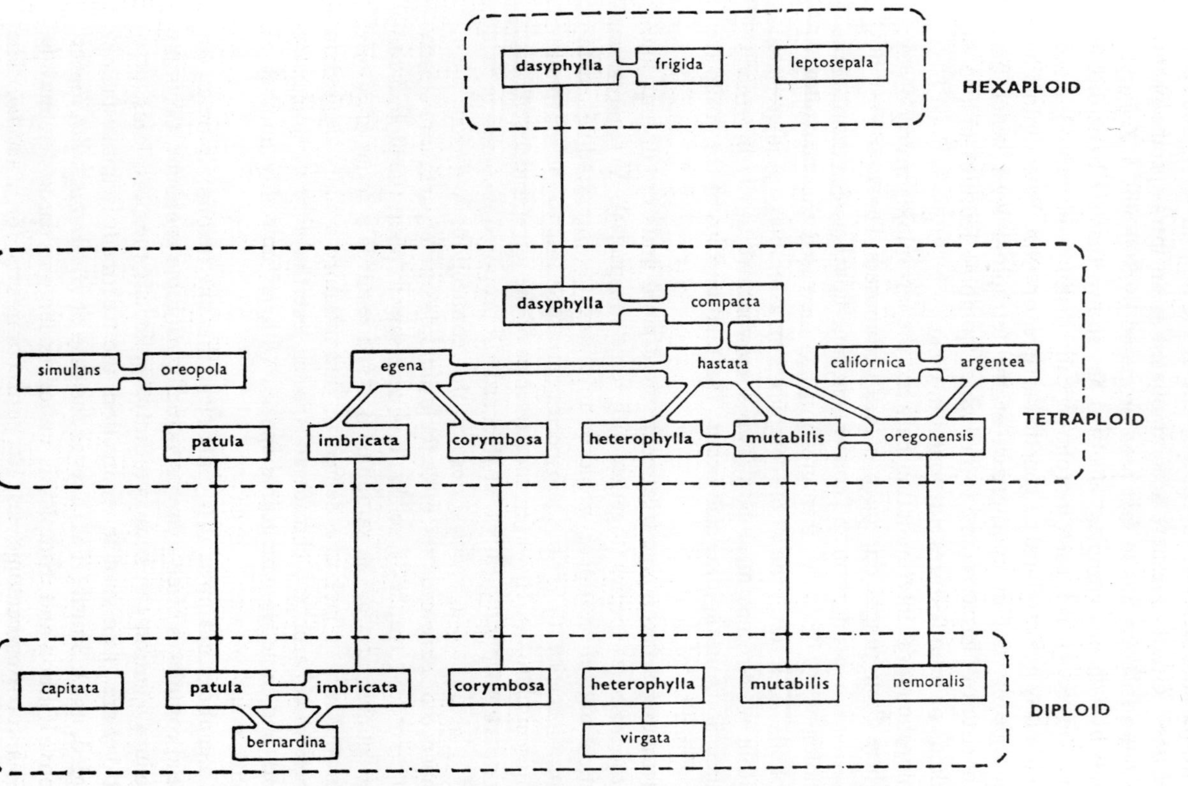

FIG. 17. Polyploid pillars in the *Phacelia magellanica* complex. Chromosome levels or grades are enclosed in broken lines. Taxa which occur at more than one level are joined by vertical lines. Channels indicate intergradation between taxa. (After Heckard, 1960.)

gene-pools and therefore deserving specific rank, and taxonomic treatment based on morphological differentiation which would unite the diploid and tetraploid in the same unit.

A full assessment of the taxonomic difficulties and procedures applicable requires a consideration of the evolutionary and biosystematic significance of polyploidy in such situations. Among the factors that have to be considered are the consequences of the increased fertility that frequently occurs at the polyploid level compared with the diploid level. From what we have already seen, however, polyploidy would appear to represent in these complexes a short-term progressive advance over the diploid condition, and the closely linked tetraploids or higher polyploids can be regarded at each level as a polyphyletic grade. This is well illustrated by the recent work of Barlow (1959) on the *Casuarina distyla* group of 13 closely related species. Sexual tetraploids are formed by eight of the species (and triploid apomicts by four of them). The polyploids and apomicts do not apparently characterise particular morphological groupings: "On the contrary, they occur as conditions within the species, and these forms are at present morphologically indistinguishable from their respective diploids." Species evolution has here been entirely at the diploid level; polyploidy has arisen within the species a number of times but has little evolutionary significance and is not amenable to taxonomic recognition.

The whole polyploid superstructure may be regarded in many of these cases as ornamentation or embroidery on the basic diploid pattern. Conventional taxonomy is not adequate to reflect all the details.

Cytogeography

The importance of chromosomal features in understanding the distribution of plants has led to the development of a new line of study termed cytogeography or chromosome geography (Darlington, 1956). A general discussion of the field is given by Reese (1961), who considers what rules govern the relationship between the chromosomes of a plant and its geographical distribution (and ecology). The most important karyological changes from a geobotanical point of view are polyploidy and to a lesser extent other numerical changes, and structural hybridity. We have already alluded to various relationships between chromosomes and geography and ecology, such as the incidence of polyploidy in different floras, and the polyploid nature of present day taxa as one of the main lines of evidence for estimating the relative age of taxa. Favarger & Contandriopoulos (1961) regard it as the only objective method and discuss the role of cytology in the understanding of endemic and corresponding (vicariant) taxa (cf. Löve, 1954b). These topics will be considered further in Chapter 13.

Chapter 7

TAXONOMIC EVIDENCE

III. Phytochemistry

INTRODUCTION

Within recent years there has accumulated a vast literature, much of it not readily accessible to the taxonomist, dealing with the systematic distribution of particular chemical constituents in plant groups. The attention of taxonomists has been drawn to a certain proportion of this work by a few workers such as Bate-Smith, H. Erdtman, Mirov and Darnley Gibbs. Recent reviews cover only a small amount of the evidence and it needs to be stressed at the outset that the number of chemical features of potential systematic value may run into hundreds. Yet most of the data so far available are to a large extent ignored by taxonomists.*

In any attempt to produce a natural classification which takes into account as many characters from as many lines of evidence as possible, it is quite clear that plant chemistry and biochemistry will provide a wide new source of characters. Due to the development of modern and often simplified techniques, the application of chemical tests can in some cases be a relatively easy matter and is likely to be increasingly used.

These chemical characters may have a particularly high taxonomic value when they can be shown to be stable, unambiguous and not easily, if at all, changeable. In most instances much further information on these aspects is required. There is a tendency, however, to regard chemical characters as more "fundamental" than other kinds, especially morpho-anatomical ones. They have also been referred to as more important, more basic, and more privileged. All these concepts have little meaning unless precisely defined and their purpose declared. Probably what is felt subconsciously is that the smaller and more cryptic the character the more important it is for indicating relationship. This line of argument has already been discussed and rejected in connection with anatomical characters, where we saw that endomorphic features are no more reliable in general than exomorphic ones. At the protein

* Since this chapter went to press, R. Hegnauer (1962) has published his *Chemotaxonomie der Pflanzen*, **1**: Thallophyten, Bryophyten, Pteridophyten und Gymnospermen. (Cf. review in Nature, **197,** no. 4867, p. 527 (1963).)

level it could be argued that since these substances are the immediate secondary products of the genetic material (controlling the development of phenotypic characteristics), a knowledge of them would allow taxonomists to arrange plants according to their true relationships. The connection between genes and characters is so complex, however, and the developmental pathways so involved and susceptible to both internal and external factors about which so little is known, that to look forward to protein chemistry as a panacea for all taxonomic ills is unrealistic. It seems likely that the finer the level of analysis employed (and this if pursued would ultimately take us outside the fields of biology and into theoretical nuclear physics), the simpler and more unified the pattern would become. In other words, to relate all plants to a basic pattern of which they ultimately represent variations would reduce the concept of relationship to a taxonomically valueless level.

When considered for the purpose of delimitation and diagnosis of groups and for their consequent arrangement in a phenetic system, chemical characters may be important for the reasons mentioned above, namely their stability, unambiguous appreciation, and unlikelihood to change. This is, however, not always so, and chemical characters appear to be susceptible to all the "failings" of other kinds of evidence, such as parallelism, convergence, inconstancy, environmental modification, reversibility, etc.

To consider that chemical characters are more indicative of relationship than, say, morphological characters, as is often proposed, is simply loose thinking. Chemical characters will show chemical relationship in the same way that morphological characters show morphological relationship. If it can be established *ex post facto* that chemical characters show a high correlation with other features, to this extent they will be more important for the purposes of marking overall relationship than characters (morphological, cytological, etc.) not showing such high correlations. It is pertinent to note in this connection that the role of biochemical characters in the classification of certain groups of micro-organisms (to which flowering plant taxonomists are frequently directed for example) is by no means clear despite the obvious practical relevance of this kind of evidence in these groups. For example, Barnett (1960) has proposed that, in the yeasts, an improved and stable classification could be achieved by the use of biochemical criteria exclusively. In a short but extremely valuable criticism of this viewpoint, Roberts & Thorne (1960) reject the proposition that the natural relationships of the yeasts can be disclosed simply by means of the enzymes contained in their cells, even if the biochemical pathways and enzymatic organisation are studied. They consider that "while biochemical processes are the basis of life, they are not synonymous with it, and to ignore life's outward manifestations, however subjective they may be, can only lead to an imperfect conception of the taxonomic whole: no one, in attempting to distinguish between a lion and a tiger, would ignore morphological criteria in favour of enzymatic".

Classification of the bacteria involves the same kind of problem although again biochemical characters for diagnosis have greater practical value than morphological ones. In flowering plants the same arguments hold, with the exception that morphological markers are much more useful in practice. The general position has been summarised by Pirie (1955), who states that "many different types of information can be, and should be, woven into the classificatory scheme, and there is no obvious advantage in attaching overriding importance to any one type".

The attractiveness of chemical characters from a phylogenetic point of view stems from the possibility that in many cases the chemical evolution of a character has gone in a particular way. In some cases it may be stated that a certain compound must of necessity have been derived from another compound. This kind of evidence is of very great help in mapping the course of evolutionary relationship, but it does not tell us *how* a particular taxon has been derived. Chemical features evolve at different rates and are not necessarily synchronised with other characters. They can, however, suggest the direction of change and avoid possible errors of interpretation, but they are only one facet in a complex mosaic.

HISTORICAL

The connections between the chemistry of plant constituents and taxonomy has already a long history. A treatise on phytochemistry was published by Rochleder in 1847, but the characterisation of certain taxonomic groups by their essential oils, e.g. *Rutaceae*, *Labiatae*, or by pharmaceutically important substances, e.g. *Digitalis*, *Atropa*, etc., was clearly appreciated not only in the early days of scientific taxonomy but in fact much earlier when plants were classified largely according to their economic and medicinal uses.

There is an extensive literature reflecting the large number of investigations into the occurrence of pharmacologically important substances in plants. Much of this work is of a highly specialised nature and of little relevance to taxonomy, but more systematic surveys would in many cases provide valuable data for taxonomists. The main drawback to the use of this kind of information in taxonomy is in fact its incomplete and sporadic nature, although it is clear that the chemical constituents of plants may frequently be characteristic of genera or even species.

Within recent years a welcome sign has been the number of systematic surveys which have been undertaken by chemists who have interested themselves in the taxonomy and relationships of the plants containing particular chemical substances. Prominent amongst these is H. Erdtman, who has discussed some of the general principles of chemical taxonomy.

The fact that certain compounds or chemically related substances may be characteristic of genera or even families is now well established. According to Erdtman (1956), these substances must have been formed by certain metabolic processes in the plants which were retained when the group in question underwent further evolution and differentiation. This type of situation could have considerable importance in the study of the evolutionary history of groups, since knowledge about the principles and direction of chemical evolution might contribute to the understanding of the relationships of present day groups. This aspect has been discussed by Bate-Smith, who considers that chemical *processes* are probably the essential events distinguishing certain groups of plants from others, rather than the presence or absence of particular constituents (1959).

Chemical Constituents as Comparative Data and their Evolutionary Interpretation

The generalisation has been made that the higher the position in the *Cormophyta* the more complicated the chemistry (Weevers, 1943), but it is probably premature since a number of the same natural products have been found in micro-organisms. N. A. Sörensen, for example, has recorded polyacetylenes in the *Compositae* and *Basidiomycetes*!

As occurs with other kinds of evidence, the value of individual chemical constituents varies from substance to substance and from group to group. Certain substances are so widely distributed in the flowering plants that they are of virtually no taxonomic value (e.g. glucose, oleic acid). Some products are closely connected with ubiquitous metabolic processes and are therefore widely distributed—for example xanthine, which is an ubiquitous product of nucleo-protein metabolism. Caffeine has a similar structure that might have arisen from xanthine by a single mutational step; it occurs in apparently unrelated Angiosperm groups such as *Parietales*, *Centrospermae*, *Rubiales*. In such cases the product is not correlated with taxonomic position, the chemical resemblance being due to parallelism, as happens with other characters. And as with morphological or anatomical characters *combinations* of products can often be of more value in assessing taxonomic position than single products and are more diagnostic. Such combinations are less liable to be the result of parallelism or convergence although this cannot be altogether discounted.

A study of the distribution of those products (often the end points of metabolism) which show a positive correlation with other kinds of character may be of importance in making a natural classification. They will often give a lead in deciding difficult cases where we have to choose between several alternatives on the basis of other evidence (as with cases of floral reduction).

Chemical characters can be considered under three headings:

1. Directly visible, e.g. starch grains and raphides.
2. Plant products, e.g. alkaloids, flavanoids and terpenoids.
3. Serology and electrophoresis of proteins.

1. *Distribution and Description of Directly Visible Chemical Substances*

Very few chemicals in plants are directly visible and it is not surprising that these have attracted the attentions of anatomists (cf. Solereder, 1908) and their distribution is rather well known. The main two groups are starch grains and raphides.

Starch grains. Starch develops as grains in the plastids. The grains may be single or compound; they vary considerably in shape and frequently show layering (probably due to the alternation of different chemical layers). Their differentiation and specificity of occurrence in relation to genera and lower taxa has been discussed by Reichert (1913), who lists 350 species.

The form of starch grains has been used as an aid to the classification of the *Gramineae* together with chromosome number and other characters discussed above. Typical genera of the tribe *Hordeeae* have simple grains, while *Nardus*, *Lolium* and *Parapholis* have compound grains and appear to be out of place—a conclusion confirmed by the study of other cryptic characters. A review of this topic is given by Tateoka (1962).

Raphides. These are bundles of calcium oxalate crystals contained in large cells and may be observed in suitable sections. Gibbs (1958a) notes that the ability to store oxalate in this form appears to be restricted to certain groups of plants and may therefore have some value as evidence of relationship. This, as he says, had been pointed out a century previously by Gulliver in a paper on raphides as "natural characters in the British Flora". Raphides occur in about 35 Angiosperm families. In the Dicotyledons they are known in several systematically separate groups and Gibbs (1958b) considers that they must have arisen more than once. He has also considered their occurrence in the *Monocotyledons* where he finds that they are absent in the presumedly primitive orders (*Alismatales*, *Hydrocharitales*, *Scheuzeriales*, *Potamogentonales*, *Triuridales* and *Najadales*), but that they do occur in the *Liliales* and in many other orders following Kimura's recent system (1956). If Kimura's arrangement is correct, he says, then raphides arose in the Monocotyledons independently from the Dicotyledons and after they had diverged from them. Further, they would have arisen once only in this group and later dropped out in at least one branch.

The presence or absence of raphides has been employed by some authors as a taxonomically valuable feature. Bremekamp, for example, has employed them within the large family *Rubiaceae* together with testa structure, degree of development of albumen and floral type for a more natural classification into genera. All members producing raphides are put in the *Rubioideae*. Raphides are found in several families of the *Centrospermae* and also in the *Cactaceae* (a family whose position is controversial); in this case the presence of "nitrogenous anthocyanins" only in the *Centrospermae* and in the *Cactaceae* together with similarities in embryology strengthen suggestions of affinity of the two groups.

Dormer (1961) has studied the crystals which are to be found in the inner tissues of the ovary wall of many *Compositae*. The crystals are apparently of calcium oxalate and show a variety of different forms, some of which have a restricted systematic distribution. His study was confined to members of the *Cynareae*, but the results suggest that these characters may be of considerable taxonomic value in other parts of the family. In *Centaurea* Dormer (1962) finds that crystal structure (prismatic or curvilinear) cuts across the conventional classification of the genus, but shows a rather close correlation with the pollen types established by Wagenitz (1955).

2. *Chemical Test Characters; Plant Products*

Because of the labour involved in studying many of the natural plant products, work has been largely done on those of economic importance, often pharmacological. Many of these products are formed at the end of long biosynthetic pathways and may be regarded as metabolic by-products. They are seldom formed as a consequence of a fundamental biochemical process if one is to judge by their frequent sporadic, indeed erratic, occurrence in the different taxonomic groups. Heslop-Harrison has termed them "luxury diversifications"—simply expressions of biochemical virtuosity. Their occurrence is not, it would seem, necessary to the functioning of the plant, so that variations on the theme of their presence are found.* This variation may be at all taxonomic levels and even within species. Little enough is known about the infra-specific variations shown by these products, and of their susceptibility to environmental modification (cf. Schratz, 1960). Their presence may also be affected by ontogenetic changes in the individual plants. In the genus *Digitalis* (*Scrophulariaceae*) which is one of the most intensively investigated groups from a pharmacological point of view, the occurrence of glycosides such as Digoxin has been shown to vary considerably between different species; their presence does not appear to be correlated with taxonomic position; and variations within the individual species are well known and of considerable economic importance.

The ecological relationships of the intraspecific variation of these products has been investigated in some cases. One of the best known cases is Daday's much-cited analysis of the distribution of cyanogenetic glucosides in *Trifolium repens*. When the leaves of plants containing the glucoside and the appropriate hydrolytic enzyme are bruised, HCN is formed. The occurrence of cyanogenesis and its genetic control has been known for many years, but it was not until Daday showed the distribution of the pairs of alleles controlling the presence of the enzyme and the glucoside had particular distribution patterns that the possible adaptive nature of these characteristics was suggested (Daday, 1954a, b, 1958). Daday showed that both glucoside and enzyme were more frequent in plants growing in warmer areas (cf. Morley, 1959).

Heslop-Harrison (1960) puts forward the suggestion that plant products which appear to have no adaptive significance or function may in fact

* Some chemists would dispute this viewpoint and suggest that these products do have a function although its nature has yet to be elucidated.

contribute in some way to adaptation of populations; hence any ecological correlation demonstrated may well indicate adaptive significance. On the other hand, they may be the result of direct environmental modification, particularly in the case of non-essential plant products. The possibility exists that some of the products valued by man may in fact act as buffers for other processes which are more important to the metabolism of the plant.

Clinal variation in chemical characters is also known. It has been reported by Barber and co-workers in *Eucalyptus* (Barber, 1956).

The number of substances that has been extensively studied from a systematic point of view in the Angiosperms is quite small. Best known are the alkaloids, glycosides, phenolic substances, water-soluble carbohydrates and essential oils.

Alkaloids

Although not susceptible of precise definition, alkaloids are easily identifiable by their physical and chemical properties (Rowson, 1958). Usually they show considerable pharmacological activity and for this reason investigations into their distribution are highly incomplete, having been concentrated largely on plants of pharmacological importance. Also, much of what is known is incomplete and in need of revision. The genus *Pinus*, for example, was assumed to be alkaloid-free until two alkaloids were recently detected in the leaves of one species.

Despite the fact that our knowledge of the distribution of alkaloids, their role in physiology and biosynthetic pathways, are so incomplete, sufficiently detailed data are available in certain cases and they are of much taxonomic interest.

Distribution. Alkaloids appear to be absent from most plants and the details of their restriction to certain taxonomic groups is valuable. For example, such families as the *Labiatae* and *Rosaceae*, as far as is known, are devoid of alkaloids. According to Rowson there seems to be little correlation between the "phylogenetic" position of a plant and either the presence or absence of alkaloids in it, or their chemical nature when present.

A remarkably high specificity is shown by some alkaloids, a classic example being *morphine*, which is known only from *Papaver somniferum*, the opium poppy. Most alkaloids (and other secondary substances such as heterosides, essential oils) occur in nature accompanied by other principles of a similar type. Thus in *Rauwolfia* (*Apocynaceae*), about 20 alkaloids are known to occur. Within a species, however, plants from different provenances of the same species may differ considerably in their alkaloid complexes: in *Aconitum napellus* variation in aconitine and other alkaloids has been detected between plants from the Tyrol, Auvergne and Haute-Marne (cf. Dillemann, 1960). These differences may be due to racial differentiation within the species or to environmental modification. Chemical races, genetically determined, certainly do occur in many groups and it is likely in some of them that they do not necessarily correspond with morphologically recognisable infra-specific taxa.

Within genera containing alkaloids, species which are alkaloid-free are rare; small quantities or only traces may be found, but complete absence is unusual. An example of the latter is *Lobelia erinus*, which contains only traces of alkaloids while the other species investigated are much richer in them.

One of the best examples of the value of alkaloid distribution in supporting a taxonomic treatment is in Manske's work on the *Papaveraceae*, all the members of which contain several types of alkaloids, of which protopine is common to all species. Similarly, all members of the *Fumariaceae* contain protopine and this is taken by Manske (1944) as decisive evidence of the close relationship of the two groups. This is contrary to Hutchinson's view that the *Papaveraceae* is quite distinct from the *Fumariaceae* on general taxonomic evidence. On the other hand, there is evidence that each genus within the *Papaveraceae* sensu lato has its own characteristic combination of alkaloids and these combinations differ in different parts of the family. The *Fumariaceae* possess a combination of isoquinoline derivatives different from anything found in the *Papaveraceae* sensu stricto. Here we appear to have a situation in which chemical evidence can be used to support two courses of taxonomic action.

The story does not end here. Protopine was thought to be confined to the above groups by Manske, but it has since been found in *Nandina domestica* in the *Berberidaceae*. This is interesting since Hutchinson regards the *Fumariaceae* as allied to certain members of the *Berberidaceae*, such as *Bongardia* which itself is similar in some respects to *Nandina*. On the other hand, Gibbs (1954) supports the view that *Nandina* might be better considered a monotypic family separated from the *Berberidaceae*. He raises the possibility that *Bongardia* may also turn out to contain protopine.

Although we started this consideration of alkaloids in the *Papaveraceae* by saying that it was an example of taxonomy being confirmed, it is evident that *which* taxonomical classification receives confirmation is debatable. We have not yet heard the end of this particular tale. Premature conclusions must not be drawn, nor can we agree with Manske that the nature of the contained alkaloids alone can decide the issue of relationship between the *Papaveraceae* and the *Fumariaceae*. This is not to deny the close *chemical* relationship of the groups in terms of alkaloids, but even this can be interpreted in different ways according to which aspect of the protopine distribution is considered. What is shown is that much further work is needed from both a chemical standpoint and from other branches before a satisfactory assessment can be made. We also need information on the alkaloids of the *Hypecoideae* which form a morphological link between the *Papaveraceae* and *Fumariaceae*.

A detailed survey of the occurrence and known structure of alkaloids in the *Veratreae* (*Liliaceae*) and its implications regarding the taxonomy of the group has recently appeared (Kupchan, Zimmerman & Afonso, 1961). The most significant generalisation arising from these studies was that the alkaloids strongly supported the classification of the groups arrived at by the use of other evidence. The elaboration of different alkaloids by different genera (*Veratrum* and *Schoenocaulon*) agrees with their wide taxonomic separation. Other chemical differences showed a correlation with their geographical distribution, suggesting that further studies might be of value in tracing the chemical evolution and migration of the *Veratreae*. It is emphasised that further data must be accumulated before any firm generalisations can be made.

The techniques of paper chromatography and electrophoresis which are being used to analyse the mixed or total alkaloid fractions of species will lead to a much more detailed knowledge of their composition, and consequently allow the breakdown of these products into taxonomically significant fractions. It is the restricted or unique occurrence of compounds which often provides the taxonomist with more valuable information than do the more widespread substances. Alkaloids and other products can therefore be used as analytic or synthetic characters according to their pattern of distribution in taxonomic groups and the details of the information about them.

Paper electrophoretic analysis of total alkaloids in genera of the *Apocynaceae* has lent support to Pichon's separation of species of *Catharanthus* from species of *Vinca*. The patterns shown by electrophoresis were sufficiently clear to allow a distinction between the two groups of species studied (Paris & Moyse, 1957; Farnsworth, 1961). Another example of the possible value of alkaloid distribution at generic level is in the *Genisteae* group (*Leguminosae*) where taxonomic opinion is still divided as to generic limits. A chemical analysis of several taxa was carried out by Nowacki (1960) and some interesting correlations with their systematic position were revealed. As frequently happens, however, the incompleteness of the data detracts from their potential taxonomic value.

The total alkaloid content or the proportions in which they are found may be of taxonomic significance at the species level, but it must be stressed that variation in such features may be the result of environmental factors, age of the plants sampled, etc. Chemical characters must not be regarded as privileged and exempt from all the safeguards to which other kinds of character are subjected. The fact that they may be obtained only after complicated analytic techniques does not make them more valuable than, say, exomorphic features or chromosome counts. Some may turn out to be so, others not.

Cyanogenetic glycosides

HCN usually occurs in plants as glycosides (so-called cyanogenetic or cyanophoric). About 14 cyanogenetic glycosides are known and are of restricted occurrence. They may be present in some parts of a plant and absent in others. They may, for example, be absent from leaves (the organ usually tested) yet be found in the seeds (e.g. in *Anthemis*, Greshoff, 1909). Their presence may also vary according to the time of the year and according to the age of the plant or organs.

Tests may be made on herbarium specimens where positive results may be valuable; but negative results may be less significant, since a positive reaction might be obtained from fresh material. The tests may be simple, using small amounts of ground up material and testing with sodium picrate paper in a test tube. Details are given by Gibbs (1954, 1958b). Gibbs warns that since there are many cyanogenetic glycosides all yielding HCN, the evolution of HCN may not provide a clue to relationship.

Phenolic constituents (including phenolic glycosides)

These are best considered in conjunction with the glycosides just mentioned. The chemical terminology of these substances can be highly confusing and a table is given of their classification in Bate-Smith (1962). According to Bate-Smith (1958), little thought has been given to the phenolic constituents of plants with the exception of the anthocyanin pigments of flower petals. They are, however, present in all of the higher and most of the lower plants. He suggests that the lack of interest in them is due largely to the fact that they have no apparent function and no explanation can be given for their great diversity. It is largely as a result of his work that a large amount of information is now available about plant phenolics as taxonomic guides. These results have been published in series of papers to which the reader is referred for detailed information (Bate-Smith, 1954, 1956, 1958, 1959, 1961, 1962; Bate-Smith & Lerner, 1954; Bate-Smith & Metcalfe, 1957; Bate-Smith & Ribereau-Gayon, 1959). Only some of the main points of this work will be considered here.

Phenolic constituents are known to be present in the leaves, flowers and seeds of plants. Bate-Smith has shown that there is an interesting correlation between the presence of leuco-anthocyanins* and the woodiness of the plant.

Flower colour pigments. A great deal of work has been done on the chemistry and genetics of flower colour pigments (Lawrence & Scott-Moncrieff, 1935; Scott-Moncrieff, 1939; Lawrence & Price, 1940). Anthocyanins are mainly responsible for the wide range of colours, from reds and purples to deep blue, found in flowers and some fruits, and also, together with carotenes, for the autumn hues of leaves. For a survey of the chemistry of these pigments, see Geissman (1962).

A detailed example can be taken from Hall's monograph of the genus *Tulipa* (1940). The colour of the tulip flower is largely determined by two classes of pigment: (*a*) a ground colour, white or yellow, situated in the

* The term leuco-anthocyanin was applied by Rosenheim to colourless compounds which may become visible when boiled with mineral acid, giving brilliant red pigments having the characteristics of anthocyanidins.

plastids in the mesophyll cells; (*b*) anthocyanin—a soluble sap pigment of various shades of crimson or purple, in the cells of the cuticle except in pure yellow or white flowers. The visible colour of the flower is therefore produced by the anthocyanins modified by the background of plastids: colour intensity depends not only on anthocyanin concentration but, in flowers of lighter shades, upon a mixture of unpigmented with pigmented cells. Or lighter shades may be caused by presence of a co-pigment or by variation in the pH of the sap. Flower pigments (ivory or yellow) may also be found in the cuticle of some tulips and this modifies the anthocyanin colour.

The taxonomic significance of pigmentation lies largely in the restricted occurrence of anthocyanins. Species which are considered "good" on general taxonomic evidence normally contain only a single anthocyanin with no admixture of a secondary pigment. Hybrids of species characterised by different anthocyanins can give rise to pigments found in neither parent, due to chemical recombination.

Too much stress should not be placed on the diagnostic value of flower colour, since exceptions to the general rule that plastid colour is constant in any species occur. Hall cites the case of *Tulipa aitchisonii*, a diploid species which contains both white and yellow variants. Amongst tetraploids derived from it, *T. stellata* and *T. chitralensis* have white backgrounds while *T. stellata* has yellow plastids. White and yellow ground variants occur in *T. kaufmanniana*, and *T. greigii* is typically full yellow in ground colour though other forms with palest primrose or ivory white occur.

Albino forms in which anthocyanin pigments are absent are more common. In the Near East pure yellow forms of scarlet tulips of the *Oculus-solis* group are found. The yellow-flowered *T. batalinii* differs from the scarlet-flowered *T. linifolia* only by the absence of anthocyanin.

Two additional groups of pigments, which are in some respects similar to the anthocyanins with which they have been classified in the past, are the red-violet betacyanins and the yellow betaxanthins. They occur in the cell sap. Their distribution in the *Centrospermae* is discussed by Dreiding (1961).

Flower colour is, of course, of great selective significance in relation to insect and bird pollination (see p. 435).

Phenolics in vegetative tissues. Bate-Smith has concentrated on the phenolic constituents of leaves since petal pigments are highly variable genetically and, apart from their specific diagnostic value just discussed, are frequently not closely correlated with the systematic position of plants. Phenolics in the vegetative tissues are, on the other hand, found to be of considerable systematic value.

The commonest phenolic constituents of leaves are: (1) *flavonoid* compounds: the *flavonols* myricetin, kempferol and quercetin, together with leuco-anthocyanins yielding cyanidin and delphinidin; (2) *phenylpropane* derivatives: the hydroxycinnamic acids, *p*-coumaric and caffeic acids. These compounds, as well as a number of less commonly occurring phenolics, are closely related chemically. The first group all have 15 carbon atoms in their basic skeleton ($C_6C_3C_6$); the second group have nine (C_6C_3). Further details of their chemical structure are given in the papers of Bate-Smith (1958, 1962). (Other valuable classes of phenolic compounds from a systematic viewpoint are the methoxyacids, flavones, coumarins, isoflavones, methylenedioxy compounds and anthraquinones.)

Leuco-anthocyanins. These are present in all classes of vascular plants, but their distribution in the Angiosperms is irregular. The presence of leuco-anthocyanins and their relative strength may be simply tested and the identity of the anthocyanidins produced may be established by paper chromatography. Unlike those from petals, the anthocyanidins produced by leaves from the leuco-anthocyanins are nearly always cyanidin or delphinidin.

The correlation shown between the presence of leuco-anthocyanins in petals and woodiness was confirmed by a more extensive study by Bate-Smith of their systematic distribution in leaves. As a rule the herbaceous members of generally woody families possess leuco-anthocyanins, while whole families (of the *Gamopetalae*), whether woody or not, lack them. The tendency for leuco-anthocyanins to be associated with woody habit also occurs within woody and herbaceous families, e.g. *Leguminosae*. Bate-Smith took as a basis for habit classification the details given by Metcalfe & Chalk (1950) in their systematic survey of Dicotyledon anatomy. Five groups recognised were (*a*) wholly woody, (*b*) woody habit dominant, (*c*) woody plants and herbs in various proportions, (*d*) herbaceous habit dominant, (*e*) wholly herbaceous. It was found that 61 per cent. of the members of the woody families examined contained leuco-anthocyanins but only 15 per cent. of the members of herbaceous families.

Bate-Smith comments on the small number of woody plants in the last six orders of the *Metachlamydeae* (in Engler's sequence) containing leuco-anthocyanins: they are to be found only in the *Apocynaceae*, *Asclepiadaceae*, *Rubiaceae* and *Caprifoliaceae*, while no leuco-anthacyanins are found at all in the arboreal members of the *Oleaceae*, *Verbenaceae*, *Scrophulariaceae* and *Bignoniaceae*. On the other hand, the herbaceous *Primulaceae* and *Plumbaginaceae* are rich in leuco-anthocyanins.

A most interesting situation is found in the *Leguminosae*, where the woody *Mimosoideae* and *Caesalpinoideae* nearly all contained leuco-anthocyanins (and flavonols), while the *Papilionoideae*, which is largely herbaceous but also contains trees and shrubs, has members with and without leuco-anthocyanins. He finds that the trees and herbs in this subfamily are more or less equally divided as to those whose leaves contain leuco-anthocyanins and those which do not. This distribution agreed largely with the subdivision of the group into tribes, although there were certain exceptions. The irregularities of distribution occurred in the tribes *Galegeae* and *Hedysareae*, and Bate-Smith discovered that Dormer's division of the *Papilionoideae* into two groups—those that possessed a foliar pulvinus and those that did not—broke down in just these two tribes. He found that there was a high correlation between Dormer's division into pulvinate and epulvinate and the presence or absence of leuco-anthocyanins (Dormer, 1946; Bate-Smith, 1958).

It should be remarked here that Gibbs (1954, 1958b) has made a survey of the reaction of families to the HCl/Methanol test and finds that they agree closely with the distribution of leuco-anthocyanins. Species giving positive results agreed so closely with those which possessed leuco-anthocyanins (and vice versa) that he began to wonder if reactions to the two tests were due to the same substances. Bate-Smith (1958) believes that the positive HCl/Methanol reaction is due to the presence of leuco-anthocyanins in the wood. Certain exceptions were, however, recorded by Gibbs, as in the *Caprifoliaceae*. He also noted that raphides never seem to occur in his HCl/methanol-negative series, apart from the *Rubiaceae*.

Flavonols. These are even more widespread than the leuco-anthocyanins and again it is their absence that is more interesting from a systematic viewpoint than their presence. Woody families were found to contain a higher proportion of species containing flavonols than herbaceous families—90 per cent of species in the former, 58 per cent. in the latter.

Phenylpropane derivatives (methoxy and hydroxy acids). Hydroxy acids (caffeic and ellagic) are more frequently present than not in the Dicotyledons. Ellagic acid calls for special comment. It appears to represent the missing trihydroxycinnamic acid which does not occur in plants (Bate-Smith, 1962). It is regarded as an excellent clue to relationship by Bate-Smith since it can only occur in certain "chemically-derived" situations (see below). Methoxyacids (ferulic and sinapic) are more often absent than present. There is a very high correlation between the presence of these methoxyacids and the presence of tannins.

Tannins. These are phenolic substances which are recognised by their characteristic appearance in microscopic sections when fixed in formalin-acetic-alcohol and stained with iron salts. Bate-Smith & Metcalfe (1957) examined 500 species belonging to 175 families for tannins and also analysed the leaves of the same or closely related species for the presence or absence of leuco-anthocyanins. In 83 per cent. of cases presence of tannins was found to coincide with the presence of leuco-anthocyanins and it was therefore suggested that the histological reaction for tannins was most commonly due to the presence of leuco-anthocyanins. As would be expected, most tanniniferous families are woody or probably allied to woody families; most of the non-tanniniferous families are herbaceous. The occurrence of tannins has a certain taxonomic value since some families are highly tanniniferous, some contain little or none and some give mixed results.

A certain correlation has also been found between the presence of tannin and the evolutionary position of the family as estimated by Sporne's advancement index (cf. Chapter 2, p. 68); 0 represents the most "primitive" condition, 100 the most "advanced" condition).

	Index No.
Completely tanniniferous families (82)	14-75 (14 over 50)
Almost non-tanniniferous families (13)	32-93 (4 below 50)
Non-tanniniferous families (38)	36-100 (8 below 50)

There is therefore a tendency for the capacity to synthesise tannins to decrease as the advancement index increases. This could be interpreted as meaning that the capacity to synthesise tannins is a primitive character which tends to be lost with increasing specialisation. The absence or reduction of tannin production is often, although not invariably, associated with herbaceous as opposed to woody habit. Too much should not be read into these results, however, since Sporne's advancement index is only a rough guide and of debatable significance. The association of leuco-anthocyanins and tannins with the woody habit, however, can scarcely be taken as confirmation of Hutchinson's division of the Dicotyledons into woody and herbaceous lines but rather as a reflection of it.

From a taxonomic viewpoint (as opposed to phylogeny), Bate-Smith & Metcalfe note that there is a marked correspondence between the capacity for Dicotyledons to synthesise tannins and current concepts regarding the composition of families. At the order level there is less agreement and it might be recalled that Metcalfe regards Angiosperm taxa above the family level to be heterogeneous as regards their wood anatomy. This suggests that our present concepts of orders would repay a critical study (cf. Chapter 3).

Restricted occurrence of phenolics in certain taxa. So far, we have been largely concerned with the overall distribution of a small number of phenolic constituents, i.e. as synthetic characters. There is also evidence that the restricted distribution of rarer products is of taxonomic interest, i.e. as analytic characters.

Phenolics of restricted occurrence are largely found in the anthraquinones, nitrogenous anthocyanins and isoflavones. Substances which are neither widespread nor greatly restricted in occurrence are the flavones, flavanones, catechins and the coumarins (Bate-Smith, 1958), but it is always difficult to know how far the distribution patterns are valid owing to technical difficulties of recognition and incomplete sampling. A few examples will, however, be given of apparently significant distributions.

In the *Myrtaceae* nearly every member examined contained ellagic acid. This has been found in ten other families including the *Rosaceae*, *Casuarinaceae*, *Euphorbiaceae* (part), *Theaceae* and *Ampelidaceae*. The *Rosaceae* has been studied in detail by Bate-Smith (1961) with respect to phenolic constituents which can be determined by paper chromatography. The overall pattern is that of a woody family: the leaves of the woody members contain leuco-anthocyanins (leuocyanidin) and abundant flavonols (quercetin and kaempferol). With the exception of the *Chrysobalanoideae* and *Potentilla anserina* (*Rosoideae*), the trihydroxy representatives of these phenolic classes are absent. All members of the *Rosoideae* except the tribe *Kerrieae* contain the trihydroxy acid, ellagic acid, which is absent from the other subfamilies. This tribe is therefore unique in the *Rosoideae* and its constituent three genera are consistent among themselves. *Potentilla anserina* contains myricetin and leuco-delphinidin (the trihydroxy compounds found elsewhere in the family only in the *Chrysobalanoideae*) which supports its separation as a section or even genus.

The sections of the genus *Prunus* sensu lato (often treated as separate genera) contain certain uncommon phenolic constituents in addition to the typical phenolic pattern of the family. In examples such as these the chemical data may provide useful evidence suggesting the more natural positioning of critical or difficult species or higher taxa, where to draw dividing lines, etc.

In the Monocotyledons, with the exception of the *Gramineae* and perhaps the *Bromeliaceae*, phenolics show a rather uneven distribution. Their diagnostic value may therefore be high in some cases as in the *Iridaceae*, *Amaryllidaceae*, *Liliaceae*, etc., where the diverse phenolic distribution pattern follows generic and infra-generic classification to a considerable extent. In the genus *Iris* for example, sections *Pogoniris* and *Evansia* have a certain biochemical similarity—none of the species contain leuco-anthocyanins or flavonols; the *Sibirica* and *Californica* groups consistently contain both compounds, while Subsect. *Spuria* contains only leuco-anthocyanins. Two isoflavones, iridin in *Pogoniris* and tectoridin in *Evansia*, have been found nowhere else in the plant world.

Phenolic constituents of wood. The phenolic constituents of wood have also been extensively studied. A special study of Conifer heartwood has been undertaken by Erdtman and his collaborators. It has been found that the *Pinaceae* and *Cupressaceae* can be separated on chemical grounds and within the genus *Pinus* there are distinct chemical differences between subgenus *Haploxylon* (with needles in 5's) and *Diploxylon* (with needles in 2's and 3's). Intraspecific variation is found in the relative proportion of the phenolics between local races of *P. sylvestris*. Unusually large amounts of pinosylvin phenols have been isolated from several specimens of heartwood from the Pyrenees. In Sweden there are traces of clinal variation in phenol pinosylvin content. Details can be obtained from the papers of Erdtman (1956, 1959).

A new class of natural products related to the flavonoids has recently been discovered. They are termed *biflavonyls* and are extensively distributed in the Gymnosperms. A survey of their systematic occurrence has been made by Baker & Oliis (1961) which shows certain interesting features. All families in the *Coniferales* studied, with the exception of the *Pinaceae*, contain species from which biflavonyls have been isolated. The only non-Gymnospermous plant known to contain a biflavonyl is *Casuarina stricta*!

In a survey of the lignin chemistry of the genus *Acer* covering about a third of the known species, it has been found that certain species are sufficiently different in their chemistry to warrant their separation off as a distinct genus

(Towers & Gibbs, 1953). It has been proposed that they be placed in the genus *Negundo*, but the evidence from other sources such as floral anatomy conflicts with this view (Hall, 1951, discussed above under floral anatomy).

Leuco-anthocyanins in seeds. Data concerning the occurrence of leuco-anthocyanins in seeds are limited. A limited chromatographic survey was made by Bate-Smith & Riberau-Gayon (1959) who found that "presence or absence of leuco-anthocyanins in the testa reflects more consistently the character and relationships of the family as a whole than does their presence in the leaves". Thus herbaceous members of the *Geraniaceae*, for example, differ from woody members in the absence of leuco-anthocyanins in the leaves, while the seeds are consistently positive. The seeds of the *Leguminosae* were found to contain leuco-anthocyanins whether or not the leaves of the species concerned possessed them. The absence of them from epulvinate members of the *Papilionoideae* is regarded by Bate-Smith as the local suppression of a synthetic capacity originally possessed by the whole plant.

Summary of the role of phenolic constituents in taxonomy. Leuco-anthocyanins in general are correlated with woodiness; methoxycinnamic acids with the herbaceous habit. This is of little taxonomic value since the wider the presence of a character the less its value for group-making. There is no suggestion of dividing flowering plants into taxonomically meaningful groups simply according to the presence or absence of leuco-anthocyanins any more than because they are woody or not. The unusual occurrence of a phenolic constituent in a group serves as a pointer to possible taxonomic discontinuity. If other differences are correlated, there may be a basis for division. The chemical character by itself is no more a sure guide to taxonomic separation than any other single character or line of evidence. As taxonomic characters in a natural system, phenolic substances must be regarded for correlation purposes as equivalent to any other characters. Their physiological significance is of no importance.

Phenolic constituents are more important in assessing evolutionary relationships. They are regarded by Bate-Smith as chemically privileged. The chain of biosynthetic processes which give rise to their formation appears to be irreversible beyond a certain point. Absence of a particular phenolic compound does not, however, necessarily mean complete absence since other phenolics may be present. The pathway may be diverted, or suppressed (cf. Leguminous seeds and leaves), or the compound may be used up so fast that it cannot be detected by ordinary techniques. Bate-Smith stresses that it is the pathways or processes which are important, not the end-products. The absence of the end-product does not mean necessarily that the pathway is not there.

The evolutionary significance lies in the possibility of establishing the necessary chemical evolution of the phenolic constituents in terms of enzymatic processes. If it can be established that certain processes are irreversible, in other words once the enzyme is lost it cannot be regained, then we are in possession of information concerning the direction of change. As in other demonstrated directional changes, the value of this information is more in terms of its negations than positive affirmations. In other words, we may be able to say that a group containing a certain compound cannot have been derived from another which contains a more highly modified compound. What we cannot often say is which particular group did in fact give rise to another: we can usually only postulate the general direction of change.

Water-soluble carbohydrates

Little work has been done on the taxonomic value of water-soluble carbohydrates but a recent study by MacLeod & McCorquodale (1958) on the sugars, oligosaccharides and water-soluble polysaccharides in the seeds of 22 species of grasses suggests considerable taxonomic implications. Both quantitative and qualitative determinations of the compounds were made. They found, for example, that the *Bromeae* form a very natural tribe, containing fructosans but no raffinose (oligosaccharide). In this respect *Brachypodium sylvaticum* stands apart (as it does cytologically), and has already been recognised as the basis of a new tribe by Hubbard despite the hairy appendage on the top of the ovary which it shares with other members of the *Bromeae*. All the *Festuceae* contain soluble β-glucosan in greater or lesser amounts. But *Festuca* and *Lolium* (the latter previously placed in the *Hordeeae* because of its sessile spikelets) are distinctive in containing an unusual trisaccharide. This supports the close relationship of these two genera suggested by their ability to hybridise.

Recently Percival (1961) has analysed by paper chromatography the types of nectar found in the Angiosperms. In this survey 10 different nectar types are distinguished and certain tendencies in the types and distributions of nectars are revealed. These show a relationship both to systematic position and to method of pollination, but as Percival points out, the evidence is still insufficient for taxonomic conclusions to be drawn.

Terpenes

The terpenes found in Conifers have been the subject of extensive study, especially by Mirov on the Pines. When the trunk of a pine tree is cut, a quantity of oleoresin is exuded: that part which is volatile with steam is turpentine which consists mainly of cyclic terpenes. Terpenes are end-products of metabolism and once formed are not utilised by the plant (Lvov, 1954). Mirov (1956, 1958) considers they have certain taxonomic advantages over other plant constituents: the turpentine composition of most pine species is relatively simple and enough can be obtained from an individual tree to allow its variability within the species to be studied.

An analysis of the turpentine composition of 94 out of approximately 100 known *Pinus* species has been completed largely by Mirov and his associates (Mirov, 1961). They found that the turpentine composition of each valid species was specific and little modified by the growing season or transplanting. Although some widely separated species (from a taxonomic viewpoint) contain almost identical major turpentine components, they can still be separated by certain specific properties of the turpentine, frequently in its high boiling fractions (Mirov, 1954).

It is found that the chemical composition of the turpentine is not always correlated with the systematic position of the species. The reasons suggested by Mirov (1958) are that: (1) systems of classification of the genus differ considerably according to the author concerned; (2) chemical and morphological characters may have followed different evolutionary pathways. Further, the chemical evolution of the pines may not be closely synchronised with that of other features. As Mirov says, the biochemical relationships of the species existing, say, in the Jurassic was different from that obtaining today. As a consequence we now have patches of an old biochemical pattern and patches of an old morphological pattern, and it is often difficult to fit them together in the present systematics of the genus. These relict patches may, however, be of considerable value in elucidating the possible evolutionary relationships of contemporary species. This is well illustrated by *Pinus jeffreyi*, a "relict" Californian species. It resembles *P. ponderosa* of the *Australes* group in certain morphological features and forms fertile hybrids with it in nature. Chemically, however, *P. jeffreyi* is related to a widely different group, the *Macrocarpae*, comprising *P. coulteri*, *P. sabiniana* and *P. torreyana*. Again it forms fertile natural hybrids with the former species. Mirov interprets this situation as follows: *P. jeffreyi* occupied an intermediate position between the two groups, the *Macrocarpae* being the older and the *Australes* the more recent. During the course of evolution the links between these groups disappeared and only one species, *P. jeffreyi*, remains as a testimony to the former relationship. There is chemical evidence supporting its interpretation as an older species than *P. ponderosa* (Mirov, 1938, 1958). The relationships may be shown diagrammatically as below.

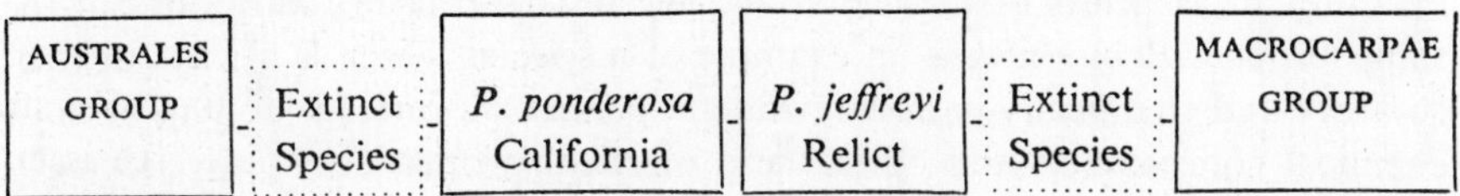

Despite the high specificity of the turpentines, apparent deviations do occur for the following reasons:

1. When a taxonomic species contains several physiological or chemical races, each with its own particular turpentine.

2. When one species has been included mistakenly in another. *P. engelmanii* has been treated in the past as a synonym of *P. ponderosa*, but it has been shown that their turpentine composition is very different and they are currently regarded as distinct species.

3. When chemical mutants occur within a species which is otherwise

homogeneous. These have been described in *P. sylvestris*, *P. monticola* and other species and are possibly quite common.

4. When natural hybridisation takes place between species. Mirov (1956) has made a study of the composition of the turpentine of hybrids between *P. contorta* and *P. banksiana* which cross in nature in Alberta, Canada, forming hybrid swarms. The turpentine in the F_1 hybrids produced artificially contained the terpenes of both parents with the more complicated compounds (bicyclic terpenes) of *P. banksiana* predominating over the simpler compound (monocyclic terpene) of *P. contorta* in an approximate 3:1 ratio. Samples taken from the area of overlap between the two species were, as would be expected, heterogeneous morphologically, but the chemical composition of these natural hybrids did not always correspond with their morphological characteristics. Plants from the hybrid zone which appear to correspond morphologically with *P. contorta* were found to be chemically *contorta*-type; or chemical hybrids possessing the two kinds of terpenes in differing proportions; or chemically pure *banksiana*-type. None of the plants easily identifiable as *P. banksiana* possessed the *contorta* compound. Morphologically intermediate pines were often also intermediate chemically. Some were chemically pure *banksiana*, but none were chemically pure *contorta*. Mirov concludes that when all the reasons for the lack of correlation between the taxonomy of a pine and its chemical composition are understood, the specificity of turpentine appears to be a valid character. It would be unwise, however, to rely too heavily on this kind of evidence without corroboration from other sources (including chemical ones).

Variation in the chemical composition of pine populations is well known. Variation may also be detected by studying their variability in physical properties, such as the optical rotation of turpentine. Some species vary little throughout their geographical range while others vary widely. Variability of optical rotation depends on (1) different amounts of laevo-rotatory and dextro-rotatory terpenes (turning polarised light to the left and right respectively); (2) different amounts of the laevo- and the dextro-rotatory antipodes of the same terpene. *P. merkusii* is an example of a species which is always dextro-rotatory and chemically uniform, while *P. ponderosa* varies widely in both its chemical composition and optical and physical properties (Mirov, 1958).

The two closely allied Mediterranean species, *P. halepensis* and *P. brutia*, may be readily separated by their terpenes. Two terpenes, β-pinene and Δ^3-carene, are found in *P. brutia* but are absent in *P. halepensis*. Furthermore, the turpentine of *P. halepensis* is dextro-rotatory and that of *P. brutia* laevo-rotatory. Critical pines allied to *P. brutia* in the Black Sea region are also laevo-rotatory. As a reflection, presumably, of their chemical differences, *P. halepensis*, but not *P. brutia*, is attacked by the scale insect *Matsucoccus*. Other examples will be found in Mirov (1961).

The geographical distribution of the chemical components of the turpentines is now known for most of the genus. A notable feature of the biochemical geography of the genus is that the same or essentially similar turpentine components are found in the pines of western America and eastern Asia, on both sides of the Pacific, suggesting that the origin of the genus was in the land mass (no longer existing today) that lay between the northern parts of America and Asia (Mirov, 1959). Eastern American pines are similar chemically to south-west European pines in possessing relatively simple turpentines. This supports palaeobotanical evidence that most European species of the genus came to Europe via north-east America, Greenland and Spitzbergen, while *P. sylvestris* and *P. peuce* possess turpentines similar to those of Asiatic pines and probably came to Europe from the East across Asia. This kind of chemical analysis can clearly contribute much to the elucidation of natural relationships and migration routes when considered along with all other evidence.

The classification of the genus *Eucalyptus* (centred in Australia) has proved difficult on gross morphology. A remarkable feature of eucalypts is their essential oils, and it is not surprising that they have been the subject of systematic study with a view to determining their taxonomic value. Baker & Smith (1920) found in such a survey that the oil composition for a given species was more or less constant and therefore regarded the oils as having equal weight to other, morphological, characters that had been employed in the taxonomy of the genus. As a result of this dictum they made alterations in rank when differences in oil composition were found between taxa, and conversely they united species when their oil composition proved identical despite the large morphological differences they showed—a very dubious practice!

Baker & Smith were able to divide the genus into large groups containing different combinations of oils and found that these groups were correlated with bark structure and leaf venation. In fact, they generalised that progressive morphological changes in *Eucalyptus* were correlated with concomitant changes in the chemical composition and yield of essential oils. For example:

(1) Obtuse "feather" venation indicated low yield of oils, with pinene as the essential component.

(2) More acute venation with marginal veins indicated a slightly higher oil yield, with cineole and pinene as main components.

(3) "Butterfly" venation indicated a very high oil yield, with phellandrene and piperitone as main constituents.

It is unfortunate that Baker & Smith went beyond their evidence in constructing a very optimistic phylogenetic tree showing the origin of all species from other contemporaneous forms. Penfold & Morrison (1927) discovered that there is variation in oil composition between trees which are indistinguishable morphologically. These chemical differences were not due to soil, climate or other ecological factors but to inherent differences in the trees, and they introduced the terms "physiological form" and "chemical variety" to cover these variants. These physiological forms were found to be widespread among the commercial oil-yielding species and it is now generally agreed that there is more variation in oil composition than Baker & Smith allowed for. This variation may also follow clinal patterns as has been reported by Barber and co-workers (Barber, 1956).

Chemical variation in species and races has been intensively studied in various commercially important groups. Schratz (1960) considers that such chemical variants are widespread in nature. He proposes distinguishing between quantitative and qualitative infra-specific chemical variation. The quantitative variation is difficult to treat taxonomically without extensive study to show how much of it is due to morphogenetic, ontogenetic and environmental factors. Even when this can be elucidated the taxonomic characterisation of the variants would be difficult, due to the number of chemical constituents accumulated by different genotypes being under the control of polygenic systems. All this applies not only to essential oils but to other secondary constituents such as glycosides where in *Digitalis purpurea*, for example, glucoside composition appears to be controlled by the age of the plant, not by temperature or season. Plants from diverse provenances differ in their morphological and anatomical characters, but these differences do not agree with the various chemical types of glucoside found. In other cases there is no appreciable morphological differentiation accompanying chemical differentiation (Schratz, 1960; Dillemann, 1960; Jaminet, 1960). Qualitative differences in chemical composition are, however, easier to treat taxonomically since the different genotypes are easier to delimit as a result of the discontinuities in the variation pattern and the fact that such differences are often less modified by external factors.

There is much debate as to the desirability of the recognition and naming of the various kinds of chemical variants, especially when they are independent of morphological, anatomical and cytological differences. Proposals for an official policy have been published in the journal *Taxon* by Lanjouw, Mansfeld and Tétényi (1958), and by Tétényi (1960, 1962).

Tétényi (1962), in a discussion of chemical races within species, considers that the essence of their chemical differentiation "is not the difference between the accumulated secondary products of metabolism, but the phasic deviation of the ontogenetic changes of metabolism."

Fats and oils in seeds and fruits

A wide range of fats and oils are synthesised by plants which store them particularly in their seed and fruit coverings. Only a limited number of these products have been analysed chemically and these especially in plants of economic importance, but what is known from them suggests that consistent surveys of whole taxonomic groups might give results of taxonomic value. There is a very high level of specificity of fatty acid composition in the *Palmae*, for example (Hilditch, 1952).

The fascinating story of the chaulmoogric acid series in the *Flacourtiaceae* and its role in the treatment of leprosy is another example where chemical investigations instigated for economic or medical reasons have provided results of taxonomic importance.

The oils of various members of the *Flacourtiaceae* have been employed for a long time in tropical Africa, India and China for the treatment of leprosy, and analysis has shown that the most effective ones contained glycerides of the chaulmoogric acid series. Moreover they appear, so far, to be confined not only to this family but to certain tribes of Gilg's classification of the family, thus lending support to this taxonomic disposition (for a summary see Gibbs, 1945).

At the species level an example may be taken from a study of the volatile oils found in *Brassica* seeds by Hemingway *et al.* (1961). Variation is shown in the cultivated mustard species *Brassica juncea* in morphology, especially leaf shape. There is also variation in the type of volatile oil produced on hydrolysis of the seed glucosides. Three conditions have been described:

(1) production of allyl isothiocyanate only;
(2) production of 3-butenyl isothiocyanate;
(3) production of a mixture of the two.

An investigation of 96 seed samples from a wide range of sources grown in Britain included all previously described forms, and revealed an apparent correlation between the oil variation and geographical distribution.

B. juncea is an allopolyploid of *B. nigra* × *B. campestris*. *B. nigra* produces only allyl isothiocyanate while *B. campestris* produces only 3-butyl isothiocyanate. It is thought that this may explain the variations in oil production. When the correlated evidence of morphology, geographical distribution and chemistry is considered, there are some grounds for believing that the origin of *B. juncea* may have been polytopic, the natural hybrid having arisen at different geographical points and subsequently doubled to give the allopolyploid.

Amino acids

The techniques of paper chromatography have been applied extensively to amino acids, but the number of acids which have been characterised is quite small, around 50, and the occurrence of perhaps half of these may be so general that they are of no particular systematic value (Alston & Irwin, 1961).

Refined techniques have made the identification of individual components feasible and these may have potential taxonomic significance, but so far the results have been disappointing although the restricted distributions shown by some of these acids may prove valuable (Fowden & Steward, 1957).

Alston & Turner (1959) suggest that paper chromatographic analysis of biochemical patterns may give an insight into the evolution of a group in the same way as patterns of chromosomal rearrangement have been applied. It should be possible, using improved techniques, to obtain sufficient biochemical characters to allow "standard chromatograms" or "biochemical profiles" to be produced as a supplement to standard descriptions of individuals, populations or species.

It has been recently shown that there appears to be a correlation between chromatographic pattern and hybrid expression: hybrids may show biochemical recombinations of parental components and be intermediate in their exomorphic characters (Alston & Turner, 1959; Hillis & Orman, 1962).

Chromatographic analysis may be employed in taxonomy without actually identifying the substances which distinguish the chromatograms of one species from another. An example is the recent work of Riley & Bryant (1961), whereby nine species of the *Iridaceae* have been shown to differ chromatographically by comparing the patterns of extracts made from small portions of their root tips.

3. *Proteins*

Protein analysis is very much in its infancy and results obtained so far have not been encouraging from a systematic viewpoint. Technical difficulties are such that it is unlikely that comparative protein studies will afford a ready means of providing taxonomic evidence. Serological methods, on the other hand, are comparatively rapid and may be systematically valuable. They are regarded as an objective means of assessing biochemical similarity between plants and animals, but again there are serious technical and interpretative difficulties.

The early serological work carried out by Mez and his Königsberg school in the nineteen-twenties which produced spectacular phylogenetic trees of the whole plant kingdom (including fossils!) is now largely of historical interest. It is fully reviewed by Chesters (1937). It was largely unsuccessful for two reasons: firstly, it employed such comparatively crude serological methods as were available at the time; secondly, there was no third criterion for deciding between the serological and morpho-anatomical results when they conflicted. Taxonomists, not unexpectedly, were hesitant about preferring such novel methods to the well-tested ones with which they were familiar. Little faith was placed on the antibody-forming mechanism of a mammal to make their distinctions for them, especially when the Berlin-Dahlem school of serology came to different conclusions from the Königsberg group after employing modified techniques.

The great advantage of the serological method is its ability to distinguish antigenic substances which are indistinguishable by other chemical means. However, when considering the taxonomic use of such phenomena the following points must be borne in mind:

An antibody against a given protein derived from one species in a group may either: (1) precipitate with another protein in that group in a quantitatively identical manner—*prima facie* evidence that the two proteins are identical; (2) it may "cross-react", i.e. precipitation may occur, but more antibody will be required to produce a given amount of precipitate, and the interaction may be less efficient in other ways. The current explanation is that the population of antibody molecules is not homogeneous. When antisera which contain antibodies against a number of unrelated antigens are used, cross-reactions of this type cannot readily be distinguished from "homologous" reactions. In the work recently carried out by Gell, Hawkes & Wright (1960) on the application of immunological techniques to the taxonomy of *Solanum*, the analysis has been based on the presence or absence of any reaction at all—no attempt has been made to measure the amount of precipitate produced, as was done by Mez and his associates. There is a serious risk of error if taxonomic conclusions are based on measurements of the amount of precipitate produced when crude plant extracts, containing numerous different antigenic proteins, react with their corresponding antisera.

In this type of work it is important to distinguish between different antigens in the extracts: the more antigens we can differentiate the more data for determining relationships will be available, and presumably the firmer our conclusions. For this it is necessary to use serological methods which will distinguish between different antigens and antibody systems in the same total reaction. Such methods have been employed by Gell, Hawkes & Wright through the use of gels instead of liquid media to bring about precipitation. Crude plant extract is injected into a rabbit so that antibodies form in their blood, producing the antisera used for the experiments. These methods depend on the fact that antigen-antibody systems in a mixture form separate bands of precipitate when allowed to diffuse towards one another in a thin sheet of agar gel. Thus a characteristic "spectrum" of precipitin lines is formed for every species or group of species. Furthermore, one species may be directly compared with another by seeing how many of these lines join up when protein extracts from each are placed side by side and allowed to diffuse towards a common antiserum. A comparison can them be made by noting how many antigens common to a pair or group of species are present, and how many antigens are not common to the pair or group. Further information may be derived by noting the density of the lines.

Conclusions reached in this way can be compared with those obtained by classical taxonomic methods. When they agree, they serve to reinforce one another. If they disagree, the results can be checked by cytogenetic methods, although even then it has to be decided whether the relationships being sought are the same—whether they are intended to assess overall affinity or some particular aspect such as crossability.

These methods have been applied by Gell *et al.* (1960) to the tuber-bearing *Solanum* species (potatoes) and it has been possible to arrange the 38 species studied into groups which show a fairly close agreement with orthodox taxonomic classification and with cytogenetical relationships. It threw little light on the relationships of the South American species studied because the antisera used could not distinguish between them as they did with the Mexican species.

A revival of interest in comparative serology is largely led by workers at Rutgers University and some of their results may be found in the *Serological Museum Bulletin*. Seed proteins of genera of the *Magnoliaceae* and *Ranunculaceae* have been studied serologically and the results are largely consistent with current systems of intergeneric relationships in the two families.

The grasses have received their customary attention and a recent study has been undertaken by Fairbrothers and Johnson (1959) on the precipitin reaction as an indication of relationship in the subfamily *Festucoideae*. Antisera to proteins of a species give intense reactions with proteins from species estimated to be closely related, but little or no reaction with those species judged to be distantly related. Examples of the results are that *Festuca rubra*, *Dactylis glomerata* and *Poa pratensis* show strong serological correspondence with *Bromus inermis* antiserum: six species of *Bromus* were found to be sero-diagnostically separable from one another and different degrees of serological correspondence were found between certain of them. The interpretation of such results is not easy.

In general the rather close agreement between serological, morphological and in some cases cytogenetic evidence (as in *Solanum*) augurs well for the inclusion of comparative serology as a technique in modern taxonomy. With direct chemical determination of the proteins an extremely laborious and difficult matter, it seems that refined immunological techniques are the best that we can use. But one must assume that many proteins remain undetected by these methods (cf. the failure to distinguish between the 22 South American *Solanum* species studied). What exactly the varying degrees of serological correspondence tell us about relationship is very difficult to judge. If serological relationship could be shown to be a more or less accurate reflection of overall genetic relationship, then clearly these methods would be of the greatest importance. That this is so, however, is very far from being established. That the methods are "objective", as is often claimed for them, is even less obvious since it appears that "subjective" assessment of degrees of correspondence cannot be avoided. Results so far indicate that serum diagnosis may take its place along with other disciplines in providing data for establishing relationships—phenetic and phyletic. But like the other disciplines serum diagnosis has to be interpreted.

Parasitism

A small number of workers have attempted to apply knowledge derived from fungal parasite relationship to the relationship of their hosts. Parasite relationship is according to Savile (1954) a valuable "biochemical" tool if properly used. Taxonomic mycologists have been able to work out with a reasonable degree of probability the evolutionary trends of some genera, notably in the *Uredinales*, and these may in turn suggest relationship or comparative age in the host plants. It has been suggested that a new rust form becomes adjusted regularly "to some new group of host plants which is phylogenetically younger than the host plants of its origin" (Leppik, 1953); i.e. the fungi have evolved along with their hosts; there has been no return of "advanced" rusts to "primitive" hosts. The evolution of the fungi is in this sense irreversible.

This approach has been applied (through a study of some of the smuts) to *Carex* and related genera where it has revealed a previously unrecognised natural subgenus of *Carex* and suggested improvements in the taxonomy of the group (Savile & Calder, 1953; Savile, 1952); it was also found that within each major host group a smut tends to be restricted to a few closely related species.

Savile (1959), employing inferred evolutionary trends in the teliospores in *Puccinia*, arranged the host plants in the *Saxifragaceae* in an approximately chronological sequence. As Savile points out, the value of such a procedure depends on obtaining a natural classification of the parasites and accurate identification of the hosts (voucher specimens being a necessity). Even then, the parasite relationship may only give a limited amount of information about the host relationships, or none at all in some instances.

It must be reiterated that interpretations depend entirely on the validity of the evolutionary trends in the parasites: reliance on spore morphology without satisfactory supporting evidence is likely to provide a very shaky basis for phylogenetic speculations. If parasite classification is influenced by the classification or "phylogeny" of the host, then conclusions on host relationships are invalid. Circular reasoning must be avoided.

Summary and Conclusions

The major part of the preceding discussion on the role of phytochemistry in taxonomy may be summarised as follows:

1. Chemical characters may be employed at all levels of the taxonomic hierarchy.

2. They should not be regarded as *a priori* more important than other characters, either for the purposes of analysis or synthesis.

3. Living material is necessary for many of the chemical tests employed. This means that many species cannot be analysed since they are not in cultivation (a drawback in all experimental work).

4. Some techniques are still rather crude. For example, colour tests may not distinguish between closely allied products.

5. Chemical characters are often affected by environmental factors such as temperature, mineral deficiencies. Alkaloids are particularly susceptible, especially with regard to the quantity produced.

6. Little is known about the ways in which many products are formed, the metabolic pathways involved, and whether they are reversible or not.

7. Little is known about the adaptive significance of plant products.

Chemical adaptation may obscure relationships in the same way as morphological adaptations do.

8. Larger and more complex molecules are not necessarily the most advanced. Chemical reduction, related to metabolism, may occur in plants as widely as morphological reduction. It is more helpful to regard products in relation to those already in the plant and with reference to the way in which they are formed.

9. Chemical parallelism: similar end-products may arise in groups which are unrelated taxonomically, especially when formed by mutation from ubiquitous metabolic products (the reaction being modified by mutations affecting the enzyme relationship). End-products that have been most studied are not those that control morphogenesis.

10. The possibility exists that chemical and morphological features may have followed different evolutionary pathways.

11. Chemical variability is known to occur quite widely, such as chemical mutants within species.

12. The protein content of a plant may change during its life and differ from organ to organ. It is therefore essential in comparative work to compare organs of the same physiological age. Many proteins are not detectable by methods so far employed.

13. Comparative chemistry and serology are aids to establishing consanguinity, not (at least above the genus level) of phylogeny proper, which involves the tracing of phyletic lines back into the past.

Chapter 8

FIELD, HERBARIUM AND LIBRARY

INTRODUCTION

For practical reasons, the classification of the world's flora is primarily based on herbarium material and the literature associated with it. Despite its limitations, a herbarium has certain advantages over living collections. It is usually only in the herbarium that we can compare all the related species of a genus in the same place, in the same state and at the same time (Davis, 1961).

The pioneer and consolidation phases of taxonomy (cf. page 3) are an essential part of classification. The importance of these phases cannot be overemphasised, because it is upon them that further research is founded. Many botanists (biosystematists among them) are under a serious misconception about the amount of basic work that needs to be done. The success of biosystematic research—at least in terms of taxonomic results—largely depends on having a satisfactory (or at least, workable) basic classification to serve as a frame of reference. We cannot interpret classification in evolutionary terms, or attempt to improve it, until we have a basic natural classification to improve and interpret. It is not generally realised by those who are not "formal" taxonomists how very far we still are from achieving this basic classification, particularly at the genus and species level.

We may distinguish between different kinds of herbaria according to the purposes which they serve. Except for a very few large University herbaria, the *major herbaria* are mainly national institutions and adequately cover the flora of the world, although specialisation in the flora of one region or another (associated with the preparation of Floras) is the common practice. They serve the dual purpose of research and identification.

Although the national herbaria of the British Commonwealth are mostly arranged by Bentham & Hooker's system, those elsewhere usually follow the Englerian system as set forth in Dalla Torre and Harms' *Genera Siphonogamarum* (1900-07). In each genus it is usual for the species to be grouped under various geographical regions. Within these divisions the species may be arranged according to a relevant Flora or monograph, which facilitates research, or alphabetically, which serves as the most convenient filing system but is inconvenient for revision. Type sheets are often kept in distinctive folders, and are either placed with the rest of the material of the species (the most convenient for research), or kept together in a separate collection (the safest practice).

Where to place cultivated material—often poorly represented in national herbaria—is difficult to decide. For general purposes, it is probably simplest to store it in separate covers at the beginning or the end of each genus. Wherever applicable, it should be co-ordinated by a cross-reference to the wild material (present in the herbarium) from which the cultivated plants originated. Liaison between herbarium and botanic garden seems to be one of the weakest links in many national herbaria.

Regional or local herbaria make little attempt to cover the world's flora but concentrate on that of their own and adjacent countries. Many museums and most University herbaria come into this category, and furnish important collections for the preparation of county, state or country floras. Small University herbaria present a rather special case, since they are primarily used for teaching and postgraduate research. Ecological collections often form part of University herbaria, and the many recent studies in population analysis (often part of Ph.D. research in taxonomy) have resulted in the accumulation of mass gatherings (cf. p. 265) that may or may not be incorporated into the herbarium. Such mass collections provide vital data on variability, but they tend to be so bulky that the national herbaria, primarily concerned with the regional variation of the world's flora, can seldom afford space to accommodate them unless only limited parts of the plant have been collected. It seems possible that University herbaria, which work on a much narrower geographical basis than national ones, may at least be able to share some of the burden of storing mass-gatherings on which their research students have worked. When once these collections have been thoroughly analysed and the results published, however, it is a moot point whether any herbarium can afford to incorporate more than a representative sample. Now that so much cytological investigation is done in University departments, it is essential that the material (voucher specimens) on which these studies are based should be incorporated in their herbaria. The University of Michigan Herbarium has initiated the experiment of incorporating photographs of fossil plants into the modern plant collection (Graham, 1962)—a development which might well be considered by national herbaria.

Working herbaria are designed expressly for identification. They consist of the species from a particular area arranged within their genera, and represented by only one or very few "representative" sheets, so that the collection can be quickly consulted; in critical groups it will, of course, be necessary to examine the main collection in order to verify the determinations. Working herbaria often form part of a major herbarium (as the European reference collection at the British Museum (Natural History), London), or they may be all that small institutions possess. Universities, for instance, may find that a working herbarium of their country's flora adequately serves their needs for identification.

The increase in size of national herbaria poses numerous problems. The largest,* Kew, with over six and a half million specimens (including *c.* 300,000 type specimens of vascular plants) adds about 50,000 a year. As the world contains about a quarter of a million species of Angiosperms, this vast collection is still grossly inadequate to deal with even the broad regional variation of species. Can governments be expected to go on expanding their herbaria indefinitely? Sooner or later some international co-operation (let alone national co-operation) on policy will have to be faced. At present, collections are duplicated in numerous herbaria (even in the same country or city) so that, from an international viewpoint, we are not making the most of our resources, either in material or staff. Although local herbaria are necessary for identification, more centralisation is needed for research.

When a new herbarium is being built up, it is essential to consider very carefully what its main uses are likely to be, not only in terms of the present staff (whose interests may not be continued beyond their period of service) but in the future.

Collecting Herbarium Material

Since the primary classification of Angiosperms is very largely based on herbarium material, it stands to reason that the success of classification depends on the quality and quantity of the material for study. Good collecting is the corner-stone of the whole edifice. We need to enrich our herbaria, not to fill them with poorly annotated scraps that waste space and can only tantalise, mislead or confuse taxonomists in their efforts to achieve a stable basic classification.

The methods of collecting adopted will vary from group to group, and must be modified according to the climate and facilities available. The following points will be found to be of wide application, especially in temperate floras.

1. *How and What to Collect*

Two main approaches are possible: (1) to collect as much as one possibly can—a method desirable in little-known areas but only when the collector is able to cope with the tiring work involved; (2) to collect certain groups, either those in which a taxonomist is particularly interested, or of which herbaria require better representation. In this connection, expert advice should be sought from herbarium staffs. It is more valuable to collect a limited amount of material really well, than to collect a great deal in a cursory manner. Common plants, one might add, are often the least well represented in herbaria (especially if difficult to press, e.g. thistles), so that the herbarium often provides a false picture of their distribution. Good material of common

* Accurate estimates of the size of herbaria are difficult to make. Leningrad runs Kew a close second, if it does not actually exceed it in numbers of specimens. The British Museum (Natural History) herbarium also approximates to this size.

plants may, therefore, add considerably to our knowledge of them, particularly if they are collected to show as much of their life history as possible.

Herbarium specimens should, as far as possible, be representative of their populations. Collect to show the range of variation, but if a particular variant is exceptional the fact should be recorded—shade forms, for instance, or depauperate or grazed individuals can mislead the taxonomist in the herbarium if the label does not draw attention to their conditions of growth. Whenever possible specimens should be collected in flower and fruit. It is very helpful to show the variation of the individual: the leaves of juvenile and fertile shoots (as in *Ulmus* and *Quercus*) can differ greatly. In herbaceous plants the rootstock should be collected whenever possible—and the corm, bulb or tuber if it has one. Fleshy underground parts can be sliced open to facilitate drying. Bulbs can be killed by plunging them in boiling water or immersing them in alcohol. Tall, bulky herbs (like *Heracleum*, *Cirsium*, *Verbascum*) will need to be cut up into sections, so that two or three herbarium sheets may be needed to represent an individual. Sterile rosettes, as well as flowering plants, should be pressed if the species is biennial. In the press, paper pads or cardboard can be used to even out the pressure on leaves and flowers. When specimens are bulky, however, the corollas should be pressed separately so that they do not curl up while drying.

The use of "flimsies"—folders of absorbent tissue paper or even newspaper—can improve the pressing of delicate specimens and speeds up the drying process. The specimen is arranged in the flimsy and need not be handled again until it is dry, the flimsy being moved into dry papers with the specimen inside. Flowers of showy Monocotyledons (*Iris*, *Crocus*, *Tulipa*) are best taken to pieces and their parts pressed separately. Succulent plants (e.g. *Sedum*) and flowers in which the form of the corolla is very important in identification (as in *Orchidaceae*) are best preserved in "spirit". A suitable preservative is made up of industrial spirit (11 parts), 40 per cent. formalin (1 part), water (8 parts) and glycerine (1 part); it may be decanted into screw-topped vials. Material preserved in this way should supplement normal herbarium material, and bear the same number. The same applies to all supplementary collections (fruits, barks, bulbs, etc.) and photographs, all of which must be co-ordinated with the normal herbarium material by bearing the same serial number as the latter.

The polythene bag has largely replaced the vasculum as the most practical collecting container, but a square of oilcloth, folded to the size of the specimens contained in it, is equally effective. If pressing has been delayed, water may be sprinkled inside to raise the humidity and keep the plants fresh. Small polythene bags will serve to separate very small plants from large ones, and so prevent them from getting lost. Many specimens, however, are better pressed when collected, especially if they have evanescent petals. It is very convenient to collect straight into a portfolio containing flimsies or newspaper, and temporary labels may be slipped in as the plants are pressed.

Specimens collected during the day in a portfolio or press will be found usually to have lost much of their rigidity by nightfall. Then is the time to go over them carefully, spreading out crumpled leaves or petals, turning a few leaves or petals so that some show the lower, some the upper side, pushing apart organs which accidentally overlap, and so on, before drying is completed. After a strenuous day in the field, this often proves a tedious, irksome task but the pleasing and instructive specimens thus made justify such extra care. How much labelling is done in the field will depend on the circumstances and the visual memory of the collector. As little as possible should be left to memory.

Field notes. The data accompanying the specimen are as important as the specimen itself, badly labelled material being more of a hindrance than a help. The first essential is that specimens should be properly localised, but the main objective is to record data of taxonomic value that will not be apparent from the specimen. Herbaria often provide books of herbarium labels, but for extensive collecting we have found that large, ledger-sized notebooks are more time-saving to write up in the field; if ruled to accommodate the following data in vertical columns they enable the information to be readily typed on to herbarium labels later*:

(1) *Number of plant.* This should preferably be a straight serial number so that the specimen can be referred to merely by the collector's name and number (e.g. *K. H. Rechinger* 3274). Another commonly used method is a serial number prefixed by the year of the collection as an integral part of the collector's number (e.g. *G. Brown* 1962/173). The number may safely be attached to the specimen by writing it on a jeweller's tag.

(2) *Name of plant* (as far as known—even if only to family).

(3) *Province*—or whatever major divisions of the country you wish to use.

(4) *Locality.* This should be as precise as possible, but should always include a locality shown on an easily available map. Grid references are certainly very helpful, especially for little-populated areas. Some collectors distribute their material with the locality marked on a small map attached to the sheet.

(5) *Altitude*, preferably in metres. The altitude at which the gathering was actually made should be given here, not the altitudinal range (if known, this can be referred to in the "observations").

(6) *Habitat*, which should include the type of terrain and preferably some indication of the community (e.g. *Quercus coccifera* maquis, *Festuca rubra* salt pasture, *Calluna* heath, lowland Dipterocarp forest).

(7) *Observations.* This should include any data that may help the taxonomist in the herbarium. For example, flower colour, scent (if any), duration and habit (e.g. biennial or perennial, height if a tree or shrub), observed

* On the other hand, if a series of small numbered notebooks is used and information transferred from them, the loss of one will not be as serious as the loss of a large notebook containing the data of many days collecting.

pollinators, etc. Notes on frequency and variability can be given here, and remarks on properties and the vernacular name if known with certainty.

(8) *Date of collection.*

If all herbarium material was labelled with data of this kind, the consolidation phase of taxonomy would be a great deal nearer completion than it is at present.

In addition to these minimal data, however, the collector may often wish to record other data of an ecological or phytogeographical nature, even if it is not always practicable to include it all on the herbarium label stuck on the sheet. A separate notebook will be required for information of this type, since detailed records on population variability may be of the greatest help to the monographer. Additional information of this kind can be conveniently given on a supplementary label, folded and placed inside the paper "capsule" often attached to herbarium sheets and used for unmounted material, drawings and floral dissections.

2. *Pressing and Drying*

The specimens (unless very bulky or spiny) are best laid in flimsies pressed in between sheets of absorbent drying paper (the Kew size which is frequently adopted measures 16 in. by 10 in., convenient for a mounting paper size of 16½ in. by 10½ in.) in lattice presses. Corrugated cardboard (preferably the kind which is flat on both surfaces) can be inserted between the drying paper to even out the pressure and to facilitate the aeration of the press; corrugated aluminium sheets are sometimes used in America.

The essential thing is to dry the plants as quickly as possible before the tissues have time to decay. It is therefore necessary to change the drying paper every day—at least for the first few days—and to see that the paper is really dry before it is used again. Presses may be stood in the sun or near gentle artificial heat. When a large collection is made (each gathering represented by several duplicate sheets), considerable difficulty may be experienced in drying the paper—various tips will be found in Davis (1961). It is advisable to take plenty of drying paper on a collecting trip; nothing is more thwarting than being unable to collect material because of a lack of dry paper.* When it is possible to take material back to the herbarium within 24 hours, much of the above procedure may be abbreviated. Many institutions possess large "base" presses, often of special design (cf. Kruschke, 1955), or facilities for speeding up the process of drying material (drying chambers, drying machines, etc.).

When the specimens are quite dry, they may be set aside in their flimsies which should bear the number of the gathering in pencil. They can be tightly

* If presses can be kept *extremely* thin, the paper usually need not be changed at all in the presence of gentle heat or wind; by this means half a dozen sheets of plants can be dried in a day or two. In the tropics the generous use of thin corrugated metal sheets greatly facilitates drying.

bundled up and packed in boxes with plenty of naphthalene for transport home. Further hints of collecting, both in temperate and tropical regions and with reference to special groups, will be found in Steyermark (1947), Lawrence (1951—with a bibliography on p. 260), Burtt (1951), Davis (1961); many National herbaria (including the British Museum) put out booklets on the subject.* Reference should be made to Lawrence (1951, Chapter 11) for information on mounting techniques and on the preservation of specimens against insect attack. A good bibliography is given by Porter (1959, Chapter 4).

Curating techniques vary considerably, often according to personal preferences. Variables include the size of mounting paper, its quality, the kind and quality and labelling of the species and genus folders, whether or not the specimens are mounted (i.e. affixed to the sheets), whether they are stored directly on the shelves of herbarium cupboards in loose covers, in portfolios with tapes, or in boxes.

Since most botanists, unless they are building up a private herbarium, will have to conform to the norms and practices of the institution in which their material is preserved, we have not gone into any detail. A botanist faced with the task of organising or reorganising a herbarium is well advised to seek guidance from other institutions.

Our own preferences, for the record, are that the material should be mounted on good quality white paper (lightweight board), the weight varying with the kind of specimen; the size approximately 16½ in. by 10½ in. The labels should be affixed near the bottom right-hand corner of the sheet and a capsule at the top for loose fragments, dissections, etc. The sheets should be grouped in white or cream-coloured species covers, and these in turn in buff or coloured manila folders (genus covers). The covers should be appropriately labelled and numbered.

3. *Mass Gatherings*

The use of herbarium materials can be greatly extended by the method of mass collecting described by Anderson (1941). These gatherings usually consist of 30-50 individuals (or selected parts of individuals) and should comprise random population samples. For small plants mass gatherings may well consist of the entire plant (as even ordinary herbarium material may for annuals like *Euphrasia*, *Erophila*, etc). For large plants considerations of space necessitate the selection of some critical portion such as the leaves, inflorescence or fruits, depending on the group being studied. This selection of parts, however, must only be made after a careful preliminary investigation, because there may be no opportunity later to study the variation in those parts which have *not* been collected, thus running the risk of missing characters of taxonomic importance. Anderson claims that a mass gathering, properly

* Among the most useful of these is the British Museum (Natural History) publication, *Instruction for Collectors*, no. 10, *Plants*, 6th. edit. (1957).

made and filed, requires little more space than a few sheets of ordinary herbarium specimens. This will, however, very much depend on the species being studied.

Mass collecting brings population studies into the herbarium, so that we may describe and analyse geographical trends, delimit and interpret centres of variation, establish and analyse variation patterns in different genera, and tackle various other problems that cannot be solved by the general run-of-the-mill herbarium material. The latter, however, still provides the broad framework in which population samples must be studied. The range of variation found throughout a widespread species, even as it is represented in a great herbarium by ordinary material, is probably considerably greater than that found within a few local populations. Taxonomic conclusions based on mass gatherings should therefore be drawn with extreme caution unless they cover the major parts of the species' range. On the other hand, we may well avoid describing limited material as a new species if we have mass gatherings of its allies to show their range of population variability. Mass collectings should be particularly useful for detecting introgressive hybridisation (cf. p. 476).

Reference should be made to Anderson's papers for a detailed account of mass collecting procedures. The problem of storing mass gatherings is referred to on p. 260.

IDENTIFICATION

After making the collection, the next step in taxonomic procedure is identification. This is an integral part of all taxonomic work, even when it needs to be provisional. It pre-supposes a classification. As Mayr *et al.* (1953) have pointed out, routine identification takes up time that a systematist might otherwise spend on taxonomic revision. Other taxonomists, and not only they but also ecologists and economic botanists, are thereby in a position to parasitise this work (cf. Blackwelder, 1962). Ironically, it is only through monographic work that satisfactory identification is possible. On the other hand, progress in taxonomy and biology in general is closely linked with identification. To teach botany (and ecology in particular) in a country without a practical Flora must be a thankless task. Van Steenis (1957) has stressed the need for taxonomic revision before floristic work, but though this is no doubt the ideal procedure it is difficult to see how it can be achieved in practice. The vicious circle could only be broken by having sufficient herbarium staff to deal with both revision and routine identification.

When a large collection comes in to a herbarium, the first step is to sort it into families. For this purpose reference will often have to be made to family keys or synopses. Some of the most useful works for family identification are J. Hutchinson's *Families of Flowering Plants* (1959), Engler & Gilg's *Syllabus der Pflanzenfamilien* (1936) and the family key in L. H.

Bailey's *The Standard Cyclopedia of Horticulture*, vol. 1 (1917),* in addition to the family keys provided in many regional Floras, notably the *Flora Zambesiaca*, **1**: 55-78 (1960).

Dissection of herbarium material will often be necessary in this and other stages in identification and revision. A flower boiled up in water for a minute or two, or soaked in a detergent solution (e.g. Teepol), may be carefully dissected (from outside inwards) in a drop of water placed on a slide on the dissecting stage, preferably under a binocular microscope. The parts may then be mounted on a white card, this often being achieved by covering the card with a suitable paste (e.g. Gloy) and pressing the slide bearing the dissected floral parts (from which the excess water has been drawn off with blotting paper) down upon it. If the slide is carefully lifted off, the parts should stick to the card, though some slight rearrangement may be necessary. The dried dissection is then placed in a "capsule" stuck to a corner of the herbarium sheet, the collector's number being written on the card. As fine structure is more readily seen during the process of dissection, it is usually preferable for descriptions or drawings to be made during dissection rather than from the finished product†. Permanent dissections, however, enable rapid comparisons to be made and usually obviate the necessity of dissecting other flowers on the sheet—an important consideration when material is scarce. Harder floral parts, like the involucral bracts of *Compositae*, may be arranged on the card in sequence and covered with cellophane (stuck to the back of the card) or Scotch tape; the latter, however, lacks permanence as an adhesive.

The identification of the genus is often the most difficult step in determination for the exclusively English-speaking student, since much of the literature is in Latin or German. If the provenance of the plant is known and a Flora is available, we shall probably use the latter for generic (and specific) identification. If a Flora cannot be used (or we cannot identify our genus in it), we must turn to Engler and Prantl's *Die natürlichen Pflanzenfamilien* (1887-1914; second edition, 1925—still in process of publication) in which the keys are in German, or to Bentham and Hooker's *Genera Plantarum* (1862-83) in Latin. A. Lemée's *Dictionnaire descriptif et synonymique des Genres des Plantes phanérogames* (1925-43, Suppl. 1951) contains a key (in French) to the Monocotyledons and Dicotyledons in volumes 8a and 8b respectively. Keys to genera in most of the Monocotyledon families (and those of only the smaller Dicotyledon families) will be found in Hutchinson's *Families of Flowering Plants* (ed. 2, 1959).

Very detailed bibliographic information on Floras and floristic papers is given in S. F. Blake & A. C. Atwood's *Geographical Guide to the Floras of the World* (part 1, 1942; part 2, Western Europe, 1961), an invaluable work

* A key to cultivated genera will be found in the same volume.

† Organs often shrink considerably in the dried state, and enlarge again when wet. Measurements should therefore be made on material in a comparable state.

which every taxonomic institution should possess. A fairly comprehensive survey of European Floras and similar works is given by Lawalrée (1960) and a list of standard European Floras will be found in Heywood's *The Presentation of Taxonomic Information* (Leicester University Press, 1958) and its Supplement (1960d). A survey of the major taxonomic and floristic papers published in each European country since the war, giving extensive bibliographies, is given in Heywood & Pichi-Sermolli's *A Survey of Taxonomic and Floristic Research in Europe since 1945* (Webbia, vol. 18, 1963). A list of the major North American Floras is given by Porter (1959:31), who also cites many useful manuals for identification. For China, Japan, Formosa, Korea, Manchuria and Mongolia, E. D. Merrill & E. H. Walker's *Bibliography of Eastern Asiatic Botany* (1938), with Walker's *Supplement* I (1960), will be found extraordinarily useful, both on account of its wide coverage, which includes major works on the plants of India, Burma and the Philippines, and its excellent subject indexes.

To identify the species, several ways are open to us. Four sources may be tapped: (1) Floras, (2) Monographs and revisions, (3) Herbaria, (4) Illustrations. The first of these has already been referred to, and of course can only be used when we know where the plant originated. Monographs and revisions are only available for certain groups, and are sometimes only traced with difficulty. The largest series of monographs is that published in Engler's *Pflanzenreich* which contain Latin keys to the species. References to others will be found at the end of the family accounts in Lawrence's *Taxonomy of Vascular Plants* (1951), and (for European genera) Janchen's *Catalogus Florae Austriacae* (1956-63), which is a particularly valuable source of information. Such modern Floras as Hayek's *Prodromus Florae Peninsulae Balcanicae* (1924-33) and Soó & Jávorka's *A Magyar Növényvilág Kezikönyve* (1951) are very helpful in this respect.* Many of the major herbaria keep a card index of such revisions (that in Leningrad filling 144 drawers), but the publication of such a compilation remains a desideratum.

It is possible, if we have access to a good herbarium, to "match up" the specimen directly with named herbarium material. Whether this is feasible or not will depend on the size of the genus and whether we know the genus's provenance or not, since to match up an unknown *Carex*, *Astragalus* or *Senecio* can take many hours. Herbarium material, within each genus, is usually arranged according to geographical regions, so that if we know where a plant comes from our task will be much easier. For the majority of genera, up-to-date monographs are not available, and the absence of some diagnostic part of the plant may prevent us from "running it down" in a key; in such cases comparison with herbarium material is the obvious procedure.

* If the genus occurs in south-eastern Asia, reference to the subject indexes of Merrill & Walker's *Bibliography*, cited above, will usually lead to any available monograph or revision of world-wide scope as well as to papers exclusively concerned with Asiatic species.

It is necessary, however, to warn a student of certain snags attendant on "matching up" his specimen with herbarium material: (1) in critical genera it is very easy to misidentify the species; (2) the specimens with which we have matched our plant may themselves not have been correctly identified; (3) the species cover may have become a "dustbin", specimens matched up with the extremes of variation so that eventually two or more species may be included within a single species cover; (4) the name in general use in the herbarium (or in the literature) may not be correct for nomenclatural reasons.

Illustrations are usually referred to when specimens have not been seen, but the danger of misidentification applies to illustrations as much as to herbarium material. To find out whether a species has been illustrated, reference should be made to the *Index Londinensis* (1929-41) and (for cultivated plants) to F. J. Chittenden's *Curtis's Botanical Magazine Index, 1787-1947* (1956). In the majority of cases, herbarium material and illustrations can be used to confirm identifications made by using Floras, monographs and revisions.

When it comes to naming *cultivated* material, we often find that no record is available of the plant's wild origin, so that we do not know which Flora to turn to. For such plants we may refer to L. H. Bailey's *Manual of Cultivated Plants* (ed. 2, 1949), A. Rehder's *Manual of Cultivated Trees and Shrubs hardy in North America* (ed. 2, 1940), C. K. Schneider's *Illustriertes Handbuch der Laubholzkunde* (1904-12), and the Royal Horticultural Society's *Dictionary of Gardening* (1951-56)*. It should be borne in mind, however, that our specimen may only be closely allied to the species actually included in these works, so that it is advisable to check identification by reference to monographs, herbarium material (if available) or illustrations. Comparison with other cultivated material is probably the least reliable method of identification, since plants are widely grown in gardens under the wrong names. There are, of course, monographs of many cultivated genera, many of which are listed by Stearn (1942). It is worth bearing in mind that plants that have been in cultivation for a very long time often differ considerably from their wild counterparts.

There are several different roads leading to the identification of species (wild or cultivated). As the late Sir William Wright Smith was fond of telling his students: "It doesna matter, laddie, how you identify a plant as long as you get it right!"

Identifications of herbarium material should be signed and dated for the convenience of later botanists working over the same material. The usual practice is for a "*Determinavit*" slip to be attached to the sheet, bearing the name of the plant (with authority), the author's name and the date of identification. On no account should earlier determinations by other workers be crossed out or removed. If the author approves a previous identification, he

* To these must now be added Krussmann's handsome *Handbuch der Laubgehölze* (1962).

may attach near the label a small slip bearing his name and the date after the printed word "*Approbavit*". Sometimes an exclamation mark is used to show that a determination has been approved, or a new "*Determinavit*" slip is added. Specimens named according to citations in a monograph or revision are often given "*Fide*" or "*Teste*" slips bearing the identification and a reference to the relevant work.

Documentation

The steady flow of taxonomic literature presents the taxonomist with one of his biggest problems, and even with various forms of documentation he may have great difficulty in finding out what has been published (cf. Heywood, 1962). No taxa below the rank of species are listed in the *Index Kewensis*, so that infra-specific taxa cannot be traced there or taxonomic work that has not resulted in the descriptions of new species or genera. Many documentary works are mentioned elsewhere in this chapter, so we shall only draw attention here to those not otherwise dealt with.

The *Gray Herbarium Card Index* (in continuous publication) includes all new names and combinations (including infra-specific taxa) relating to New World plants. The *Index Nominum Genericorum* (being published by the International Association for Plant Taxonomy) aims to index all generic names with their place of publication and to cite their type species. J. C. Willis's *A Dictionary of Flowering Plants and Ferns* (ed. 6, 1948) is a valuable handbook that lists alphabetically all accepted families (with the characters of each) and genera, the family of the latter being given in every case; it also defines many morphological terms and includes a conspectus of the Bentham & Hooker and Englerian systems.

Many taxonomists learning their craft have found Lindley's *Glossary of Botanical Terms* (1848), Asa Gray's *Structural Botany* (1879 and later editions) and Le Maout & Decaisne's *A General System of Botany* (transl. Mrs. Hooker, 1876) very helpful for the definition and illustration of technical terms; these works belong to the heyday of phytography. Lawrence (1951) is more up to date and provides helpful illustrations of indumentum types. Moll's forbidding *Phytography as a Fine Art* (1934) presents a highly complex system by whose aid very detailed descriptions can be drawn up in a more or less mechanical manner. The art of omission is little in evidence.* B. D. Jackson's *Glossary of Botanical Terms* (ed. 4, 1928) is one of the better glossaries, but the use of the same term in different senses by different botanists makes this a particularly difficult field to cover. We still await a satisfactory glossary, let alone agreement on the definition of terms—an achievement which would not,

Sprague's review of Moll's book (1936) is worth quoting. "Written between the ages of seventy-two and eighty-two years by a blind man suffering from gradually increasing deafness, with the aid of three successive secretaries and the use of Braille, it represents a triumph of will-power and endurance over grave bodily disabilities."

however, free us from the necessity of interpreting the terms used in earlier literature. A step in the right direction has been taken by the publication by the Systematics Association in *Taxon* (**9** (8): 245-257: 1960) of a list of works relevant to descriptive biological terminology—a most useful compilation—and proposals for the standardisation of terms for flat shapes (*Taxon*, **11** (5): 145-156: 1962).

Taxonomists should be grateful for a multilingual botanical dictionary: Davidof & Baxteevas' *Botanical Dictionary: Russian-English-German-French-Latin* (Moscow, 1960).* (See also Artschwager & Smiley: 1925.)

Abstracting Journals exist on the national, regional and international levels, but we only list the latter here:

Biological Abstracts (new taxa being listed under the heading "General and Systematic Botany")—Philadelphia, U.S.A.

Excerpta Botanica (*Sectio* A), useful but incomplete—Stuttgart, Germany.

Plant Breeding Abstracts (particularly valuable for biosystematic information and for literature on cultivated plants)—Farnham Royal, England.

Bulletin Signalétique (*Sect.* 17). Published by *Le Centre de Documentation du Centre National de la Recherche Scientifique*, Paris.

Referativnyi Zhurnal, seriya Botanika—the most comprehensive abstracting journal, Moscow.

In addition to these, mention should be made of two very useful American publications which list titles only:

Bulletin of the Torrey Botanical Club.

Taxonomic Index, published as a regular part of *Brittonia*.

A more extensive account of taxonomic literature than that given in this chapter will be found in Lawrence (1951, Ch. 14) and Porter (1959, Ch. 3). The information relevant to taxonomy is extremely diverse and often includes material originally published without any taxonomic aim in view, particularly that of phytochemistry. The taxonomist should read widely outside his own immediate field. If not he may fail to grasp the implications or be totally unaware of much that would make his own work more efficient and save it from that narrowness of outlook, aridity and mechanical treatment which unfortunately characterises much taxonomic literature.

Taxonomic Revision

The success of biosystematic work is largely dependent on having a fairly satisfactory working classification to start with. Such a classification can withstand the test of time remarkably well provided certain general principles and procedures are followed. The importance of this phase in taxonomy

* An *Index Polyglottum*, giving the equivalent in most European languages to the commonest terms used in Floras, is being prepared by Webb in connection with *Flora Europaea*.

cannot be overestimated, since it is on this foundation that further improvement and progress are based.

As Bentham pointed out, "The preparation of a revision or monograph is still recognised as the best exercise for the young botanist"—or at least for the embryo taxonomist. It is only by revising a group of species that a botanist can learn to solve the practical problems of classification, find his way about its extensive literature, gain information from unpromising herbarium material, chart distribution and seek its explanation, gain experience in understanding the code of nomenclature and set out his results in a clear, useful manner—in short, to handle properly the tools of his craft. The revision of small groups of species in the Mediterranean flora has been the basis of a short, extramural course in the practice of taxonomy given at Edinburgh University during the last few years.

Given adequately preserved and labelled normal herbarium material (i.e. not mass gatherings), there are only a limited number of activities which a taxonomist can perform. The "born" taxonomist carries these out more or less intuitively, but we have endeavoured to break down the process into a series of progressional stages. A considerable amount of back-tracking goes on, which serves as a check on the progress of the work. The stage at which the literature is consulted will depend very much on the published accounts (how extensive and reliable they are) and the purpose of the revision. If we assume that a genus is being revised, the stages may be summarised as follows:

(1) Assess the scope of the problem by consulting the literature. We need to know if previous revisions exist, and to have some idea of the number of species involved and the distributional range of the genus. For monographic work, a card index of all species listed in the *Index Kewensis* is usually compiled at the beginning of the investigation. Herbarium material must then be assembled, preferably from as many herbaria as possible.

(2) Study individual specimens and select characters which show discontinuities within the whole group.

A picture of the variation is more easily obtained if the character expressions are tabulated for individual plants or herbarium sheets; these may be referred to at later stages in the revision and in the preparation of keys, descriptions and graphical methods of presentation.

(3) Group individuals together according to the highest correlation (maximum co-variation) of characters which have been selected and which show more or less continuous variation between these individuals. These groups will represent the major "*species*", and often possess a characteristic facies. Graphic methods to portray variation can be used at this stage (or earlier) to check taxonomic conclusions.

(4) Seek discontinuities which may separate the resultant groups from one another; that is, the *diagnostic characters* are established.

(5) Next study variation in those differential characters that show

incomplete separation *between* the groups, and any other patterns of character variation that occur *within* the groups. This may disclose further discontinuities within the groups first established under (3).

(6) Map the geographical distributions of the groups, at the same time seeking ecological as well as geographical regularities of variation within the groups; these may reveal *infra-specific* patterns that deserve subspecific or varietal recognition. The taxonomist should decide on his specific and infra-specific criteria within the group and apply them as consistently as he can.

(7) The species which have been recognised by the above processes can then be grouped on the basis of their similarities, so that natural series, sections, etc. can be built up. A natural classification of the genus should thus be achieved. In some genera it may prove impossible to recognise satisfactory sections, the best that can be achieved being groups of closely allied species which may constitute species aggregates or series. In many cases it is possible to recognise major groupings at an earlier stage of the revision (corresponding to sections, subsections, etc.) which can be broken down into species at a later stage.

(8) For the purpose of communication, names must be attached to the groups in accordance with the *International Code of Botanical Nomenclature*, synonyms listed, keys and descriptions drawn up, etc. Naming involves the scrutiny of original descriptions and, often, examination of type material. Additional methods of presentation (scatter diagrams, maps, etc.) may be used to describe the patterns of variation (cf. Chapter 9).

(9) The taxonomic characters can be arranged (or considered) in series. The interpretation of these in terms of evolutionary trends may be attempted with some degree of probability, but very seldom with certainty. The same applies, but with even greater force, to the phylogeny of the group. This can only be discussed in terms of possibilities.

The order in which these processes are carried out, the degree of emphasis given to them, and the amount of back-tracking required, will depend on several variables. These include the complexity of the group, the purpose of the revision, the literature already available, the amount of available material; and the working methods and personality of the botanist concerned.

So much attention is drawn to the variation of plants, and ways devised to portray it, that it seems surprising that the other major variable in taxonomy —the Taxonomist—has received so little experimental study. Yet it is this factor which partly explains the discrepancies between taxonomic accounts in different Floras and revisions. The taxonomist should try to gain some understanding of the species concepts of the authors of monographs and Floras, and to understand the ways in which they use the ranks.* This can be most

* When a monograph or revision comes to hand, the systematist should take one or two species with which he is familiar and use them to test the keys and descriptions. By this means he will become better equipped to identify unknown material with its aid and to assess the quality of the work.

helpful when it comes to assessing earlier work, and when, for example, taxonomists from Western Europe are working with revisions carried out by many taxonomists in the Soviet Union—or vice versa (cf. p. 102). A word of warning here: although taxonomists (if communicative about their methods) sometimes give quite lucid accounts of their species or genus concepts, these do not always correspond to what they do in practice. Splitters, in particular, often fancy themselves as upholders of a conservative species concept, and only recognise pulverisation when it is not in their own group.

As we have pointed out, the stages of taxonomic revision must be greatly affected by the purpose for which the revision is being done. In the preparation of a flora, for instance, it is extremely important to have as complete a list of recorded species as possible. It is therefore customary to survey the floristic literature at an early stage, often before we start the revision at all. The revision, in fact, will be partly concerned with checking the records given in the literature. When writing a Flora, time is limited—we cannot afford to follow every fascinating train of investigation that offers itself. To neglect literature and names in the early stages of writing a Flora can waste a lot of time, since some of our taxonomic and nomenclatural problems may already have been thoroughly studied and clarified. In a regional Flora the most important objectives are to delimit the species and to provide a practical means of identifying them. We cannot go into generic or family limits with the same thoroughness as we can the delimitation of species. To do so, we should need to write a Flora of the World!

If we are revising a genus throughout its entire range, floristic accounts will matter less to us in the early stages of the work. We can, in fact, afford to be more independent of the literature in the early phases than we can when writing a Flora, since all herbarium material and literature will have to be thoroughly surveyed before we can complete the work. Indeed, by deliberately soft-pedalling the literature, a more independent judgment may be achieved which can be compared later with the findings of previous authors. It is very easy to be influenced by other taxonomists' opinions, and one must always remember that all taxonomists make—and must make—mistakes, especially in the pioneer phase of taxonomy. That is the price of progress.

Whether we are revising a group for a Flora, or writing a full-scale monograph (which should contain a synthesis of all available information), the following points should be borne in mind:

(1) Classification does not begin and end in the herbarium—it only starts there on the basis of what has been collected and observed in the field, and of necessity. As our knowledge increases, detailed population, biosystematic and other data may be obtained and as far as possible or desirable incorporated into a taxonomic classification. It is therefore essential that classification, based primarily on herbarium material, should not be pushed too far. Unless field notes are exceptionally complete, or we have studied the group in the field ourselves, or have mass gatherings available for population

analysis, we can seldom go with confidence further than the recognition of geographical subspecies; varietal rank, when used, is often a neutral, non-committal category, used to draw attention to a variation pattern whose nature is not understood, although it may be accorded to "subspecies within subspecies". Mayr (1959b), for example, does not consider the sub-specific category suitable for describing the population structure of a species. To encumber the literature with a hierarchy of infra-specific categories serves no useful purpose and often hinders progress. Variation at this level needs population analysis, collateral cultivation and other biosystematic techniques if it is to have a really sound biological basis (cf. Chapter 12). Below the level of the subspecies, the herbarium taxonomist treads on very shaky ground. He should avoid naming single-character variants that do not appear to have a population basis.*

(2) Great care must be taken not to assume, *ab initio*, that the species we have recognised in the herbarium (often from limited samples that are not even contemporaneous with one another) are valid representatives (statistically) of natural populations in the gene-pool sense. The temptation to treat them as such, however, is in many cases too strong, especially when we are exhorted to think in terms of populations, not typology; yet no amount of mental gymnastics will alter the facts or our state of knowledge of them. It is a common pitfall in taxonomy to regard species and infra-specific taxa, known primarily from normal herbarium material, as gene-pool populations, and to discuss their variation, distribution and evolution in these terms. Taxonomists should not impute a gene-flow relationship between the individuals within a herbarium taxon—i.e. they should not regard them as representatives of a population in the interbreeding sense, although they may well turn out to be so when studied later in the field and bred in the experimental plot (cf. Chapter 11). Even with adequate population samples, gene-flow is a likely inference, not an established fact. Two sympatric biological populations, if very closely allied, may easily be mistaken for one such population in the herbarium, especially if material is limited. Conversely, if only the extremes of variation are known to us, the Mendelian population (cf. p. 353) may be inadvertently referred to two taxa in the herbarium.

(3) If we cannot understand the basis of the variation we meet in herbarium material, we can at least be sure that we understand the structure of the organs we are describing. We can learn much about the morphology of a group from living cultivated material. Even if breeding experiments and extensive transplanting cannot be undertaken, individual plants of several species can generally be grown. This can give an insight into ontogenetic development and reveal characters that are not apparent in herbarium material. Whenever possible, living flowers should be dissected. Care must be

* This may, however, be necessary when classifying cultivated material, wherein a single character may be of economic value, but for this the *International Code of Nomenclature for Cultivated Plants* should be followed.

taken not to regard cultivated plants as necessarily typical of their species; like herbarium material, they only represent a small part of the natural variation, and may even represent a single clone. Horticulturists are particularly prone to suffer from this typological viewpoint.

(4) When revising a genus in the herbarium, attention should be paid to the information on the labels. Geographical and ecological data can provide the taxonomist with valuable clues, and so can differences in flowering time. These may represent barriers which may separate (or help to separate) closely allied taxa in nature. On old "mixed" sheets, where more than one gathering has been mounted on one sheet (a practice discontinued in most herbaria now), the labels may sometimes have become attached to the wrong specimens. Genuinely mixed gatherings, however, in which two allied taxa have inadvertently been collected together, provide valuable evidence for the constancy (or breakdown) of distinguishing characters.

(5) In general, it is most advisable that like should be compared with like. We must not, for instance, compare juvenile with adult foliage, or a grazed inflorescence with a normal one if we are seeking diagnostic chracters to separate taxa and for the construction of keys. This is particularly important with regard to leaf shape, which generally varies according to its position on the stem (especially in herbaceous plants) so that we should usually compare leaves in comparable positions.

This applies, however, only when the plants have been growing under fairly similar conditions. In specimens which have been stunted by grazing, exposure or lack of water, the sequence of organs laid down under more favourable conditions will be modified, leaf forms that would normally appear late in the "leaf spectrum" (cf. p. 333) being formed at a lower node on the stem; conversely, in shady or wet conditions the juvenile leaf stages may continue to be formed at later stages in the spectrum. The taxonomist must have his wits about him and may be helped by information on the labels and by his own field experience. Simple experiments on cultivated material can provide useful information about the phenotypic variation he is likely to find in the group.

Great care must be taken when comparing material collected at different times of year (see the example in *Ononis* cited on p. 340). This applies particularly to plants growing in arid regions. Perennial species of *Alyssum* or of *Phlomis* Sect. *Dendrophlomis* (*Labiatae*) can look remarkably different in habit in their spring and summer states. This is largely due to the shedding of leaves on the main stems and to the production of short axillary shoots or pseudo-rosettes whose leaves differ considerably from those produced in the spring. The form of inflorescence can alter strikingly between the flowering and fruiting states. Such seasonal variations have not infrequently resulted in the description of "paper" species.

It is, of course, impossible to distinguish with certainty in the herbarium between hereditary variation and variation produced by the environment, but the experienced taxonomist must attempt to do so if he is to avoid giving taxonomic rank to modifications. Modifications, as well as inherited variations (as in ecotypes) will tend to be associated with certain ecological conditions wherever they may occur, whereas variation showing a regular geographical pattern is almost certainly inherited. It is for this reason that geographical subspecies can often be recognised with confidence in the herbarium.

(6) Specimens that are intermediate between previously recognised taxa naturally are of great importance to the taxonomist, especially in considerations of status. By studying their distribution we may gain some idea as to whether we are dealing with a cline or with hybridisation between taxa. If hybridisation is involved, there is a likelihood that it may be accompanied by a reduction in fertility. This may be inferred by a failure to set seed, or by the production of sterile pollen (as in *Verbascum* hybrids). Sterile pollen can be readily recognised by heating the grains for a minute or two in a drop of acetocarmine, sterile grains appearing shrivelled and without a darkly stained nucleus, thus allowing the percentage of sterile grains to be calculated. We must, however, compare the results with those from supposedly uncontaminated taxa, because hybridisation is not the only cause of sterile pollen production. Hybrids may also reveal their presence by heterosis—as detected by Woodson (1947) in *Asclepias*.

Two well-established species should not be combined on the strength of one or two intermediate specimens. It is the extent of the intermediates that matters—whether or not they obscure the limits of the taxa to such a degree that their recognition as separate species (or subspecies) is no longer convenient. The more distinct taxa are morphologically, the less justification there is for combining them on the strength of putative hybrids. Very often we can only recognise a specimen as an intermediate—reason unknown. Similarly, species should not be maintained on the strength of one or two experiments indicating intersterility, if they grade into one another on a wide geographical basis.

(7) In order to attach the correct names to specimens, they must be compared with original descriptions and with type material. It must be pointed out that later specimens are very unlikely to match exactly either the type specimen or the description. These differences will in part be genuine (due to population variability), in part be due to the slightly different use of technical terms by different botanists—sometimes even to misobservation. The botanist cannot expect his material to match the type in every detail, so that he must very carefully assess and interpret the degree of difference. When material is very limited, mistakes are bound to be made. Some guidance may be obtained by applying the law of homologous series (p. 82), i.e. by observing the range of variation found in better known taxa. Experience shows, however, that the range of variation observed increases with the increase of material available for study; as species become better known, botanists tend to attribute more, not less, variation to them. The taxonomist

cannot be a prophet, but he should show some foresight, particularly in the pioneer phase of taxonomy.

On the other hand, a strong case can be made out for describing "small" species in the pioneer phase, particularly in groups where differentiation is correlated with geographical isolation. The inconvenience caused by "lumping" later (if further material necessitates it) may be less dangerous than the confusion caused by initially failing to separate different taxa from one another. Faced with this dilemma, the taxonomist must rely upon his experience and judgment. Only time can show whether he has done his work well, and whether posterity will consider him as a good taxonomist or a bad one.

(8) It is possible that a revision leaves us with some material for which no name can be found. This may represent a new species (or subspecies) for which a description and new name will have to be provided. However, several possibilities indicate the need for caution: (*a*) the plant may have been already described and reduced to synonymy, in which case the original name will have to be reinstated; (*b*) it may have been described under a generic synonym and lost sight of; (*c*) it may have been described very recently, so that it has not yet appeared in the *Index Kewensis* whose supplements appear every five years; (*d*) it may be a species from another country that has not yet been recorded for the area we are dealing with—an occupational hazard when writing a Flora*; (*e*) it may be an extreme genetic variant of a species already familiar to us, particularly when material is very limited; (*f*) it may have been growing under abnormal conditions (perhaps in dense shade, or grazed by goats); (*g*) it may be a hybrid between two previously described taxa. It is obvious that one must go very cautiously before describing a species as new. The *Index Kewensis* must be systematically checked in case a forgotten name is already available.

(9) In any revision, it is very desirable that all names should be taken into account. If their status is uncertain (as is often the case when a species is represented by very limited material, or if the type has been destroyed), this should be frankly recognised. Names should either be accepted, reduced to synonymy, mentioned in observations, or listed as belonging to taxa imperfectly known or doubtfully recorded. Synonymy, if carefully checked, is not a sterile encumbrance but an index of knowledge. Sometimes the most important information about a plant, the best descriptions and the best or only illustrations have been published under names now treated as synonyms. Good synonymy saves this from loss.

If taxonomic progress is to be made, it is essential that ignorance should not be hidden. Decisions taken on limited material will often need to be altered in the light of further knowledge. If we maintain a species provisionally, rather than reduce it to synonymy, it should be made clear that its status is in doubt. Only by drawing attention to the weak spots in a classification can

* It may also, of course, be an escape from cultivation not yet recorded as naturalised.

we hope to put it on a firmer footing when more information becomes available.

Typification and Nomenclature

1. *The Type Method*

In the following account of typification and nomenclature, reference is given in parenthesis to the relevant articles of the *International Code of Botanical Nomenclature* (adopted by the Ninth International Botanical Congress, Montreal, 1959; publ. Utrecht, 1961).

The correct application of plant names is determined by nomenclatural types. A *typus* is that constituent element of a taxon to which the name of the taxon is permanently attached, whether as a correct name or as a synonym (Art. 7). If a species or genus is given a new circumscription which excludes the type, then it can no longer bear the same name. The nomenclatural type of a family name is a genus, of a generic name a species, of a specific name a single specimen—or, in the case of small herbaceous plants, one sheet of specimens belonging to a single gathering (Art. 9). (The possibility of a mixed gathering should be borne in mind.)

It is only names that have types, not species, so that typification is an entirely nomenclatural procedure. The type is usually the gathering with which the name is first associated, so that it need not be typical of the species in terms of population variability; it may, in fact, even be an extreme variant. It is therefore no more important *taxonomically* than any other specimen; its importance lies entirely in fixing the application of the name. If a species is based on a single gathering, it is very unlikely that the next gathering of it (particularly if it comes from another area) will closely match the type. Lack of exact agreement with the original description or type specimen certainly does not mean that the new material necessarily represents a different species. Allowance must be made for population variability. If species were only represented by type specimens, distinguishing species would be simple; it is the additional material that usually makes taxonomy difficult! In the case of a new species being described, it is of course very desirable, if we are in a position to do so, to choose an "average" specimen as the type.

The Code (Art. 7) recognises four kinds of types. The choice of a specimen by the author, if made when he published his species (or other taxon) is final. That specimen is the *holotype.*

In the absence of a specimen, the type may be a figure or description. It is nowadays essential that a holotype should be designated and its place of deposit stated when a new taxon is described; it is recommended that holotypes should be preserved in a permanent public institution. Duplicates of the holotype are known as *isotypes.* (In the early days of taxonomy there was no type method in the modern sense; due to the doctrine of fixity of species, all plants might be considered typical of the species.) Until very recently taxonomists often cited several specimens when describing a new species, without selecting one as the holotype. These are known as *syntypes*, and from among them a taxonomist may wish to choose one specimen as the *lectotype*. If a holotype was originally designated and has been lost or destroyed, an isotype (if it exists) should be chosen as the lectotype. The choice of a lectotype is particularly necessary if the syntypes (in the absence of a designated holotype) cover more than one taxon, as often happens in critical groups. In such cases our guide to the choice of lectotype should be the author's original description of the taxon, and any indication he may have left for any particular preference. The lectotype should be chosen, if possible, from among the specimens actually seen by the author when he described the species, not from duplicates of them in other herbaria, since there is always a risk of mixed gatherings or of labels misplaced when specimens are distributed.

When once a lectotype has been chosen and published as such, it must be retained unless it can be shown that the choice was unsuitable. At present there is no published index of lectotypes, so that it is rapidly becoming very difficult to keep track of them. Except in monographic work, it would therefore seem advisable not to choose a lectotype unless it is necessary for nomenclature reasons, as would be the case, for instance, if an author has inadvertently included two taxa among his syntypes.

In some cases, no original specimens exist; in that case the type of a species is the original description or figure or one given by a previous author (Art. 9, note 7). This quite often happens with Linnaean species, when no Linnaean specimens are extant and the synonymy takes us back to earlier literature. Other things being equal, when choosing a lectotype preference should be given to a specimen over pre-Linnaean descriptions or figures.

When all the type material is irretrievably lost or destroyed, a *neotype* may be chosen. This is not really a type but is a specimen chosen to serve as a standard reference point; it simply represents a later taxonomic opinion, not an act of the original author. Such a specimen should match the original description and if possible come from the *locus classicus*. Neotypes should never be designated unless there is a special reason for doing so, and then with great care, the justification for a particular choice being recorded. In most instances it suffices to state that the type is not known to exist. A guide for the determination of types appears on p. 64 of the Code.

A *paratype* is one of the original cited specimens other than the holotype or isotype(s). Although it is therefore nomenclaturally irrelevant (at least while the holotype is extant), it is taxonomically just as important as the holotype, a paratype (like a syntype) representing the author's species concept in so far as it is covered by his description.

When the rank of a taxon is changed (species to variety) or it is transferred to another genus, it automatically retains the same type, since the epithet is permanently associated with the type. Sometimes it is found that the name of a taxon is invalid for nomenclatural reasons, often because it is a later homonym. In that case, if no other epithet is available, it is given a new name (*nomen novum*) with a reference back to the original description under the other name which is cited as its synonym. The type of the *nomen novum* is automatically that of the old one (Art 7, note 4). However, if the new name is accompanied by a Latin description or diagnosis, it is not necessary that the original type should be retained; indeed, it often gives one a chance to choose a better and more representative specimen, since the plant can be published as a new species instead of being given merely a new name (Art. 72).

As discussed on p. 277, original descriptions and type specimens must be carefully interpreted in the light of the taxon's variability. Failure of a specimen to agree with the type does not necessarily mean that it does not belong to the same species.

The typification of genera, sections, etc. generally rests upon the choice of a particular species as a lectotype. It is often a time-consuming business to find the right name and authority for groups between the genus and the species, and it is essential that the type method should be strictly applied in order to avoid mistakes and the use of cumbersome "emended" authorities. Sectional nomenclature has been very laxly used in both Monographs and Floras. Pfeiffer's *Nomenclator Botanicus* (1873-74) will be found very useful for tracing the original descriptions and authorities.

Type specimens are widely scattered in numerous National Herbaria and, except for early collections of great historical importance (e.g. Willdenow in Berlin, de Candolle in Geneva, Linnaeus in London, Lamarck in Paris), can usually be sent on loan to taxonomic institutions.* The addresses of the world's herbaria (with additional information) will be found in the *Index Herbariorum*, part I, ed. 4 (1959). Part 2 of the same work (Collectors, A-D, 1954; E-H, 1957) shows where particular collections exist. Information on the whereabout of type collections will be found in Alph. de Candolle's *La Phytographie* (1880) and A. S. Hitchcock's *Location of type specimens, list* 2 (Washington, 1935—30 pp., mimeographed); cf. Lawrence (1951—p. 328) for additional references. Several contemporary Floras cite the whereabouts of type material under each species (e.g. *Flora URSS*, *Flora of Tropical East Africa*). The International Documentation Centre (Sweden) has now made available micro-fiche editions of several herbaria of great historical importance, including the de Candolle, Linnaean and Burser herbaria. These editions are technically excellent, but since interpretation is left to the user, they should be employed for typification with the greatest caution.

1a. *Typification of Linnaean Taxa*

As modern botanical nomenclature starts with the publication of Linnaeus's *Species Plantarum* in 1753, typification of Linnaean species is an important part of Angiosperm nomenclature. Unfortunately for us, Linnaeus

* Other herbaria are restricted by testament or national law (e.g. the British Museum (Natural History) London).

and his contemporaries had no concept of nomenclatural types. It is the efforts of later taxonomists, trying to fix the application of Linnaean names, that have resulted in the typification of Linnaean species. To Linnaeus the species was a subdivision of the genus (his basic working unit) comprising all specimens, illustrations and descriptions falling within the differential limits of the specific phrase name (Svenson, 1945); these elements on which the species is based are known collectively as the *protologue* (cf. Code, Art. 64).

Linnaeus considered his *Species Plantarum* to be a compendium without descriptions. As noted on p. 89, each species was given a binomial and, when the genus possessed more than one species, a polynomial or phrase name intended concisely to distinguish it from the other species of the genus, in short, to be a diagnosis; references were given to the polynomials in his earlier works, e.g. *Flora Lapponica* (1737), *Hortus Cliffortianus* (1738), *Flora Suecica* (1745), *Hortus Upsaliensis* (1748) etc., to Caspar Bauhin's *Pinax* (1623) and to figures by earlier authors. He cited the locality (sometimes erroneously) and frequently the habitat (often from personal observation if a Swedish species). The following extract from the *Species Plantarum* will serve to illustrate the form of the work.

CARLINA

acaulis 1. Carlina caule unifloro flore breviore. *Hort. cliff.* 395. *Hort. ups.* 252. *Mat. med.* 378. *Roy. lugdb.* 135. *Sauv. monsp.* 293. Carlina acaulos, magno flore albo. *Bauh. pin.* 380. Chamaeleon albus. *Clus. hist.* 2. p. 155. *Cam. epit.* 428. *Habitat* in Italiae, Germaniae *montibus.* ♃

Herbarium specimens are cited but rarely and then indirectly by mentioning the name of the collector of a hitherto undescribed or little known species, as *Loefl.* (i.e. Loefling) under *Echinops strigosus*, *Osbeck* under *Helicteres angustifolia*, *Gmelin* under *Cacalia hastata*, etc., but in the majority of cases probably exist. Most of them are in the Linnaean Herbarium (at the Linnean Society's rooms, Burlington House, London) or in the Hortus Siccus Cliffortianus or the Hermann Herbarium at the British Museum (Natural History); others are in Uppsala, Stockholm and Paris (cf. Stearn, 1957: 103-124). Thus Linnaeus's *Carlina acaulis* can be satisfactorily typified by a Hortus Cliffortianus specimen agreeing with the Clusian woodcut cited above. It must be stressed that the existence of a specimen supposedly representing a given species in the Linnaean Herbarium does not automatically make it the type of that species, many of the specimens there being irrelevant to typification.

In his great task of bringing order out of chaos, it is not surprising that Linnaeus made mistakes. His species concept was often a very broad one. The circumscription of a plant in the first edition of the *Species Plantarum* (1753) was sometimes changed in the second edition (1762). Research often shows that the specimens now extant in his herbarium do not belong to the same species as those described or figured by his predecessors—or even by himself prior to 1753. In very many cases—possibly in the majority—Linnaeus's concept will be found to cover more than one species by modern standards. This is not because he believed in polymorphic species; his species concept was essentially unitary, and such confusion was due to ignorance or error. The specimens extant may conflict with the synonymy; in such cases the locality given in the *Species Plantarum* may help to typify the species, but too much reliance should not be placed upon it as such data are not always reliable. If there are several specimens of a taxon, they may in fact represent different species. Many of his specimens were not in the herbarium when the first edition of the *Species Plantarum* was published, and if this can be shown to be the case such a specimen should not be taken as the type; those sheets which bear the Linnaean trivial name and the number are most likely to be original specimens. The most important single element in the protologue of a Linnaean species is the phrase name (cf. Stearn, 1961b). Reference should be made to Savage's *Catalogue of the Linnaean Herbarium* (1945).

Such uncertainties add greatly to the difficulty of typifying a Linnaean species, often a time-consuming task. Naturally when a succession of botanists have consistently applied a Linnaean name to a particular taxon, the transfer of that name to a different taxon is likely to cause inconvienence and confusion; hence they are reluctant to change accepted usage when a Linnaean specimen proves at variance with this. In such cases there may in fact be reason for typifying the Linnaean name not from this specimen but from an earlier figure or description consistent with general usage. Often, however, there is no escape. It is only relatively recently that careful attention has been paid to Linnaean typification, and unpleasant surprises still come to light. Now that a microfiche of the Linnaean herbarium is available (numbered according to Savage's *Catalogue*) we may hope for wider attention being paid to the interpretation of Linnaean names, though it must be borne in mind that the majority of specimens cannot be accepted as type material.

In the synonymy of the *Species Plantarum* Linnaeus often cites his *Hortus Cliffortianus* (1738). This important work lists under polynomials and mononomials the plants cultivated on George Clifford's estate near Haarlem Holland, during the years 1736-37 when Linnaeus had charge of his collections. For the *Species Plantarum* Linnaeus often copied the phrase name directly from the *Hortus Cliffortianus*. In such cases, and especially if the reference to the "Hort. Cliff." immediately follows the phrase name and no change in the polynomial has been made, a specimen in the Clifford Herbarium at the British Museum (Natural History) may often be chosen as the lectotype, since it is the basis of the phrase name and usually older than any specimen in the Linnaean herbarium.

When the species concerned is one of Central or Southern Europe and no specimen can be found in Linnaeus's own Herbarium or the Clifford Herbarium, Linnaeus may have based his account on a specimen in Joachim

Burser's *Hortus Siccus* at Uppsala. This 16th century collection was arranged according to C. Bauhin's *Pinax* (1623) so often cited by Linnaeus in synonymy and, owing to the close co-operation between Bauhin and his former student Burser, it accurately represents with specimens the names used by Bauhin. Sometimes Linnaeus cites directly Burser's herbarium (of which there are 23 volumes); often his localities give the connection. Thus under his *Cucubalus quadrifidus* in the *Species Plantarum* **1**: 415 (1753) Linnaeus gives a brief diagnosis, cites Bauhin's *Pinax* and an illustration in works of Clusius and L'Obel, and then says "Habitat in Styriae monte ad oppidum Eisenertz". This precise indication of habitat appears in none of the printed literature cited by Linnaeus. The specimen in Burser's herbarium (XI. 63) corresponding to the Bauhin synonym is labelled 'prope oppidum Eisenertz Styriae in monte' (cf. Juel, *Studien in Burser's Hortus Siccus*, 72; 1923) and is thus clearly the source of Linnaeus's information; it belongs to the same species as that plant portrayed by Clusius and L'Obel. Unfortunately it is also the same as that later named *Silene alpestris* by Jacquin. It becomes increasingly evident that many Linnaean species are to be typified by Burser specimens. Among these may be mentioned *Antirrhinum dalmaticum* L., the basionym of *Linaria dalmatica* (L.) Miller, which is discussed by J. F. Alex in *Can. J. Bot.* 40: 295 (1962), and *Isopyrum aquilegioides* L., discussed by G. de Beer and W. T. Stearn in *Bull. Brit. Mus. Nat. Hist.* **2**: 195-202 (1960).

When Linnaeus's phrase-name is taken directly from *Roy. lugdb.* or paraphrases this, and no specimen represents it in the Linnaean Herbarium, then the type is likely to be a specimen in Adriaan van Royen's herbarium at the Rijksherbarium, Leyden. An example is *Mimosa glauca* L., the basionym of *Acacia glauca* (L.) Moench, discussed by H. C. D. de Wit in *Taxon,* **10**: 50-54 (1961).

Many north temperate plants were first described in Linnaeus's *Flora Lapponica* (1737), which is often cited in the synonymy of the *Species Plantarum.* Two-thirds of the species described in *Flora Lapponica* are in Burman's herbarium in Paris (cf. Alston in *Proc. Linn. Soc. Lond.* **168**: 102; 1957), but if a specimen cannot be traced the description in the *Flora Lapponica* will probably have to be taken as the lectotype. When no authentic specimen can be chosen, one must fall back on a previously published figure or description. The beautiful woodcuts in Clusius's *Historia Plantarum* (1601) often provide satisfactory lectotypes when no Linnaean specimen can be traced.

Some of Svenson's views (1945) on the typification of Linnaean species are rather extreme. He does not agree with the widespread view that a new phrase name in the *Species Plantarum* means that the diagnosis was based on a specimen before him.

In the *Species Plantarum* occur a number of new phrase names specially drafted for that work. Sometimes a new phrase name is merely an earlier phrase name changed slightly in wording but without any change of meaning or addition of information, having been redrafted in order to bring it into line with the phrase name of an added species in accordance with the rules laid down in Linnaeus's *Critica Botanica* (1737). This is a matter Svenson (1945) has emphasized. When, however, the new phrase name contains information which Linnaeus could not have taken from earlier descriptions, then it must have been based on a specimen or else an illustration seen by him on which the details given were evident. To typify a Linnaean species it is necessary to go over every element which may have contributed to this. Svenson describes Linnaean species as aggregates of synonyms, herbarium specimens, illustrations and living plants, for Linnaeus always examined living material whenever he could and was a keen gardener. Neither Svenson nor Stearn admits that a Linnaean specimen automatically takes precedence over a cited figure; a good illustration may provide information not evident in a poor specimen and have received more attention from Linnaeus.

Typification of Linnaean species is an extremely tricky matter, not to be embarked on lightly. Anyone attempting it should first read Stearn's introduction to the Ray Society's facsimile edition (1957) of Linnaeus's *Species Plantarum* (1753). Taxonomy will be best served if expert advice is taken on all but the most straightforward cases of Linnaean typification.

After the difficulties of typifying Linnaean species, typifying Linnaean genera is relatively simple. It is necessary to see which species listed in the *Species Plantarum* (1753) best fits the generic description in Linnaeus' *Genera Plantarum* ed. 5 (1754), which should be compared with that in the first edition of 1737. Sometimes there may be significant changes. Often, however, the description has come essentially unchanged from 1737, when the number of species known to Linnaeus was much lower than in 1753; they are listed in his *Flora Lapponica* (1737) and *Hortus Cliffortianus* (1738). Very often we shall find that there Linnaeus's generic description is based on a single species well known to him in the living state. *Veronica*, for instance, contains 27 species in the *Species Plantarum*, but only *V. officinalis* perfectly fits the generic description and is therefore chosen as the type of the genus—*fide* Pennell (1930).* An unofficial supplement to the Rules of Botanical Nomenclature, 1935, gives a list of the types of Linnaean genera, but these should not be uncritically accepted (see Stearn, 1959, 1960; Jones, 1960). (Mistakes are not infrequent; e.g. *Delphinium peregrinum* L. (not *D. consolida* L.) should be the type of *Delphinium*, and *Anemone coronaria* L. (*not A. nemorosa*) the type of *Anemone*.)

In Stearn's index to the *Species Plantarum* at the end of volume 2 of the Ray Society's facsimile (1959), the names of the species chosen by various authors as lectotypes of Linnaean genera are printed in italic and accompanied by symbols, e.g. $ for the *North American Flora*, indicating where they were so

* *Zanthoxylum* provides an example of a genus of which the type is a species not included in the *Species Plantarum* (1753), an anomalous situation discussed by F. R. Fosberg in *Taxon*, **8**: 103-105 (1959).

designated. The matter is discussed in the preface to this facsimile and also in Stearn (1960).

As a guide to the selection of types of Linnaean taxa, the following references are among many that should be found helpful:

CLARKE, C. B., On certain authentic *Cyperaceae* of the Linnaean Herbarium. *J. Linn. Soc. Bot.* **30**: 299 (1894).

HEDGE, I. C., Studies in E. Mediterranean species of *Salvia*: IV. *Notes R.B.G. Edin.* **23** (4): 559-567 (1961).

HELLER, J. L. & STEARN, W. T., An appendix to the "Species Plantarum" of Carl Linnaeus. Appended to the Ray Society facsimile of Linnaeus's *Species Plantarum*, **2** (London, 1959).

HEYWOOD, V. H., The nomenclature and typification of *Ballota hispanica* (L.) Benth. *De Flora Lusitana Commentarii*, fasc. 12: 199-204 (1959).

JONES, K., The lectotype species selection of Britton and Brown and the standard species of Hitchcock and Green. *Taxon*, **9**: 175-189 (1960).

PENNELL, F. W., Genotypes of the *Scrophulariaceae* in the first edition of Linné's *Species Plantarum. Proc. Acad. Nat. Sci. Philad.* **82**: 9-26 (1930).

SAVAGE, S., A catalogue of the Linnaean herbarium. 225 pp. London, 1945.

STEARN, W. T., An introduction to the "Species Plantarum" and cognate botanical works of Carl Linnaeus. Prefixed to the Ray Society facsimile of Linnaeus's *Species Plantarum*, **1**. (London, 1957).

STEARN, W. T., The background of Linnaeus's contributions to the methods and nomenclature of systematic biology. *Systematic Zoology*, **8**: 4-22 (1959).

STEARN, W. T., Notes on Linnaeus's Genera Plantarum. Prefixed to Historiae Naturalis Classica facsimile of Linnaeus's *Genera Plantarum*. (Weinheim and Codicote, 1960.)

STEARN, W. T., A new photographic record of the Linnaean Herbarium. *Taxon*, **10**: 16-19 (1961).

STEARN, W. T., Botanical gardens and botanical literature in the eighteenth century. *Catalogue of botanical books in the Collection of Rachel McMasters Miller Hunt*, **2**, xliii-cxl (1961).

STEARN, W. T., Three Prefaces on Linnaeus and Robert Brown. (Weinheim, 1962.)

SVENSON, H. K., On the descriptive method of Linnaeus. *Rhodora*, **47**: 273, 302, 363, 388, (1945).

2. *Nomenclature*

For the accurate naming of plants, it is necessary to follow the *International Code of Botanical Nomenclature*, last revised at the Ninth International Botanical Congress in Montreal (1959) and published in 1961. The object of this section is to give an account of the general principles and rules adopted in the Code, with reference to the relevant Articles of the rules; many of the

Articles are supplemented by Recommendations which tend to obscure the rules.

The declared object of the Rules is to put the nomenclature of the past in order, and to provide for that of the future. The rules aim at the certainty of names, and at the avoidance of rejection of names which may cause error or ambiguity.

Principle of Priority. Each taxon can have only one correct name—the earliest name that is in accordance with the rules. Most of the "new" names for familiar plants are of course really old names that have been "unearthed" in the literature. In Britain, Bentham and Hooker recognised "*Ruppia maritima* L." and *Ruppia rostellata* Koch (1824). But examination of the Linnaean type of *R. maritima* has established that Koch's *R. rostellata* is identical with the true *R. maritima* L. (1753). The name *R. rostellata* must therefore be replaced by *R. maritima*, which is the earlier name. The first valid name for the species which Bentham and Hooker were wrongly calling "*R. maritima*" is *R. spiralis* Dumortier (1827).

Limitations of the principle of priority. So far as the flowering plants are concerned, modern botanical nomenclature begins with Linnaeus's *Species Plantarum* of 1753.* A name is not treated as *effectively* published unless it is printed and distributed (e.g. in a book or journal). Mechanically reproduced handwritten names on herbarium labels are only effectively published if they were distributed prior to 1953 (cf. Arts. 29-31: rather tricky). A name is not *validly published* until it is (*a*) effectively published, (*b*) accompanied by a diagnosis of the plant (which since 1 January 1935 must be in Latin), or by reference to a previously published diagnosis (Art. 36).† A very short diagnosis will serve to validate a name provided the distinguishing characters are clearly stated. A name is said to be *illegitimate* when it is not in accordance with the rules; *legitimate* when it is.

Synonyms and Homonyms. Synonyms are different names for the same species, genus, etc. It is common practice to distinguish between nomenclatural and taxonomic synonyms. *Nomenclatural synonyms* are different names based on the *same* type; they may be in different taxa of the same rank (e.g. species in different genera) or at different ranks within the same taxon (i.e. as varieties or subspecies of the same species). Since they are based on the same type their identity is not in question, though there may be a difference of opinion as to their taxonomic position. *Taxonomic synonyms* are based on *different* types, so that their identity is entirely a matter of taxonomic

* Rusts, Smuts and Gasteromycetes date from 1801 (Persoon, *Synopsis Methodica Fungorum*); the rest of the Fungi date from 1821-32 (Fries, *Systema Mycologicum*). The Mosses (*Sphagnaceae* excepted) date from 1801 (Hedwig, *Species Muscorum*).

† The Code apparently uses the term diagnosis here to cover either a differential diagnosis (the sense in which we use "diagnosis" in this book) or a description including diagnostic characters.

judgment. These two types of synonyms are now often referred to as *homotypic* and *heterotypic* synonyms, respectively.

Citation of author's names. For a plant's name to be accurate and complete, it should be followed by the name of the publishing author. This enables one to trace the original description and to ascertain its type and date of publication. (For this purpose reference should be made to the *Index Kewensis*, published in quinquennial supplements.) If an author takes up an unpublished name for a new species and validly publishes it with his own description, "ex" is usually inserted between the first man's name and his own (e.g. *Capparis lasiantha* R. Brown ex DC.).* (This should not be done if the botanist who validates the name does not retain the original type; in that case "ex" should be omitted, or a new name chosen for the taxon. The rules allow "ex" to be omitted for brevity of reference; in that case the name of the validating author must always be cited, not that of the author who originally proposed the name.)

Transference. A species may have been described in one genus and transferred to another. When this has happened, the name of the original author is cited in brackets, followed by the name of the author who made the transfer (e.g. *Sisymbrium murale* L. is now treated as *Diplotaxis muralis* (L.) DC.). By means of the "double citation", the original description (and thus the type) can be traced. A variety may be raised to specific rank: *Crocus sativus* L. var. *vernus* L. becomes *Crocus vernus* (L.) Hill. Similarly a species may be sunk to varietal or subspecific rank, and show it by a "double citation": *Thalictrum arenarium* Butcher has become T. *minus* L. subsp. *arenarium* (Butcher) Clapham. *Thalictrum arenarium* Butcher is called the *basionym* of Clapham's subspecies—i.e. it is the name-bringing or epithet-bringing synonym with which the type is associated. Since 1 January 1953, new combinations are not validly published unless the basionym is clearly indicated, with authority, direct reference and date of publication (Art. 33). It sometimes happens that an author making a transfer erroneously applies the epithet in its new position to a different species which he may even describe (cf. *Seseli crithmifolium* (DC.) Boiss. discussed in Davis, 1953: 117). In such cases it is the identity of the basionym, not the taxonomic intentions of the author who made the combination (often without seeing the type), which must decide the application of the name.

When the *rank* of a taxon is changed, it is the earliest legitimate name in the *new rank* which is valid. *Magnolia virginiana* L. var. *foetida* (1753) is now treated as a species; its correct name is *M. grandiflora* L. (1759), not *M. foetida* (L.) Sargant (1889), because *M. grandiflora* is the earliest available name at specific rank (Art. 60).

When a species is transferred to another genus *without* change of rank,

* In this example Robert Brown recognised the plant as distinct and coined a name for it, but de Candolle first published Brown's name with a description and thus validated it.

the original specific epithet must be retained (as with *Diplotaxis muralis* (L). DC., cited above), unless, in its new position, it becomes illegitimate, as happens when the genus to which the species is transferred already contains a different species with the same epithet. In such cases a new name must be found for the transferred species, in order to prevent the creation of homonymy which can, of course, only lead to confusion; later homonyms are illegitimate. E.g. *Spartium biflorum* Desf. (1798) was transferred by Spach to *Cytisus* in 1848. But, as there was already a species called *Cytisus biflorus* L'Héritier (1791), Spach had to give a new name (*Cytisus fontanesii*) to Desfontaines' species.

Attention must be drawn to another point which is not as well understood as it should be. A name cannot be taken up if it was superfluous when first published (Art. 63). (Thus *Chrysophyllum sericeum* Salisb. (1796) is illegitimate, because Salisbury cited *C. cainito* L. (1753) as a synonym, thereby rendering Salisbury's name superfluous.) In the case of a taxon transferred from one genus to another and possessing an illegitimate name in its original genus and no legitimate name in the genus to which it is transferred, it is possible to retain the originally illegitimate epithet under the second genus by publishing it there as a new name and not as a new combination. Thus the name *Talinum polyandrum* Hook. (1855) is illegitimate, being a later homonym of *Talinum polyandrum* Ruiz & Pavon (1798). Transferred to *Calandrinia*, Hooker's species bears the name *Calandrinia polyandra* Benth. If this is done nowadays, a new type may be chosen if the original species was based on unsatisfactory material, but a Latin diagnosis (or description) must then be given.

Note that all the transfers we have mentioned depend upon the taxonomic acumen of the botanist. It is his judgment which decides what rank a taxon should be given, and to which genus it should be referred.

Rejection of names. We can consider this under the heading of nomenclatural and taxonomic rejection.

Nomenclatural rejection

(*a*) A name cannot be rejected because it is inappropriate (e.g. *Scilla peruviana*—the name of a Mediterranean species).

(*b*) A name must be rejected if it is a later homonym (Art. 64). The South African *Impatiens capensis* Thunberg (1794) is a later homonym of *I. capensis* Meerburgh (1775)—i.e. they are different species with the same name. *I. capensis* Thunberg must therefore be rejected and another name found for that species (*I. duthiae* L. Bolus, 1921).

(*c*) A name must be rejected if it is used in different senses, and has become a source of persistent confusion or error (Art. 69). Certain widely accepted family and generic names which would otherwise be illegitimate have been specially conserved, and form Appendices II and III of the Code. There is, as yet, no such provision for the conservation of specific names. Confused specific names, however, can sometimes be rejected, but this should not be done unless there is a very strong case for it. Thorough research into the author's original description and specimens, etc. may establish the author's intentions and make further confusion unnecessary. If this, however, results in a reciprocal name switch, confusion is bound to result, and it seems to us a weakness in the Rules that such names are difficult to get rid of—though you may drop them, someone else is at liberty to reinstate them. It should be possible to reject names likely to become persistent sources of error and not wait until they have become such.

Taxonomic rejection

When two previously recognised species are united, the earlier epithet must be chosen for the united species and the other name treated as a taxonomic synonym (e.g. *Myosotis collina* Hoffm. (1814) is a taxonomic synonym of *M. hispida* Schlecht. (1791), and must therefore be rejected). Such changes depend on taxonomic opinion as to whether the two species are the same, and the names, therefore, taxonomic synonyms. There is nothing to prevent a botanist from taking up the name *Myosotis collina* Hoffm. if he considers Hoffmann's plant specifically distinct from *M. hispida.*

A name must be rejected if it was based on a monstrosity (Art. 71). The Code unfortunately does not define what is meant by a monstrosity, but presumably excludes double-flowered forms which typify such species as *Ranunculus asiaticus* L.

From the above account it should be clear that many name changes are due simply to the application of the Rules—especially the rule of priority and its limiting clauses. Other name changes are essentially taxonomic, and may be beyond dispute, or may still be a matter of opinion.

Some other matters treated in the Code

(1) When describing a new species, a name must not be chosen for it that has been used in the genus before, even if botanists now agree that the first name is a synonym for a different species and therefore no longer used. This is to avoid confusion if the plant represented by that synonym should some day be reinstated as a "good" species. It is permissible, however, to use a name for a new species that already exists as a *nomen nudum*—i.e. has never been validated by a description and therefore has no standing under the Rules; the same applies if the name has only appeared in synonymy ("pro syn."), and thus has no validity.

(2) As soon as a variety or subspecies is described within a species, there is automatically a "type" variety or subspecies. This bears the same epithet as the species, but is not followed by an authority (e.g. *Arenaria acerosa* Boiss. var. *glabra* Boiss. is automatically balanced by var. *acerosa*. This epithet is given priority over such infra-specific epithets as var. *typica*, *originalis*, etc., which are now illegitimate (Art. 24)).

Similarly, within genera the subgenus and section which include the type species of the genus automatically bear the same name as the genus (without authority). We must, for instance, use *Ranunculus* Sect. *Ranunculus*, not Sect. *Euranunculus* (Art. 22).

(3) *Orthography*. Plant names should be spelt as they were originally published unless the original spelling was definitely wrong or can be shown to be a misprint (Art. 73). "*Onopordon*" for instance, should be *Onopordum*, because Linnaeus originally spelt it that way.

Species named after people should, in the masculine genitive, end in *-ii*, except after vowels and *-er* (e.g. *wilsonii*, *farreri*, *drucei*). This is retroactive. It is recommended (Art. 73F) that all specific and infra-specific epithets should be spelt with a small initial letter—a practice which brings botanical nomenclature into line with zoological nomenclature.

(4) *Citation of synonyms and misidentifications*. Synonyms are best cited chronologically and as originally published (even if spelt wrongly), though the authority can be corrected in accordance with modern practice; to omit a double citation, for instance, can cause confusion.

Some taxonomists have preferred to cite the basionym first, but this is bound to add to the chances of overlooking an earlier name. If synonyms are cited chronologically, enough information should be given to indicate why a name has not been adopted if it appears to have priority over the accepted name. This will be necessary, for instance, if it is a homonym. The following synthetic example (in which *Planta turcica* is the basionym) should make the points clear:

Herba turcica (Boiss.) Bornm. (1940)
Syn. *Planta dubia* L. (1753), *nom. confus.*
Planta nigra L. var. *anatolica* DC. (1800)
Planta kurdica Benth. (1850), non *Herba kurdica* Hooker fil. (1865)
Herba squalida Boiss. (1861), non L. (1753)
"*Herba orientalis*" sensu Doe (1865), non DC.
Planta turcica Boiss. (1867)
Herba perplexa Siehe (1910), *nom. nud.*
Herba anatolica (DC.) Stapf (1920)
Herba anatolica var. *turcica* (Boiss.) Hall (1925)
Herba neglecta Brown (1959)

The alternative method, used for example in Y. I. Barkoudah's *Revision of Gypsophila* (1962), would be:

Herba turcica (Boiss.) Bornm.
Basionym: *Planta turcica* Boiss. (1867)
Homotypic synonym: *Herba anatolica* var. *turcica* (Boiss.) Hall (1925)
Heterotypic synonyms:
Planta dubia L. (1753), *nom. confus.*
Planta nigra L. var. *anatolica* DC. (1800)
Planta kurdica Benth. (1850), non *Herba kurdica* Hooker fil. (1865)

Herba squalida Boiss. (1861), non L. (1753)
Herba perplexa Siehe (1910), *nom. nud.*
Herba anatolica (DC.) Stapf (1920)
Herba neglecta Brown (1959)

Misapplied name: "*Herba orientalis*" sensu Doe (1865), non DC.

As a refinement, the types of synonyms and their whereabouts may be cited.

A few taxonomists delight in the intricacies of nomenclature and practise it as almost an end in itself. As has been observed in a review of St. John's extraordinary book (1958) on the "case method" in nomenclature, "It is a subject riddled with snobbism, prejudice and intrigue." To most systematists, however, nomenclature is a time-consuming necessity that comes between them and the plant. Nevertheless, it is one of the tools of the taxonomist's trade and for that reason its principles must be mastered. To practice nomenclature, however, without reference to the status or identity of taxa is asking for trouble. The greatest care must be taken to see that the names we are treating as synonymous do in fact refer to the same taxon.

The nomenclatural treatment of hybrids is dealt with in Chapter 14.

Chapter 9

PRESENTATION OF DATA

"The proper aim is not to name species but to know them"
G. F. Ferris, 1928

Introduction

The value of any taxonomic revision, however well it has been done, depends to a very large extent on the way the taxonomist presents his findings. He must consider very carefully the *purpose* of his publication. Very often there is some conflict between what the taxonomist would *like* to present, and what exigencies of space *enable* him to present. The more acute this conflict, the more important it is for the taxonomist to decide not only in what order to present his data but what to emphasise, restrict or omit. Time and again one finds revisions or Floras in which the space has not been used to the best advantage.

In this chapter we cannot do more than emphasis the salient points in presentation. Every taxonomist should read the classical essays by Bentham (1874) and by J. D. Hooker (1855) which still remain unsurpassed for clarity and wisdom. More up-to-date accounts on presentation will be found in Hitchcock (1925), Van Steenis (1954, 1957), and Mayr, Linsley and Usinger (1953); the latter, although primarily a zoological work, contains in many ways the most comprehensive account of taxonomic presentation.

Although there is a big overlap in the format and material of monographs and Floras, it seems best to consider their aims and content separately before we go into more detail on the information they contain.

Monographs and Revisions

A monograph, in the true sense, should synthesise all the information that is known about the group. Revisions are less ambitious, since they give less detail and largely confine themselves to taxonomic and geographical information, though cytological data are increasingly used; revisions may be done on a regional basis (as part, for instance, of a major Flora) or throughout the world range of the group. In either case the object of a revision (and an integral part of a monograph) *is to delimit the taxa clearly (particularly species), to group them in a natural manner and to provide means of identification.* A monograph goes further, being nowadays often concerned with evolutionary interpretation, generally in the light of combined morphological, geographical and cytogenetic data. The introductory chapters in such a work are, to students of evolution, often the most interesting part of the monograph, (cf. Constance, 1951) and may include accounts of cytogenetics, taxonomic characters (their variation and evolutionary interpretation, geographical distribution and ecology), and a discussion on the probable evolution of the group. Aspiring monographers should study E. B. Babcock's *The Genus Crepis* (1947) and T. H. Goodspeed's *The Genus Nicotiana* (1954). Harling's beautiful *Monograph of the Cyclanthaceae* (1958) shows how much can be done with a difficult tropical group for which biosystematic data are almost lacking.

As examples of shorter works from which a great deal can be learnt on taxonomic method and presentation, the following could be cited: Stearn (1938) on *Epimedium* and *Vancouveria*, Nordlindh (1943, 1946) on *Compositae-Calenduleae*, Lee (1948) on *Swainsonia*, Shan & Constance (1951) on *Sanicula*, Hedberg (1955) on *Sibthorpia*, Rollins (1955) on *Lesquerella*, Hara (1957) on *Chrysosplenium*, Leenhouts (1959) on *Canarium* (*Burseraceae*), K. R. Stern (1961) on *Dicentra*, and Raven (1962) on *Oenothera* Sect. *Chylismia*. The reluctance of modern taxonomists to take up *very large* groups, however, remains a serious impediment to taxonomic progress. The brunt of this broad, basic work is still borne by the authors of *Das Pflanzenreich*, Wimmer's monograph of the *Campanulaceae-Lobelioideae* (1957) being a fine recent example.

Descriptions are of course an essential part of monography, though in a revision these may be short or even absent provided a good key and critical notes are provided. Specimens must be cited, but how extensive this enumeration should be will depend on the space available. Cited specimens enable other workers to check the basis of a monographer's delimitation of species, and will often enable duplicate specimens to be identified in herbaria not examined by him. On the other hand, it must be admitted that if the monographer has done his work properly and determined the material in different herbaria, the citation of a large number of specimens becomes less necessary. Often a compromise can be achieved, the monographer citing specimens which are available in several herbaria and which cover the geographical range of the taxon. It would be a great help in all monographic and revisional work if type material were always cited—its whereabouts and whether the author has seen it or not—since this is a vital reference point in taxonomic research.

The value of monographs, revisions and Floras is largely dependent on the amount of original critical observation they contain. The taxonomist should make it clear whether the specimens he cites have been examined or not. Uncritical citation of synonyms may lead to a repetition of errors. The monographer should accept nothing on trust that he can confirm personally.

Both monographs and revisions must take the *names of all the taxa* in their group into account. Where it is not possible to adopt these names or reduce them to certain synonymy, they may be cited as doubtful synonyms or listed at the end as belonging to taxa of uncertain status (e.g. "*species non satis notae*"); in any large genus there are sure to be some such names which, due to the loss of type material, cannot be applied with certainty. As Hooker (1855) wrote a century ago, "The difficulty of determining synonymy is now the greatest obstacle to the progress of systematic botany."

No attempt should be made to hide lack of knowledge. On the contrary, these *lacunae* must be emphasised so that other workers will know where information is still required in order to clarify the taxonomy of the group.

Van Steenis (1957) has drawn attention to the psychological effects of a bad monograph. Excessive "splitting" is apt to set a pattern, so that later workers, unfamiliar with the group, often describe plants as new that will not fit into the narrowly circumscribed species accepted by the monographer. "The aftermath of such pseudo-monographs is either to bring its author into disrepute or the genus or family respectively. Non-systematists derive from such works even a repugnance towards the whole of taxonomy as a branch of science. . . . Monographs must be useful and monographers should have the responsibility for this clearly in mind. Good monographs are the only means of bringing progress; this holds for the past and holds for the present and future. They must be constructive and solve problems. Taxonomy is not in need of mere compilations without creative spirit; there is no sense in publishing such works. . . . I have come to the definite conclusion, based on long experience, that the majority of "difficult groups" are not so intended or created by nature, but that the difficulties have been created by the monographer."

As Raven (1962) says, the modern monographer should transcend the bare bones of keys and formal description. If he wants his work to be widely read by biologists, it is up to him to make it interesting reading.

Floristic Works

These cover a wide range of treatments, from the most comprehensive Flora to the bare species list. The standards required today, with much attention being paid to typification and nomenclature, distribution and ecology, in addition to the comparative morphology which was the main preoccupation of nineteenth-century Floras, means that "creative" regional Floras (as van Steenis calls them), cannot be written in a hurry and require a team of taxonomists; the increasing accumulation of synonyms adds to the burden. As examples of comprehensive regional Floras we may cite the *Flora Malesiana*, *Flora of Tropical East Africa*, *Flora Zambesiaca* and *Flore du Congo Belge*, all in the process of publication. Van Steenis points out that Floras are best written for a natural botanical region (as the area covered by *Flora Malesiana*), since the maximum number of related taxa will be found in such an area. Political considerations, however, result in Floras still being largely written for individual countries (or parts of countries) and in such cases the taxonomist must be very careful to review the flora of neighbouring territories if he is to avoid mistakes in identity.

The majority of Floras, designed for ready identification, contain considerably less information than the comprehensive Floras referred to above. As Bentham pointed out, *the principal object of a Flora is to afford the means of determining any plant growing in the area circumscribed.* This is usually achieved by a combination of keys (or synopses in older Floras) and descriptions. These descriptions, Bentham says, "should be clear, concise, accurate and characteristic, so that each one should be readily adapted to the plant it relates to, and to no other; they should be as nearly as possible arranged under natural divisions, so as to facilitate the comparison of each plant with those nearest allied to it". Bentham's *Flora Hongkongensis* (1861) and *Flora Australiensis* (1863-78) set the standard for the British Commonwealth. Many works which are called Floras do not come up to these basic requirements: they may contain a key but no descriptions (e.g. Rechinger, *Flora Aegaea*), or descriptions but no keys (e.g. Post & Dinsmore's *Flora of Syria, Palestine and Sinai*, 1932-33). The very useful manuals which provide artificial keys to the species but little other information are known in Russian as *Opredelitel'*, a term which can be translated into English as "*Determinator*" . . . a useful title. A good Czechoslovak example is Dostál's *Klíč k Úplné Květeně ČSR* (1958).

Although there is some overlap between them, Floras (in the broad sense) can be considered under two headings, since their content must be largely determined by their purpose:

(1) *Research Floras*, which include those comprehensive works we have already referred to (*Flora Malesiana*, Fawcett & Rendle's *Flora of Jamaica*, etc.). Although these can, of course, be used in the field, their bulk and detailed information probably make them most valuable for further revision and identification in the herbarium. They are in the nature of regional revisions and stand between the shorter Flora and the monograph. They are self-contained, and provide an excellent basis for further taxonomic research.

(2) *Concise or Field Floras*, which are much shorter than the above, descriptions usually being brief or (in Determinators) absent. Types and herbarium specimens are not cited, and synonymy is reduced to a minimum. On the other hand, there is usually rather detailed information on distribution and ecology. Good examples of Floras of this type, dealing with countries whose flora is well known, are Clapham, Tutin & Warburg's *Flora of the British Isles* (ed. 2, 1962), Fernald's edition of *Gray's Manual of Botany* (1950) and Binz & Thommen's *Flore de la Suisse* (1953). Heywood's guide for the contributors to the *Flora Europaea* (1958, 1960d) contains many useful suggestions for Floras of this kind.

Van Steenis (1954) does not mince his words about the content of a satisfactory flora. "Some modern Floras are so full of non-phytographical text, on economics, ethnobotany, anatomy, history, etc., that the descriptive text is suffocated by a mass of other information and is treated as of secondary rank.* Apparently the means at the disposal of the authors and/or their aims vary to extremes. Whatever the excuse, I maintain that everybody is responsible for his own writings, and in descriptive botany it is the amount of original critical observation rather than the repetition of the errors of others, which is decisive for the standing of Floras. I emphasise that neither science nor botanists are at present in need of any non-critical Flora; these mediocre works only add to the difficulties we already have to face."

A third kind of Flora is that known as the Excursion Flora, in which space is saved by combining keys and descriptions, as in Rothmaler's *Excursionflora von Deutschland* (1958), or by omitting, or at least abbreviating, certain kinds of information, such as distributional, ecological or ethological, as in Clapham, Tutin & Warburg's *Excursion Flora of the British Isles* (1960).

The content of a fairly comprehensive Flora will of course vary according to the amount which is known about the flora of the area concerned, the scope of the work, the personality of the author, etc., but the following is a fair summary of what one expects to find in such a Flora, though the emphasis may vary:

(1) *Introduction.* This must clearly delimit the area covered, and show (usually with the aid of a map) the divisions which are being recognised in the Flora. Ideally these divisions will be floristic or phytogeographical, but until the Flora has been written there may be insufficient evidence to delimit areas by botanical criteria; in such cases they may be geographical. Various other kinds of information can add greatly to the value of a Flora: topography, climate, geology, vegetational cover, etc. Botanical exploration and a list of collectors is often included.

(2) *Taxonomy.* A Flora is not the place to propose a new classification. The families should follow an accepted system (allowing for some modification) which for convenience may be one that is already used in the area. Short family descriptions will be followed by keys to the genera. The generic description may be followed by references to relevant revisions which will enable the reader to study the group further if required. Keys to species are of prime importance and will be discussed further below.

The species should be cited in as natural an order as a linear format allows, and may be grouped under their sections, etc., if these are satisfactory. It is convenient to number the species within each genus (and the genus within each family), so that the same numbers can be referred to in the keys. The accepted name of the species must be followed by its authority, place of publication and date. This may be followed by references to other works

* A remarkable example of such a Flora, containing a great deal of everything except taxonomic revision, is Tackholm & Drar's *Flora of Egypt* (1941-54); even Hegi's otherwise excellent *Illustrierte Flora von Mitteleuropa* (1906-31; 2nd edition in progress from 1935), suffers from an excess of non-taxonomic information.

where the species is helpfully treated, and to previous illustrations or to an illustration given in the present Flora. Citation of all references (a practice adopted in many monographs) tends to be self-defeating—not knowing which are the useful ones, the reader is deterred from looking up any of them.

Synonymy follows. This need not be as extensive as that in a monograph, but should be as complete as possible for the area under review and for adjacent territory.

The specific description should at least cover the species as it exists within the area of the Flora—that is, including its known variability within that area. Descriptions should be concise and diagnostic, and serve as a check on the key to the species. If the species contains more than one infra-specific taxon, these should be keyed out after the species description. Infra-specific taxa may require short descriptions and a separate synonymy.

The description of the species (or infra-specific taxa) is often followed by the symbol for the *life form* (using Raunkiaer's system*), the *flowering time* and *chromosome number*. The latter should only be cited if it has been counted for material known to have come from the area covered by the Flora (cf. Heywood, 1958). Alternatively, counts based on material from outside the area should be indicated as such if included at all.

Ecological data often follow the species description, though they may be placed after the geographical data. The altitudinal range (in metres) should be given whenever it is known. The amount of ecological data that can be given will depend very much on the tolerance of the species and the size of the area covered by the Flora; if a large area is covered, habitats may vary so widely from place to place that they become very difficult to summarise.

In research Floras *type material* should be cited—locality, collector's number (if any) and herbaria in which the types are kept. It is essential to indicate whether type material has been examined or not. *The Flora of Tropical East Africa* is meticulous about the citation of types, even those synonyms being given—a refinement of doubtful necessity.

The *geographical distribution* of species is extremely important in a modern Flora, and should be as accurate as one can make it for the area covered. It should be summarised, using the geographical or botanical divisions adopted in the Flora or a grid system; more precise data may be given for rare and endemic species. Whether specimens are cited or not will depend very much on the purpose of the Flora and how well the flora is known. It would not, for instance, normally be helpful to cite specimens in a Flora of Britain (which will be largely used in the field and where the species are taxonomically well understood), but would be very helpful for a pioneer Flora of Afghanistan. Specimens must be cited economically, and in a consistent, geographical sequence; the locality (preferably including the altitude) should precede the

* Raunkiaer, C., *The Life Forms of Plants and statistical Plant Geography*. Oxford Univ. Press, 1934.

collector's number. If specimens are numerous and only a few can be cited these should be chosen to cover the distributional range of the taxon. Whenever possible they should be good representative specimens and readily accessible. It is certainly helpful if the distribution outside the area of the Flora can be summarised, at least for the species in the broad sense. This is a time-consuming job if it is to be done accurately, and the author must decide for himself how much time he can afford to spend on it. It seems reasonable to allow for a decrease in accuracy the further one gets from the area whose Flora is being compiled.

In a modern Flora one of the most helpful parts (at least for further research) is the *observations* which may end the accounts of many species. These give the taxonomist scope to show his insight, and can be the most interesting part of a Flora. This section is the place to discuss variability (genotypic and phenotypic), differentiation from allied species (or from ones with which there has been confusion) and biological facts of particular interest, such as pollination or dispersal. It is helpful for phytogeographical studies if the affinities of endemic species are given and an indication of the ally's range outside the area of the Flora. New taxa, new combinations and nomenclatural discussions are best published in a separate series of papers in step with the preparation of the Flora.

As an appendix to a Flora or monograph, it is a useful practice to list herbarium numbers (with page references) so that the herbarium taxonomist can quickly check the identity of specimens. A bibliography, and abbreviations of references and collector's names, may also be provided. An efficient index is of course essential, and must distinguish between accepted names and synonyms.

The publication of *Species Lists*, except for areas where the Flora is very little known, serves a very limited purpose and, if not backed by careful revision, frequently leads to the repetition of errors in identification or nomenclature. Species lists are, however, an important stage in building up the knowledge of the flora of an area. They allow other workers to find out what is already known, what previous collections have been made and where they may be found. In many cases the time would be better spent on taxonomic revision and in the preparation of critical regional or national Floras. As such lists, however, will continue to appear and provide data (of varying reliability) for Floras, it is worth citing the minimum requirements for them as set forth by Blake & Atwood (1961) after examining several thousand such publications: "The essential features of even the barest list of plants include a title accurately describing its contents; the most definite possible statement of the geographical area covered, with reference, in the case of small or obscure localities, to the distance and direction from some locality that can be found on any reasonably detailed map . . . and a statement of the material on which the list is based, whether the author's collections or observations alone or all published or unpublished information. . . ."

Comprehensive species lists or catalogues, giving a large amount of detailed geographical, ecological or phytosociological information, but without keys or descriptions, are often compiled for counties, provinces, etc. Some County Floras of Great Britain are good examples of such works (e.g. Dony, *Flora of Bedfordshire*, (1953)) and two continental examples are Chassagne's *Inventaire Analytique de la Flore d'Auvergne* (1956-57) and Braun-Blanquet & Rübel's *Flora von Graubünden* (1932-35).

Having outlined the general content of monographs and Floras, we will now go on to discuss in more detail some of the more important features of presentation.

Keys and Synopses

Modern Floras are usually provided with *dichotomous keys* which are of two main types: *indented* keys and *bracket* keys (cf. p. 302). The former have the advantage over the latter in giving a visual presentation of the group, so that the reader can more readily obtain a picture of the taxa; on the other hand, indented keys take up more space and are more expensive to print than bracket keys. Both are based on the same data—on the diagnostic (and sometimes differential) characters which in large or critical groups will need to be tabulated before the key is constructed.

The first aim of the key (*clavis*) *is ease and certainty of identification.* Naturalness will often have to be sacrificed to convenience. This means that related species do not necessarily come out next to one another in the key—they may be widely separated by some technical character which is easily seen and described (e.g. glandular versus eglandular hairs) but which cuts across the relationships of the group as assessed by overall resemblance; the affinities of the species, however, will be catered for in the enumeration of the species and by grouping them as far as possible into natural sections, etc. If the subgenera and sections are natural and clear cut (as in *Iris*, *Trifolium* or *Primula*), there is no reason why these should not be keyed out before the species, particularly if the genus is a large one in the area; but if the sectional distinctions are ill-defined, unnatural or difficult to observe, it is better to key out all the species without reference to sectional classification.

Dichotomous keys consist of contrasting propositions (a couplet), each statement of which is a *lead*. The leads should be simply numbered and if possible begin with the same word. The characters used *must contrast*. This sounds obvious enough, but it is extraordinary how often one comes across keys which do not conform to this simple requirement. Characters may be drawn from any part of the plant, but the most constant and readily observable characters should be chosen in preference to those which are variable or observed with difficulty (e.g. embryo structure). Vegetative characters (if constant), such as opposite versus alternate leaves, or even types of indumentum, have an advantage over characters of the flower and fruit in that they are visible for a much longer period in the life of the plant, so that the key will "work" for such features (provided the genus is recognised) even when the plant is in the sterile condition. It will therefore often be more convenient to use such characters in the early dichotomies of the key rather than details of flower and fruit which can be increasingly employed in the later dichotomies. Separate keys are sometimes provided for flowering, fruiting and sterile material (as in *Salix*).

Qualitative characters should usually be used in preference to quantitative ones. Measurements which show some degree of overlap frequently break down still further when more material is available, so that they should only be used if better characters cannot be found. In critical groups, and when overlapping subspecies are being keyed out, several characters often have to be used in the ultimate dichotomies in order to allow for some breakdown in characters correlations. Such vague terms as "large" and "small" should not be used unqualified in keys, since the reader, trying to name a specimen in an unfamiliar genus, cannot be expected to appreciate the usage of such relative terms; they should be replaced (or supplemented) by a range of measurements.

The taxonomist must reconcile himself to the fact that polymorphic species may have to be inserted two or three times in the key; this is preferable to the "either . . . or if not then . . ." type of lead that can bemuse the reader into following the wrong trail in the key.

The construction of a key is an intellectual challenge. We may quote from van Steenis (1957): "It has been the opinion of Bentham that the inability to give a key to the species is itself always a sure warning signal of the absence of a sound synthesis and I believe this is unanimously agreed on by all good taxonomists. A key represents the proof of the pudding, because the author of the group is challenged to crystallise the essence of his work in it and to show in which way he has found the hierarchial value of differential characters." Although it is questionable whether all "good" species can be satisfactorily keyed out (at least in so far as some species are recognised in temperate floras), the claim we have quoted has practicality to recommend it and may be sound policy for "alpha" classification.

Good accounts of the mechanics of key construction will be found in Hitchcock (1925), Stearn (1956: 251-3), Mayr, Linsley & Usinger (1953), Lawrence (1951: 255), Z. P. Metcalfe (1954) and Sneath (1957b). We shall give here only an example of indented and bracket keys (adapted from Clapham, Tutin & Warburg's *Flora of the British Isles*) though variants exist. The figures in parenthesis in the bracket key enable it to be worked backwards if one has "gone astray" in a long key or wants to extract the characters of a species. In very long indented keys it is advisable to give a page reference at the end of the first lead of a couplet if the second lead is widely separated from the first (as in the key in Hubbard's *Grasses*: 1954).

Examples of keys, using the British species of *Primula*

(a) *Bracket key*

1. Leaves mealy beneath; flowers lilac or purple . . . 2
— Leaves not mealy beneath; flowers yellow 3
2. (1) Leaves crenulate; flowers lilac, heterostylous; corolla lobes distant; calyx teeth ± acute; capsule cylindrical, much exceeding calyx . . . . 1. *farinosa*
— Leaves not crenulate; flowers blue-purple, not heterostylous; corolla lobes contiguous; calyx teeth obtuse; capsule ovoid, slightly exceeding calyx . . . . 2. *scotica*
3. (1) Scape distinct; pedicels finely pubescent; limb of corolla concave, rarely exceeding 20 mm. diam. . . . 4
— Scape 0 or nearly so; pedicels with shaggy hairs; limb of corolla flat, usually more than 30 mm. diam. . . . . 5. *vulgaris*
4. (3) Calyx uniformly pale green; corolla with folds in the throat; fruit ovoid, enclosed in the calyx . . 3. *veris*
— Calyx with midribs conspicuously darker green than the remainder; corolla without folds in the throat; fruit oblong-ovoid, exceeding calyx . . . 4. *elatior*

(b) *Indented key*

1. Leaves mealy beneath; flowers lilac or purple
  2. Leaves crenulate; flowers lilac, heterostylous; corolla lobes distant; calyx teeth ± acute; capsule cylindrical, much exceeding calyx . . 1. *farinosa*
  2. Leaves not crenulate; flowers blue-purple, not heterostylous; corolla lobes contiguous; calyx teeth obtuse; capsule ovoid, slightly exceeding calyx . . . . 2. *scotica*
1. Leaves not mealy beneath; flowers yellow
  3. Scape distinct; pedicels finely pubescent; limb of corolla concave, rarely exceeding 20 mm. diam.
    4. Calyx uniformly pale green; corolla with folds in the throat; fruit ovoid, enclosed in the calyx . 3. *veris*
    4. Calyx with midribs conspicuously darker green than the remainder; corolla without folds in the throat; fruit oblong-ovoid, exceeding calyx . . 4. *elatior*
  3. Scape 0 or nearly so; pedicels with shaggy hairs; limb of corolla flat, usually more than 30 mm. diam. . . . . 5. *vulgaris*

A *synopsis* is defined by Hitchcock (1925:104) as "a segregation of groups of the same category into successively smaller divisions according to characters common to its members". It often has the format of a key (being a "synoptic key"), but usually does not account for exceptions or distinguish between all species and thus does not enable individual plants or species to be identified*. There is, however, no clear dividing line between synopses and keys. The synopsis is now most used to summarise a classification, particularly above the species level. Fine examples are found in the synopses used in Bentham & Hooker's *Genera Plantarum* (1862-83) and in Willkomm & Lange's *Prodromus Florae Hispanicae* (1861-93). Examples of synoptic keys to Orders are given in Lawrence (1951, p. 372, 439).

Diagnoses and Descriptions

Diagnoses. A diagnosis may be defined as a "brief listing of the most important characters or character combinations that are peculiar to a given unit and by which it can be differentiated from other similar or closely related ones. The direct comparison of a species (or other taxon) with other specifically mentioned species (or other taxa) is usually called a *differential* diagnosis" (Mayr *et al.*, 1953).

A diagnosis is of very great practical value in that it makes it very much easier for someone without the material of a newly described form to appreciate its characters and to tell whether material he is trying to name belongs to that taxon or not. It would be a very great help if all new species were published with a Latin differential diagnosis preceding (or following) the detailed description. Great care must be taken, however, to see that the species with which the new taxon is compared in the differential diagnosis is, in fact, the most closely allied to it on overall resemblances. If a new species is wrongly placed—particularly if it is described in the wrong section or even genus (as can easily happen in *Cruciferae* or *Compositae*), the diagnosis will put the reader off the scent. If a species appears to be equally closely allied to two or three species, the difference from each of these may be cited in the diagnosis. It is not necessary—or even desirable—to cite all apparent differences in the diagnosis, but only the most certain differences which serve to distinguish it from its allies known at the time. If the plant has no close allies, this should be stated and comparison made with some well known species in the genus. In Floras without dichotomous keys (as in Boissier's *Flora Orientalis*), diagnoses are often of more help than descriptions in identifying species.

When a differential diagnosis is not given, a similar purpose may be served by underlining or italicising those characters in the description which differentiate the species from its allies.

Descriptions. Plant descriptions can for convenience be considered under three headings: original descriptions, redescriptions and short descriptions.

The *original description* should be fairly complete, although there is no

* An example of a synoptic key is given on p. 150.

need to stress the generic or family characters in the description of a species It should be possible to identify the species from it even if the type material is later destroyed. In little-known floras we must therefore bear in mind the discovery of unknown species with which the new species may be later confused. As Mayr *et al.* (1953: 158) note, the so-called "detailed description" of a species is in fact highly selective. "How much subject matter is included in a description depends on the group concerned and the state of knowledge of that group. Excessively long descriptions obscure essential points; short descriptions omit pertinent data." The style should be telegraphic and concise, consisting largely of nouns and adjectives in the nominative case. Descriptions should follow a conventional sequence (cf. Heywood, 1958: 5-6). It is usual to describe the habit and vegetative parts of the plant (from below upwards) before the inflorescence and floral parts; the latter are described from the outside inwards. The underground parts and the habit of the plant are described first. The indumentum, if of the same type throughout most of the aerial parts of the plant, is often described after the habit to avoid repeating it for each organ. Then follows the description of the leaves (arrangement, stalked or sessile, shape—including toothing, base, apex—dimensions, petiole, stipules), although if the plant has only basal leaves they are described before the scape. The stem descriptions should include height, thickness, branching, etc. The general form of the inflorescence precedes the description of bracts, pedicels and bracteoles. Flowers may be described first in general terms—size, colour, position, scent. The organs of flowers are described in the following order: calyx, corolla, stamens, staminodes, ovary; fruits and seeds are described last. In the case of gamopetalous bilabiate corollas (as in *Labiatae*), tube and upper and lower lips must be separately described, e.g. *Tubus corollae cylindricus superne sensim ampliatus, 4-5 mm. longus, extra breviter barbatus, intus glaber; labium superius profunde bifidum, porrectum, 3 mm. longum; labium inferius trilobatum 3 × 4 mm. lobo mediano reniformi crenato purpureo-maculato lobis lateralibus triangularibus duplo longiore.*

Measurements should always be given in the metric system, and must include the *whole range of material definitely included in the taxon* by the author. If there is reason to believe that certain measurements are exceptional they may be placed in brackets (e.g. (3-) 5-9 (-11) mm.), but this should not be attempted if the material is very limited. The taxonomist must always try to convey the habit and facies of the plant—the most difficult part of any description.

If a description is in Latin (as it should preferably be), a separate differential diagnosis is not necessary to validate the name under the Code since the diagnostic characters will be included in the description. However, the publication of both a Latin differential diagnosis and a Latin description are to be strongly recommended for reasons of convenience; one or other are in any case compulsory under the Code.

As models for detailed descriptions (and differential diagnoses), the reader cannot do better than study the descriptions (with their accompanying figures) in the recent numbers of *Curtis's Botanical Magazine* and *Hooker's Icones Plantarum.*

Redescriptions (*emended descriptions*). If an original description is seriously incomplete or faulty, it is an important element of revisional work to re-describe it on the lines of the detailed description outlined above. Ferris (1928) in an entomological context, has stated that ". . . a distressingly large percentage of the named species. . . cannot be recognised positively or even at all, on the basis of the existing literature. It is more important, for the advancement of our study, to redescribe such forms than it is to describe new species. The redescription of such forms should be regarded by the student as an essential part of his work upon any group which he may elect to study. The fact that a species has been named should make no essential difference in the way in which it is treated. . . . The proper aim is not to name species but to know them." The quotation is equally applicable to Angiosperm taxonomy.

With regard to measurements, and to the proportion of certain variations within the population, a description is never final. Nevertheless, for practical purposes a line must be drawn somewhere. With so much basic description still to be done in taxonomy, it would be wasteful to publish redescriptions of species again and again when good descriptions are already available. It is a monographer's business, however, to provide redescriptions of species in a comparable form, since he should be in a better position to do so than anyone else.

The *short description* is an art in itself. As Bentham (1874) put it, "Any tyro with a little practice can draw up long descriptions of *specimens*, fairly detailing every organ; but selecting the characters necessary to give a good idea of a *species* in a short description requires a thorough knowledge of the subject and a methodical mind."

In a modern Flora short descriptions usually serve as a check on the key, to verify identifications. In poorly known floras such descriptions should be long enough for us to know if we have before us a species already recorded for the area, or another not yet recorded. In well-known floras the latter consideration scarcely applies. Long descriptions are, however, frequently given in Floras, but one doubts if they are often read; they would be more appropriate in monographs or revisions. Five to ten lines should be sufficient for the identification of most species, the length varying according to the complexity of the group, the number of closely allied species and the probability of others being discovered. Descriptions, which should exclude characters shared by all members of the genus (and which will appear in the generic description), may be shortened by omitting subgeneric or sectional characters if a synopsis of subgenera and sections is interpolated in the enumeration of species. The order will follow that given under "original descriptions" above, but certain parts may be left out. Except for the size of the plant, measurements should usually be given only when they are of diagnostic value—an economy which greatly shortens the time required to prepare the description. The shorter the description is, the more difficult it is to write and the greater the skill required of the taxonomist. The very short description is virtually a diagnosis.

As an example of the art of short descriptions, the reader is advised to study J. D. Hooker's *The Student's Flora of the British Isles* (ed. 3, 1884). Combined with synoptic headings, Hooker's diagnostic descriptions serve the purpose of a key. (Hooker considered that keys to species "promote very superficial habits among students"!)

Before leaving the subject of plant description, a general point requires consideration. It is very desirable that terminology should be applied as consistently as possible. Taxonomists must bear in mind that terms have changed their meaning with time; they are used in different senses by different botanists and even by the same botanist in different taxonomic groups (as with indumentum). Many terms are synonymous, overlap with others or are unnecessary. Until we get a glossary of descriptive terms that are internationally agreed upon (if that is possible), we must allow for some inconsistency in the use of terms. The important thing to aim at, at present, is the consistent use of the same term within the same taxonomic group, particularly within the genus.

CITATION

Authority and reference (place and date of publication). A plant's name, to be complete, must be followed by the correctly cited authority, usually printed in a different type from that of the name. The author's name (authority), whether abbreviated or not, should be separated from the reference by a comma, e.g. *Rumex undulatus* Schrank, Baier. Fl. **1**: 625 (1789). The date of publication (which must be accurately ascertained) is usually cited in brackets. It is recommended in the Code (Art. 46D) that the preposition "in" should now be restricted to cases when a name with a description supplied by one author is published in a work by another author, e.g. *Digitalis leucophaea* Sibth. in Smith, Fl. Graecae Prodr. **1**: 439 (1809). It is also frequently used for names published in journals instead of independent books, e.g. *Achania stylosa* Schrank in Flora (Regensburg) **2**: 449 (1819), a practice which has much to recommend it because it prevents confusion between references to books and journals. Schrank, for example, was the author of a *Baiersche Flora* and a *Flora Monacensis* as well as articles in the periodical *Flora*.

Much space would be wasted by citing every book and periodical by its full title: hence botanists generally abbreviate such titles unless they consist of a single word, e.g. *Darwiniana*, *Rhodora*, *Watsonia*, etc. Unfortunately there is no international agreement on abbreviations. Thus under one system *Bulletin of the Bernice P. Bishop Museum of Polynesian Ethnology and Natural History* is abbreviated to *Bishop Museum Bull.*, under another to *Bull. Bishop Mus., Honolulu.* Following the *World List of Scientific Periodicals*

(3rd ed., 1952), the word *Journal* is often abbreviated to the letter *J.*, e.g. *J. Jap. Bot.* for *Journal of Japanese Botany*, and *Jahrbuch* or *Jahrbücher* to *Jb.* e.g. *Bot. Jb.* for *Botanische Jahrbücher*, also cited as *Engler*, *Bot. Jahrb.* and *Bot. Jahrb. Engler*. A partial list is given by Lawrence (1951: 307-317). More comprehensive is L. Schwarten and H. W. Rickett's "Abbreviations of titles of serials cited by botanists" in *Bull. Torrey Bot. Club*, **85**: 277-300 (1958), **88**: 1-10 (1961) which lists the abbreviations used in indexes to American botanical literature in the *Bulletin of the Torrey Botanical Club*. Some of the abbreviations used here, e.g. *Va.* for *Virginia*, *La.* for *Louisiana*, are not, however, likely to be readily understood outside the United States; in general, such geographical names should be abbreviated as little as possible, because periodicals are usually filed geographically in libraries. *America* or *American* is better abbreviated to *Amer.* than to *Am.*, which is very like *Ann.* for *Annals*. Very short words such as *Acta* and *Actes* should be left unchanged, not abbreviated to *Act.* or *A.* for example.

When preparing a Flora it is well to compile at the beginning a standard list of abbreviations based, for example, on the *World List* or Schwarten and Rickett's list. When, however, a revision is intended for publication in a given journal, reference should be made to its instructions for contributors, or its editor should be consulted about its standard procedure when not evident from study of the journal itself.

It is now common practice for volume numbers to be published in bold type, parts in arabic numerals in parenthesis,* and the page or illustration in arabic numerals cited immediately after a colon. The volume may be preceded by a series number cited as a large roman numeral, or as a small arabic numeral after "ser.". To take an example: ser. 2, **4** (2): 47, f.3 (1892). If the whole reference is in brackets, it is often convenient to place the date after a semicolon to avoid double brackets. A "Guide to the citation of botanical literature" will be found in the Code (1961: 66-68) based on the editorial procedure of the New York Botanical Garden.

For both nomenclatural and bibliographical reasons the date of publication of a name should be stated correctly; indeed all citations should be dated. This is not as simple as it seems. The date on the title-page of a book is not necessarily the date on which it was actually issued. Most of the book may have been published in parts before or after the date given. Moreover the date, though correct, may not be precise enough; often two names for the same plant have been published in the same year, sometimes in the same month and in a few instances on the same day! To ascertain the precise date of publication of an old work is often a very difficult task, necessitating reference to contemporary reviews, to library accession lists, to publishers' advertisements, etc. and demanding much time in the search for information and experience in assessing its validity. A useful rule is that three pieces of evidence should lead to the same conclusion. Fortunately the dates of publica-

* It is unnecessary to cite parts if the volume shows a single serial pagination.

tion of many botanical works have been investigated, but as stated in *J. Soc. Bibl. Nat. Hist.* **4**: 1 (1962) "bibliographical information is so widely scattered, and may occur in places so little expected, that systematic search for data relating to a particular problem can lead to much loss of time and small success; it is all too easy to overlook relevant material". Catalogues of such papers will be found in *J. Soc. Bibl. Nat. Hist.* **1**: 1-30 (1936), **2**: 1-17 (1943), **3**: 5-12 (1953), 165-174 (1957), **4**: 1-19 (1962), *Flora Malesiana* I, **4**: clxiii-ccxix (1954), Blunt & Sitwell's *Great Flower Books* (1956), and *Taxon*, **10**: 70-91, 111-121, 132-149 (1961).

Citation of specimens. We have already considered the choice of specimens for citation (p. 298). If the whole field data are cited, they are preferably placed in the following order: country, province, locality, altitude, habitat (or community), observations (e.g. flower colour), date, collector's name and number. In a Flora this layout, which would be desirable for the description of a new species, must be cut down, so that often only the locality (preferably with the altitude) and collector's name and number are cited; the other information is synthesised elsewhere in the account of the species. The author may either state that all specimens cited have been examined unless otherwise indicated (generally by "n.v."—*non vidi*) or that specimens examined are followed by an exclamation mark.

In the citation of *type specimens* it is probably best to cite the specimen(s) as originally published by the author, though additional information written on the sheet may be interpolated in brackets. After the collector's number, the taxonomist should cite the standard abbreviation, as given in Lanjouw & Stafleu, *Index Herbariorum*, part I (4th ed., 1959), for the herbarium in which the material exists, making it clear whether it has been examined or not (e.g. holo. K, iso. E! means that the holotype is at Kew, and an isotype in Edinburgh, but that only the latter has been seen).

Meticulous care in the citation of type specimens, and the pains taken to examine them, will be rewarded by greater stability of nomenclature, and makes readily available a piece of information which is fundamental to further research. Furthermore, it makes clear how much original work the author has done, and how much he has taken on trust.

Mapping

The distribution of taxa, particularly at the genus and species level, has come to play an increasing part not only in plant geography but also in descriptive taxonomy and in evolutionary interpretation. In this section we shall deal only with the mapping of taxa; character mapping is dealt with under the section "Geographical-morphological Method" on p. 313.

An excellent essay on species mapping, from which our account is largely drawn, has been given by Stearn (1951); the reader should turn to this for more detailed information and additional references. The symposium, *The cartographical Presentation of biological Distributions* (Royal Geographical Society, 1954), will also be found of value.

Constructing any map depends on finding where the taxa occur, and th s necessitates being quite clear about the morphological units. It is very im portant that maps should be based on correctly identified material. Published records, unsupported by material examined, are accepted at the cartographer's risk. If the group is a critical one, or the records have not been verified by a reliable monographer, such published records are best ignored. On the other hand, in mapping well-known and distinctive species, it may be worth including important records of unexamined material; this becomes particularly tempting in common plants (including trees), since they are often very poorly represented in herbaria. Remarkable errors in identification can, however, be made; in Turkey, for instance, the distribution of even *Cedrus libani* and *Picea orientalis* has been confused by early workers!

How a taxon is mapped depends on how much we know about it and how much needs to be put on paper. We can consider mapping procedures under three main headings, since other methods are largely variations or combinations of these.

(1) *Blocking in of specially delimited areas*. This was the method adopted in most of the maps published before 1957 in the *Biological Flora of the British Isles* (*The Journal of Ecology*), the areas being the vice-county divisions adopted by Druce in his *Comital Flora of the British Isles* (1932). The main objection is that the area of the species is exaggerated; we cannot tell whether a plant is rare or common. This may be improved by only blacking in the areas where the species is common, and shading them where it is scarce.

The current refinement of this type of method is the grid map adopted in *The Atlas of the British Flora* (edit. Walters & Perring, 1962). The basis of the scheme is to indicate by means of a dot (or circle for pre-1930 records) the presence of each species in every 10-kilometre square of the National Grid in which it occurs. Because of the use of modern data-processing equipment, which includes rapid mechanical map-production, it is possible (on request) to trace the source of the records mapped. Naturally the accuracy of grid methods will increase, the smaller the grid. They can only be used effectively in countries which are floristically very well worked, and are probably more suitable for phytogeographical than taxonomic studies, since the most closely related species do not necessarily have a distribution which can be shown on the same map.

(2) *The dot method.* Here each station is indicated by a dot. In contrast to the first method we have described, the dot method is likely to underestimate the distribution. As Stearn puts it: "The appeal of a dot map lies in its honesty. It makes clear how much the author knows; it may also indicate how much remains to be learned." A concentration of dots may well represent over-collecting, absence of dots under-collecting, but this effect is lessened if care is taken to plot localities rather than specimens. For a species of limited

occurrence the dot method is the most suitable; for species which occur abundantly, hatching may be preferred. For plants which are common in some areas, rare in others, a combination of the two methods is preferable, as we find effectively employed in Hultén's *Atlas of the Distribution of Vascular Plants in N.W. Europe* (1950). Various refinements can be introduced in the dot method: a special mark for the type locality, or for localities where the species is extinct (rings or crosses); large dots may be used for plants which have been cytologically examined, small dots for those whose chromosome number is not known. Different symbols are often used for mapping different chromosome numbers (see maps in Böcher *et al.*, 1953; Jones *et al.*, 1961).

Pictorialised dot maps may also be constructed, so that information is given about the geographical variation of taxa. White (1962) has produced distributional maps of *Diospyros* species in Africa with drawings of the variable organs (tied up with particular dots) given round the margin of the map, together with maximum, mean and minimum values for each organ on all the material of each taxon examined (Fig. 18). Hedge (1960) has mapped the variation in leaf shape in *Salvia candidissima* in Turkey by plotting the length/breadth ratios on the map. It is also possible to "pictorialise" the dots in the same way as is often done on scatter diagrams*, so that the character combinations are indicated for each locality (Fig. 19). This method may show up possible allopatric introgression or topoclines, but can only be a rough guide, since local population variability cannot be adequately shown on the map†.

Any map attempts to express three dimensions in two. The information presented by the map may be greatly increased if ground above a certain altitude is stippled—corresponding, say, to the tree line. Distributional gaps may then be clearly seen to be correlated with altitudinal preferences.

In all cartographical and graphical presentation, care must be taken not to obscure the picture by too much detail.

(3) *Boundary line method.* In many monographs the range is indicated by boundary lines. This can be very useful when several overlapping ranges need to be presented on the same map, and when the area is large and the scale of the map consequently small. By using different kinds of boundary line (e.g.: - - - - - - - -; · · · · · · · · ·; + + + + + + + +; ————————; etc.), it is easy to

* The method of producing these "pictorialised" dots (metroglyphs) is described on p. 478.

† The latter can be shown by means of "pie" diagrams (extensively used in pollen analysis maps) to portray the percentage of plants possessing a particular character in a population. A simple example is given in New (1958: 461) in which "pies" are drawn on the map to show the proportion of sculptured to non-sculptured seeds in populations of *Spergula arvensis*. The proportional occurrence of a particular character is blacked-in on the disc in the manner shown in Fig. 14 (p. 166). Up to four characters, however, can be clearly shown simultaneously on a "pie" by assigning each quarter of the disc to a different character; the percentage occurrence is then blacked-in for each segment instead of for the whole disc.

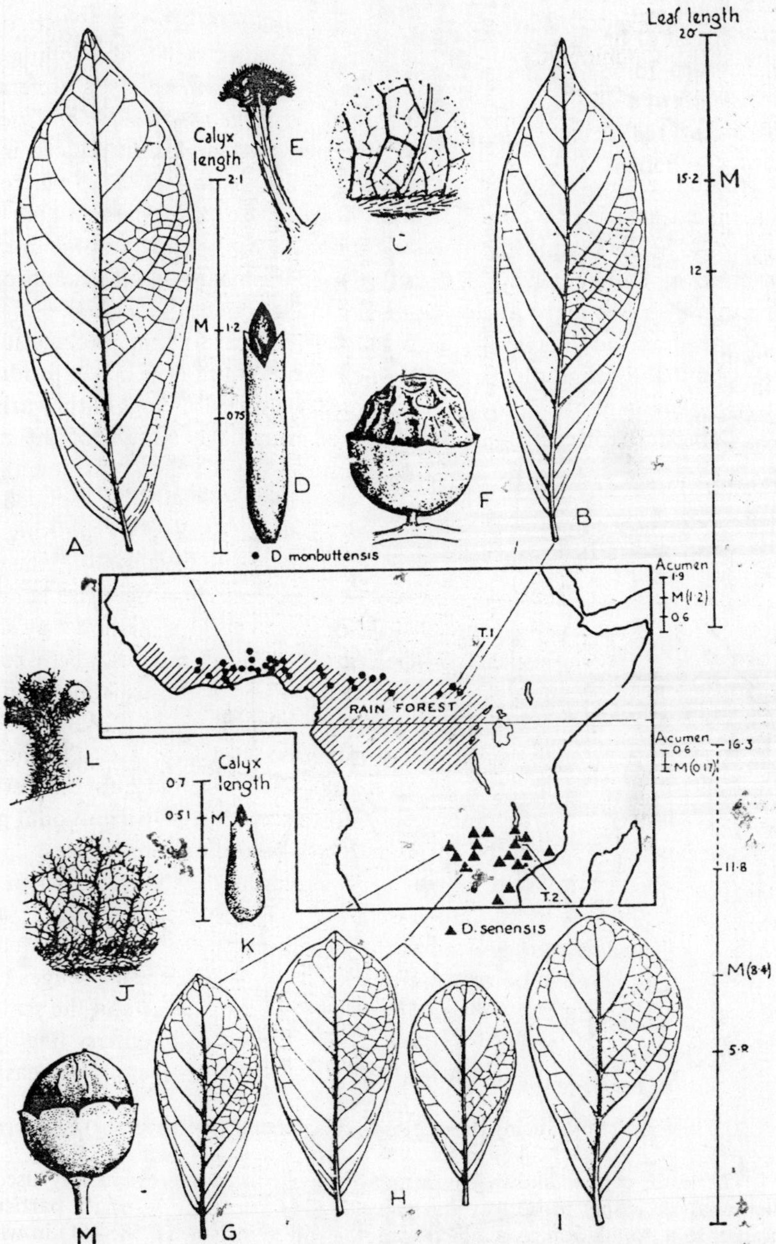

FIG. 18. Pictorialised distribution map of *Diospyros monbuttensis* and *D. senensis*, showing main similarities and differences, and maximum, minimum and mean lengths of calyx, acumen and leaf on all material studied. The organs illustrated belong to specified herbarium specimens. (From White, 1962.)

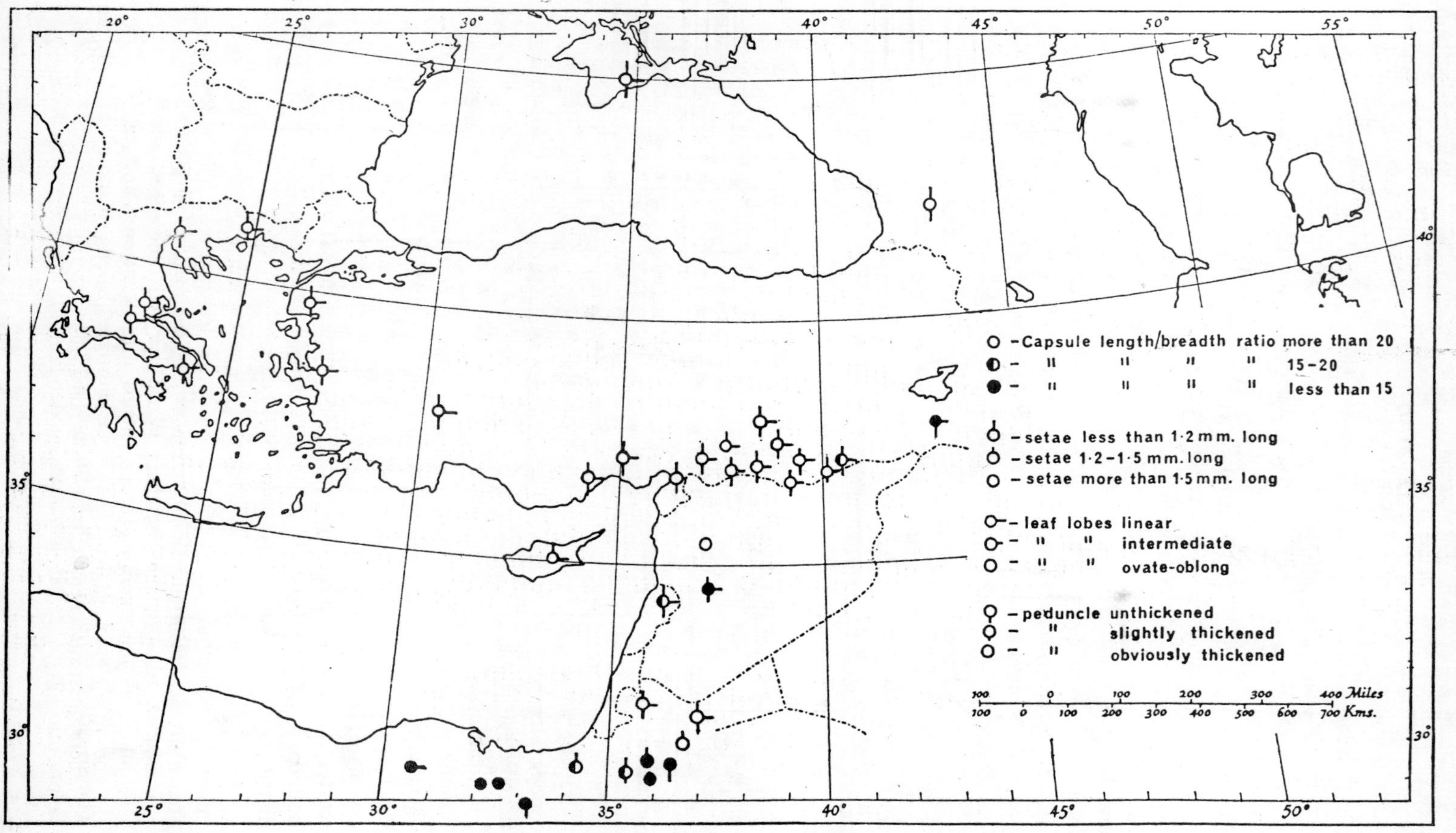

FIG. 19. Pictorialised dot map of *Roemeria hybrida*, based on herbarium material and showing intermediates between subsp. *hybrida* and subsp. *dodecandra*. Typical subsp. *hybrida*

chart several species on one map. It is convenient to place the same numbers that are used in the enumeration of species against the boundary lines of those species on the map. This enables the reader to turn straight from the text to the map, instead of having to study a long caption first, and in many cases will do away with the necessity of using different kinds of boundary line.

As Stearn remarks, "The disadvantage of this boundary line method is that it gives no information of the occurence of species within the area enclosed. It tells the systematist what species are to be expected from a certain area but it does not tell him from what localities they are known; it gives no indication whether a species is evenly spread over the area or occurs only at isolated localities. Moreover, the drawing of the boundary line itself is essentially subjective. A great variety of patterns can be made by drawing a line through or round the outermost of a number of scattered dots representing the localities from which a species is known." Just how arbitrary this can be is shown in Fig. 20. The smoother the outline, the more it pleases the eye

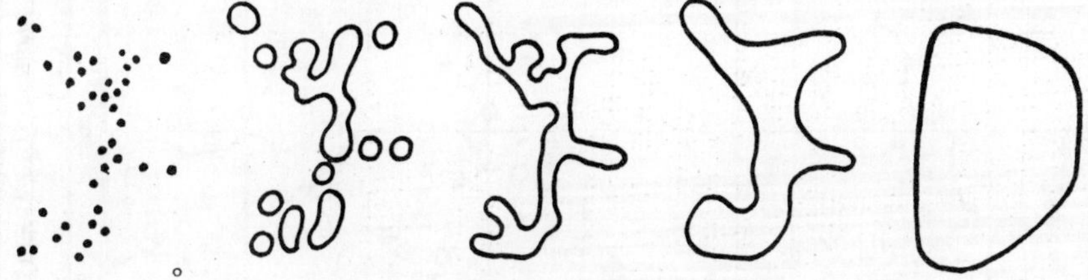

FIG. 20. Relationship between spot localities and outline areas. (After Stearn, 1951.)

—but the more it departs from the facts. In many cases a consideration of topography (e.g. alpine/lowland species) or ecological preferences will help to indicate how the boundary line may best be drawn.

An effective compromise can be achieved by combining dots with boundary lines (cf. maps in F. C. Stern's *Study of the Genus Paeonia*, 1946). The method has the advantage of combining the accuracy of the dot method (so far as known localities are concerned), with the ease of observation associated with a boundary line which focuses attention when the dots are scattered or several species are plotted on the same map (Fig. 21). Used alone, boundary lines are generally of more scientific use for generic or sectional distributions than for the mapping of species.

Advice on the type of projection to use (which varies with the area mapped) and techniques employed in distribution mapping will be found in Stearn's review (1951).

THE GEOGRAPHICAL-MORPHOLOGICAL METHOD

The so-called geographical-morphological method, entrenched in German taxonomic literature, is a natural extension of the morphogeographical

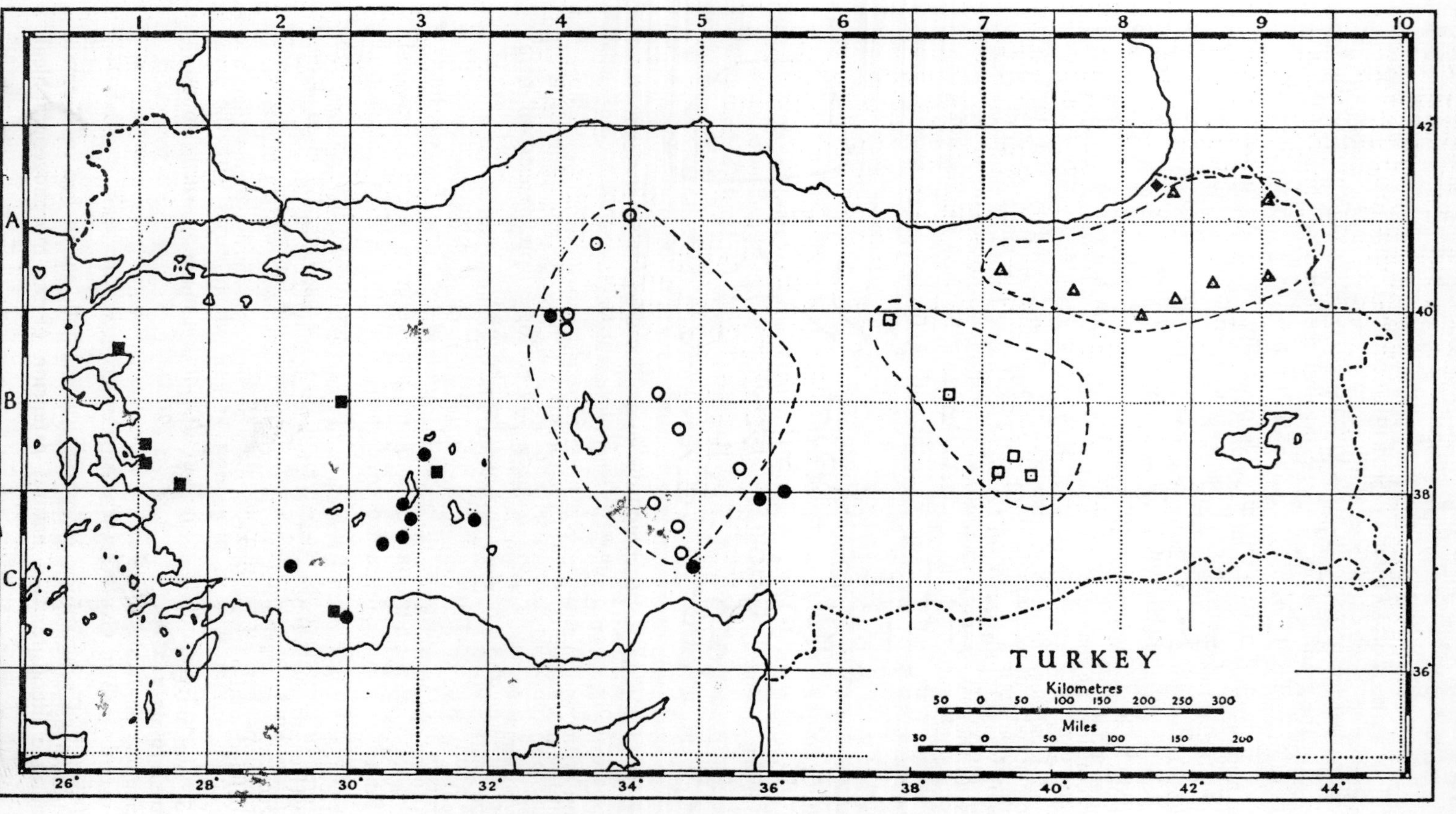

FIG. 21. Distribution in Turkey of *Saponaria* Sect. *Silenoides*.

Key to symbols: ■—*S. chlorifolia*. ●—*kotschyi*. ○—*S. prostrata* subsp. *prostrata*. □—*S. prostrata* subsp. *anatolica*. △—*S. prostrata* subsp. *calvertii*. ◆—*S. cerastioides*. (After Hedge, 1961b.) The map shows a grid used for citation of specimens in a Flora of Turkey.

species concept that was firmly established by A. Kerner, Korschinsky and Bunge. The method is based on the fact that morphology and area are the expression of a taxon in nature and are an outcome of its history. Like Willis's *Theory of Age and Area* (1922), it involves a space-time correlation.

The approach was first used in practice in Bunge's monograph of *Acantholimon* (1872). In this remarkable work he mapped character variation by the ingenious use of different coloured type for the names of species on his distributional map. On the basis of morphological and geographical considerations, Bunge recognised three main evolutionary lines within the genus and, by considering character correlations, distinguished between resemblance due to ancestral relationship and that due to convergence. This Russian taxonomist, who monographed some of the most difficult genera (*Echinops, Cousinia*), was certainly quick on Darwin's trail, and helped to establish the interest in the geographical aspects of taxonomy that has remained a strong feature of the Leningrad school of systematics.

The geographical-morphological method was first clearly expressed, however, by Wettstein (1898) in his book *Grundzüge der geographisch-morphologischen Methode der Pflanzensystematik*, and put into practice in his monograph of *Gentiana* Sect. *Endotricha* in Europe (1896). His method is based on the knowledge that species are adapted and confined to particular environmental conditions.* Wettstein claims that species of common origin, of recent formation, frequently inhabit contiguous, although perfectly distinct, areas; if they are found in areas that are far separated, or in the same area, their affinity is less. The method is thus descriptive and interpretive. It enables us to enquire into the evolutionary history of the group most successfully, of course, when the climatic and geological history of the religion is known. Wettstein (1935) considered the method, although only applicable to the lower categories, to be relatively objective—a claim which many botanists would question.

The method is primarily interpretive; the taxa are usually recognised before it is applied, and a study of their distribution allows inferences on their origin to be drawn. It may, however, provide evidence for deciding whether to apply specific or subspecific rank to a particular situation. Wettstein himself, though recognising the concept of geographical subspecies, did not give them trinomials—that was a policy adopted later in the *Rassenkreis* theory (Rensch, 1929) that has penetrated zoological taxonomy far and wide. The geographical-morphological method has been much written about but little-defined, and appears to be more a way of considering plants than a

* S. A. Cain (1944) points out that Wettstein adopted a neo-Lamarckian hypothesis in his revision of *Gentiana* Sect. *Endotricha*. His maps show the species replacing one another geographically or altitudinally, and he maintained that they arose over the whole area of the ancestral population in direct response to habitat influences—a view taken by his father-in-law, Kerner, but later rejected in favour of hybridisation as a major factor in the build-up of new species.

taxonomic method. This elusive pimpernel is, in fact, built into the corpus of modern taxonomic procedure. When we consider the distribution of taxa, particularly for evidence of their probable evolutionary history, we are adopting some aspect of the geographical-morphological method. The concept is an integral part of Babcock's *The Genus Crepis* (1947), and of numerous other monographs and revisions (e.g. Hedberg, 1957).

Despite the amorphism, or rather polymorphism, of the "method", certain techniques of distribution mapping associated with it and devised

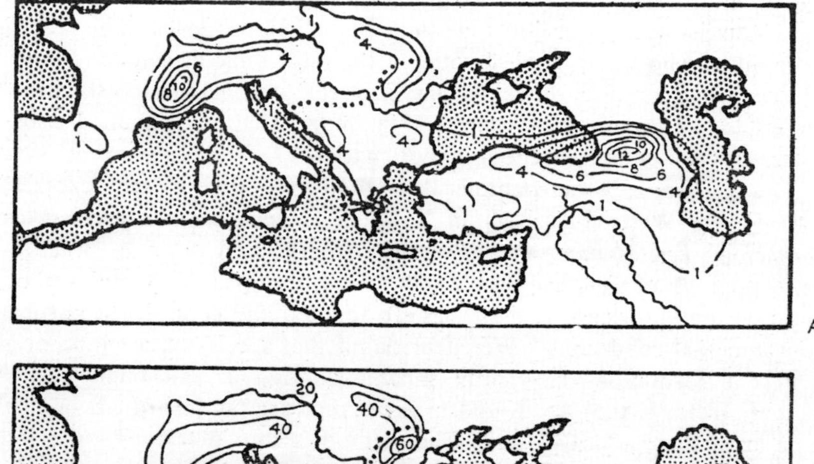

FIG. 22. *Isoflors* (*A*) and *isochars* (*B*) for the species of *Alchemilla* Subsect. *Calycanthum*. Explanation in text. (After Rothmaler, 1955.)

during the last twenty-five years are so valuable both for factual presentation and for interpretation that we give them in some detail here. The first two techniques are described by Rothmaler (1955), who gives the most comprehensive account of the geographical-morphological method in its modern form.

The first procedure involves the drawing of *isoflors* (called *Isoporien* by Rothmaler), which enclose areas with the same number of species within a higher taxonomic group (e.g. section or genus). The strongest species concentration is found in one or more focal points. The area where the largest diversification of the group occurs is known as its centre of diversity (cf.

S. A. Cain, 1944). In the construction of isoflors, it is not the distribution of the individual species that is important, but only the number of species occurring together at any one point. In order to produce the isoflor map, a grid is drawn on the map and the number of taxa noted which occur in each quadrat. We are thus able to draw contours linking quadrats with the same taxon-frequency. An isoflor map is exemplified by Rothmaler's map of *Alchemilla* Subsect. *Calycanthum* which we reproduce in Fig. 22A.

The second procedure involves the drawing of *Isopsepheren*, a term which we can anglicise to *isochars* (Fig. 22B)*. These are based on the largest possible number of contrasting character expressions found in the taxa being studied. Here, again, a particular number of characters falls to the share of each quadrat of the map, and those with the same number of characters are joined together by lines—isochars. This map of character distributions shows still more clearly where the centre of the greatest diversity of the group occurs. In these character-centres a particularly strong development of the group has taken place, so that we can term them development centres. Rothmaler claims that primary and secondary development centres can be distinguished, although he admits that the primary centre need not be the same as the centre of the origin of the group; the terms are relative.

By studying together both isoflor and isochar maps, we can see where the largest number of species and the highest number of characters occur; where these coincide (if they do) is the primary development centre. If the concentration of both values in another area is less (though more or less proportional to the values in the primary centre), then this is a secondary development centre. A large number of species with a low number of characters indicates an area of secondary development and the recent formation of numerous, weakly differentiated species. If, on the other hand, a large number of characters is associated with a small number of species, then we may assume that in this area very heterogeneous elements, perhaps relicts of different origin, occur together.

Applying this interpretation to the distributional maps of *Alchemilla* (Fig. 22A and B), Rothmaler comes to the following conclusions: the Caucasus, with 12 species and 80 characters, has been the primary centre of development; the Alps, with 10 species and 40 characters, have been the secondary centre of development; and the Balkans, having 4 species but 60 characters, are a relict area of heterogeneous elements. The distributional facts must of course be interpreted in the light of climatological and geological history. We must also bear in mind that some areas have been better collected, or their flora more intensively worked, than others—a consideration relevant to *Alchemilla*.

The fact that such methods of interpretation have been less used in N.W. Europe and America than in Germany appears to be due to the widespread reluctance (ably justified by S. A. Cain, 1944) to equate centre of diversity with centre of origin, particularly in areas strongly effected by the Pleistocene

* We are indebted to Dr. W. T. Stearn for suggesting this term.

glaciation—a limitation frankly admitted by Rothmaler. Furthermore, the methods of mapping are laborious and only confirm what is often apparent to a thoughtful monographer without them. The main object of methods of presentation, however, is to present data to other people, not to ourselves, so that any clear way of doing it is to be encouraged. The field deserves further exploration, but the evolutionary interpretation of isoflors and isochars obviously needs considerable caution.

In order to show the distribution of a particular character, an area can be gridded and for each quadrat the percentage of specimens showing a certain character can be noted. Contours, known as *Isosemen* (isophenes of Huxley) are drawn through quadrats with equal numbers. The procedure was used by Pearson (1938) to demonstrate and interpret the distribution of melanism in the Tasmanian bush opossum. The distributional pattern which emerged was largely explained in historical terms.

Woodson (1947) enlarged this concept in his classical study of the polymorphic North American *Asclepias tuberosa*. By normal taxonomic methods he was able to recognise three overlapping subspecies (found to be interfertile) differing from one another mainly in the form of their leaves. As part of his study of leaf shape, Woodson plotted average leaf angles (leaf apex and leaf base) for each quadrat and drew lines through squares of the same average angles; these lines were called *phenocontours*. He then plotted the variability for each quadrat, in terms of standard deviation, and drew phenocontours through these. It was found that the maximum variability occurred in the area where two subspecies came in contact; in this belt (approximate to the southern extension of the Pleistocene ice sheet) an increase in leaf size was interpreted as an indication of hybrid vigour. Woodson's studies show how effectively biometrical studies can be applied to abundant "normal" herbarium material (supplemented by population studies and crossing experiments), and provides the most impressive example of phenocontour mapping which successfully clarifies a taxonomically difficult situation. As he is not concerned with centres of origin, Woodson's results do not suffer from the hazards that may accompany the interpretation of isoflors and isochars.

A recent refinement of character mapping has been devised by Ehrendorfer (1958) and used for his studies on the *Galium canum—G. graecum* complex in the Levant. The contours (Fig. 23) are primarily based on normal herbarium material, although he has superimposed results from population studies upon them. The procedure is particularly applicable to cases where one taxon grades into another over a large area, as is the case with allopatric introgression.

The stages in the preparation of such a map are illustrated by a simple hypothetical example in Fig. 24a-d.

(*a*) A spot map is drawn up of all specimens under consideration. (Fig. 24a).

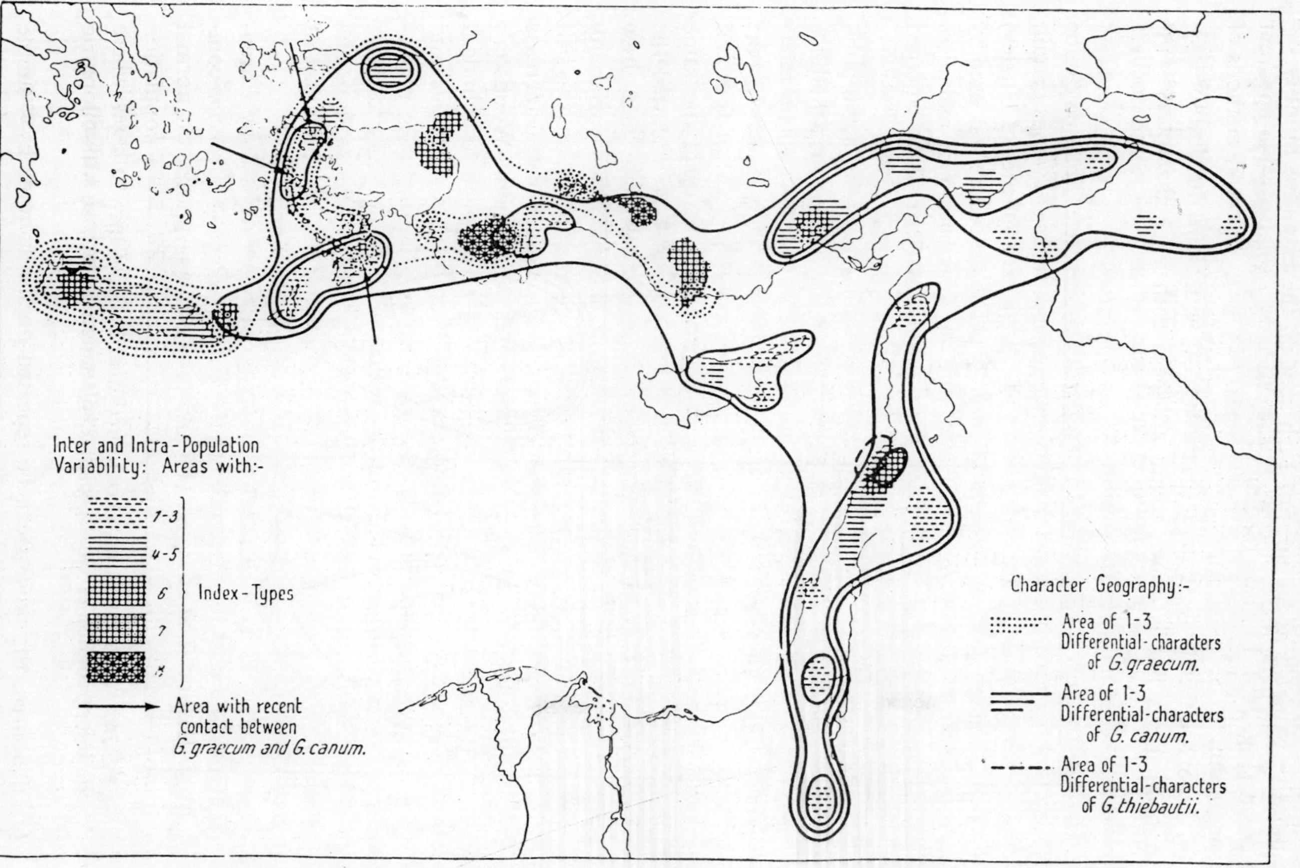

FIG. 23. Distributional map showing the intergradation and variability in the *Galium graecum*—*G. canum* complex in the Levant. Explanation in text. (After Ehrendorfer, 1958.)

(*b*) For each specimen the number of diagnostic character expressions of one putative parent is added up. For each spot on the map, this number is put down (Fig. 24b). For easy reading of the map, four characters are about the maximum.

(*c*) Lines are then drawn round: (*a*) those spots with all (n) the characters; (*b*) outside this a line is drawn round all spots with $n-1$ characters; (*c*) and so on for $n-2$, etc. (Fig. 24c).

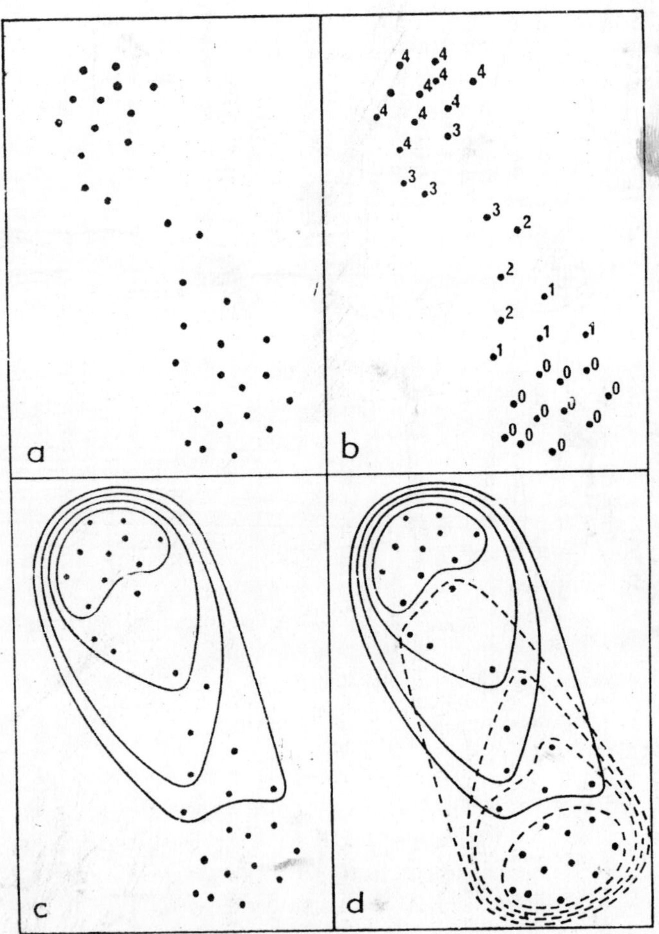

FIG. 24. Preparation of a contoured distributional map to show the intergradation of two species. Explanation in text. (J. Cullen.)

(*d*) If desired, the process can be repeated on the same map, using the characters of the other putative parent, the contours being drawn in dotted or broken lines (Fig. 24d). Reference should be made to Ehrendorfer's paper

(1958) for details of this and the following procedure superimposed upon it.

If these lines are thought of as contours, then the high points represent the pure species; the slopes represent the declining influence of the species in terms of numbers of characters.

Ehrendorfer has superimposed upon this map the results of population studies based on a hybrid index (cf. p. 479). For each population sample he calculated the hybrid index value for every plant (on an $0-8$ scale and using the same characters that he employed in the contour maps). In order to do this, he worked out for each character the number of index values (1, 2, or 3, etc.). Then if n_a is the number of index values for character A, and n_b for character B etc., then his index value is formed by: $1+(n_a-1)+(n_b-1)+(n_c-1)\ldots$ Reference should be made to the example given in Ehrendorfer's paper: 1958, p. 238. By using different cross-hatchings, these different index values are superimposed on the contour map (somewhat at the sacrifice of visual clarity) to show the areas of least and greatest variability (Fig. 31). The same result could be produced by plotting hybrid numbers (Gay, 1960; cf. p. 479) on the map and drawing contours round those.

Ehrendorfer's method is ingenious and suggestive. In the case of *Galium*, its interpretation in terms of an ancient pattern of introgression resulting in a "fossil" hybrid complex seems extremely probable. It remains, however, a probability which cyto-genetic investigation may help to substantiate. In such cases the possibility of divergence from a variable centre must always be borne in mind.

Although in this section we have stressed the newer procedures of distribution mapping (and particularly character mapping), it should not be forgotten that accurate portrayal of the distribution of related species (and higher groups), as described on p. 309 is a fundamental part of the geographical-morphological method. The newer techniques may be considered as elaborations and extensions of the basic mapping of individual taxa.

Illustrations

It is obviously desirable that, wherever possible, species should be illustrated, especially when they are first described. In a monograph the aim may be to illustrate fully every species (as in Babcock's *The Genus Crepis*), whereas a Flora may have to content itself (if illustrated at all) with drawings of critical parts.

Black and white, pen-and-ink drawing is certainly the most generally satisfactory method of illustration at reasonable cost. The technique can only be acquired with practice and skill, and we cannot do more than mention a few important points here.

Drawings should be larger than the size they will appear in print; it is usually satisfactory to allow for a reduction by $\frac{1}{3}$. If they are larger, irregularities of line appear on the printed page, whereas if they are further reduced, fine detail tends to get lost. The thickness of line and the type of shading used (if any) need to be carefully considered in relation to the amount of reduction employed in the preparation of the block. The scale of the figure should be precisely indicated, and, if possible, the specimen on which it is based.

Useful books on the subject are J. L. Ridgway's *Scientific Illustration* (1938), W. Kuhl's *Das wissenschaftliche Zeichnen in der Biologie und Medizin* (1949), H. Graham Cannon's *A Method of illustration for zoological papers* (1936) and L.N. Staniland's *The Principles of Line Illustration* (1952). The reader should also turn to the excellent summary of this skill in Mayr *et al.* (1953, Ch. 9), and read W. H. Fitch's articles on botanical drawing reprinted in Wilfred Blunt's admirable book, *The Art of Botanical Illustration* (1950).* He may learn most, however, by studying the work of the masters of plant illustration. For accuracy of detail, the black-and-white drawings in Baillon's *Histoire des Plantes* (1866-95), Sargent's *The Silva of North America* (1891-1962), recent volumes of *Hooker's Icones Plantarum*, S. Ross-Craig's *Drawings of British Plants* (1948-), and Mason's *A Flora of the Marshes of California* (1957) are unsurpassed.

Photographs of type specimens serve the useful purpose of indicating the habit of the plant (often so difficult to convey in words) and vegetative form, though floral detail is seldom adequate unless dissections are made and separately enlarged. It is possible to supplement photographs of specimens by pen-and-ink drawings of floral structure (with a scale attached); these drawings may even be photographed on the sheet together with the whole plant e.g. as in *J. Linn. Soc. London, Bot.* **51**: tt. 26, 31 (1938), *Bull. Brit. Mus. Nat. Hist., Bot.* 2: t. 12 (1960). Heslop-Harrison (1954) has published photographs of labellum mounts for populations of *Dactylorchis* which give a very good impression of the variability in the form and marking of the labellum in natural populations.

The use of X-ray photography for studying and presenting the details of plant morphology in herbarium material is described by Slepyan & Kishkovsky (1960a, b) and Nitzelius (1963).

Tabulation, Symbolic and Graphical Methods

Comparative data are of course the basis of visual methods of presenting variation patterns. In population analysis, where large random samples are available to the taxonomist, statistical treatment is an essential tool (cf. Chapter 12). On normal herbarium material, which does not represent a random sample and suffers from various limitations, statistical methods are of more limited application. We may quote Mayr *et al.* (1953: 153) on their

* A simple, easily made aid to botanical drawing is described by E. H. Walker in *Phytologia*, 1: 165-166 (1935).

use: "In practice, taxonomists should realise that there is no particular virtue in statistics *per se*. Statistical analysis is one of several tools which are available and which may or may not be used to advantage in a particular problem. Taxonomic studies at the alpha level utilise in general only the simplest statistics, e.g. size, including range, if several specimens are available, and simple proportions or ratios. Population analyses are impossible because there is no uniformity of sampling. Furthermore, comparisons between intra-specific populations to determine overlap cannot be made as long as the diagnostic differences of the species and their delimitations are still uncertain."

We cannot get more out of statistics than we put into them. When applied to herbarium taxonomy any fairly elaborate statistical treatment should only be used when a preliminary survey suggests that such treatment is going to serve a useful purpose. It is possible, for instance, to substantiate and confirm subjective taxonomic judgment by statistical means and to make the basis for that judgment clearly apparent to others who have not studied the material. The majority of taxonomists, however, are not statisticians. The visual perception which is an important part of taxonomic methodology means that graphical methods of expressing statistical data, or the use of symbols, are more readily appreciated by most taxonomists than a full statistical treatment—even if the latter were possible. The main purpose of this section is to outline the most useful visual methods of presenting qualitative and quantitive data.

(1) *Tabulation*. Before any morphological variation can be presented visually, it must be tabulated. The usual procedure is to list the herbarium sheets (citing the locality and collector's number), and to tabulate the measurements (or index value) of each significant character for each sheet, usually in vertical columns. The range of measurements may be divided into arbitrary categories, as in the construction of a hybrid index (see page 479). Punched cards may also be used, but need to be specially prepared for each genus.*

It is of course important that characters should be measured in a logical and consistent manner. Hedberg (1961, fig. 1) illustrates the way in which he measured leaves and flowers in *Canarina;* Woodson (1947, fig. 3) shows how the apical taper and the angles at the leaf base can be measured in *Asclepias tuberosa*. An index of labellum shape has been devised by Heslop-Harrison (1951) to distinguish between critical species of *Dactylorchis*. A Leitz eyepiece micrometer, with a scale calibrated in tenths of a millimetre, will be found extremely useful in measuring indumentum and seeds.

Tabulation is a basic taxonomic procedure, and can be utilised when drawing up keys and descriptions, and even for the presentation of raw data

* Punched cards have a wide application as a taxonomic tool, both for recording data and storing information (cf. Phillips, 1941, 1948; Forest Products Research Bulletin, 1952).

as well as for more sophisticated methods of presentation. In the accompanying table of *Erysimum linifolium* in Spain, the specimens are arranged geographically, leading to the recognition of subspecies and varieties (Heywood, 1954b). For leaf-width, four arbitrary categories have been chosen:

N=0·5-1 mm. n=1-1·5 mm. b=1·5-2 mm. B=2-..(4) mm.

Variation in *Erysimum linifolium*

Collection		*Character*						
		Height in cm.	*Basal leaves*	*Leaf-width*	*Flower-length in mm.*	*Petal-limb length in mm.*	*Sepal-length in mm.*	*Style-length in mm.*
1. Orense	*Winkler*	23	toothed	b–B	13–14	7–8	7	2–2·5 (imm.)
2. Asturias	*Durieu*	25–35	toothed	n–b	14	8–10	5–8	1·5–2
3. Asturias	*Bourgeau*	*c.* 30	± toothed	n–b	14–16	10–11	7	2–2·5
4. Asturias	*Levier*	30–35	± toothed	n–b	15	7–9	7	2–2·5
5. Asturias	*Lacaita*	45–50	toothed	n–b	15–16	9–10	7	2–2·5
6. Salamanca	*Lacaita*	20–30	toothed	N–n	12–15	7–9	6–8	2–2·5 (imm.)
7. Salamanca	*Lacaita*	15–30	± toothed	n	14–15	8–10	7	2–3
8. S. Gredos	*Graells*	*8–16*	± toothed	b (–B)	14–16	10–12	8–9	2–3
9. Toledo	*Bourgeau*	30–35	± toothed	b–B	13–16	8–9	8	2–3
10. Morena	*Joau*	20–30	toothed	b–B	14–15	7–8	7–8	2–3 (imm.)
11. Morena	*Lacaita*	30–40	toothed	n–b	13–16	7–11	7–8	2·5+ (imm.)
12. Jaén	*Ell. & Sand.*	25–35	toothed	*N*	15	8–10	8	*5–6*
13. Jaén	*Lacaita*	15–40	± toothed	N–n	15–18	10–12	8–10	*4–6*
14. Jaén	*Wilmott*	25–35	toothed	N–n	14–15	8–9	7–8	*4–6*
15. Granada	*Ell.& Hubb.*	(15) 25–35	± entire	N–b	14–15	10	7–9	*2–4*
16. Almeria	*St. Lager*	25–35	toothed	*B*	15–17	11–12	9–10	*3–4*
17. Baza	*Bourgeau*	20–25	toothed	b	*18–20*	12–13	8–*11*	*3–4*
			± entire	n–b				
18. Vél.-Rub.	*Ell. & Sand.*	(5)–15	toothed	*B*	14	9–10	9	—
19. Minas	*Lofthouse*	15–30	toothed	n–b	15–16	9–11	7–8	*2–3·5*
20. C. Rayo	*Wil. & Loft.*	10–15	barley toothed	n–B	14–16	8–10	8	*2·5–3·5*
21. Cazorla	*Lacaita*	15–20	*entire*	n–b	15–16	9–11	7–8	*2·5–4+*
22. Cazorla	*Wilmott*	40–80	*entire*	b–B	15–16	8–9	8–9	*5–6·5*
23. Cazorla	*H. & D.: 182*	30–65	*entire*	*N*	14–15	8–9	6–8	*2–4*
24. Cazorla	*H. & D.: 314*	20–30	*entire*	n–b	14–15	8–9	6–8	*2·5–4*

N.B. All measurements have been made from dried material.

All significant variations have been shown in italics. Several features may be observed from the table, including the following topoclinal variation in only one character—style length which increases from north to south; plants from Jaén are distinctive in the very narrow leaves and long styles; the Cazorlan plants, while fitting into the general lines of variation shown by Baetic plants have additional characters which make them quite distinct. No statistical treatment is attempted in such limited samples.

(2) *Formulae.* A method of portraying variation by the use of formulae is exemplified in the same paper by Heywood (1954b) for *Sisymbrella aspera*

(sensu lato). In studying the variation of this species it became evident that the important taxonomic features were as follows:

P = pedicels long, 8-12 mm.	p = pedicels short, 3-8 mm.
S = siliques long, 25-40 mm.	s = siliques short, 15-20 mm.
G = stems ± glabrous.	g = stems to a greater or lesser degree asperous.
U = seeds 1-seriate in each locuius.	u = seeds 2-seriate in each loculus.
F = flowers large, 7-12 mm.	f = flowers small, 5-7 mm.
L = siliques smooth.	l = siliques scabrous, asperous.

Thus the typical formula for subsp. *aspera* is psgufl, and for subsp. *boissieri* PSGuFL*. But, in fact, in a series of specimens the following variations of these formulae resulted:

Subsp. *aspera*

P	s	Gg	u	f	l
P	–	Gg	u	f	l
p	S	g	u	f	l
p	sS	G	u	f	l
p	S	g	u	f	lL
p	s	g	u	F	l
p	s	g	u		lL

Subsp. *boissieri*

Pp	S	gG	u	F	L
p	S	gG	u	f	L
P	S	g	u	F	L
P	S	gG	u	F	L
P	S	g	u	F	Ll
P	S	G	u	F	l

These examples have been taken from Spanish material only. French material of S. *aspera* subsp. *aspera* adheres closely to the formula psgufl, from which it can be claimed that subsp. *boissieri* and *aspera* are distinct separable taxa; although there are transitions in the characters given above, their different combinations serve to distinguish the two subspecies.

Similar symbolic methods have been used by Chaytor & Turrill (1935) to portray variation of the polymorphic *Clypeola jonthlaspi* (*Cruciferae*) and by Ehrendorfer (1951) for *Galium incanum*, a numerical index being used in the latter case.

The hybrid index method (p. 479) may be used to portray variation irrespective of hybridisation, when it is best termed a biometric index.†

(3) *Histograms*. The use of histograms in taxonomy is to express the variation of individual characters on a frequency basis. This may be done for at least three different reasons: (1) to show the variation in individual organs within previously delimited taxa; (2) to show the frequency distribution of characters in different (often geographically isolated) populations; (3) as a means of delimiting taxa which show some morphological overlap in the least arbitrary manner.

Histograms cannot, of course, express the character correlations of individual plants. Good examples of the first and second of these uses will be found in Hedberg's revision of *Canarina* (1961) and of vicarious taxa in the Afro-Alpine flora (1957). The third use of the histogram is particularly

* A variant with uniseriate seeds also occurs, subsp. *pseudoboissieri*: PSgUFL.

† We are grateful to Dr. J. McNeill for this suggestion.

helpful in deciding where best to draw the line between two species or sub-species where some overlap in the range of characters occurs. By the use of blocking-in and cross-hatching it is possible to present two or more histograms on the same base line (cf. Heslop-Harrison, 1953b, fig. 7; Borrill, 1956, fig. 2). It is also possible to elaborate a histogram by using hatching to show the frequency of another character within each frequency class (Fig. 25).

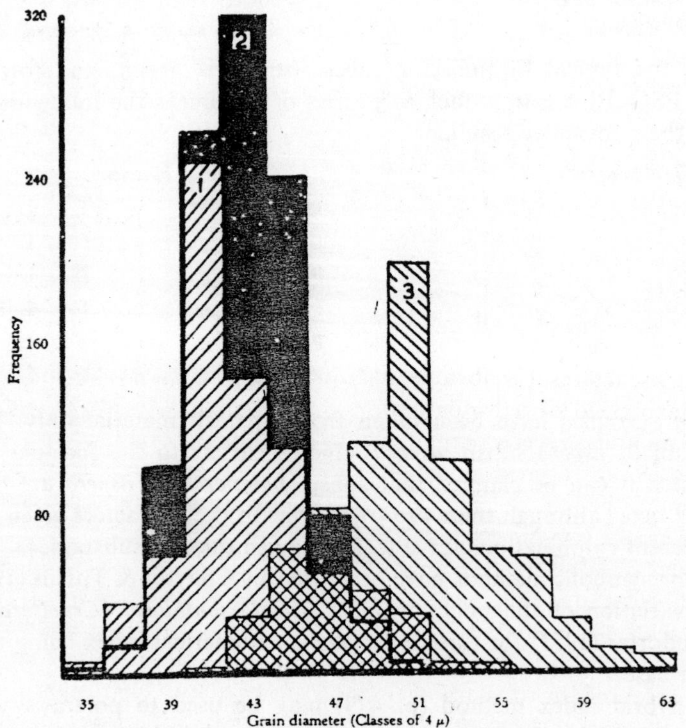

FIG. 25. Cross-hatched histogram showing pollen diameter in three species of *Glyceria*. Grain diameter (classes of 4 μ). Means: 1, *G. declinata* 41·1μ. 2, *G. fluitans* 42·1μ. 3, *G. plicata* 48·3μ. Difference between 1 and 3 significant at $P=0{\cdot}05$. (From Borrill, 1956.)

Gillett (1960) tabulates the differences found in the Kew material of *Indigofera astragalina* and *I. hirsuta* in order to decide whether they can be maintained as separate species or not, a recent revision having, as a result of statistical study, treated them as conspecific. As Gillett points out, "before handling the characters mathematically, it is necessary to enumerate the material as regards each character, using three or more grades wherever possible. If a bimodal type of distribution appears, genetic factors are almost certainly involved; a normal unimodal curve may well be due to chance. If only two grades are used for each character, we have no means of telling which type of distribution is involved, nor can we tell whether we have chosen the best dividing line as regards this character."

Gillett starts by tabulating the maximum length of the peduncle in mm. as follows:

	0–9	10–14	15–19	20–24	25–29	30–34	35–39	40–49	50–100	*Totals*
hirsuta in Australia	—	—	—	—	2	2	1	12	11	28
hirsuta in India, etc.	—	—	—	—	—	—	1	5	11	17
hirsuta in Angola, etc.	—	—	—	—	—	2	—	—	6	8
hirsuta elsewhere	—	—	—	1	2	5	18	49	142	217
astragalina in India, etc.	2	26	16	8	—	—	—	—	—	52
astragalina in Angola, etc.	3	12	8	4	—	—	—	—	—	27
astragalina elsewhere	6	25	25	8	—	—	—	—	—	64
intermediates in Angola, etc.	—	—	—	6	2	5	—	—	—	13
intermediates in W. Africa, etc.	—	—	1	1	1	—	—	—	—	3
atypical from Australia	—	—	—	—	—	—	—	1	1	2
Totals	11	63	50	28	7	14	20	67	171	431

From these figures it is obvious that there is a bimodal distribution (which might be expressed as a histogram); the same applies to breadth of pod, maximum number of seeds per pod, and colour of hairs on dorsal side of pod. In order to test the correlation of these factors, the 431 specimens measured for maximum peduncle length (with 19 exceptions) were divided into *Indigofera hirsuta* (maximum peduncle length over 25 mm.) and *I. astragalina* length under 25mm.). From his results it is apparent that breadth of pod, etc. is well-correlated with length of peduncle and that the two taxa can be conveniently separated as different species. In discussing why the earlier author came to the opposite conclusion, Gillett finds that the latter completely neglected the breadth of the pod, one of the two most important criteria. Secondly, the earlier worker's "method of dealing with the four criteria which he does investigate are faulty. Instead of first tabulating the total variation of each character and then letting such differences as may exist appear from the facts, he divides his material into two groups only for each character choosing the point of division arbitrarily." Peduncle length, for instance he divides at 50 mm., thus ignoring the significance of the bimodal curve. He "devotes no fewer than 22 pages to tabulating his measurements, calculations of probability based upon them, and a discussion of his results. But all this vast edifice is worthless when it is based on erroneous assumptions." Gillett's instructive paper demonstrates all too clearly that "ponderous, erudite and laborious mathematical methods may be just as fallible as the most superficial examination if care is not taken to eliminate logical errors in their use".

(4) *Scatter diagrams.* In recent years the scatter diagram has been increas-

ingly used in taxonomy, particularly in its pictorialised form in population analysis.

The plain scatter diagram shows the *relationship of two characters to one another;* correlation may be positive, negative or absent (cf. Cazier & Bacon, 1949). Each axis of the diagram gives the range of variation in each character, and the individual is plotted where the values for each character intersect. By the use of different symbols it is possible to show the relationships of three characters; this is a simple form of the pictorialised scatter diagram (see below). Hedberg has used the method effectively to show the variation in vicarious taxa of the Afro-Alpine flora, the results being based on normal herbarium material. In Fig. 26 the ratio of leaf length to leaf width (horizontal

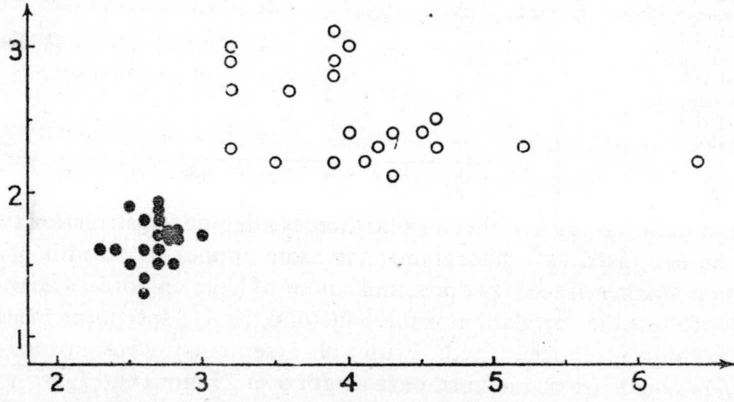

FIG. 26. Scatter diagram illustrating the relationship between three characters in *Bartsia longiflora* and *B. macrophylla*. Explanation in text. (From Hedberg, 1957.)

axis) in *Bartsia longiflora* and *B. macrophylla* is plotted against the ratio of corolla length to calyx length (vertical axis); in this example the plants with stems below the inflorescence less than 2·2 mm. wide are marked by open circles, and those which are more than 2·2 mm. broad by dots. The open circles all belong to material from Ethiopia, Mt. Nyiro, Elgon, Aberdare, Kilimanjaro and Mt. Meru (*Bartsia longiflora*), and those with dots to the Virunga volcanoes and Ruwenzori (*B. macrophylla*). The diagram is based on mean values from measurements of three flowers and leaves of each collection, and of all flowering stems present* (Hedberg, 1957). It is obvious that there is no overlap in these two characters plotted, and that the ranges of the two taxa are distinct. Hedberg would assign specific rank in such a situation. On the other hand, his scatter diagram of *Philippia trimera* (Hedberg, 1957, fig. 4) shows that each mountain population differs in at least one character, though

* Mean values should be used with caution as they may obscure important variation between individual plants or organs.

the differences are slight, allowing only subspecific status to be accorded. By using different symbols, Hedberg has been able to show to which of the six mountain populations each spot belongs.

When three characters are involved, it is also possible to use triangular charts. The actual means are not plotted in such cases, but their percentage contribution to the sum of the characters (cf. Mayr, Linsley & Usinger, 1953, p. 149).

Anderson (1949, 1953, etc.) appears to have been responsible for the widespread use of the *pictorialised* scatter diagrams in population analysis. Though the method was designed to demonstrate introgressive hybridisation, there is no reason why it should not be used to show variation due to any other cause within populations and taxa and to distinguish between them. The technique enables a visual impression to be obtained of the correlation between several different characters. This is usually achieved by attaching radiating arms to the dot, in the manner explained under "introgression" on p. 478 (see Fig. 41 on p. 477).

(5) *Polygonal graphs*. The polygonal graph (polygraph) originally described by A. H. Hutchinson (1936) and elaborated by Davidson (1947) offers another way for the simultaneous portrayal of several variables and, though usually used for population analysis, can also be used for demonstrating the differences between taxa.

To construct the polygonal graph one draws a "key in the form of spokes radiating from a central point, the centre being represented by a point or circle-shaped area. Each radius represents a single character, and the distance along it expresses the variation of that character (e.g. calyx length)." This is illustrated in Fig. 27. In practice not more than 12 radii can be conveniently used. In population analysis individual plants can be scored for each character (forming a spider's web effect), but when distinguishing between taxa it is often sufficient (and much clearer) to plot three lines corresponding to the extremes and the mean. If the sample has been adequate, it is likely that most specimens collected later will fall within the area bounded by the extremes. Related taxa can be compared by examining the form of their polygonal graphs. Hybrids will tend to have polygraphs intermediate between those of the parents.*

There seems no reason why polygonal graphs should not be drawn on spokes standing erect on a base line (instead of radiating from a focal point); they might be easier to compare if constructed in this way, or more characters could be shown. This is, in fact, similar to a method used by Wilmott (1950b) for the identification of critical taxa that differ from one another in diverse combinations of small difference. Wilmott tried out his method on *Euphrasia* material determined by an expert on the genus. Although Wilmott's presenta-

* An elegant modification of the polygraph method, portraying data on variability as indicated by the coefficient of variation, has been employed by Brandt (1961) in a study on *Veronica prostrata*.

tion portrays the variation in previously delimited taxa, it might well prove useful in the identification of obligate apomicts, as in *Alchemilla*, where the taxonomist has great difficulty in grasping the complexity of character combinations.

A refinement of this method in population analysis is described by Löve & Nadeau (1961). "It is concerned with the particular characters and their relationship within the population rather than the correlation of these characters in each individual. By plotting each arm from the individual

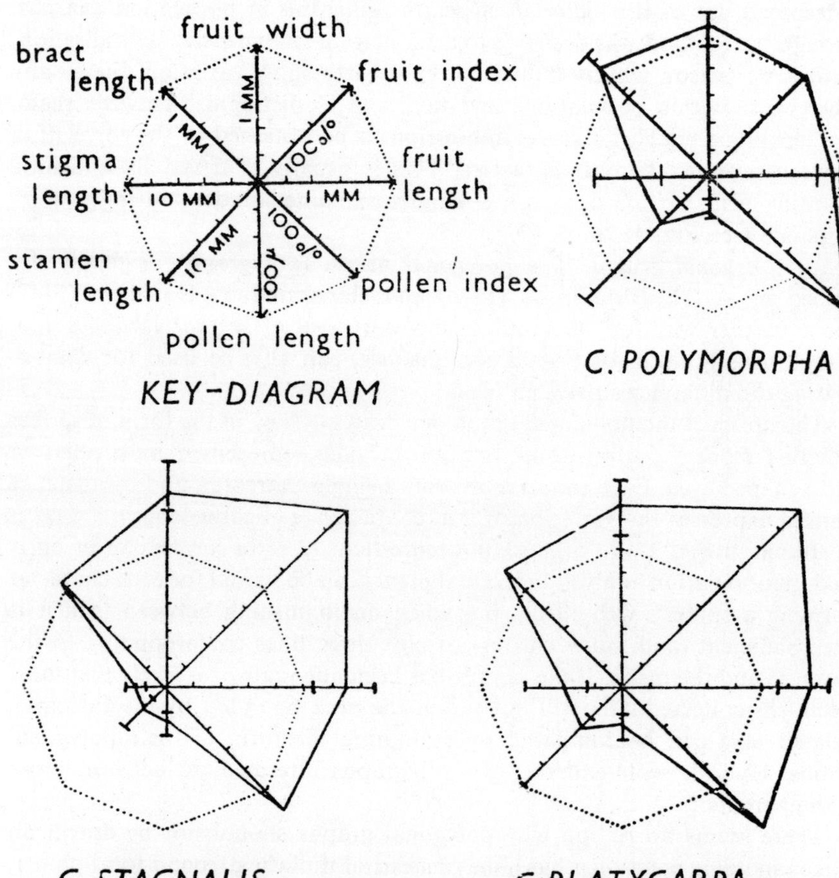

FIG. 27. Polygons of *Callitriche platycarpa*, *C. polymorpha*, *C. stagnalis* and key-diagram. The thick continuous line linking the radii indicates the mean value for each character or character index. The extreme limits for each character are indicated by transverse lines across the relevant radii. The dotted lines and calibration markers on the radii are measuring guides. Fruit index is 100 length × width; pollen index is 100 length/width. Note the intermediate position of *C. platycarpa*. (After Savidge, 1960.)

polygraphs separately on a polygram of this sort, it becomes necessary to enlarge the point of each character-variable to a circle (or other shape of area) in order to accommodate all arms linked to it. The size of the areas then automatically becomes a measure of the degree of increase in the frequency of any particular variable over other variables of the same character. At the same time the density of the lines connecting any two points becomes a visible measure of the degree of correlation between these two points." In Fig. 28 we reproduce Löve and Nadeau's polygram of a natural hybrid

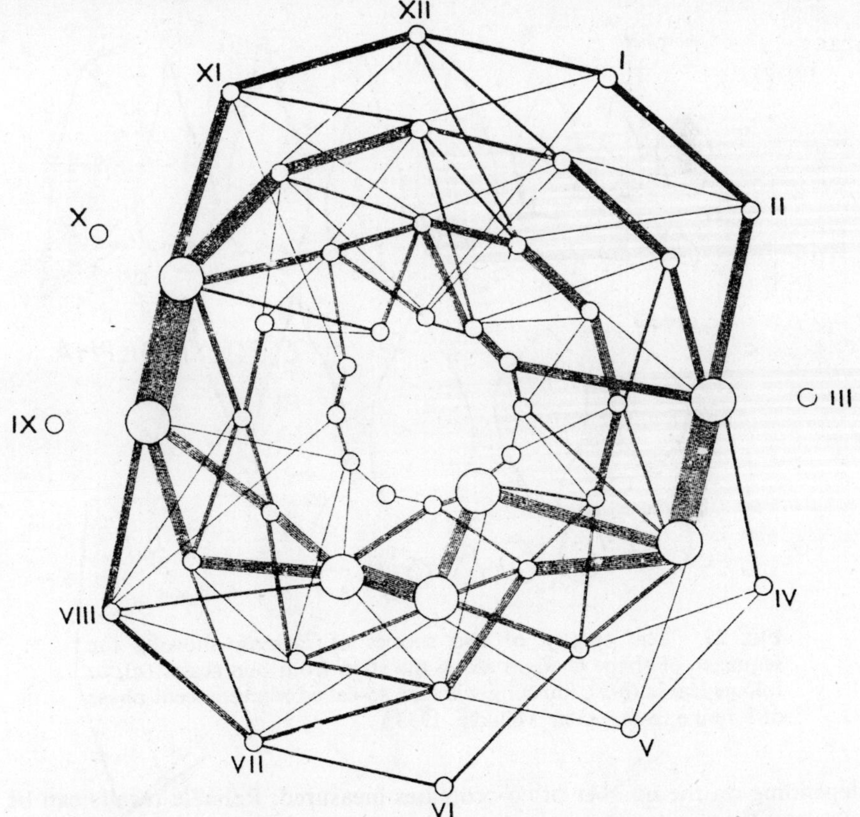

FIG. 28. Polygram illustrating a natural hybrid population between *Empetrum eamesii* subsp. *eamesii* and subsp. *hermaphroditum*. For explanation see text. (From Löve and Nadeau, 1961.)

population between *Empetrum eamesii* subsp. *eamesii* and subsp. *hermaphroditum*. By the use of their "key" diagram, it is possible to draw conclusions on character correlations.

(6) *Portrayal of form by co-ordinates*. An accurate way of defining and comparing fairly simple leaf shapes (as in *Ulmus*) by means of *rectangular co-ordinates* has been worked out by Melville (1937, 1960a). The leaf outlines

are drawn on tracing paper and projected in a photographic enlarger to cover a decimal scale, from which co-ordinates of the breadth of the half lamina can be read off directly in terms of percentages of the lamina lengths. For simple leaves (as in *Ulmus*) it is sufficient to determine co-ordinates at 10 per cent. intervals along the leaf margin and at a few points of inflection near the apex and base. For more complex leaf shapes (as in the sinuate-dentate leaves of many oaks), a larger number of co-ordinates is needed. Using this procedure, leaf shapes may be described and compared, the accuracy

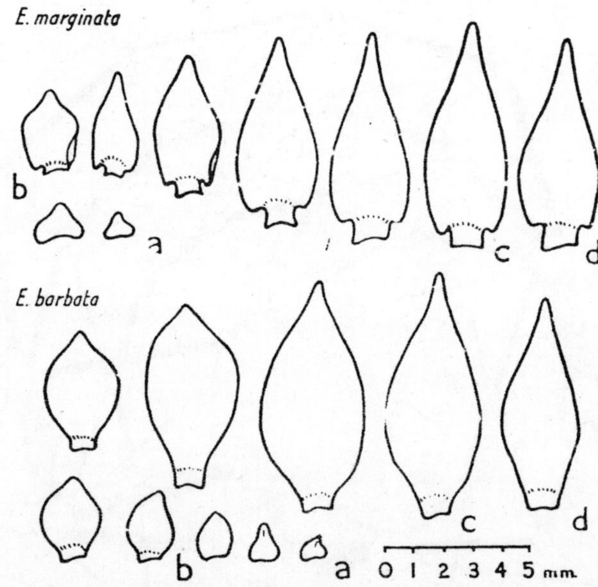

FIG. 29. Leaf spectra of two species of *Epacris*, showing the sequence of shape changes along the stem from bud scales (*a*), to foliage leaves (*b-d*), finishing with the so-called rejuvenescent phase of Krenke (*c-d*). (After Melville, 1953.)

depending on the number of co-ordinates measured. Reliable results can be obtained by plotting the average leaf shape of ten comparable leaves on an individual. Similarly, average leaf shape can be calculated for a population. For palmately divided leaves (as in *Ribes*), *polar co-ordinates* must be used (Melville, 1951, 1960a: 89).

Melville (1960a) has analysed the leaf shapes of F_1 hybrids in several woody genera. In none of them were the hybrids exactly intermediate between the parents, the hybrid interaction being in every case complicated by some other phenomenon. The simplest type of interaction was an intergrading of characters along the leaf axis, one parent being partially dominant towards the leaf base and the other towards the apex, as in *Prunus* × *fontan-*

esiana. Heterosis was found to complicate the leaf shapes of hybrids within *Rhododendron* and *Ribes*.

(7) *Spectra*. The cycle of leaf shape changes differs in different species. Although individual leaves may resemble one another from species to species, the leaf shape cycle, in relation to its position on the stem, may be quite distinctive The cycle of leaf shape can be treated, from the systematic standpoint, as a single character and is termed the *leaf spectrum* by Melville

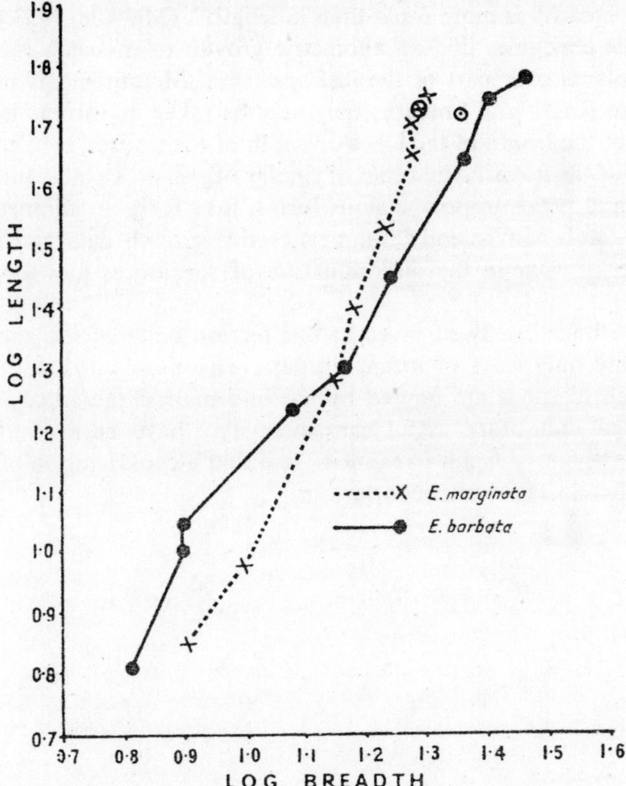

FIG. 30. Logarithmic plottings of leaf lengths and breadths of two *Epacris* species. The circles represent the last-formed leaves in the sequence illustrated (i.e. Krenke's so-called rejuvenescent phase). Further explanation in text. (After Melville, 1953.)

(1953) who shows how such spectra can be graphed. An example is given for *Epacris* in Figs. 29 and 30.

In Fig. 30 the lengths of individual leaves are plotted against their breadths on a logarithmic scale. The points representing the leaves formed after the bud scales and before the final "rejuvenescent" phase fall nearly on a straight line. This indicates that the rates of growth, in length and breadth, bear a constant relationship to one another, irrespective of size. "When the straight line relationship holds, growth is said to be allometric and conforms with the allometry equation: $y = bx^k$, or expressed in its linear form: Log $y =$ Log $b + k$ Log x. Here b is a constant related to the initial difference between the length [x] and breadth [y]. When k is unity the growth rates are equal and the graph rises at an angle of 45°; if k is greater than 1 (length plotted vertically), the graph rises steeply and growth in length is more rapid than growth in breadth. Similarly if k is less than 1, the graph rises more gently and growth in breadth is more rapid than in length". (Melville, 1953.)

Melville concludes that an allometric growth relationship is maintained in many plants over part of the leaf spectrum. Measurements need not be confined to length and breadth; they may be taken at various levels of the blade, or of the length of the lobes or depth of the sinuses, etc., according to the shape of the leaves. Any series of similar organs of a plant, such as bracts of sepals and petals in polypetalous forms, may form an allometric growth series. A graph conveniently summarises this growth data and its slope is a character of value in the discrimination of species, as it is for *Epacris* in Fig. 30.

The methods briefly surveyed in this section on symbolic and graphical methods are only ways of arranging data. Taxonomic deduction from the use of such methods are limited by the information presented. It must be remembered that other useful characters may have been overlooked (or ignored if they are difficult to express), and that a consideration of them may lead to different taxonomic conclusions.

Chapter 10

MODIFICATION OF THE PHENOTYPE

PLASTICITY OF THE PHENOTYPE

Much of the phenotypic variation encountered by the taxonomist is the result of the plastic response of the individual to factors of the environment. Such phenotypic plasticity has tended to be underestimated in taxonomic and evolutionary work in plants. From a practical point of view the problems it poses are most acute in certain groups such as aquatic and marsh plants and in certain regions such as the tropics, where van Steenis notes that "using it for taxonomic delimitation has given rise to a multitude of 'paper' species and varieties without any taxonomical value".

Development of the individual follows a more or less orderly pattern and sequence to produce an organism of specific structure. These developmental activities are nowadays explained in terms of gene action and the production of specific growth substances, and come within the province of developmental physiology and morphogenesis.

An important difference between plants and all but the smallest animals is that growth is restricted to certain parts of the plant body. These are permanently meristematic regions—regions of indeterminate growth, resulting in the continuous formation and development of organs. Such regions provide good and extensive material for morphogenetic studies.

This difference is related to the sedentary life of plants, with large surface areas providing for nutrition. The stationary habit also makes individual plants more susceptible to environmental fluctuations, and their plasticity provides morphogeneticists with the possibility of studying the effects of the environmental factors and presents taxonomists with numerous problems.

Although the plant body as a whole is relatively variable and not subject to the same determinism as in animals, the various organs form a clearly organised whole. Individual organs such as leaves, carpels, etc. have a much more precise form than, say, the number of branches of a tree, but it is quite evident that the overall shape of a tree, despite the variation in the number and position and relative length, etc. of the axes, is clearly recognisable and characteristic for each species. There is in fact some kind of homeostasis or regulatory process which keeps development within certain norms (cf. Mather, 1955).

The number of organs, branching and size, tend to be more easily modified in development than form and pattern. This is largely due to the period of growth of the meristem and the degree of cell elongation at a later stage of growth. These features are easily modified by factors such as temperature, water, light and nutrition as discussed below.

Genotype and Phenotype

As we have seen, living plants, like all living systems, are today regarded as organised systems which maintain and adjust themselves, through their capacity for homeostasis (cf. Waddington, 1957), to the environmental flux. A distinction is made between the *genotype* which remains unaltered (except under certain circumstances) and the outward appearance or *phenotype* which alters in structure in response to changes in the environment, these alterations sometimes being partially reversible. There appears therefore to be a fixed factor—the genotype—and a variable expression—the phenotype—in response to varying environmental conditions, and a homeostatic relationship co-ordinating the system.

It is only recently that the dichotomy between heredity (genotype) and environment, and the effects of each on the phenotype, which has been overstressed in the past, has been questioned. The effects of heredity and environmental factors always operate jointly and frequently it is not possible to disentangle them. As Waddington has expressed it (1957), "organism and environment are *not* two separate things each having its own character in its own right, which come together with as little relationship as a sieve to a shovelful of pebbles. It has been shown, on the contrary, that there is a close integration between the two: the environment can modify the channel which the form of the organism can take." Again, as Sinnott in his recent review of morphogenesis says (1960), a visible trait (i.e. morphological feature) is the developmental reaction of a specific (and constant) genetic constitution to a specific environment.*

The response of the different features of the plant to different environmental conditions or factors varies considerably. Some characters, such as leaf arrangement or floral structure, remain more or less unchanged under a range of different conditions. These are frequently referred to in taxonomy as "good" characters because of their relative constancy and are regarded as "genetically fixed". Other features such as leaf shape, stem height, time of flowering, etc. develop differently under different environmental conditions. The term *phenotypic plasticity* is applied to these variations shown by the

* Since this subject strikes at the very roots of taxonomy, some readers may wish to pursue it in depth. Important surveys are given by Mather (1955), Sinnott (1960) and by Waddington in his fundamental though difficult work *The Strategy of the Genes* (1957) and in a later work, (Waddington, 1962). For an interpretation of the relationship between genotype, phenotype and environment in terms of the information theory, a recent paper by Schmalhausen (1960) should be consulted.

phenotype in response to environmental fluctuations. The same genotype may, according to the environmental factors prevalent, give rise to different phenotypes. This plasticity may be so wide as to overlie or even obscure the genotypic variability. In fact, the distinction between genetically determined characters and environmentally induced characters is a false one, since both types of character are the result of genetical inheritance and environmental control. In other words, each genotype has a range of plasticity which is genetically determined. In some cases the range of expression is narrow—the plants are termed *stenoplastic*—or broad—the plants being called *euryplastic*. It is the inherited response which produces the character or phenotype, not just the environmental factor. The adaptive or non-adaptive nature of these differences is discussed below in connection with ecotypes.

The phenotype can be regarded in the light of what we have just discussed as the product of the interaction of all genes with each other and with the environment. It is the phenotype which is exposed to natural selection, and the phenotype that the taxonomist must be primarily concerned with, not the individual genes, chromosomes or genotypes, for the latter obviously can only be perceived through their phenotypic effects.

The factors which affect the plasticity of the phenotype are clearly of great concern to the taxonomist. It is, for instance, very necessary to find out if variation is due to genetically inherited differences or to the particular range of environmental factors which produce largely reversible modifications from the same genotype. The fact that similarity in phenotype may be due either to environmental modification or to genotypic differentiation makes such similarities unsure guides of genetic relationship. They are usually disregarded by evolutionists due to their possibly non-genetic nature. Taxonomists may be misled by phenotypic variation in their selection of taxonomic characters. It is not of course possible to determine the nature of the variation—whether genotypic or environmentally induced—in the herbarium or even in the field, and comparative cultivation is normally necessary. It may be possible to argue in some cases by analogy and extrapolate from experience of other situations, but there are always attendant dangers on such indirect methods. Nonetheless, taxonomists have frequently to attempt it.

Growth Substances

The regulation of growth and organised development is largely achieved through the agency of growth substances. Recent morphogenetic studies suggest that many of the features of form and structure on which the taxonomist relies (such as symmetry of flowers, inflorescence structure, number of floral parts, nodal anatomy, etc.) are probably determined by complex patterns of balance between gene-controlled cellular states at primordia; the latter are activated or deactivated through hormonal control causing the initiation of particular developmental pathways. The precise mode of action of these systems is not yet known, but it is suggested that profound alterations in the phenotype can be caused by small changes in auxin balance or concentration and the time and rate of diffusion of these substances. For a detailed discussion of these problems see Wardlaw (1952) and Sinnott (1960).

The determination of the particular pathway along which a tissue will develop probably takes place at a very early stage of growth in the primordium, but whether the normal developmental pathway is continued to its particular end may depend on substances which enter the developing primordium from the apex (Heslop-Harrison, 1959, 1961). In *Cannabis sativa* the effect of treatment with 2, 3, 5-triiodo-benzoic acid (TIBA) is to modify the pattern of emergence of the perianth members and stamens to form flowers of a tubular form but without deflecting their developmental pathways. To quote Heslop-Harrison (1959), TIBA simulates "the effects of the genetical changes which, in the course of evolution, have brought about such transitions as that from polypetaly to sympetaly".

Major structural and functional alterations may be induced in intact flowers by externally applied auxins, such as the suppression of the corolla and androecium in species with diclinous flowers and the hypertrophy of the calyx and gynoecium; and the production of female or inter-sexual flowers in dioecious hemp. In the latter case the ontogenetic development of the presumptive stamens appears to have been diverted towards the pathway producing carpels, the auxins having possibly influenced some determining process in the flower primordium.

Heslop-Harrison concludes that since floral morphogenesis can be influenced by applied auxins, we would expect that environmental factors which alter auxin metabolism should have similar effects on the structure of flowers. That structural alterations induced by environmental factors parallel those caused by auxins does not necessarily mean that native auxins have been responsible, but the likelihood is there.

Differences in auxin concentration at the apex of the plant can alter the form of branching which may constitute an important taxonomic difference between species, as in *Aster novaeangliae* and *A. multiflorus*. The former has one main stem whereas the latter is much branched; this has been shown to be correlated with the amount of auxin present in the plant (Delisle, 1937).

The effect of growth substances on flower induction and sex determination has been well studied and documented. Details can be obtained from the account by Sinnott (1960) and two symposia held in London recently (*Journ. Linn. Soc. Lond.* **56**: 153-302 (1959) and *Proc. Linn. Soc. Lond.* **172**: 90-127 (1961)).

The Investigation of Phenotypic Variability

The techniques for the study of the individual plasticity of the phenotype are, according to Clausen *et al.* (1940):

(*a*) Artificial modification of the environment by means of greenhouses allowing factors to be controlled, ranging to full-scale growth chambers or phytotrons (see below).

(*b*) Transplant experiments. In deciding whether variations are heritable or non-heritable and reversible, a dual approach is employed:

(i) Cultivation of plants of uniform heredity in varied environmental conditions so as to show the effect of different environments on the same genotype.

(ii) Cultivation of plants of different heredity, i.e. different genotypes, in uniform environmental conditions so as to neutralise the effects of the different environments on the phenotype.

The attention of taxonomists has therefore become increasingly directed in the past few decades to the study of the factors which affect the development and plasticity of the individual. These factors are, as indicated, usually distinguished as (*a*) environmental, (*b*) genetic. Van Steenis (1957) recognises four sets of factors: (*a*) intrinsic, (*b*) climatic, (*c*) edaphic and (*d*) biotic, each of which is further subdivided. Sinnott (1960), in regarding the factors from a morphogenetic viewpoint, distinguishes three sets:

(*a*) Physical, e.g. light, gravity, etc., which originate chiefly in the internal environment.

(*b*) Chemical—including auxins.

(*c*) Genetic—regarded as part of the internal environmental.

Physical Factors and Modifications

1. *Light*

(a) *Intensity*. Differences in light intensity can cause quite marked variations in form, habit and anatomy. They can affect features such as stem elongation and internode length, branching, leaf shape and thickness, flower colour, etc. The phenomenon of sun and shade leaves is well known—those developing in strong sunlight tending to be smaller and thicker with a more highly differentiated palisade than those which develop in the shade. Other features which may be intensified by light concentration are degree of lobing or dissection, prominence of venation, hairiness, glaucousness—all features which may be used for taxonomic discrimination.* Heslop-Harrison (1953b) illustrates *Rubus* plants of the same genotype grown in full light, medium shade and deep shade: differences are shown in leaf size and stem thickness, hairiness, shape of stem section and prickle form, all of which are important characters in *Rubus* taxonomy (cf. Brainerd & Pieterson, 1920). Shade forms have in some instances been treated, both in temperate and tropical floras, as distinct taxa until they were shown to be modifications of normal-leaved forms.

* A list of 17 morphological and 14 physiological features affected by light availability is given by Daubenmire (1959), whose account of environmental modification of plants should be consulted.

Light intensity is partly responsible for the differences found between mountain and lowland plants (hypselomorphosis). Mountain forms are frequently dwarfed, short-stemmed, few-flowered and often show increased anthocyanin pigmentation. Normally white-flowered species may develop pink or purple flowers at high altitudes, e.g. *Ptilotrichum spinosum* in Sierra Nevada, Spain, at 2500 m. (Similar genetically fixed variants are also known.)

Seasonal dimorphism can be sufficiently striking to mislead the taxonomist into thinking he is dealing with more than one species. A spectacular example of such dimorphism, mainly controlled by illumination, has been described from Israel (Jardeni, 1938). *Ononis leiosperma* (*Leguminosae*) undergoes an exceptionally complete metamorphosis which affects the root system, stems (which become thorny during summer) and leaves.

Epiphytes may be included under the heading of light effects, although in some instances rooting conditions may be the deciding factor. Van Steenis (1957) cites cases where terrestrial specimens are much more vigorous than epiphytic forms. *Wightia borneensis* (*Scrophulariaceae*), for example, is sometimes a smallish tree, sometimes an epiphyte, sometimes a liana. The effects of altitude may be due to a number of factors such as exposure, extreme temperatures, poor soils, insolation, etc., and it is not easy to disentangle exactly which is causing a particular phenotypic response. Another effect of shade is to protract the juvenile stage of some species (Ashby, 1948). The quality of light (different wavelengths) may also affect several developmental features.

(b) *Light duration: photoperiodism*. The relative length of the alternating daily period of exposure to light (photoperiod) and to darkness (nyctoperiod) may modify morphological features of the plant, especially young, not fully developed leaves. The anatomical and structural features of the leaves may also be affected. Such vegetative responses have been greatly neglected in photoperiodism while control of flowering has been regarded as the essential response. The effect on vegetative development may, however, be important (Went, 1957). It has been suggested by Wareing that the major significance of the photoperiod for woody species in relationship to habitat adaptation is its effect on dormancy. In *Ulex europaeus* the plant passes through a leafy phase of seedling ontogeny and a spiny adult stage although there is not a clear distinction between them (Milliner, 1961), and the juvenile-adult transition may be significantly altered by differences (daylight as well as other factors).

More attention has been paid to the physiological effects of photoperiodism on time of flowering, the number and type of flowers produced and differentiation of the sexes. As will be evident from the examples discussed here, it is difficult, and not always practicable, to draw a line between the plastic effects of photoperiodism and genetically fixed responses.

Species and races differ genetically in their range of photoperiodic reactions. This can be of considerable adaptive significance, in that it may be a mechanism which ensures flowering at the most favourable time for a particular environment. Different strains of *Trifolium subterraneum* vary in their flowering time: they have become adapted to the length of the photoperiod in the range of habitats occupied (Morley, 1959). In this case early flowering is a mechanism which allows the plant to avoid the periods of drought.

Photoperiodism is seldom a simple process and is interrelated with the somewhat parallel phenomenon of thermoperiodism. The time of flower initiation in *Trifolium* has been shown to be determined by dark inhibition, a high temperature, a light-independent process and a vernalisation process. In different environments different factors in this process complex may be limiting (Morley, 1959).

Very many photoperiodic ecotypes have been described, particularly in tree species (cf. Vaartaja, 1954, and Chapter 12). They are also found in herbaceous species and grasses such as *Bouteloua curtipendula*, where northern and southern races were shown by Olmsted (1944) to be different in their light reactions; some were short-day plants, some were long-day plants.

A notable feature of some North Temperate species, especially crop plants, is that they can adapt their flowering period to conditions in South Temperate areas: flowering takes place six months earlier or later.

Photoperiodism may also determine to some extent bulb or tuber formation which can again be regarded as an adaptation to the environment. Plants growing under short day lengths, such as primaveral species may, for example, be stimulated to produce an accessory means of propagation.

2. *Temperature*

Although temperature plays a vital role in the physiological processes of the plant, there is little experimental evidence concerning the plastic response of individual plants in their natural habitats to different temperatures or temperature ranges. There is, however, a fair amount of evidence relating to cultivated plants. The different processes of the plant during its life cycle each require a particular range of temperatures, and plants are normally adapted to the seasonal variations of temperature within the overall extremes found in the environments where they grow. As with phenomena such as photoperiodism, it is often difficult to make a clear distinction between plastic responses and genetic adaptations to temperature conditions.

The kind of features affected by temperature are largely periodic phenomena, such as germination (cf. cold treatment of seeds), shoot regeneration, leaf fall, flowering, fruiting (cf. Hiesey, 1953), and are often intimately associated with other climatic factors such as light and moisture (see sections above and below). Examples of obvious plastic modifications similar to those produced by other factors are rare, and apart from damage caused by extreme temperatures (frost damage, scorching, etc.) seldom have appreciable consequences for the taxonomist. The adaptation of plants to temperature conditions is to some extent expressed in life form classifications, such as that of Raunkiaer (1934) based on the way in which the perennating buds are protected during the dormant season (see p. 146).

3. *Water*

It is well known that the amount of water available to a plant can have considerable and dramatic effects on its form and structure. At one extreme, restriction or scarcity of water may cause xeromorphic modifications such as reduced leaf surface, smaller cells with thicker walls, etc. Leaf number may also be affected, but this can be a drought effect causing the leaves to drop prematurely. These effects have been confirmed by experimental work.

In desert areas, life form can be strikingly modified by rainfall. In Egypt, *Diplotaxis harra* (*Cruciferae*) behaves as an annual in very dry seasons (if it appears at all), but persists as a suffrutescent perennial in favourable years.

Unless herbarium material is very limited, such modifications seldom cause the trained taxonomist much difficulty. On the other hand, the effects of abundant or excess water in the habitat, as in the case of aquatic or marsh and swamp plants, have provided many problems of interpretation for the taxonomist. Stem and leaf structures may differ widely in different depths of water.

Heterophylly in aquatics and amphibious plants. Many aquatic or amphibious plants show the phenomenon known as heterophylly—where different types of leaf (in shape or structure) occur on the same plant. Heterophylly may be due to a number of quite different causes: it may be primarily developmental (called *heteroblastic development* by Goebel, 1897), it may be a simple response to submergence in water, or it may be a response to other conditions not primarily associated with the aquatic environment.

Heteroblastic development results in the formation of adult and juvenile leaves. In *Sparganium erectum*, *Alisma plantago-aquatica* and *Sagittaria sagittifolia*, for example, the aquatic leaves are juvenile. These plants may be retained in a juvenile state by submersion in deep or swiftly flowing water or in low light intensity, but they usually remain in a sterile condition. Flowering is associated with the adult or aerial leaves. In the above three species, at least, the placing of adult plants in flower into deep or swiftly flowing water will not cause a reversion to the juvenile state without a great deal of conditioning.

In nearly all plants (not only aquatics) submergence in water will cause changes in leaf form. In many plants such as *Poa annua*, a terrestrial species, or adult *Sagittaria sagittifolia*, the differences are not great, while in *Hippuris vulgaris* or *Myriophyllum* sp. (typical aquatic plants) the differences are very considerable. In the water crowfoots, in particular *Ranunculus peltatus*, the differences are so great that some states have been described as separate species, the terrestrial state being called *R. bauhinii*, the deep-water state

R. elongatus, the submerged non-heterophyllous state *R. submersus* and the early spring state *R. truncatus*. These states are formed sequentially.

In the water crowfoots the situation is further complicated by genetically controlled heterophylly. Two distinct types of leaf occur—one, capillary, found submerged or terrestrially, and the other, laminate, found floating or submerged. In the submerged state initiation is controlled by a photoperiodic-temperature effect. The morphology of the terrestrial state is also under photoperiodic control. When *R. aquatilis*, for example, is cultivated terrestrially in a heated mist propagation unit in summer, it looks like the normal terrestrial state, but in winter it looks like the normal submerged state.

Extensive studies on the effects of submergence were carried out by H. Glück (1905, 1906, 1911, 1924), and numerous papers have been published by other authors. Many of the latter need to be treated with considerable reserve since their results did not take into account sufficiently varying factors of the environment other than water itself, and too many uncritical generalisations have been made.

It has been suggested that the wide plasticity of such plants may be associated with their having become secondarily adapted to the aquatic habitat after earlier adaptation to a terrestrial way of life (Heslop-Harrison, 1953b). Takhtajan (1959a) goes further in suggesting that aquatic plants are neotenous derivatives of terrestrial plants as a consequence of heteroblastic development. While this may be true in some instances, for example *Sagittaria rhombifolia* and *S. natans* which could be regarded as neotenous derivatives of *S. sagittifolia*, a better example, perhaps, occurs in the series *Sparganium erectum*—*S. emersum*—*S. minimum*. In this case the seedling stages are very alike. Whereas *S. minimum* flowers at the seedling stage, *S. erectum* develops into a tall semi-terrestrial plant. The adult or mature leaf of *S. minimum* is virtually the same as a seedling leaf of *S. erectum*.

One of the classic studies on the variation in *amphibious* plants is that of Massart (1902) on *Polygonum amphibium*, which is adapted to both terrestrial and aquatic habitats. Three morphologically different variants have been recognised (and named)—the "typical" terrestrial form, the aquatic form and the xeromorphic form which differ from one another in the form and structure of the stem, branching, leaf shape, indumentum, etc. Massart transplanted samples of the terrestrial and xeromorphic forms to dry sand, moist sand and water, and in due course the plants developed the phenotypic modifications found in the corresponding natural habitats. Massart's studies were incomplete, however, in that he did not show that the forms, which he demonstrated could be interconverted, were genetically identical. He did, however, emphasise the great developmental plasticity shown by these plants and their consequent ability to grow in a range of habitats.

Tropical examples are known although, as van Steenis (1957) notes, it is seldom possible to confirm them by experiment or make adequate field observations. One such case is found in *Limnophila* (*Scrophulariaceae*)—*L. indica* is typically aquatic with finely dissected submerged leaves and more or less simple toothed leaves, while in the terrestrial form the pinnatifid leaves are often absent.

As already mentioned, precocious flowering may be induced by a drying environment in aquatic or swamp plants. An elegant example is cited, again by van Steenis (1957), from the inundated rice field in the tropics. Rice seedlings are planted in mud and have a growth period of some three to four months. During the first two months weeding is carried out only once or twice, and at the end of the growth period the rice fields are drained to allow the crop to ripen. Any weeds growing in the rice fields have to adapt themselves to the short growing period if they are to survive. This is effected by dwarf habit and precocious flowering—ecologically induced neoteny. Plants which are normally perennial behave as annuals and this phenomenon, together with the production of flowers and fruits by vegetatively juvenile plants, has misled taxonomists on several occasions until the rice culture adaptations were recognised. This recalls the specific adaptations to flax cultures found in several species.

Neoteny may be unrelated to such conditions, however, and flowering by juvenile stages, even seedlings, may have considerable taxonomic significance. (Cf. discussion of neoteny on p. 157.)

4. *Wind*

Persistent wind action may cause various phenotypic modifications. It may result in desiccation and consequent premature abscission of leaves and other organs. It may result in dwarfing in habit, especially of tall erect plants: it is one of the factors involved in altitudinal modifications. Such dwarfing can cause considerable changes in the pattern and number of branches (cf. Daubenmire, 1959, for a detailed treatment). Deformation (wind-training) is a well-known phenomenon—the effect of strong wind pressures from a single direction is often to produce oblique, unilateral growth, a flattening of the trunk and branches in the direction of the prevailing wind, or similar adaptations. These are seldom likely to cause difficulties of interpretation taxonomically.

Very small differences in wind speed may cause considerable differences in cuticular pattern (Juniper, 1959), and the papers by Whitehead (1951, 1962) should be consulted.

5. *Soil*

The effects of different soils on the phenotype have been less well studied than those of other factors. It is often not known whether responses observed are due to genotypic adaptation to particular soil types (see Chapter 12) or simply to plastic modifications.

It is well known that plants growing on poor soil may be dwarfed and few-flowered, but in well-studied species they are unlikely to mislead the taxonomist into describing them as distinct, although this has happened in the past and is still a factor to be guarded against in poorly known floras.

The best documented instances of the interactions between plants and different soils are the experiments undertaken by Marsden-Jones and Turrill at Potterne for the British Ecological Society. For these experiments, which extended over a period of 13 years, special enclosures (55 × 10 × 3 feet) were constructed, each filled with about 50 tons of one of the following imported soils: sand, calcareous sand, clay, chalky clay, as well as Upper Greensand Potterne soil. The beds were arranged near each other so that all were exposed to the same climatic conditions; likewise care was taken in the spacing of the plants to ensure that there were no appreciable differences in shading or other microclimatic effects.

Plants of known wild origin were employed, in most cases as *clones*, although seedlings were sometimes used, as in *Anthyllis vulneraria.* Cloning has obvious advantages in that the vegetative divisions or *ramets* are genetically identical with one another and can be so selected that they are virtually uniform in size and appearance before being used in the tests. Eleven different species and varieties were employed, and 26 ramets planted in each of the five different soil beds at Potterne and a further 26 ramets at Kew. They were scored at intervals for a series of features.

The reactions varied considerably as might be expected. The greatest plasticity was shown by *Plantago major* in which rapid and considerable changes were recorded. Marsden-Jones & Turrill noted that "phenotypes have been produced within two years, from seedlings and ramets of a single clone, which have been classed as varieties and subspecies in a recent monograph of the genus [Pilger in *Das Pflanzenreich*]". There was a remarkable difference in size of all organs, number of crowns produced and in reproductive capacity between ramets grown on sand and those on clays.

Other species showed very little plasticity on the different soils. Details of these experiments can be obtained from the original papers (Marsden-Jones & Turrill, 1930, 1933, 1935, 1937, 1938, 1945). The following essential requirements in such work are listed by Turrill (1948):

1. Preservation of good voucher herbarium specimens which allow comparisons to be made and identifications checked.

2. Description and analysis of the environmental conditions as far as possible.

3. Description of structural and behaviour characters at frequent intervals for each individual seedling or ramet, all of which should be numbered.

Other phenotypic modifications, in specialised soil conditions, are mentioned by van Steenis (1948), notably *fumarole* plants where steam escaping under pressure has the effect of reducing the size of plants and making them more compact in habit, and *solfatara* plants where there is again dwarfing in habit, largely due to the soil sterility, extreme temperature conditions, poisonous fumes and acidity.

Halophytes. In maritime habitats numerous modifications are found, often paralleled by genotypic adaptations (genecotypes). They include dwarfing, prostrate habit, increased hairiness and succulence, etc. It is often extremely difficult to decide which environmental factor is responsible for a particular effect in such conditions (soil, exposure, wind, etc.). In halophytes, both coastal and inland, soil conditions are a major factor, although again they operate together with other features of the environmental complex. This can be illustrated by reference to the genus *Salicornia*.

Variation in size, habit and degree of branching is very extensive in *Salicornia* species. In *S. ramosissima* the variation ranges from simple dark, purplish-red plants, not more than 5 cm. high, to bushy green plants over 30 cm. high and up to 15 cm. in diameter. The former plants occur in dense stands on bare relatively dry mud in the upper part of salt marshes. In cultivation the F_1 generation obtained from such plants is usually indistinguishable from the large bushy plants. It would appear that the small red form occurs where there is intense competition in open habitats (usually from other *Salicornia* plants). The large green plants occur in open habitats where there is much less competition and probably where the water content of the soil remains high. Where the plants are subjected to trampling or high-velocity currents, pseudo-prostrate forms often occur. These often appear to have the lower branches spreading at right angles to the main stem but in many cases this appears to be due only to physical action. However, these factors are sometimes genetically controlled and it is possible to obtain F_1 plants from such a population which are genetically prostrate with widespreading lower branches, and plants which are erect with widespreading lower branches.

Another aspect of phenotype plasticity in this group is the occurrence of almost identical forms of different species in some habitats. Populations of *Salicornia* occurring on bare soil, particularly at the base of sea walls, are usually composed of several species, including *S. fragilis* (or *S. lutescens*) and *S. obscura*, tetraploid and diploid species respectively. Under these circumstances it is often impossible to identify a high proportion of plants without resorting to microscopic or submicroscopic characters such as anther size.

6. *Biotic*

Plants may be modified during their growth period by a number of biotic factors. The following classification is adopted from van Steenis (1948):

Phytomorphosis

Fungal and bacteriological infections.

Zoomorphosis

Ant habitation (Myrmecomorphosis)

Galls (cecidio-morphosis)
Grazing and browsing (pascuomorphosis).
Anthropo-morphosis
Trampling
Fire (pyromorphosis)

Infection by fungal parasites can have rather striking effects on the hosts. In the normally dioecious *Silene* (*Melandrium*) *dioica*, infection of the anthers by the smut, *Ustilago violacea*, results in the development of the ovary in the male flowers—evidently the ancestral condition in the genus. In the *Liliaceae*, the oblong corolla of *Muscari* Sect. *Leopoldia* (sometimes treated as a separate genus) is transformed into a globular one when infected by *Uromyces vaillantii*, thereby simulating Sect. *Botryanthus*. *Agrostis tenuis*, parasitised by the smut *Teletia decipiens*, is so greatly dwarfed and its inflorescence so modified that it was described as *A. pumila* by Linnaeus and has long persisted in Floras as a variety of *A. tenuis* (Philipson, 1937). *Trachyspora intrusa* (*Uromyces alchemillae*) changes both the habit and the leaf morphology of *Alchemilla vulgaris* agg. Virus infection can produce erect fastigiate forms, particularly in the *Leguminosae*.

Galls formed as a result of attack by insects, mites, etc., are of common occurrence and have led to a limited amount of taxonomic confusion. Van Steenis (1948, p. xxxvi) cites several examples of tropical species and even genera, based on the galled fruits of other species. Usually, however, the effects of galls are easily recognised, particularly when they distort the vegetative parts. Such abnormalities should not, of course, provide the basis for new taxa. When a plant is almost constantly galled in nature, there is good reason to mention this morphological phenomenon in its description, since it becomes a characteristic of the species. The Cretan *Salvia pomifera*, for instance, is nearly always covered with large ovoid galls, a feature which provided Linnaeus with the specific epithet. An impressive systematic account of European plants infected with galls is given by Houard (1908-1913).

The most widespread and taxonomically most misleading biotic effects are those of animal grazing. Grazing or browsing depends to a considerable extent on the life form of the species concerned, as Daubenmire (1959) points out: palatable annuals may be virtually eliminated from a habitat in a short period if grazing prevents them setting seed and propagating. Apart from such extreme damage, grazing, especially in perennial plants, affects the habit and degree and form of branching, often to such an extent that the plants are largely atypical of their species with regard to general facies. When the top of an inflorescence is grazed, as in many *Labiatae*, the bracts or upper leaves of the remaining portion may become abnormally enlarged. Time of flowering may also be affected. It is sometimes extremely difficult in herbarium material to recognise that specimens have been grazed, since the main shoot may be eliminated and grown over by lateral shoots, leaving little trace. Floral and fruit characters are not appreciably affected, however. In a few instances the habit of some plants simulates grazing damage and has led to taxonomic misinterpretation.

Trampling by man can alter the habit of plants, causing nanism and poor development. Often, however, genotypic adaptations to such conditions are found.

The increasing use of selective weed killers means that the phenotypic effects of chemical "drift" are often very noticeable in populations adjacent to those to which the weed killer has been directly applied, and could seriously affect population analyses. The effects include stem distortion, increased dissection of the leaves, and a tendency for the main lateral veins of the leaf (of Dicotyledons) to become parallel.

Interaction of Factors

The preceding discussion has focused attention on single factors of the environment and their effects on the plasticity of individuals. As we have seen, however, it is often difficult in practice to disentangle the various factors. This is probably due to the characters studied being the complex end points of various processes, and the difficulty is one of sorting out which processes were in fact involved. We also have to remember that the various environmental factors do not operate singly in nature but form part of an intricate and dynamic environmental complex. An excellent discussion of the problems involved is given by Daubenmire (1959). Geiger's book on microclimates (1950) will be found very useful in this meeting ground of taxonomy and autecology.

It is possible today to investigate in great detail the effect of these factors under experimental conditions through the use of growth cabinets, temperature-controlled rooms, growth rooms, greenhouses and phytotrons employing various degrees of control (Hudson, 1957).

Such controlled experimentation has reached high levels of sophistication and is a basically valid procedure, but as Daubenmire (1959) points out it is often performed under conditions that produce results of little practical value. He notes that plants grown under artificial and unnatural environmental conditions cannot be expected to react to individual factor variations in the same way as they would in nature. As a result the data obtained from greenhouse and even phytotron cultivation may not be a valid representation of field responses. Responses in the greenhouse may in fact be reversed in the field.

It is clearly important to know what range of phenotypic variation is possible in response to the variation in single factors and in combinations of factors, but if the results are to have more than a theoretical significance they must be related to what actually happens in nature. Phenotypes may be produced under experimental conditions which are not known in the field, especially in highly plastic species. Examples are found in species of water buttercup already referred to. This is largely explained if it is remembered that the phenotypic expression of most genes is modified to a certain degree by environmental factors, as Clausen & Hiesey (1958a) point out. The expressions of certain genes do not appear above or below certain environmental thresholds: such genes are termed latent. When they are expressed these latent genes may produce striking phenotypic effects. Among the examples that may be cited are petal colour in *Primula sinensis* var. *rubra* which is genetically controlled. Baur (1914) found that the petals were red when grown at low temperatures (15° to 20°C) but they became white at higher temperatures (above 30°). More spectacular results were reported by Lawrence (1945) in *Deschampsia caespitosa:* plants grown in California from seeds collected in northern latitudes in Sweden and Finland showed a high development of pseudo-vivipary although it was not evident in native habitats. The genes controlling pseudo-vivipary are apparently present in some individuals and are revealed by growing them at the more southern latitude.

Environmental conditions may also affect the *degree of expression* of an individual gene in a multiple gene system (Clausen & Hiesey, 1958a, b).

As we shall see later, genecological studies have demonstrated how closely adjusted genotypes may be to the environmental conditions where they are native. We must not lose sight of this when interpreting the results of experimental cultivation or modification of habitat.

Chapter 11

VARIATION WITHIN POPULATIONS

INTRODUCTION

Morpho-geographical taxonomy, although indispensable for evolutionary and genetical studies as a prime source of information and as a reference system, is no longer adequate for many such investigations into the nature of variation, the operation of natural selection and evolutionary processes. The study of evolutionary relationships forces us to study not only the morphology as expressed by the phenotype but the hereditary bases of the morphological characters, in other words the genotype. The plasticity of the phenotype and the nature of its genetic control under different environmental conditions the pattern of variation shown within populations, and the role of the reproductive system in affecting the type of variation pattern are also of the greatest importance.

For these reasons taxonomy has been extending its activities into the fields of cytology, cytogenetics, genetics, population studies, experimental comparative cultivation and similar activities, many of which are nowadays grouped under the general title of *Biosystematics*.

Two points must be stressed at the outset. Firstly, taxonomy, as has already been discussed, has a functional or practical aspect. Its basic task of organising information about organisms and expressing the results in a morphological framework with formal nomenclatural recognition has not been superseded, nor is it likely to be in the foreseeable future, so long as man relies on visual information. Secondly, the morphological framework will be continually susceptible to revision as a result of evidence from these newer fields. Some people feel strongly, however, that the temptation must be resisted to alter the morphologically based taxonomy every time some piece of "experimental" information comes to hand. In other words, what is new and different is not necessarily a better basis for classification: a broad view must be taken of all evidence. None the less, since a vast amount of the data from cytogenetics, etc. will not be susceptible to morphological treatment or expression, or will in fact cut across the morphological-taxonomic frameworks possible, a second activity for taxonomy will be to devise means of integrating and relating such data to formal taxonomic schemes. Alternative classifications tressing, for example, cytogenetic relationships irrespective of morphology, will become necessary. These special classifications, such as the coenospecies-ecospecies-ecotype scheme, come within the scope of biosystematics but must be attached in some way to orthodox taxonomic schemes.

The Population Concept

It is a tenet of modern taxonomy, although stated in many different ways, that species and lower taxa should be formulated around natural populations. This implies that the inherent variability of species is accepted as a phenomenon requiring study and appropriate treatment, not just as an inconvenience in that individuals deviate from a single pattern or type.

Even though a species is originally described (as indeed the vast majority have been) from one or a few specimens, it is realised that these specimens represent only a very small percentage of those plants that bear these same or similar characteristics in the field. This is well put by Dobzhansky (1950): "the classic way to study a species, or a part of it, is to determine the modes or averages for as many traits as possible, in as large samples of its representatives as practicable. The resultant system of averages is taken to be a common property of the species as a whole. . . ."

The greater part of descriptive taxonomy even today is not able to deal adequately with these problems due to the inadequacy of samples available. Heslop-Harrison (1952b) goes further and suggests that the failure of taxonomists to define populations rather than individuals is due partly to their not adopting methods necessary to acquire knowledge of population structure, and partly to the difficulties of presenting the data in an easily assimilable form for others.

If we consider the case of a taxonomist describing a new species, we can see how far these strictures may be justified. It is legally permissible to provide no more than an indication in Latin of how the proposed species differs from other ostensibly related species—in other words, a diagnosis. The description of the samples regarded as representing the new species may be as detailed or as brief as the individual taxonomist wishes. Measurements of particular organs may be included or not, an indication of the range of variation found in the samples may be given, or, as is more frequent, restricted to those features regarded as differential vis-à-vis related species.

The only check on the validity of the description of a new species is in terms of the nomenclatural aspects laid down by the *International Code of Botanical Nomenclature*. The taxonomic validity of the species may only be tested, in many cases, by the passage of time—when it is possible to collect it again, compare it, discover how representative the original samples were, whether the diagnostic characters hold or break down through intergradations in later samples.

As a result it is perfectly possible to describe species in such a way that other taxonomists cannot interpret the unit which it is proposed to describe or even estimate the probability that any individual specimen belongs within that unit or not (Heslop-Harrison, 1952b).

In general, however, the whole basis of taxonomy is reproducibility—in other words the description of a taxon permits further samples to be accommodated in it. Of course, later collections necessitate in most cases a modification of the original description (but not, we must add, of the name) until, in theory, the point is reached at which all variations, all deviations from the original specimens and description, have been allowed for. Then complete reproducibility of identification is achieved for the named taxon in question. This is the traditional and inevitable basis of modern "typological" taxonomy. The history of taxonomic investigation of a flora is a series of such stages of improved knowledge.

From randomly collected samples "types" are *recognised* on the basis of comparison and contrast, and *described* as new taxa in terms of their resemblances amongst themselves and differences from other groups of organisms. Later collections are compared with these "types" or descriptions, or with that combination of specimens and literature plus experience which we have previously referred to as the typological picture (p. 11), in order to see which of them they agree with sufficiently for inclusion within them.

Population taxonomy, in the sense in which it is widely employed today, rejects to a large extent this procedure, although it cannot escape from it entirely. In sexually-reproducing organisms the evolutionary unit is the *local breeding population*, and this is the basis of the units of classification of Biosystematics. The populations have therefore to be studied in the field and sampled statistically in such a way that the samples give a valid representation of their variation. Even this is insufficient because statistical sampling methods are not necessarily valid for biological populations, particularly in outbreeding organisms where each individual is in some degree unlike every other genetically and with a different evolutionary potential* (cf. Mayr, 1959a). The system of averages mentioned above is, according to Dobzhansky, a statistical fiction, and a biologically realistic description of a species or a race should at least in theory, he says, "indicate the frequencies in its gene pool of gene alleles and chromosomal variants". It is more important in living organisms to consider diversity and variability than types and averages. This is a consequence of the realisation that with gene heredity, a population cannot be simply regarded as a series of characters and averages, but as a space-time phenomenon.

Populations defined by biosystematists obviously have a different significance from those defined by taxonomists and will be discussed later, but it follows that taxa will have defined geographical distributions, since they consist of one or more populations. Species and subspecies are regarded by taxonomists today as populations, or series of populations of individuals,

* This point was commented on long ago by Johannsen: "We must work at the study of heredity *with* mathematics, not treat it *as* mathematics."

with all the dynamic changes that this implies. As Mayr has recently pointed out, it may be noted that the population concept came into evolutionary and genetic thinking from systematics and not vice versa.

That any named taxonomic species may not in fact be *known* to represent a population is not a criticism of taxonomy but a reflection of our stages of knowledge of species. If in fact it turns out not to be a population but random specimens belonging to different populations, then it may be rejected or remodelled in time. What is important is that the inductive leap from a taxonomic species to a biological population should not be made without confirmatory evidence. It is very necessary to stress this point (discussed further on p. 275) since it is one of the commonest pitfalls open to taxonomists today.

When a taxon is known to represent a population, its characters have to be considered in relation to the individual members and in relation to the population as a whole. There are, therefore, *individual* characters—those possessed by some individuals, and *population* characters—those shared by the majority of the population individuals. Variation in terms of populations can therefore be divided into:

(1) *Individual variation*—i.e. between individuals within the population. Included here is plasticity of the individual itself (phenotypic plasticity).

(2) *Collective or group variation*—i.e. between populations.

Population Definitions. As "population" is becoming a key word in taxonomy, it is important to define what is meant by a population as precisely as possible in any particular context. Populations are defined in different ways for different purposes. That they are groups of individuals is common to all these definitions. The common-sense or general definition, used largely by taxonomists, is: *any group of individuals considered together at any one time because of features they have in common.* In practice this comes to mean a group of plants growing together which look similar, or a series of such plants as they occur over a particular area.

In population genetics, the importance of the cohabitation of the particular group of individuals is that they can (in sexual organisms) exchange genes. It is therefore a breeding population, and defined as a group of individuals between which there can occur a greater or lesser degree of crossing (or potential crossing) with consequent gene flow. It is formed of an indefinite number of individuals. Two points may be noted: firstly, the possibility of interbreeding, secondly, genetic similarity. Populations in this specialised sense are not defined on the basis of morphological characteristics which the individual members share, but because they grow in breeding contact and thereby share a common gene pool. Populations are regarded in this sense by many biosystematists. The morphological characteristics are secondary and serve to recognise the population (and this not always), not to define it. Populations of this nature are often termed Mendelian and are defined by Dobzhansky as "*a reproductive community of sexual and cross-fertilising individuals which share in a common gene pool*". The largest and most inclusive Mendelian population is the "biological" species. Groupings above the species level such as subgenera, genera, etc. do not normally possess common gene pools.

Sirks (1951) distinguishes between plurispecific and unispecific populations. The *plurispecific* population comprises all the plants (and animals and micro-organisms) that live together within the same area. The area may be a square yard or a square mile; a wood or an ant heap. The relationships here are largely sociological. The *unispecific* population he defines as "a group of individuals all belonging to the same species, and living in the same area". Such a broad distinction is of little meaning for taxonomy but if used in a narrower sense it is valuable, since a population may be of a recognised taxonomic species or it may be of a group of plants which are being considered for specific treatment and may in fact contain individuals which will be assigned to more than one species. In a sense, therefore, to talk of a unispecific population may be begging the question, especially if, as Sirks himself says, the micro-evolutionist should avoid using such taxonomic terms as species because they cannot be precisely and accurately defined and should restrict his units to populations and individuals.

The third use of the term is made by the statistician for whom the population is an abstraction. Mather (1946) defines it as "the hypothetically infinitely large series of observations or individuals of which those observations or individuals actually obtained form a sample". There is a remarkable similarity between this definition and the taxonomic concept of a species and our state of knowledge of it at any particular time, as discussed above (p. 352). As Heslop-Harrison says, when we describe a biological population in terms of statistics we are in effect attempting to estimate the parameters of the hypothetical infinite population of which the data we use form a sample. How closely the estimated parameters of the hypothetical population will agree with those of a real biological population of the unit which it is attempted to define will depend on the representativeness of the samples and the mathematical treatment of them. There are very serious difficulties in fact in applying mathematical methods of analysis to biological populations which are units of evolution. Mayr (1959a) puts this point provocatively by asking "Is it not perhaps a basic error of methodology to apply such a generalising technique as mathematics to a field of unique events, as is organic evolution?"

Populations are always subject to change: they vary in size, sometimes noticeably from year to year, in composition and area. They may increase in size or decrease, It is remarkable in this connection that much attention is paid to extensions of the range of species and the reasons for this (for example, the remarkable spread of introduced weeds such as *Veronica persica*) but less notice is given to contractions in range, which must be very frequent, unless the effect is striking as in the case of so-called relict species whose habitat is in danger of elimination.

Variations in population size in many species of *Orchidaceae* are well known: some years no trace of a species may be found in a particular habitat and its reappearance in later years is widely hailed. Sharsmith (1961) commented on some striking changes in populations of *Hesperolinon* (*Linaceae*) in N. America. In one habitat *H. californicum* was found abundantly in 1952, and was homogeneous in appearance. In 1957 she had to search to find even a few plants and these were more or less intermediate between *H. californicum* and *H. micranthum*. This phenomenon has been noted in other members of the genus and she suggests that "selection of different hybrid genotypes may take place under the varying climatic conditions of different years and that these genotypes may be maintained by inbreeding until they are displaced in subsequent years by more highly selected genotypes". Population size may therefore be a reflection of genotypic composition and the factors affecting this. One of the most important of these factors is the breeding system which is discussed at length in a later section.

Small populations may be succeeded in later years by large populations due to the high reproductive capacity of the components, as in *Atriplex* where sufficient seed may be produced by a small population to ensure the production of a large one, even allowing for normal mortality.

Mendelian populations may diverge in time into two or three more derived units; or previously separated populations may merge and eventually fuse with each other, the results often being spectacular as in cases of introgressive hybridization discussed in Chapter 14.

These dynamic changes in populations take place gradually and continuously, and, as Dobzhansky (1950) points out, the boundaries between populations may not be clearly distinguishable, leading to an unavoidable degree of arbitrariness in their delimitation, and in the delimitation of the races and species which are based on them. This point is worth stressing since it is often claimed that biological species based on Mendelian populations are objectively delimitable, compared with subjectively delimited taxonomic species. Populations may be clearly delimited spatially—frequently they are used in a local sense. Their limits may be defined by small topographic or physical or ecological discontinuities. Much attention has been paid to such local populations in experimental taxonomy and genecology since, as already said, they are regarded as the fundamental evolutionary unit. In this connection the local populations (*topogamodemes*) are defined as those individuals occupying a precise locality which form a reproductive or breeding unit. How the breeding relationships of the individuals is established is, however, a complex problem. In practice, only minute samples of populations are tested and it is assumed on the basis of their morphological characteristics that the other individuals have the same or similar genetic constitutions and breeding relationships.

Small local populations may show certain characteristics in relation (adaptation) to the local environmental conditions and may therefore constitute ecological varieties or ecotypes. These may be genetically fixed—genecotypes—or simply phenotypic modifications—phenecotypes. They are discussed in Chapter 12.

Populations vary in size, however, and may be considered on a wider, regional basis. The principle of populations replacing one another in sequence in geographical regions is the basis of the *Rassenkreis* (mosaic of races, Rensch), largely used in zoological taxonomy and nowadays increasingly so in botanical taxonomy (see Chapter 9).

When loosely defined, populations do not necessarily coincide with any particular taxonomic unit and when defined in the strict genetical sense their coincidence with taxa is even less frequent.

Within populations the assemblage of individuals with the same or essentially the same genotype is known as the *biotype*. In autogamous plants (inbreeders) the biotype may be composed of very numerous, more or less homozygotic individuals. In allogamous groups (outbreeders) each biotype generally consists of a single individual. Thus a population may consist of one or more biotypes. In inbreeders the population and the biotype may in effect coincide, which tends to cause taxonomic problems if there are several such local, uniform populations each with a slightly different morphology (see below, p. 377).

The realisation that the population, as an interacting group of individuals rather than the individual itself, is the fundamental unit of evolution has caused much attention to be focused on the genetic structure of populations. It can be said that evolution proceeds by changes in the gene frequency of populations (or changes in the frequency of chromosome arrangements or numbers) and that when isolation occurs between populations of the same species that kind of evolution known as speciation may result.

Only rarely does the individual play a primary part in speciation, and even then it is only temporary unless the individual concerned gives rise to an increasing number of progeny, thus building up a population more or less like itself (cf. speciation through polyploidy, Chapter 6).

The genetic structure of a population, the way in which the various alleles of the different genes are distributed throughout it, is largely determined by how far and in what way the breeding system departs from being at random (Crosby, 1954). This is considered in detail in a later section.

Causes of Variation in Populations

The basis of variation between the individual members of a population is (*a*) external environmental modification, (*b*) mutation, (*c*) genetic recombination. The pattern of this variation is largely determined by the reproductive (breeding) system, and is utilised in adaptive evolutionary processes by natural selection.

Phenotypic plasticity in response to environmental factors has already been discussed in detail (Chapter 10) and clearly contributes to the differences between individuals in any population. Even in genetically homogeneous populations at the same stage of development, there will be small dissimilarities between individuals due to the slightly different intensities of the various environmental factors. Such variation in quantitative features as stem height, leaf dimensions, pollen size, etc. tends to be continuous and follow a normal distributional curve (bell-shaped curve). When there are adequate samples available this kind of variation may be treated mathematically. Observations on the various characters measured or assessed can be tabulated in a frequency distribution table:

Number of Individuals	*Length of Sepals* (mm.)
1	7–8
3	8–9
7	9–10
24	10–11
28	11–12
9	12–13
4	13–14
2	14–15

This information can also be expressed in a graph or histogram and provided the number of samples is sufficiently large the frequency distributions of the classes would form a normal distributional curve (or an approximation to it). The class containing the highest number of observations (called the mode) will correspond to samples growing under the commonest conditions in the habitat. Classes corresponding to less common conditions will contain fewer observations and fall on either side of the modal class. Such curves can be defined by the mean (M) and the standard deviation (σ or S.D.) which is a measure of the scatter of the observations around the mean. Meristic variation, as in number of parts of the flower, which has already been mentioned in Chapter 4 (p. 117) may also follow a more or less normal curve of distribution. When data show a non-normal distribution, an appropriate transformation technique may be applied (see Cochran, 1938). Detailed accounts of the statistical and biometrical methods that can be applied to biological material will be found in standard works such as Goulden (1952), Cochran (1953), Fisher (1954) and Snedecor (1956).

Such environmentally caused variation acting on a common genotype is likely to be encountered by the taxonomist in apomictic groups or in strict inbreeders (see below).

Genetic Variation

Apart from the special cases just mentioned, populations normally contain genetic variation caused by random gene mutation. This variation is shuffled from generation to generation by gene segregation and recombination, depending on the breeding system operative in the population.

(1) *Gene mutation.* In order to have the greatest possible significance for evolutionists we should follow Mayr (1942) in defining mutation in the broadest possible terms as "a discontinuous chromosomal change with a genetic effect". Adopting this definition we must consider mutation at three levels: those of the gene, the chromosome and the chromosome set or genome. The boundaries between these levels are difficult to define, and some mutations at one level are similar in their effects to those at the level above or below; the difference is mainly a shift of emphasis. Gene mutations primarily alter developmental processes, and thus have specific effects on individual phenotypes. Chromosomal and genomic mutations, on the other hand, either alter whole series of different processes, or affect the relationship between them. They are therefore most likely to have a series of generalised effects on the phenotype, and also tend to bring about a new organisation of hereditary material; this contributes to the reproductive isolation of the mutant individual and thereby to the origin of "biological" species (Stebbins, 1959b).

The only practical working definition of a gene is that of Stadler (1954): "the smallest segment of the gene string that can be shown to be consistently associated with the occurrence of a specific effect".

(2) *Genes and characters.* The effects of gene mutations on the organism—which are what immediately concern the taxonomist—are various. They may turn the zygomorphic corolla of *Antirrhinum* into an actinomorphic one, the gamosepalous calyx of a *Silene* into a polysepalous one, or the spurred corolla of an *Aquilegia* into a spurless one (Pražmo, 1960)—all apparently atavistic mutations (i.e. in the direction of the inferred ancestral condition).* At the other extreme effects can be so slight that they blend into the modification of the phenotype produced by the environment.

Other outstanding effects of single-gene differences range from a change of habit from annual to biennial (*Atropa*) to ability of chromosomes to pair (Riley, 1960). Further examples of the genetic basis of taxonomic differences are given by Clausen and Hiesey (1958a).

Mutation is believed to be at random, and all sorts of effects can, on occasion, be favoured by natural selection and therefore contribute to evolutionary change. However, most mutations are probably disadvantageous to the plant, and this will apply particularly to most mutations with a large phenotypic effect. According to Mather (1953), mutation appears to be more a process of degradation—it is retrogressive in that it wipes out or reduces the capacity of the individual to display some character, etc., rather than progressive in endowing it with novel ability. The great majority of mutations which contribute to evolution have relative slight, but cumulative, effects

*Not all such teratological phenomena are genetically induced; other examples are environmentally produced, or sub-pathological in nature. Plant teratology and its phylogenetic implications are discussed by Heslop-Harrison (1952c) and Takhtajan (1959a). A comprehensive review of the whole subject is given by Astié (1963).

(cf. Clausen & Hiesey, 1958a). In *Geum* Gajewski (1957) found that the biggest influence exserted by a single gene was to produce a recognizable difference in flower colour or a notch on the upper petal margin.

(3) *Polymorphism.* A commonly occurring kind of genetic variation in populations is that termed polymorphism. Huxley (1955) in an extensive review of the subject terms the phenomenon *morphism* which he defines in the sense of Ford's genetic polymorphism: "in which (usually sharply distinct) genetic variants or *morphs* coexist in temporary or permanent balance within a single interbreeding population in a single spatial region, and in such frequencies that the rarer [variant] cannot be due solely to mutation, or to the spread of selectively neutral mutants". Although it has not been widely adopted, Huxley's term morphism is preferable to polymorphism since the latter is often employed in a much wider sense to cover any extensive kind of variability shown by a taxon.

Morphism should be distinguished from meristic variation, gynandromorphism (although some permanent sex-differences are included by some writers), and sporadic mutation. It is widespread and probably a feature of populations of most species. Dobzhansky (1950) doubts if any sexual species is adaptively quite monomorphic and suggests as a working hypothesis that the amount of morphism present in a population or species is a function of the variety of adaptive niches which the populations are able to occupy and exploit.

A very wide range of characters is involved in morphism. Examples are flower colour forms (cf. Dobzhansky & Epling, 1942, on *Linanthus parryae* in California), flower patterning, chemical forms (*Eucalyptus*), smooth and hairy-stemmed variants in *Digitalis purpurea* in which the proportions of the two types vary according to the geographical region, leaf-marking in *Plantago maritima* (Gregor, 1938), etc.

Such features may be manifest or cryptic, discontinuous or more rarely continuous, adaptive or only presumed to be so. Their genetic basis may be genic (one or more genes or groups of closely linked genes), chromosomal-segmental (inversions or translocations) or chromosomal.

Not all variation between biotypes in populations are so well-marked; on the other hand, there are characters which are controlled more or less equally by many genes with small effects and the distribution of such features in a population will often approximate to a normal distributional curve. There is an intergradation between the two types of distributions (normal and polymorphic) in populations although they are distinct in their extreme forms.

Both environmental and hereditary variation can give rise to a normal distributional curve and normally both are involved in the phenotypic variation found in populations. They cannot be distinguished simply by inspection—only by breeding or selection experiments.

(4) *Pleiotropic gene action.* Many mutations, possibly the majority of them, affect several characteristics of the adult phenotype. This phenomenon is

known as *pleiotropy* and is the inevitable result of the complexity of the developmental processes which link the gene to the adult character. Although most, if not all, genes probably affect only one primary process of cell metabolism, each primary metabolic change occurs in cells of different types throughout the plant body. Pleiotropic genes may therefore produce different but related effects on different organs. In *Nicotiana*, for instance, the gene which produces long petioles also produces more acuminate leaf tips, longer calyces with narrower sepal lobes, corollas with more acuminate lobes, longer anthers and more attenuated capsules.* The taxonomist, writing the diagnosis of a new species, may often distinguish it from its nearest allies on characters which are in a similar relationship to one another and may be result from a single pleiotropic gene. This does not mean that characters related in this way should not be considered adequate for the differentiation of species, but it does mean that their genetic distinction is less than in species differing from one another in characters based on different genes and which may therefore prove more stable. The genes effect chemical processes which produce the morphological features used by the taxonomist. These are considered by the taxonomist in terms of individual characters, but he is usually unable to analyse their genetic basis. Nor can the individual effect of a gene be determined apart from the genetic environment in which it operates; in a new genetic environment its effect may be different.

Pleiotropic effects may have a considerable influence on the selective value of a mutation. These effects vary in their selective value: some are advantageous, some neutral and some even harmful. Because of pleiotropy the action of natural selection is likely to be "indirect": characters which have no selective advantage in themselves can become established through the secondary effects of genes which have been selected and which are therefore valuable to the organism for completely different (often physiological) reasons.

Gene mutations which contribute to evolution do not (as the early Mendelian geneticists believed) produce a new adaptive type at one step. They are the bricks which must be combined in particular ways in order to produce adaptive variation.

Recombination and the Reproductive System

Gene recombination is as important as mutation in the production of adaptive variation in higher plants. The recombination of genetic variability is largely determined by the breeding system, and as Barber (1956) puts it, "the whole origin of sexual differentiation is to be sought in this evolutionary principle". A whole series of mechanisms promoting cross-fertilisation and genetic systems for self-incompatibility have evolved; these have their selective advantage in promoting gene recombination. There are, however,

* See Stebbins (1959b) where the striking differences in *Nicotiana tabacum* are illustrated in fig. 2.

situations where this may be disadvantageous, so that natural selection may favour self-fertilisation or apomixis.

Cross-fertilised populations contain a large store of variability hidden in the form of recessive genes in the heterozygous condition. Most of this variability is non-adaptive, and because it is continually being brought out by recombination it reduces the immediate fitness of the population. On the other hand, this hidden store of constantly changing variability acts as an "insurance": it increases the probability that genotypes will arise that will be adapted to new environments which changing conditions may require the population to face.

It is a generally accepted premise that every population comes to strike a balance between the immediate fitness but evolutionary fixity that comes from inbreeding or apomixis, and the reduced immediate fitness but evolutionary plasticity that outbreeding confers. A compromise between long-term and short-term advantages is therefore found in natural populations (cf. Mather, 1953; Stebbins, 1959b).

There is plenty of evidence to suggest that genetic systems involving self-fertilisation, apomixis and other devices promoting immediate fitness are always secondary and have been derived from "flexible" cross-fertilised systems. Self-pollinated and apomictic populations often occupy temporary or pioneer habitats where their immediate fitness helps in the rapid colonisation of relatively uniform, newly available habitats.

Of all the additional approaches which are available to taxonomy today, knowledge of the breeding system is one of the most important, for not only does it affect the pattern of group variation but the evolutionary potentialities of the group concerned. The breeding system determines the spread of genetic variability in the population and this in turn is usually reflected to some extent in the morphology of the population.

The breeding (mating, reproductive) system can be defined as the relative frequency of cross- and self-fertilisation in a population or taxon, which largely determines the average degree of heterozygosity. The way in which reproduction is organised (cross-fertilisation, self-fertilisation, etc.) in a species population or lower taxon determines how similar or how different the gametes will be which come together in fertilisation. For obviously the gametes produced by different individuals will, when brought together in fertilisation, produce an individual which is heterozygous to at least some degree. If self-fertilisation is the rule (i.e. the gametes come from the same individual), the individual produced will be more or less homozygous.

Populations which habitually reproduce sexually are known as *amphimictic;* those in which free exchange of genes is possible are termed *panmictic* (Wright, 1943). The local breeding population (gamodeme) is normally panmictic although it should be remarked that complete breeding freedom is rarely attained except in small populations under certain conditions. The larger the population the greater may be the restrictions on gene flow:

breeding tends to take place principally between neighbouring individuals, and in large populations the individuals at extreme ends may come to be as effectively isolated from one another as if they belonged to different populations. Other factors, such as the breeding system itself, will have an important effect on the effectiveness of gene flow between individuals in such populations. Thus a distinction is often made between the effective size of a population and its absolute or apparent dimensions, the effective size determining to a large extent the degree of gene exchange possible between the members, while the apparent size is considered without relation to gene flow between individuals. As is discussed later, marginal populations, which are often genetically quite different from major population areas, present problems of their own, both to the geneticist and the taxonomist.

Apomixis

In some groups of plants sexual reproduction is absent or plays an unimportant role. The term *apomixis* has been employed in a general sense to cover all those types of non-sexual reproduction which substitute or replace the sexual method. A full survey of apomictic phenomena in higher plants is given in Gustaffson's classic work (1946, 1947).

It is not always easy to demonstrate that a given species or race is apomictic, but when it has been possible to do so there is no difficulty in recognising apomixis in plants as distinct from other reproductive modes in that pollination does not operate except in a stimulatory capacity, or is not effective in a genetic sense. There are numerous cases, however, where sexuality and apomixis occur in the same species and where it is not practicable to separate plants showing them, as in *Poa pratensis* where there appear to be transitions between complete sexuality and complete apomixis. Such cases are known as *facultative* apomicts. In *obligate* apomicts such as *Allium sativum*, no sexual phase is known.* There are added complications possible in some cases where races or ecotypes of the same species are apomictic (facultative or obligate), while others are fully sexual, e.g. *Bouteloua curtipendula* (*Gramineae*) and *Hieracium umbellatum* (*Compositae*).

Apomictic phenomena can be divided into two main classes: (1) Vegetative reproduction, (2) Agamospermy.

(1) *Vegetative reproduction* ("vegetative apomixis"). This includes all cases where structures such as bulbils, tubers, rhizomes, stolons, etc. which are normally accessory means of reproduction take over the whole reproductive function. This appears to happen in nature in several species, such as *Acorus calamus* which in Europe is a sterile triploid and maintained by vegetative spread. Other examples of clonal hybrids include various taxa in *Mentha*, *Circaea* × *intermedia*, etc. Another well-known example is *Elodea*

* Completely obligate agamospermous apomicts may not occur, since some small degree of sexuality may be present but pass unnoticed.

canadensis, which in Europe is maintained by vegetative spread. The male plant is very rare and the female plants set no seed. Like other vegetative apomicts in the Northern Hemisphere, it is normally sexual in other parts of its range.

A well-studied case is *Ranunculus ficaria*, where some races reproduce almost exclusively by axillary tubercles in the absence of seed development. They are referred to subsp. *bulbifer*. Usually diploid plants have no axillary tubercles while tetraploids possess them in Britain, but the situation is different in other races elsewhere in Europe. The distribution of the tubercle races differs in different countries. Vegetative reproduction in this instance is important in extending the range of the species, and the more widespread race in any area may depend on ecological and geographical limiting factors.

Artificial clones are widely utilised in horticulture and agriculture, as in potatoes, yams (*Dioscorea*), *Dahlia*, etc.

A high rate of vegetative reproduction in a sexual taxon may produce the same effect as vegetative apomixis, as in *Sparganium erectum*. Similarly *Calystegia pulchra* (*C. dahurica* auct., cf. Brummitt & Heywood, 1960), a probably Asiatic species which was introduced into Britain during the last century, appears to have spread there largely due to vegetative reproduction of a few clones, although the species is not apomictic. The rare *Pinguicula vallisneriifolia*, known from only a few relict localities in south-east Spain, reproduces almost entirely by means of filiform "stolons" which root and form new rosettes; the flowers set seed normally but only a small proportion is viable and seedlings have not been detected in nature.

The above cases are examples of extra-floral vegetative reproduction. Some floral examples will now be considered. Vegetative apomixis is found for example in several species of *Allium*. This is often of the type in which the propagules replace some or all of the flowers in the inflorescence. This is known as pseudo-vivipary and also occurs in *Saxifraga*, *Agave* and notably in grasses such as *Poa*, *Festuca*, *Deschampsia*, etc. (cf. the epithet *vivipara*). These species sometimes produce no seed and are equivalent genetically to obligate seminiferous apomicts.

Pseudo-vivipary and bulbifery occur in varying degrees: in some cases only occasional propagules are produced as a result of environmental factors; vivipary may be partial and under genetic control; it may be the only method of reproduction to the exclusion of sexuality. In *Polygonum viviparum* sexual and pseudo-viviparous "flowers" occur in the same inflorescence.

It seems possible that formerly widely distributed species might have become so restricted in range as to be represented by single clones. The widespread genus *Houttuynia* (*Saururaceae*) comprises a single high polyploid species which is apparently a single apomictic clone (Babcock & Stebbins, 1938).

Knowledge of the occurrence and degree of vegetative spread in even well-known species is often lacking. It has only recently been shown fairly conclusively that the bluebell, *Endymion nonscriptus*, does reproduce vegetatively although less effectively than by seed (Wilson, 1959). It is effective on a local scale but not on a wide scale.

The genetic and evolutionary importance of the above-mentioned forms of apomixis lie in the fact that the individuals produced have the same genotype and consequently belong to the same biotype. There is no formation of new genotypes through segregation of parental genes as there is no sexual process. This restricts variation, apart from mutation, and without variation further evolution is not possible. The taxonomic consequences of such means of reproduction are considered below.

(2) *Agamospermy.* The other group of specialised processes, known as agamospermy, includes all those types of reproduction which result in the formation of seeds and embryos by means of non-sexual processes. The whole sexual structure and mechanism exists, but meiosis and fertilisation is avoided or side-tracked so that the resultant embryo formed is genotypically identical with the maternal parent. Several types of process are involved (diplospory, apospory, pseudogamy), but their details will not be considered here. Agamospermy is known from about 40 families of flowering plants. Its importance for the taxonomist is considerable and its interpretation affords many problems at the present day. Frequent failure to understand it has been responsible for much confusion in the past.

In cases of agamospermy the progeny, i.e. the individuals formed as a result of reproduction, are genetically identical with their parent and constitute therefore a clone of the same biotype. If these individuals were fixed (and we shall see immediately that they are not), it would be possible to describe the biotypes as "species" since they have many of the characteristics of taxonomic species such as constancy, (relative) lack of variation, etc. On the other hand, if an attempt is made to consider these biotypes for specific status in the light of "biospecific" definitions, it will be seen immediately that they are not eligible (see Löve, 1960c). The "biological species" concept is based on the idea of populations with gene exchange among the constituent members and barriers to gene exchange with members of other such populations. Firstly, the apomictic biotype is not, strictly speaking, a population in the gene-pool sense but one derived by the multiplication of a single individual. The very nature of the agamic biotype precludes considerations of gene exchange. Moreover, several apomictic clones have arisen as the result of the combination in their genotype of the genes from previously sexually isolated species and without apomixis they could not persist because of their sterility.

All the British plants of *Taraxacum* which have been tested have been shown to be apomictic, and cytological examination suggests that apomixis has been preceded by hybridisation and followed since then by mutation. Other species of *Taraxacum* are sexual—inbreeders or outbreeders. The number of apomictic "species" described depends very largely on the availability of a specialist in the flora concerned. For example more than 500 have been described from Scandinavia; many fewer are recorded from Spain although the total number of Angiosperms in the flora is two or three times higher. Clearly such reproductive phenomena play havoc with normal taxonomic procedures and there is still no agreement as to future treatment of such groups, largely because some taxonomists are seeking entities which in a biological sense do not exist and which taxonomically make little sense. Progress in the taxonomic studies of apomicts is discussed below (p. 381).

A further complication occurs in groups where apomixis is only partial, e.g. *Rubus fruticosus* agg., *Potentilla* spp. Both reduced and unreduced embryo sacs occur side by side in the same population. Occasionally they reproduce sexually to give rise to a new series of biotypes which segregate, giving rise to new apomictic clones, sexual offspring and new "species". As a consequence, complex reticulate patterns of variation are formed and distinctions between "populations" are blurred. The morphological distinctions between the clones may be very slight but the clones are constant (apart from mutation) and occupy areas of distribution by propagation and dispersal, simulating sexually reproducing groups.

In the genus *Potentilla*, which has been extensively studied cytogenetically (notably by A. and G. Müntzing and by Rutishauser), apomictic seed formation occurs in several species, but the degree varies in different species. In the *P. argentea* group, for instance, all the Scandinavian material tested has proved to be apomictic, while in the Mediterranean region a sexual strain occurs in which some of the embryo sacs are unreduced although most are reduced (*P. argentea* var. *calabra*). In *P. verna* pseudogamous apomixis is widespread and at least five different chromosome numbers are recorded, but there appears to be no correlation with morphological differentiation.

In *Poa pratensis*, mentioned above, 5-10 per cent. sexual seed is formed in some cases; fully sexual biotypes also occur. A detailed study of the reproduction of this species is given by Grazi *et al.* (1961). A similar situation occurs in *Poa alpina* where both pseudo-viviparous and seed-bearing biotypes are found, and in both apomixis occurs, but in the seed-bearing biotypes it is not absolute. Moreover, the average degree of apomixis varies from country to country. It is, for example, higher in Scandinavia than in Switzerland and Poland; so is the average chromosome number.

In predominantly apomictic asexual species such as these Poas there is a balance between seeds which originate through apomixis and those formed by normal sexual means (Clausen, 1954). In partially apomictic biotypes, the progeny from apomictic seed are normally more vigorous than those produced from sexual seed, so that the latter may not therefore even be detected. The evolutionary importance of this phenomenon is that it allows some possibility for apomictic biotypes to hybridise and consequent recombination may allow the production of new apomictic biotypes in one or two generations (Clausen & Hiesey, 1958b).

In the *Ranunculus auricomus* group, which reproduces by pseudogamous apomixis, a vast amount of polymorphism has been observed. This is explained largely on the basis of interspecific hybridisation, even though it has not proved possible to demonstrate experimentally that apomixis in the group is facultative. As Marklund & Rousi (1961) observe, "the occurrence of new, apparently recombinant, biotypes can scarcely be explained without an assumption that the apomicts still have some traces of sexuality".

Despite their origin it must not be thought that apomictic microspecies can show no variation. On the contrary, several kinds of variation are known. In *Alchemilla*, in the microspecies of the *vulgaris* and *alpina* aggregates, variation parallel to that shown by sexual species has been recorded. This variation can be tested and shown by cultivation to be genetic in origin: three main kinds occur: (1) geographically significant variation with indistinct morphological boundaries, e.g. *A. filicaulis* and its more lowland var. *vestita* (*A. vestita*); (2) ecologically significant minor variation analagous to ecotypic variation in sexual species; (3) sporadic variation which is apparently non-adaptive, such as adpressed hair variants of a species normally possessing spreading hairs. These may be recognised as *formae* taxonomically (cf. Turesson, 1943, 1956; Walters, 1949; M. Bradshaw, 1963).

Amphimixis

The degree of heterozygosity depends to a large extent on the relative frequency of self- or cross-fertilisation. *Endogamy* (inbreeding) is defined as pairing between closely related individuals, while the most intimate form is termed *autogamy*, i.e. self-fertilisation. The result of endogamy is to increase homozygosity, and the consequence of continued endogamy in succeeding generations is the division of a population into a number of "pure lines". The more intimate the system of inbreeding the fewer generations are needed to reach the homozygous state. This should be compared with *exogamy* (*allogamy*), outbreeding, which is pairing between less closely related genotypes —those which belong to different lineages (Darlington). Exogamy increases heterozygosity; it implies genetic lability while endogamy implies genetic fixity. The majority of flowering plants probably lie between the two extremes of habitual inbreeding and habitual outbreeding and outcross to a greater or lesser degree. It is widely considered that the primary function of the pollination mechanism of flowering plants is to permit outcrossing: this is a line of thought which stems from Darwin's view that plants "ought" to be cross-pollinated. There are, however, secondary mechanisms or devices superimposed on the reproductive systems which favour selfing and consequently endogamy, or crossing and hence exogamy.

(a) *Mechanisms favouring inbreeding*. The majority of devices which favour intimate endogamy are connected with the development and structure of the flower itself. Perhaps the best known of these are examples which illustrate the phenomenon of cleistogamy, e.g. in *Viola* where the flowers fail to open in some cases, with the result that self-pollination and fertilisation is enforced. In the example of *Viola* two kinds of flower are found on the same plant—the normal condition in cleistogamy. Other examples include *Glechoma*, *Oxalis*, *Lamium*. In true dimorphism, i.e. where different plants bear the different types of flowers, some plants, as in *Impatiens noli-me-tangere*, produce exclusively small closed cleistogamous flowers, some the usual large showy flowers, and some both.

Cleistogamy appears to be closely connected in some cases with ecological conditions. Prolonged wet weather seems to favour the production of cleistogamous flowers (as in *Lamium amplexicaule*), as does deep shade. Insect pollinators are often scarce in shady conditions which helps to explain the adoption of cleistogamous flowering. Borgström has shown how in *Viola* cleistogamy is more frequent at higher altitudes, apparently due to photoperiodic response. Huxley suggests that cleistogamy has evolved through ecological pressure causing failure of the flowers to open (pseudocleistogamy) with consequent selfing, followed by further adaptations such as reduction in the petals and stamens in size and eventual germination of the pollen *in situ* without release from the anthers.

Reviews of cleistogamy in Angiosperms are given by Uphof (1938) and J. K. Maheshwari (1962).

Self-pollination may also be considered advantageous in plants growing in very dry areas or regions of drought where conditions are not suitable for cross-pollination (Stebbins, 1957). This is well illustrated by *Hesperolinon* (*Linaceae*), where all species are ephemeral annuals growing under extremes of dryness and heat in or near chaparral in California. The flowers which are pseudo-cleistogamous (see below) come to maturity after the moist spring conditions have passed.

The occurrence of dimorphic flowers—cleistogamous and chasmogamous—has been described for several members of the *Commelinaceae*. Detailed studies have been made on *Commelina forskalaei* (Maheshwari & Baldev, 1958). In this Indian monsoon herb the aerial flowers are blue and chasmogamous while the underground flowers are white and cleistogamous. The production and development of the latter are influenced by light and moisture—they develop in greater abundance under comparatively dry conditions while in water-logged soil few are produced. Cleistogamous flowers become chasmogamous on exposure to light.

The taxonomic importance of cleistogamy is twofold. Firstly, in practical terms there is the danger of misinterpreting the taxonomic value of these cleistogamous forms. It has led, for example, to the description of a new Leguminous genus (*Martia*) based on the cleistogamous morphism of a well-known species of *Clitoria* (van Steenis, 1948). This is an extreme example, but failure to understand the position may result in the recognition of varieties or species based on cleistogamous variants that are environmentally controlled.

In evolutionary terms cleistogamy is important because of its effect on the variation pattern of the populations concerned as a result of increased homozygosity. This can lead to a lack of genetic variability which may be advantageous in a stable environment, but is of course highly disadvantageous in a changing environment since the population will lack the capacity for change to meet the new conditions.

A word of warning must be added about the danger of deducing the breeding mechanism from the morphological characteristics of the flower. Many examples are known where the mechanism is other than that suggested by the morphology. Pollinating agents may be unobtrusive, such as thrips in the case of *Erica tetralix* (Hagerup, 1953). In tropical *Annonaceae* the curious shedding of the stamens and the closed condition of the flower is thought to promote selfing but it does not always occur, due to protogyny. Many plants that have flowers constructed so that self-pollination appears likely are, in fact, self-incompatible (e.g. *Trifolium repens*).

Other mechanisms similar to cleistogamy occur. *Pisum sativum* and *Hordeum vulgare* are self-fertilised, the flowers opening only after the pollen is released and deposited on the stigma. In the genus *Hesperolinon* self-fertilisation is predominant in most species. The flowers are usually pseudo-cleistogamous—the pollen matures while the anthers are closely connivent around the stigmas and the flowers still in bud. The pollen is then shed directly on the receptive stigmas, thus presumably ensuring fertilisation before the flowers open and expand as in typical cross-pollinated species (Sharsmith, 1961: 243-4).

It is interesting to note that in 10 of the 12 species of *Hesperolinon* the flowers have characteristics such as spreading coloured petals, nectariferous areas on the cup, etc. which apparently adapt them for crossing, and they are in fact regularly visited by insects. Cross-pollination probably occurs infrequently to account for the origin of new genotype combinations which then maintain themselves by selfing. This is known to occur in *Thlaspi alpestre* (*Cruciferae*) where there is a protogynous floral mechanism permitting a low level of outbreeding estimated by Riley (1956) at 5 per cent., followed by a high level of selfing. The colonial nature of the distribution of the species in suitable habitats of limited area is associated with the largely inbreeding mechanism, causing these local populations to be largely reproductively isolated. Gene flow by means of the limited amount of outcrossing allows a certain degree of variability through the production of hybrids which would be available for selection to give new inbred lines. It is probably true to say that completely obligate inbreeding does not occur in flowering plants.

Numerous other examples are known of this kind of mechanism and some cases where it varies according to external conditions and in different geographical areas. The taxonomic importance lies in the consequences of continued inbreeding resulting in the formation of local populations consisting of one or a few biotypes due to the elimination of heterozygosity.

Such populations are similar in many ways to apomictic groups and present similar problems. The morphological variation between habitually inbreeding populations is discontinuous; if a pair of stable populations are not identical, they will be more or less clearly separable and constantly so, no matter how small the morphological differences concerned (Baker, 1959a). As with apomicts, the biological species concept is not fully applicable to such local populations. Their taxonomic treatment is discussed in a later section (p. 377).

(b) *Mechanisms favouring outbreeding*. These mechanisms are many and varied; some are found in habitual outbreeders, others in probably the majority of plants which show some degree of outcrossing. Such plants may be *monoclinous* with stamens and carpels in the same flower (e.g. *Papaver*, *Hypericum*, *Althaea*, etc.) or *diclinous* with unisexual flowers—either monoecious or dioecious.

The main outbreeding systems range from mechanical devices of the flower to prevent self-pollination (*synanthesis*) and temporal features having a similar effect, to dioecism and self-incompatibility.

Complete separation of the sexes (dicliny) naturally promotes outbreeding. The simplest case is *monoecism*, where the different sexes are borne on the same plant. This does not guarantee outbreeding, but rather favours it. According to Baker it is wasteful of gametes if it is not combined with some feature to prevent selfing. Such a device is found in *Zea mays** where flowers of different sex open at different times, thus making the outbreeding system more effective. Variations of the monoecious system are polygamy: *andromonoecism* where male and hermaphrodite flowers occur on the same plant, and *gynomonoecism* where female and hermaphrodite flowers co-exist on the same plant. In *Parietaria diffusa* there is a complex pollination system with male, female and hermaphrodite flowers on the same individual plant. Monoecism is not necessarily consistent within a species: in *Mercurialis perennis*, a normally dioecious species, rare monoecious individuals occur. Plants bearing these different flower forms can mislead the taxonomist when making his specific delimitations, especially in floras which have not been intensively studied. Van Steenis cites the example of *Bauhinia castrata* which is a female variant of *B. malabarica* and notes that the correlation of specimens representing different sex forms may cause difficulties.

Dioecism, where the sexes are separate on different plants, is the most drastic system of outbreeding, since not only is outcrossing obligate but it involves the very considerable risk that fertilisation may not be realised at all (Baker, 1959b). There is, for example, a strong possibility that in any locality the neighbouring plants in the population will be of the same sex. Furthermore, only about half the plants will be pistillate and therefore potentially capable of fertilisation. Dioecism like monoecism may be incomplete: some species are

* It is interesting that by artificial selection and hybridization dioecious lines have been produced; they do not occur naturally, however, in the genus.

polygamous (*gynodioecious*) where hermaphrodite and female individuals occur; they may also both occur on the same plant. Gynodioecism is an outbreeding mechanism, since the hermaphrodite flowers may be selfed but the female flowers must be outcrossed, the different mating types contributing unequal numbers of genes to the next generation (Lewis & Crowe, 1956). The extent to which it occurs in the Angiosperms has not been fully established, but surveys have been made by Yampolsky (1922) and Baker (1948b)*. Gynodioecism is frequent in many genera of the *Labiatae* such as *Origanum*, *Salvia*, *Mentha*, *Calamintha*, *Glechoma*, etc. It is also known in a number of other families including the *Compositae*, *Caryophyllaceae* and *Saxifragaceae*. Detailed studies on the gynodioecious breeding system of species of *Pimelea* (*Thymeleaceae*) have been made by Burrows (1960). In New Zealand species studied a high proportion of females was found in populations which ensured outcrossing; a low fruit set was found in the female flowers. Widespread hybridisation in the genus is attributed by Burrows to the gynodioecious condition in many species.

Such sexual polymorphism may cause difficulties for the taxonomist since the morphological differences in the flowers may be quite marked and, if the differences are not understood, may result in the plants bearing them being described as separate species. In the *Labiatae* the female sex flowers are typically smaller than the hermaphrodite flowers; this difference may be mistaken for simple size variation and in several cases the sexual nature of the polymorphism has not been appreciated. This was the case, for instance, in the genus *Ziziphora* until recent investigations showed that sexual differentiation is quite common. There the length of the corolla tube, its degree of exsertion from the calyx and the degree of exsertion of the stamens depend on the sexuality of the plant (Hedge, 1961a). The interesting point in this case is that corolla tube length has been used by previous taxonomists as a key character in separating "species" of *Ziziphora*.

Gynodioecism should not be confused with the rare mutants causing male sterility in many hermaphrodite species. This is manifested in the abortion of pollen but has little effect on the breeding system of a population.

Within a genus both monoclinous and dioecious species may occur; for example in *Mussaenda* where *M. tristigmatica* and *M. chippii* are functionally dioecious while *M. erythrophylla* is heterostylous. In this case the floral structural differences associated with the different breeding systems do provide useful taxonomic characters although they have been overlooked (Baker, 1959a).

Dichogamy. In hermaphrodite plants differential maturation times may operate. This is known as *dichogamy* when the stigmas and anthers mature at different times. It exists in two forms—*protandry* when the pollen is released before the stigmas of the same flower are receptive; and *protogyny* when the stigmas are receptive before pollen from the same flower is released.

* See also Fryxell (1957).

Both of these mechanisms normally result in cross pollination. The difference in the time of maturity between the pollen and stigma may be quite short; indeed dichogamy may be only partial with some overlap in maturation times and this may refer to individual flowers or to individual plants. There is often some "imperfection" of this nature in pollination devices, maintaining some sort of balance between selfing and crossing. In some species such as *Zea mays* both protogyny and protandry may occur; similarly some races of *Erodium cicutarium* are protogynous while others are protandrous. Dichogamous mechanisms are known in many families and are frequent for instance in the *Gramineae*, *Juncaceae* and *Compositae*.

Other floral mechanisms which prevent self pollination. Apart from dichogamy and other cases discussed above, there are very numerous floral mechanisms and ethological factors which favour outbreeding, by adapting flowers to specific pollen vectors. These have already been referred to briefly (Chapter 4) and are discussed further in Chapter 13.

Geitonogamy. A factor which tends to be overlooked in outbreeding mechanisms is geitonogamy—the autogamous pollination that results from the transference of pollen from one flower to another on the same individual. It occurs, for example, in *Eupatorium cannabinum* (*Compositae*) between the flowers in the same inflorescence, and in various members of the *Umbelliferae* such as *Chaerophyllum*, and must be quite common in many other groups, although there is little information as to its effectiveness. It is one of the ways by which some outbreeding mechanisms may be circumvented (see Baker, 1959a: table 2).

Heterostyly and self-incompatibility mechanisms. Incompatibility systems favour outbreeding. They depend, as does dioecism, on genotypic differences but they operate in hermaphrodite flowers. Self-incompatibility is a "breeding system in hermaphrodite organisms which prevents the formation of zygotes after self-mating" (Lewis, 1954a). It is widespread amongst Angiosperms and is known from more than 3,000 species in some twenty families and probably many more examples remain to be discovered. Few species have been well analysed, largely because of the labour involved.

There are different degrees and types of incompatibility. A distinction is made between *self-incompatibility* which operates between the pollen and style of the same plant (i.e. prevents the self-fertilisation of hermaphrodites), and *cross-incompatibility* which operates between the pollen and style of different individuals, races, etc.

Three main systems of self-incompatibility are found in the Angiosperms. The incompatibility reaction may be between pollen grain and stigma, pollen tube and style, or between the male and female generative nuclei in the embryo sac.

(1) *Heteromorphic* (e.g. *Primulaceae*, *Lythraceae*, *Linaceae*, *Rubiaceae*). Here there are two or three morphologically distinct mating groups: long and short style flowers in *distyly*, long, medium and short style flowers in *tristyly*.

The pollen reaction is sporophytically controlled—the plant controls the characteristics of the pollen by means of morphological differentiation of the flower (Darlington & Mather, 1949). Genetic control is by one gene with two alleles S and s in distyly, or by two genes each with two alleles Ss and Mm in tristyly.

(2) *Homomorphic gametophytic* (*Solanaceae*, *Rosaceae*, *Leguminosae*, *Scrophulariaceae*). Here all the mating groups are indistinguishable morphologically. The pollen reaction is gametophytically controlled. Genetic control is by one gene with numerous alleles $S_{1,2,3,\ldots n}$ (multiple allelomorph incompatibility).

(3) *Homomorphic sporophytic* (*Compositae*, *Cruciferae*, *Rubiaceae*). This system combines features of the other two systems—genetic control is by one gene with many alleles but the pollen reaction is under sporophytic control.

Detailed discussions of self-incompatibility are given by Bateman (1952), Lewis (1954a, b), and Pandey (1960). Whitehouse (1950) has produced evidence to suggest that multiple-allelomorph incompatibility of pollen and carpel tissue may well have been the primary cause of the evolution of the closed carpel, and of the success which the Angiosperms achieved over their Gymnosperm ancestors. Further discussion of the implications of the distribution of self-incompatibility in the Angiosperms is given by Bateman (1954) and Lewis & Crowe (1958).

Heterostyly. In the heteromorphic system there are various accessory characters involved. In the classic case of *Primula*, e.g. *P. sinensis*, in the "pin" and "thrum" flowers morphological features are: style length and stigma position, anther height and position, size and configuration of the pollen grains, size and configuration of the stigmatic papillae.

Such features are controlled by the same S gene complex as in the *Plumbaginaceae* where the incompatibility system is allied to that of most heterostylous taxa; here there is a striking pollen-dimorphism and in some cases an equally striking stigma-dimorphism. Plants with the different kinds of pollen are found in populations of dimorphic species and those bearing each pollen-type are self-incompatible but cross-compatible. In *Limonium vulgare* there is also stigma dimorphism with cob and papillate stigmas corresponding to the two kinds of pollen. There is also heterostyly with long cob styles and short papillate styles. This culminates an evolutionary sequence in the tribe *Staticeae*, derived from a primitive condition in which there is monomorphic pollen and a capitate stigma (e.g. *Aegealitis*). Secondary monomorphism has arisen in *Armeria* and *Limonium*—this indicates self-compatibility, but the plants are still capable of some degree of outbreeding (Baker, 1953a, b, 1954a.)

In *Linum grandiflorum* only the style differs in length—it is shorter or equal in length to the stamens. Tristyly is known in *Lythrum*, *Oxalis*, *Narcissus*. The mechanism in the latter is not strictly speaking heterostyly, however, since the variation is not apparently related to the breeding group. It is known that in *Narcissus cernuus* only single posture types may be found in a given area.

In *Primula vulgaris* Crosby (1949) has studied the change-over from outbreeding to inbreeding associated with a change from heterostyly to homostyly.* The significance of this example is that it can be studied "in the act"; it has probably occurred and been completed in other species of *Primula*. Monomorphic species of the sections *Candelabra* and *Farinosae* are found near the edges of the range of distribution suggesting their derivation from the main body of heteromorphic species; and monomorphic forms of normally dimorphic species of the Sect. *Farinosae* have also been found at the margin of the ranges of these species (Ernst, 1953).

According to Baker (1959a, 1961), all these systems promoting outbreeding are reversible, permitting inbreeding. Thus morphological changes may take place in the flowers, self-compatibility alleles may arise in an incompatibility system, heterostyly give rise to homostyly (as in the example of *Primula* cited above), etc. The reverse evolution from inbreeding to outbreeding is apparently much rarer. This flexibility is of great evolutionary advantage, permitting adaptation to changing conditions. Apomictic groups, on the other hand, are correspondingly limited.

Evolutionary Significance of the Breeding System

As we have already said, Darwin believed that plants "should" be cross-pollinated. This was clearly shown, he thought, by the many and complex variations in flower structure and mechanism which he studied. In fact, the underlying aim of his work on the pollination mechanisms of the Orchids was to demonstrate that their main purpose was the fertilisation of the flowers by pollen brought by insects from other plants. Later, however, he was concerned in his work on cross and self fertilisation of plants (1876) to find out what exactly were the advantages to be gained by such a favoured process. Darwin was forced to the somewhat unsatisfactory conclusion that it was Nature's urge producing vigour and maintaining fertility. We can today understand the evolutionary importance of the breeding system better if we remember that these outbreeding systems of various kinds result in gene recombinations which play as great a role as mutation in providing the variation on which natural selection works. These gene recombinations in turn lead to the new combinations of characters on which natural selection may operate to produce in time better adapted structures and organisms. Selection may be *allaesthetic*, as in the case of the genus *Ophrys* where various flower types which arose randomly as a result of mutation and recombination in the ancestral populations were selectively perpetuated by assortative pollination (*pseudo-copulation*) and these selected types were then "perfected" to the stage at which we know them today (Stebbins, 1959b; Heslop-Harrison, 1958b).

* The degree to which the homostyle primroses outcross has been discussed by Bodmer (1958) and Crosby (1959).

The question therefore arises as to why one finds so many autogamous and apomictic species which are without doubt highly successful. There are several explanations, some of them of restricted application. For example, apomixis may be advantageous in the Arctic, where the season is too short to allow flowers and fruits to develop to maturity. There may also be a scarcity of pollen vectors, or the pattern of the population may be such that the individuals are widely spaced. Under such circumstances inbreeding may be an advantage. Selfing and apomixis give greater and more secure seed production, but against this one has to weigh the loss of genetic variability as a result of the restriction or elimination of recombination. This latter reason points to a more general explanation of the problem. The fitness of a population or of a species is judged according to the relationship between those factors which promote, on the one hand, genetic variability, and those which, on the other, lead to genetic fixity (as well as phenotypic plasticity).

In outbreeding populations variability is partly held in latent form—as a reservoir of recessive genes held in the heterozygous condition. Most of this variation is not adaptive and when brought out by recombination reduces the fitness of the population. This hidden variability is of considerable significance since it will produce an array of genotypes some of which may be adapted to meet new conditions which the species might come to face, as in changing environments. All populations have, as it were, to solve this problem of the balance of fitness by compromising in some way between immediate fitness and genetic rigidity which is advantageous in a stable environment, due to selfing or apomixis, and reduced fitness but potential evolutionary capacity though outbreeding which is of great advantage in an unstable environment.

Lewis & Crowe (1958) suggest that the answer to the paradox lies in the *critical stage in evolution*—that period in the history of an organism when it is competing in a common environment with other organisms. In this period outbreeding mechanisms are essential for survival by giving the advantages of gene recombination. It is probably a short period as indicated by Gause's principle (1934) that two or more forms with identical requirements cannot coexist in the same environment for an indefinite period of time. After establishing itself in an environment, an outbreeding system may be retained or for some reason be abandoned. Consequent inbreeding will not then be at a disadvantage for perhaps millions of years until there is some cataclysmic change in the environment (cf. also Fryxell, 1961).

Factors controlling or modifying the system

The reproductive method varies, as has been noted, from time to time and from place to place. In both cases the control may be environmental and genetical. Five major factors are recognised by Fryxell (1957) and are briefly considered here.

(1) *Entomological.* In entomophilous plants variations in the number,

kinds or activities of insect pollen vectors may bring about alteration in the breeding system. Many species are visited by a number of different species of insects, none of which predominates. These different insect species have different modes of action so that a shift in their relative proportion will effect a change in the pollination pattern of the plant species concerned. The vectors may be affected by changes in the environmental factors such as light, temperature humidity, rainfall, predators, etc. Even the shape of the area occupied by the population may alter the relationship between the vector and the plant in some cases. This has been shown experimentally in *Gossypium* spp. (Fryxell, 1956). A curious phenomenon noted by Kobel was that transfer of pollen may occur within the beehive so that there is a wider range of pollination potentially possible than would otherwise have been expected.

(2) *Geographical and climatic factors.* That there is variation in breeding habits from one geographical area to another is well known, but it is not often easy to find out which variable causes the variation observed. It may be variation in temperature, humidity, precipitation, interaction with other organisms, or combinations of these. An example is shown by *Deschampsia caespitosa*, which is fully sexual in Scandinavia, but when the same biotypes were transferred to California they became facultatively pseudo-viviparous. Native Californian biotypes are only rarely viviparous and it is suggested that aridity may be the causal factor effecting the switch in breeding system. In *Gossypium hirsutum* it has been found that the percentage of natural crossing varies according to the locality—from 1-2 per cent. to 50 per cent. This appears to be due to variations in the insect populations in the different areas, itself a response, in part, to variations in rainfall. In Peru, *Lycopersicon esculentum* exhibits a fair degree of natural crossing whereas in other areas where it is introduced as a cultigen it is essentially selfing. In this case the explanation lies largely in the presence in Peru of suitable insect pollinators which are not found in the areas where it is cultivated.

In cases where there are two flower types in a species the degree of natural crossing depends on the proportions between them. Such an example is found in *Lespedeza cuneata*, which produces chasmogamous petaliferous flowers with about 70 per cent. natural crossing and apetalous cleistogamous flowers which are selfing. The proportions of these two flower types are affected by changes in environmental conditions. A final example may be taken from *Hordeum vulgare*, which is normally a self-pollinator. At the onset of frost natural crossing will often occur—an example of seasonal variation.

(3) *Genetic factors.* It is known that the genotype can affect the rate of crossing: it is, for example, gene-controlled in *Lycopersicon esculentum* where the delicate arrangement of the anthers in a tube around the flowers is controlled by a single gene. The anthers are held by hairs and the effect is to promote selfing. The gene controls the presence of the hairs and their removal in turn displaces the anthers from the region of the stigma and so cuts down

the possibility of selfing. Such an elegant example is exceptional, however, and more often it is difficult to ascertain exactly what the relationship is between genotypic cause and effect.* Most of the examples have to do with either morphological or temporal changes in the mechanism of the flower, such as the relative size of the parts, changes in the rate of flowering, in time of anthesis, etc. Straub, for example, found that changes in level of polyploidy affected not only the size of the flower but the relative size of the parts and the position of the anthers, thus altering the form of the flower. Changes such as these may alter the type of pollination mechanism and may be adaptive as regards selection for a more efficient breeding system. Racial differentiation within a species may affect the rate of crossing. It has been well studied in *Gossypium hirsutum* (see Fryxell, 1957, for references).

(4) *Experimental modification.* It is possible in some cases to alter the breeding system experimentally by chemical, mechanical or other means. Self-incompatibility may be circumvented by various such means; and complete outcrossing may be achieved in monoecious plants by treatment of the seedlings with maleic hydrazide which causes inhibition of the staminate flowers (Wittwer & Hellyer, 1954). It should be noted in this connection that the monoecism of *Zea mays* may be readily altered by a number of environmental factors including photoperiod, temperature, soil conditions (Heslop-Harrison, 1961).

(5) *Selection.* We have already seen in the previous sections how alteration in breeding system may be considered as an adaptation to a changing environment and have a selective advantage.

Long-distance dispersal. Baker (1955, 1959b) has demonstrated the importance of the breeding system in long-distance dispersal. In the *Plumbaginaceae* floral dimorphism is found in groups near the distribution centre of the family: these members are outbreeders consequent on the floral dimorphism. In areas far removed from the distribution centre he found that self-fertilising species are prevalent. This is explained by the greater ability of the inbreeder to become established after long-distance dispersal, particularly if only a chance seed is disseminated. Clearly if it is an inbreeding individual it will have a much better chance of being fertilised in the absence of pollen from another individual and of building up a new colony, than if it were an outbreeder depending on other individuals for pollen. Stebbins (1957) says that a survey of other groups has shown that this relationship is a general one and suggests that the phenomenon should be termed Baker's Law. It provides plant geographers with a valuable new idea for studying the centre of origin of groups (Stebbins, 1959b).

* Another striking example has been recorded in *Dichanthium* (*Gramineae*): *D. fecundum* and *D. annulatum* are both facultative apomicts and the main difference between them is the sexuality of the pedicellate spikelet which, it has been shown, is apparently controlled by a single gene (Borgaonkar & de Wet, 1960).

Breeding system and habit. With increased data concerning the breeding systems of species in various floras, it is becoming possible to work out what relationships exist between the breeding system, habit or length of life cycle and habitat. As Baker has pointed out, there is a close correlation between self-incompatibility and dioecism on the one hand and the polycarpic habit on the other. This is especially true when the polycarpic habit is associated with well-developed vegetative reproduction. The advantages of self-compatibility in weedy species, notably annuals, has already been mentioned. In fact, self-incompatible annuals are relatively rare. When the only means of reproduction and persistence is by seed, as in the case of annuals, self-compatibility gives a greater guarantee of reproductive success, especially in pioneer or temporary habitats. Stebbins (1958) has remarked on the characteristic cross-fertilisation and heterozygosity of temperate forest trees such as *Pinus*, *Quercus*, *Fagus* (monoecious) and *Taxus*, *Salix* (dioecious). This whole question of integrating genetics, breeding system and ecology is discussed by Baker (1959a: see also discussion following his paper) and Stebbins (1950, 1957, 1958), to which reference should be made.

The Breeding System and Taxonomy

In the previous sections many references have been made to the implications of the breeding system for taxonomic practice and theory. Here they will be brought together and extended.

(a) *Inbreeders*. In habitually inbreeding taxa variation between populations is discontinuous because of the splitting of the taxon into numerous more or less pure lines (genetically homozygous at least for the observable characters: see later). Each line or population breeds true for many generations, and provided the genetic differences between them are reflected morphologically each line will be clearly distinguishable. They are frequently restricted to ecological niches and are consequently of restricted distribution. Because of their homozygosity they contain little capacity for evolution in response to environmental changes. Variation may be carried in a potential form—between the pure lines—and is uncovered by occasional crossing between them, forming hybrids which segregate and produce new pure lines. Some of these lines are likely to be better adapted to the new environmental conditions than the parental form. This aspect has been particularly studied in crops plants where there is often a selection, either conscious or unconscious, for uniformity in certain morphological features. Phenotypic uniformity does not necessarily mean corresponding genotypic uniformity. It is well known that in outbreeding species genotypic variability can be hidden in the form of balanced gene combinations, each of which will give essentially the same phenotypic effect under given environmental conditions. Under conditions of changing environment the variability will become uncovered by the gene combinations no longer producing the same phenotype, with a result that the population becomes split up into a number of phenotypically different forms (cf. *Lolium*, Cooper, 1951).

Similarly with inbreeders or "pure" lines, the morphological uniformity may mask genotypic diversity in a given habitat, and when transferred to another habitat or subjected experimentally to altered conditions the "pure" line will show that it is made up of a number of different phenotypes. Release of variability under such conditions is less likely in inbreeders than outbreeders, however, since the former will tend to contain less variability hidden in the heterozygous state. When it does occur, however, in both cases it is a means of achieving high immediate fitness for given conditions and at the same time retaining a considerable reserve of variability for future evolutionary change.

A consequence of continued inbreeding is the building up of population differentiation in excess of what would be found in outbreeding situations. This leads to the formation of numbers of morphologically uniform populations, all easily separable from each other and eligible for consideration as distinct taxa (Sharsmith, 1961). This phenomenon is well known to the practising taxonomist who often refers to the populations as microspecies or Jordanons after the nineteenth-century French botanist who described so many breeding lines as distinct "true" species. Jordan recognised up to 200 "species" in the genus *Erophila* as a result of a series of cultivation experiments. These "species", which he justified in extensive reasoned essays, were apparently separable from one another, constant, and showed little internal variation. Later studies, notably those of Winge (1940), showed that the chromosome numbers of various lines within the group varied from $n=7$ to $n=47$, and that the variation described by Jordan was due in part to inbreeding and occasional hybridisation between lines releasing further variation. Winge grouped the forms into four species on the basis of their chromosome numbers, but it does not appear possible in the light of later studies to recognise any overall correlation between chromosome number, morphology, ecology or distribution, although local patterns can be discerned. It is preferable to recognise one complex and highly variable species in the genus *Erophila*, namely *E. verna* sensu latissimo (containing a number of subspecies), in addition to the fairly well distinguished *E. minima*.

Jordan recognised large numbers of similar species in several other genera such as *Thlaspi*, *Biscutella*, *Iberis*, etc., and it has proved difficult to account for them all satisfactorily. It is not possible to dismiss them all out of hand, however, since occasionally they coincide with acceptable species recognised today.

The case against recognition of such inbreeding lines as species lies mainly in their relative lack of constancy due to their occasional outcrossing (although they may breed true for hundreds of generations, Stebbins, 1959b) and their large number. These are essentially the same arguments as used against the taxonomic recognition of apomict forms as species. There is no established taxonomic principle, however, dealing with the recognition of species in inbreeding groups. In general, the recognition of inbreeding helps the taxonomist to understand the variation pattern but it does not tell him what should be recognised and named.

We have already cited the case of *Erophila* where knowledge of the breeding system and chromosome situation permits us to comprehend the causes of the variation that Jordan attempted to describe, but as regards the formal taxonomy of the genus we have to return to a position not very different from that held by Linnaeus. Another celebrated example is the species *Capsella bursa-pastoris* which shows a great deal of variation in leaf shape, capsule, etc. The genetic basis of much of this variation was studied by Shull in a series of papers (see Clausen & Hiesey, 1958a: 171-175, for an extended summary) and the group was later divided into about 70 species by Almquist (1921) on the basis of these minor differences. Again the "species" can be explained as the result of extensive inbreeding, yet the taxonomy of the group remains intractable beyond the recognition of the one species, *C. bursa-pastoris*, containing a number of infraspecific taxa.

What is generally agreed, however, is that the taxonomic treatment of facultatively inbreeding populations requires a somewhat different approach from that afforded to normally outbreeding groups (cf. Stebbins, 1957). For one thing, the criterion of gene exchange between individuals of a species and restriction of gene flow between species cannot be applied rigidly to such populations. Specific boundaries must be to a large extent arbitrary, as Sharsmith (1961) notes, and the aim should be to circumscribe units in such a way as to indicate as well as possible the biological relationships within the group. In arriving at these decisions no universal criterion is applicable and each case will be judged according to knowledge available as to the distribution, ecology, population size and structure concerned. As a rough guide it may be suggested that when the inbreeding populations appear to be constant and occupy a considerable area they should be recognised taxonomically; when they only form part of a mosaic, inbreeding and outbreeding complex, their recognition might only obscure the variation pattern. Each case tends to be different and must be treated on its merits.*

Cryptic speciation may take place between morphologically similar inbreeding populations and their relative lack of genetic variability will prevent phenotypic divergence occurring, at least initially (cf. Baker, 1959a). In other words, the case for recognising morphologically similar though constantly very slightly different inbreeding populations as species may be doubtful, but when genetic isolation arises between them they become "biological" species and rather cryptic ones at that. The issue is therefore forced and species based on slender characters are described if the situation is recognised. It therefore appears that between related populations of in-

* H. Lewis (1963) regards the associated hybridisation rather than inbreeding itself as causing problems for the taxonomist.

breeders there may arise weak or strong genetic barriers. Baker suggests that it is only by allowing for the existence of these extremes that it is possible to account for the fact that amongst inbreeding annuals there exist not only cryptic genetically isolated species which the taxonomist may not recognise or even appreciate, but also taxonomic microspecies without genetic isolation which provide good material for the "splitter".

In a study of *Xanthium strumarium* recently initiated, Löve & Dansereau (1959) suggest that habitual inbreeding combined with a small amount of occasional hybridisation has led taxonomists to describe up to 50 species, but there is no evidence of sterility barriers between closely related species. They regard all previously described species of this complex as variations of a single species only, from a biosystematic point of view.

(b) *Outbreeders.* Outbreeders, unlike inbreeders, tend to show a more or less continuous pattern of morphological variation. This tends to blur the distinctions between populations; even when reproductive barriers exist there is often a possibility of some gene-flow resulting in an overlapping morphological pattern. In many cases this makes taxonomic separation difficult, since the gene-flow prevents to some extent the build up of an integrated morphological pattern which is constant and clear enough to allow the taxonomist to recognise species.

This is particularly marked in tree species which in the North Temperate region are largely outbreeding and highly heterozygous. Pollen is produced in vast amounts and often carried long distances. This has a marked effect on the variation pattern, tending to the production of continuous variation in many characters. An excellent example is the black pine, *Pinus nigra*, which ranges from Spain and N.W. Africa across south and central Europe to the Caucasus. Because of its great silvicultural importance, it has been subjected to extensive taxonomic investigation, but although scores of papers have been published about it no agreement has yet been reached as to the number of species or subspecies involved, even with the aid of microscopical wood-anatomical characters. The best means of identification today is still the place of origin! A summary of the diverse taxonomic treatments proposed is presented in a paper by Fukarek (1960) although even this is now out of date.

This phenomenon is not of course confined to tree species and most taxonomists are familiar with other examples in herbaceous species. The attitude of the taxonomist to this sort of situation is neatly summarised by Baker (1959a): "... the taxonomist who feels that he must see a break in morphological continuity before he can recognise the existence of two species keeps them under one name". For practical purposes this is clearly the best solution, for little will be served by applying names to species whose morphological boundaries cannot be discerned and the members of which cannot in many cases be identified as belonging to one or the other. This is just another example of the conflict of aims that sometimes arises between taxonomy and biosystematics.

(c) *Taxonomic relationships of inbreeders and outbreeders* The existence in nature of self-compatible and self-incompatible races within the same species is apparently unusual. Baker (1954b) cites an example within *Picris echioides* (*Compositae*) where he has found a sea cliff race in Britain which is perennial, self-incompatible, late-flowering and decumbent, and an annual or biennial, self-compatible, relatively early flowering, upright race from the edge of an arable field.

In some instances such races have been included in the same species until a knowledge of the breeding system has been obtained. As an example, Baker (1959a, 1961) quotes the genus *Primula* where in the nineteenth century several geographically and morphologically distinct taxa of Subsect. *Farinosae* were included in *P. farinosa*. Later it was shown that *P. farinosa* sensu stricto is diploid, heterostylous, and largely self-incompatible, while closely related forms constitute a chain of homostylous polyploids which must show genetic isolation from their diploid relatives. He considers that in the light of this knowledge there is no alternative to considering them separate species.

Similarly in *Armeria* (*Plumbaginaceae*) morphologically separable, dimorphic self-incompatible, and monomorphic self-compatible races were included by Lawrence (1940) as varieties of one species, *A. maritima*. Baker's experiments have shown, however, that there are strong sterility barriers between the self-incompatible European races and the circumpolar self-compatible races.

(d) *Apomicts*. For many years taxonomists have attempted to devise ways of dealing with the problems involved in the classification of apomicts, but the subject still remains controversial. In some Floras a fairly complete list of apomictic taxa is given, while in others only the major groups are described or they may be omitted altogether. The apparently endless series of descriptions of new apomicts that are published have led many botanists to regard apomict taxonomy as a harmless, if pointless, occupation best left to amateurs.

On the other hand, as Valentine & Löve (1958) point out, apomixis is widespread in the Angiosperms and occurs extensively in groups of considerable economic or ecological importance. It is therefore important that attempts should be made to find acceptable principles concerning their taxonomic recognition.

As we have already pointed out, the population structure in apomictic groups differs considerably from that found in sexual, outbreeding species. In the strict genetical sense (see p. 364) apomicts do not form populations: there is no gene exchange possible between biotypes or between clonal members of the same biotype, each of which may build up its own pseudo-population by self-duplication. On the other hand, if "population" is adopted in the broader taxonomic sense as meaning morphologically similar groups of plants growing in a particular area, apomictic populations may be found to consist of more than one biotype even though morphological differences between them are no more apparent than what would be expected in a sexual outbreeding population. (Physiological differences may, however, occur, such as flowering time, disease susceptibility, etc.)

Clausen (1960) suggests that the reason such apomictic populations contain so many biotypes when they are capable of multiplying a single successful biotype is possibly because populations rich in biotypes have a long-term selective advantage over those that possess only a few. The different biotypes may mature at different times or have slightly different ecological preferences, so that in any one season particular biotypes will be more successful than others. Similar considerations apply to inbreeders.

Valentine (1960) points out that the range of variation in an outbreeding population and an apomictic population may be much the same. As a consequence the morphological differences between the biotypes in the apomictic population may be slight, although reliable, and recognition of them may be difficult except to a specialist, if indeed recognition is desirable. The genotype composition of the population may change from generation to generation, or, as we have just noted, different biotypes may be more prominent in different seasons, and this will be reflected in the morphology of the populations. In general, the fact that each biotype is "true-breeding" means that finer discrimination may be possible than in amphimictic groups. A result of this is that very large numbers of species may be described, although the actual numbers will depend on the amount of attention that taxonomists have given to the group in question and on the nature of the apomictic mechanism involved (Valentine & Löve, 1958).

Before considering the taxonomic treatment of agamospermous apomicts, we must look briefly at the question of plants which reproduce largely or entirely by vegetative means, such as stolons, bulbils, etc.

No special taxonomic problems are posed by plants which reproduce by stolons. Occasional stonoliferous variants in otherwise "normal" populations are usually treated as *formae*, if recognised at all. When they form geographical groupings, either on a small or large scale, they may be recognised as varieties or subspecies on the basis of the stoloniferous character. The treatment of sterile hybrids or sterile triploids which reproduce vegetatively is more difficult: they are sometimes regarded as species, e.g. *Acorus calamus*, or *Poa jemtlandica* (a pseudoviviparous hybridogenous taxon which behaves in nature like a widespread species, cf. Hylander in Heywood, 1960a: 130-1); as *formae*, such as the Swedish *Populus tremula* f. *gigas*; or not given any formal recognition, e.g. triploid clones of *Butomus umbellatus* (Löve, 1960c).

Similarly pseudo-vivipary or bulbifery may occur in sporadic individuals in populations and may be under environmental or genetic control, or such characteristics may be uniform in populations.

Pseudo-vivipary is most frequent in arctic and alpine grasses (*Poa*, *Festuca*, *Deschampsia*, etc.), although it does also occur in other groups such as *Saxifraga*, *Lilium*, etc. These pseudo-viviparous forms in the grasses are very similar to the non-viviparous forms from which they are apparently derived and are usually classified with them, often as *formae* (Valentine, 1960). When the pseudo-vivipary is obligate, taxonomic treatment may be more complicated. An interesting case is *Festuca vivipara*, which has been regarded as a variety of *F. ovina* or as a variable independent species or even species aggregate. It has been shown by Wycherly (1953) that it shows almost as great a range of variation in vegetative characters as *F. ovina*, and as Valentine (1960) points out "in the taxonomic treatment of the group the whole complex of forms, viviparous, semi-viviparous and seedbearing diploid and polyploid, has to be considered". A problem here is quite simply the identity of the populations regarded as *F. vivipara*. Löve & Löve (1956) say that many arctic viviparous populations referred to *F. vivipara* (which in the strict sense they consider is apparently closely related to *F. supina*) have been found to be hybrid populations between various arctic species. Another point to be borne in mind is that *F. vivipara* may have been derived polyphyletically from different biotypes of *F. ovina* and therefore represents a *grade* group as in the case of *Deschampsia alpina* (Hedberg, 1958a). If this is a correct interpretation of such situations, the taxonomic treatment we would suggest is to regard each distinguishable pseudo-viviparous derivative as a *forma* of the non-viviparous species. It is debatable whether the assemblage as a whole should be maintained as a *grade species*. (cf. Kawano, 1963.)

The degree of pseudo-vivipary may vary from country to country within the range of the taxon concerned. In Sweden no non-viviparous *F. ovina* is found to correspond to the *F. vivipara* populations, although the degree of pseudo-vivipary varies in different biotypes.

The association of polyploidy and hybridisation with apomixis is a well-known phenomenon (Stebbins, 1950) and is one which adds to the troubles of taxonomic treatment.

Taxonomy of seed apomicts. A low frequency of agamospermy is known in several species, and it is quite possible that such a low incidence is widespread in many Angiosperms. This kind of situation presents no serious problems. But as we have already noted, very great variation in the extent of agamospermy is known, ranging from the situation just mentioned through others showing it in various degrees, to taxa in which it is obligate (bearing in mind that *strictly* obligate apomixis does not occur). A result of this is that taxonomic treatment will depend to a large extent on the nature and extent of the apomictic mechanism.

Facultative apomicts may show such a degree of variation that no useful classification of the variation into stable units is practicable, if indeed possible.

Examples of such groups in an active state of evolution can be found in *Rubus fruticosus* (cf. Gustafsson, 1946, 1947a, b), although here a fairly low degree of hybridisation has often allowed the recognition of separable biotype-clones as named taxa by specialists (e.g. Watson, 1958), *Poa alpina* (Müntzing, 1954) and other *Poa* species (cf. Nygren, 1954). Largely obligate apomicts on the other hand, such as are found in *Ranunculus auricomus*, *Taraxacum* and *Alchemilla*, allow the recognition of taxa of high stability and persistence, some of them restricted endemics, others of quite wide distribution. Hybridisation has probably been involved in the evolution of such groups in the past.

The existence of facultative apomixis ought to be stressed, according to Löve (1960c), "since the taxa produced in this way are nothing but hybrids which are perpetuated—at least for some time—by the clonal process of agamospermy. To classify such taxa as species without indicating their hybrid origin is an absurdity which would never be allowed in amphimictic groups, and this practice is the greatest cause of the disrepute into which apomict taxonomy as a whole has fallen."

Even in "obligate" apomicts, variation is possible due to mutation, autosegregation (cf. Turesson, 1956, on *Alchemilla*), and various chromosomal changes such as loss of individual chromosomes (cf. Sörensen, 1958, on *Taraxacum*) or polyploidy (Kappert, 1956, on *Taraxacum*). As a consequence, no group of apomicts is completely stable as regards the origin of new variation, although the situation in facultative groups is much more complex. Some of the new biotypes formed through the so various mechanisms may find a suitable ecological niche and be perpetuated for some time, whereupon they may be described and named (Valentine, 1960).

It is quite evident that no one system of classification will be possible for all groups of apomicts. Many attempts have been made, but as Gustaffson has noted the different classifications proposed vary widely in their practicability, utility and biological soundness.

A general approach to the taxonomic treatment of agamic complexes was proposed by Babcock & Stebbins (1938). It was based on establishing a relationship between the agamic group and the amphimictic ancestors from which it was probably derived. A detailed study of both the amphimictic and agamic members of such complexes is necessary before this method can be attempted and taxonomic divisions proposed. Babcock & Stebbins applied this method to the genus *Crepis* and Stebbins (1950) has described it as follows:

"... the needs of classification and the clearest conception of the structure, interrelationships, and phylogeny of the agamic complex are obtained by drawing arbitrary species boundaries which are related to the original boundaries of the sexual species ancestral to the complex. The number of species recognised should not be smaller than the number of sexual ancestors which are known to exist or can be reasonably inferred to have existed in the past. In most complexes, furthermore, the number of species should be larger, since groups of apomicts combining the characteristics of sexual species which are widely different from each other should be recognised as agamospecies (Turesson, 1929). The boundaries between these species must be drawn where most convenient, usually on the basis of one or a few easily recognised 'key characters' which are known to separate the ancestral sexual species".

Clearly such methods are only possible when the sexual ancestors are extant. In groups such as *Rubus*, *Poa*, etc., where the ancestral sexual species are in the main extinct or cannot be detected, other approaches have to be attempted. These include the "circle-species" system proposed by Gustafsson (1943) for dealing with *Rubus fruticosus*. These circle-species are widespread and fairly distinct, and around them the rest of the variation as circle-microspecies (groups of apomicts morphologically similar to one another), varieties, forms, etc., can be grouped. The number of circle-species so recognised may be too large for convenience but the method should, as Stebbins (1950) says, "be recognised simply as a means of bringing some type of order to a situation which from the biological point of view is incapable of resolution".

The term *agamospecies* in the above quotation from Stebbins was introduced by Turesson (1929) to cover apomictic populations whose constituents could be regarded, on morphological, cytological or other grounds, as having had a common origin. It is a grouping of apomictic biotypes that constitute a morphological unit which can be more or less adequately characterised. The individual microspecies were called *formae apomictae* (cf. Babcock & Stebbins, 1938).

This terminology was modified and extended by Turesson in 1943. He proposed the series apomict-coenospecies, agamospecies, agamotype, apomict, to parallel the terminology he had earlier suggested for amphimictic groups (coenospecies, ecospecies, ecotype, amphimict—see Chapter 12). This series as it stands is incomplete and impracticable in some respects, and as Valentine & Löve (1958) point out it does not, of course, provide any taxonomic solution to the problems of apomicts, although the terms themselves may be useful in certain circumstances.

Löve (1960c) has proposed employing the term agamospecies in a modified sense for apomictic microspecies, at a lower level than advocated by Turesson but at approximately the same level as the subspecies which Marklund (1939, 1954) recognised in the *Ranunculus auricomus* complex. (Marklund's treatment of the complex is to recognise four collective or aggregate species, *R. auricomus*, *R. fallax*, *R. cassubicus* and *R. monophyllus*, comparable to amphimictic species of other sections of the genus, and the numerous apomictic microspecies he recognised are grouped under these collective species as subspecies). The agamospecies in this sense could then be incorporated into the taxonomic hierarchy as a category at the subspecific level. If only agamospecies were included in a species, this would automatically indicate that the species itself is agamospermous only.* On the other hand, if one or more normal subspecies were included in the species, this would show that it was only facultatively apomictic and that sexual seed formation occurred. Löve & Löve have in fact employed this system in their recent Chromosome Handbook (1961).

There is much opposition to the introduction of special terms for agamospermous taxa and the use of prefixes such as "agamo-" (cf. Hylander and Merxmüller in Heywood, 1960a: 148-9), and it must be admitted that little use is made in practice of them.

The actual taxonomic treatment adopted varies from group to group and the most recent survey will be found in the proceedings of the *Flora Europaea* Vienna Symposium (Heywood, 1960a). The papers by Valentine, Walters, Pawłowski, Rothmaler and Löve should be consulted together with the discussions by several specialists. Although these were restricted to European groups (*Poa*, *Alchemilla*, *Hieracium*, *Rubus*, *Potentilla*, *Ranunculus*, *Sorbus*, *Taraxacum*), they are relevant to other areas.

* Individual plants, in such cases, could be facultatively apomictic and this could be indicated if necessary.

Chapter 12

POPULATIONS AND THE ENVIRONMENT

Introduction

In the nineteenth century a tradition was established of the study of variation, in nature and in cultivation, of plant species by botanists such as Jordan, Kerner, Bonnier and Darwin who attempted to relate this variation to natural selection. Darwin was well aware that variation must be subject to some laws, but knowledge of the mechanisms of heredity was either unformed or elementary. Mendel's classic experiments and the incipient schools of genetics started a second tradition in this study of variation, but it was not until the nineteen-twenties that the two trends were brought together and fused by the pioneer work of Turesson on the genetical basis of ecological differentiation occurring within populations of species. This work, which was aptly described as *genecology*, marked a turning-point in the study of evolutionary taxonomy and is considered in more detail below.

The technique of comparative experimental cultivation had been employed by many botanists before Turesson. It was practised by De Candolle, Boissier and other nineteenth-century taxonomists, including, of course, Darwin. In their work we find reference to varieties retaining their characters in cultivation side by side with the typical forms, and coming true from seed. More extensive and systematic trials were made by Jordan on species of the French flora, although, as we have already discussed, his aim was one which is not acceptable in its crude form at the present day, namely to divide up Linnaean species into "true" small species (Jordanons). He was able to show that many variations remained constant in cultivation and retained their characters, although we know today that his success was largely due to the choice of groups on which he worked. Many of the "species" he recognised were inbreeding lines or random biotype clusters in outbreeding species which happen to possess sufficiently distinctive morphological features to allow of their description. It is worth emphasising, however, that Jordan's researches contained much of value and focused attention on local populations and populations samples and their reactions in cultivation. Moreover, as has been mentioned earlier (p. 378), a fair proportion of his species are currently accepted after detailed taxonomic and biosystematic analyses of groups.

Turesson's Experiments

The later experimental cultures of Bonnier, Massart and others suffered from defective techniques or incompleteness. The importance of Turesson's work was that he was concerned with the differentiation of populations in different ecological conditions and the genetic, morphological and physiological nature of this differentiation. Turesson had already noted the phenomenon of the different environmental tolerances of various species, and that different populations of the same species occupying distinctive habitats frequently showed morphological and physiological traits, often of an adaptive nature. One of his aims was to see if these differences were due to modifications of the same genotype as a result of direct environmental influences or to the expression of different genotype-combinations in the different habitats. He made a series of population samples (20 or more individuals) from various parts of the species' areas and grew them in uniform conditions in his experimental garden at Åkarp in Sweden. He noted the reactions of the plants in cultivation over the years in terms of habit, height, flowering-time, etc., and found that in some cases the differences noted between populations in the field disappeared, in other cases showed inter-gradations, but in most instances persisted. These persistent differences therefore appeared to be due to genetic differences and confirmed earlier views, but he was also able to show that the genetically different races were often correlated with habitat differences and that similar selective forces in the habitat appeared to produce similar kinds of adaptations in different species. They were, as Daubenmire (1959) has emphasised, "truly ecologic races". This fundamental point was recognised by Turesson in his system of designating ecotypes by the habitat-type which they occupied, e.g. oecotypus *alpinus*, *arenarius*, *campestris*, etc. The term ecotype was applied to these products of the reaction between the genotype and the habitat: they were formed as a result of the selective action of the predominant factors of the habitat in sorting out those biotypes which were most suited to the specialised conditions, and selective elimination of the unfitted biotypes, giving a genetically adapted, heritable, ecological population (the ecotype). When similar complexes of environmental conditions exist within the area of a species, the same or essentially similar ecotypes could be expected to arise. Ecotypes could therefore be polytopic in origin. The heritable nature of the ecotypes was shown in some cases by raising progeny from seed and showing that they retained their characters.

Although Turesson's experiments may be regarded as pioneer, his detailed results are of fundamental importance, and anyone interested in this aspect of inter-population variation should consult his original papers (1922a, b, 1925, 1929, 1931). Later work,* including that by Turesson himself, has

* A reinvestigation of some of Turesson's ecotypes is being undertaken by Lövkvist (1962) who has included cytological studies of the populations.

caused many of these views to be altered, some of them quite extensively as we shall see, but it became immediately obvious that the formal taxonomic structure was too restricting to convey the details of this kind of eco-genetic variation within species. Turesson therefore proposed a new hierarchical system of terms for this purpose—ecotype-ecospecies-coenospecies. It is often forgotten that this was a *genecological*, not simply a genetical, hierarchy as is clear from the original definitions (Turesson, 1922a; cf. Heywood, 1959a):

Ecotype: the product arising as a result of the genotypical response of a species to its particular habitat. (Previously referred to as habitat-types.)

Ecospecies: "In view of the necessity of keeping the distinction between ecological and purely genetical in mind, the term ecospecies has been proposed to cover the Linnean species or genotype compounds as they are realised in nature."

Coenospecies: The total sum of possible combinations in a genotype compound.

Turesson also gave corresponding genetical analogues, but somewhat deficiently.

Phenotype: The inclusion of the different genophenes of a genotype.

Genotype: Mendelian sub-unit of the genospecies.

Genospecies: Embodies the facts of the genotypical construction of the ecospecies. The term is properly to be employed as the genetical analogue of the coenospecies.

The *genophene* mentioned in the above definition was regarded by Turesson as the range of phenotypes produced by one genotype within the limits of the habitat under which it is found in nature (Heywood, 1959a). The *ecophene* was defined as "*the reaction-types of the ecotypes called forth by the modificatory influences of extreme habitat factors*"—in other words, the characteristic phenotype produced in each particular locality. Such modifications are essentially reversible, as can often be shown by transplanting them to uniform habitat conditions where their differences gradually disappear (see below, p. 401).

Turesson later redefined the ecotype twice; in 1929 he introduced the relative incapacity to exchange genes as the major criterion in his ecotype-ecospecies-coenospecies hierarchy in an attempt to make the categories more precise. Thus the ecospecies was defined as a population which produced less vital or more or less sterile progeny when crossed in nature with other such populations. Later workers, such as Gregor (1939) and Clausen, Keck & Hiesey (1939, 1945), narrowed the concept still further so that there was a change in emphasis from the positive, ecological side to the negative, genetical aspect. The ecotype, although retained as a basic unit of the system, has come to be considered as a cytogenetic category with ecological implications, the first stage in the evolutionary sequence leading to the coenospecies and comparium.

A strict definition of the ecotype as a population with no internal barrier to gene exchange leads, however, to taxonomically unsatisfactory results, because groups which on morphological, ecological and geographical grounds are normally accorded specific status, are ecotypes in a genecological sense. Examples are common and those of *Silene vulgaris* and *S. maritima*, *S. dioica* and *S. alba* and *Aquilegia* spp. have frequently been discussed in detail (see, for example, Valentine, 1949; Valentine & Löve, 1958; Heywood, 1959a). The important point is that the ecotype serves a dual purpose: it can be regarded as a basic unit in a biosystematic classification and at the same time it is valuable to retain it as an ecological race. It is clearly not advisable to use the same term for such different purposes and we advocate retaining the ecotype in something approaching the original meaning, stressing the reaction between genotype and prevailing environment. Of the many definitions, that of Gregor, Davey & Lang (1936) comes nearest to this: "*A population distinguished by morphological and physiological characters, most frequently of a quantitative nature; interfertile with other ecotypes of the ecospecies, but prevented from freely exchanging genes by ecological barriers.*" In this definition stress is placed on the morphological and ecological nature of the ecotype, and interfertility is not the primary criterion. Moreover, absence of genetic isolation between ecotypes is generally inferred, not demonstrated experimentally except in a few cases. It is scarcely satisfactory to define ecotypes genetically and recognise them by morphological and distributional evidence as has been proposed in the instance of the "biological species" definitions by Simpson (see p. 98). Nor should the ecological-adaptive nature of ecotypes be assumed without recourse to experimentation. The problem of recognising adaptive variants is returned to later (p. 409).

THE CALIFORNIAN TRANSECT EXPERIMENTS OF CLAUSEN, KECK & HIESEY

The most detailed large-scale experiments on the genecological differentiation of populations are those undertaken in California by Hall and his associates, later continued and expanded by Clausen, Keck & Hiesey under the patronage of the Carnegie Institute of Washington. The results of their work occupy several books, monographs and papers so that only the main points can be discussed; details can be obtained from the original publications (Clausen, Keck & Hiesey, (1940, 1945, 1948); Clausen & Hiesey (1958a); Clausen (1951)).

A transect was taken across Central California, about 200 miles long and crossing mountain systems so extensive that, as Stebbins has recently noted, the climatic differences along the transect are as great as those between London and Rome. Gardens were established at 100 ft. (Stanford), 4600 ft. (Mather) and 10,000 ft. (Timberline) and were maintained free from weeds and invading vegetation. An immediate point that must be stressed is the scale of the experiments compared with those undertaken by Turesson, Böcher and other workers. The Californian transect extended from the Pacific coast across two mountain ranges and covered thereby a wide range of ecological and climatic conditions. A number of different groups were surveyed across this transect, including the much cited examples of *Potentilla* and *Achillea*. Another feature of importance is the soundness and thoroughness with which the experiments were undertaken and the diverse backgrounds of the celebrated triumvirate: Clausen a cytologist, Keck a taxonomist and Hiesey an ecologist and physiologist. The work has so far extended over thirty years, thus indicating the scope of the problems and the obvious impossibility of undertaking similar investigations for more than a limited proportion of flowering plants in the foreseeable future.

In each of the groups studied the species which extended across considerable portions of the transect were shown to comprise a series of climatic races or ecotypes forming a sequence. These climatic races were in some instances restricted to certain ecological zones within each climate, resulting in a series of more or less discontinuous ecotypes with associated morphological characteristics. There was, moreover, a certain parallelism between the ecotypes of the different species corresponding to particular climatic areas in respect of their morphological-adaptational features. This was not unexpected in the light of the results of earlier investigators such as Turesson: similar responses in morphological-anatomical characteristics are evoked by similar selective forces acting on different species, resulting in convergent ecological-genetical adaptation. As a result there was a superficial resemblance in respect of these characteristics between ecotypes of different species, so that in some cases they resembled each other more than they did other ecotypes within their own species. The parallelism in some cases was physiological rather than morphologically evident. It was complicated further by the well-known phenomenon of parallelism between ecads and ecotypes (phenecotypes and genecotypes) which is discussed below (p. 401).

In *Potentilla glandulosa* (of the section *Drymocallis*) Clausen, Keck & Hiesey recognised five climatic ecotypes in the Californian altitudinal transect (see Fig. 31). The differences between them were in both morphological features such as stature, habit, leaf area, form of the inflorescence, etc., and physiological features such as seasonal growth rhythm, time of flowering, frost resistance, etc.—features which were related to their capacity to survive in their particular habitats, although later studies showed difficulties in interpreting such relationships. The coast range ecotype (subsp. *typica*) was separated spatially from its nearest neighbour on the lower slopes of Sierra Nevada (subsp. *reflexa*); the latter overlapped altitudinally with the intermediate ecotype (subsp. *hanseni*) to some degree but was ecologically separated from it and hybridisation between the two was restricted to the zone of contact. The transition between the subalpine and alpine ecotypes (subsp. *nevadensis*) was on the other hand continuous. Thus although Clausen *et al.* called the ecotypes climatic, isolation in one case was spatial, in another ecological (in fact edaphic).

Hybrids between the ecotypes were shown to be fully fertile and could produce as a result of recombinations new genotypes which had "new" adaptive features. Clones were made of some of the F_2 plants and grown at the different stations: some plants survived and flourished at all the stations, others survived at only one or two. Variation within the climatic ecotypes was found, however, and the position is summarised by Clausen & Hiesey (1958a) as follows:

"There are no continuous trends within the subspecies with regard to mean heights of stems, but rather a random variation from population to population. Other intrasubspecific variations probably reflect differences in the relative prevalence of various biotypes colonizing a given habitat, a phenomenon to which the terms 'random fixation' and 'genetic drift' have been applied. In comparing subspecies at Stanford, it is evident that *nevadensis* has distinctly shorter stems than *hanseni*, and that the stems of *typica* are shorter than those of *reflexa*."

"In *Potentilla glandulosa* there appears to be both continuous and discontinuous variation. A complete intergradation, or cline, occurs for one character [time of flowering] within one subspecies, whereas other subspecies appear to show random variation for the same character. Further study would probably multiply such examples of degrees of discontinuity and degrees of intergradation among the various ranks of natural entities."

Similar results were found by Clausen *et al.* in the other groups with which they worked and these are summarised in Fig. 31. In the *Achillea lanulosa—borealis* complex they showed that eleven statistically different climatic races (ecotypes) could be recognised in the 200-mile transect across Central California; it is worth noting, however, that they suggest that the total number of ecological races within the three constituent species of the *A. millefolium* complex may run into the hundreds.

Rather different results were found by Lawrence (1945) in his studies on ecotypic variation in *Deschampsia caespitosa* in the same Californian transect stations. His material represented eight races from Lapland, Sweden, Finland and California. He used the same methods of experimental cultivation in uniform and distinct environments, employing clonal material, but found that although five physiologically different ecotypes could be recognised there was no corresponding morphological differentiation. Taxonomically the five ecotypes had to be included under one subspecies, since current procedure makes no allowance for the formal recognition of morphologically inseparable taxa.

As we have seen, the ecotypes recognised by the Californian workers were essentially large scale or regional phenomena. This is reflected in their frequent taxonomic recognition as subspecies. The ecotypes of Turesson were

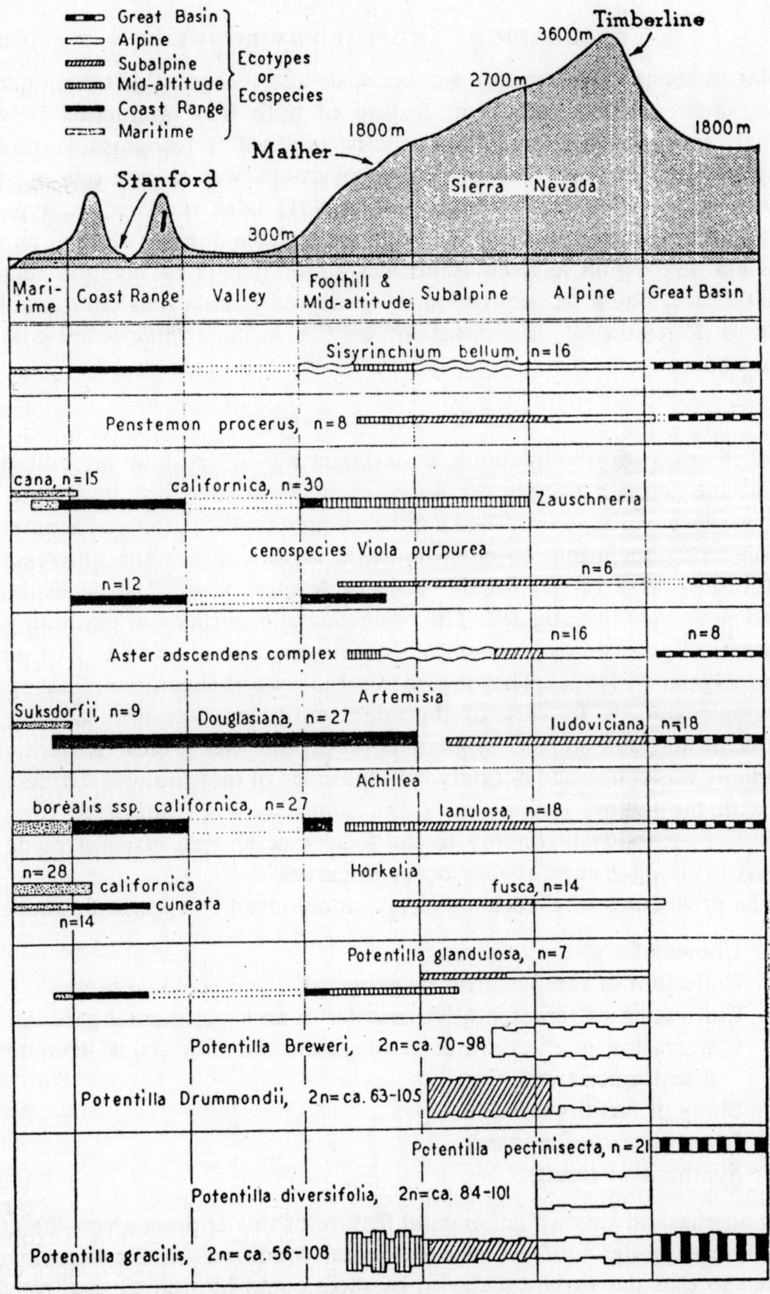

FIG. 31. Distribution of major climatic races with species complexes in a transect across central California. The width of the bar-symbols roughly indicates the degree of polyploidy. (From Clausen, Keck & Hiesey, 1940.)

similar in conception but on a smaller scale and recognisable taxonomically as varieties. It is an important feature of both that interaction between habitat and genotype could apparently produce a recognisable unit or ecotype. The difference between the two groups was largely one of scale, although it is significant that Clausen (1951) later regarded ecotypes as ecological races usually composed of a considerable number of local populations existing within a given ecological zone; that is to say, the regional climatic races could be subdivided into adapted populations comparable to those of Turesson (cf. the statement on the *Achillea millefolium* complex noted above).

The School of Gregor

Another series of investigations, undertaken by Gregor and his collaborators at the Scottish Society for Research in Plant Breeding, under the title of *Experimental Taxonomy*, shed a different light on the pattern of genotypical response of populations to environmental factors. Again the differences in their results were related to the scale or level at which the genecological phenomena were investigated. The principles of experimental taxonomy and its practical procedures were stated by Gregor, Davey and Lang in one of their early papers (1936). They regarded orthodox and experimental taxonomy as complementary. In view of the later confusions that have arisen, it is important to look at the original concepts of this group. Experimental taxonomy was concerned not only with the value of morphological differences but with the factors underlying them. Emphasis was transferred from the species of orthodox taxonomy to the local race and an attempt made "to classify evolving groups as they occur in nature".

The procedures of experimental taxonomy given by Gregor *et al.* were:

1. Choice of a group of plants.
2. Collection of samples over its geographic and ecologic ranges.
3. Cultivation of representative samples in an experimental garden.
4. Observation of discontinuous variation and biometrical comparison of continuous variation.
5. Study of fertility relationships.
6. Investigation of cytology.
7. Synthesis of results.

They emphasised that a fundamental feature of this approach was the growing of races under uniform environmental conditions in the experimental garden so that the variation shown by races could be studied unaffected by the irregularities of their natural habitat.

A species studied in detail by Gregor and his associates (Gregor, 1930, 1938, 1939; Gregor, Davey & Lang, 1936; Davey & Lang, 1939) was *Plantago maritima*, which occurs as more or less isolated, breeding populations, often of quite small numbers of individuals, in a wide range of maritime habitats, from water-logged coastal mud, through salt marsh to more or less drained coastal mud. Seed samples were collected from fairly well defined although not necessarily spatially isolated colonies, the number of seed samples (spikes) collected in any habitat depending on the density of the population. Particular stress was laid on this systematic sampling. The seed was sown in late February and the seedlings planted out in the experimental garden in June. The overwintered plants were scored the following summer for various characters such as leaf length and breadth, scape length, etc. In *Plantago maritima* the populations were most clearly distinguished from one another by quantitative characters and those which contributed to the growth habit of the plants.

The result of this work was to show that the pattern of ecotypic variation was more frequently *continuous* than discontinuous, corresponding to the gradients shown by the habitats. Gregor later adopted Huxley's concept of the *cline*—an auxiliary taxonomic principle to register continuous hereditary variation related to a gradient—and proposed that continuous ecotypic variation should be termed an *ecocline.* He suggested that several of the ecotypes recognised by other workers were parts, often endpoints, of continuous ranges of variation. He redefined the ecotype as *a range on an ecocline.*

Gregor's views on the essential continuity of ecotypic variation received support from work done on *Plantago lanceolata* by Böcher (1943) where growth habit types—distinguished as *decumbens*, *ascendens* and *erecta*—formed a series corresponding to environmental gradients, i.e. ecoclines. A pattern of clinal variation was also described by Böcher (1944) in *Veronica officinalis*, where all the races are decumbent but show conspicuous differences in size of the leaves. This increases northwards or eastwards towards more continental regions and is regarded as a *topocline* (i.e. a gradient related to topography or climate, not obviously ecological), but is also ecoclinal when followed from wooded localities into non-wooded, open areas. These relationships were elucidated as the result of comparative cultivation experiments under uniform condition and varied-environmental tests in Copenhagen. Similar investigations on *Plantago coronopus* showed a variation pattern best expressed in terms of a series of clines running N-S in Europe (Böcher, Larsen & Rahn, 1955b). As with *P. lanceolata*, Böcher concluded that certain taxonomic units correspond fairly well with some of the ecotypes (genetic races) or ranges on ecoclines. Very dwarf races occur in the N.W. of the range on exposed rocks and appear to correspond with typical ecotypes in Turesson's sense, but are connected with other "ecotypes" by a series of intermediates.

Other Patterns of Ecotypic Variation

Scores of ecoclines have been described since Gregor's original proposals, notably in outbreeding species and especially in trees. One of the most elegant demonstrations is of ecoclinal variation in *Eucalyptus* (Barber, 1955, 1956; Barber and Jackson, 1957). In a survey of 25 species of Tasmanian

Eucalyptus, the feature of taxonomic value chosen for study by Barber was glaucousness resulting from a heavy deposition of white wax on the epidermis of leaves or stems or both. This has been given high taxonomic significance by previous workers on the group: it has been employed as a diagnostic difference between species and also to separate sections and series, although it leads to an unnatural separation of otherwise closely similar or related species.

Sporadic glaucous variants in populations of normally green species, and vice versa, have been found. Such variation appears to be caused by occasional hybridisation and segregation or recent mutation and has no ecological significance today. Clinal variation in glaucousness also occurs in the group, often in the same species which show the sporadic variation just referred to.

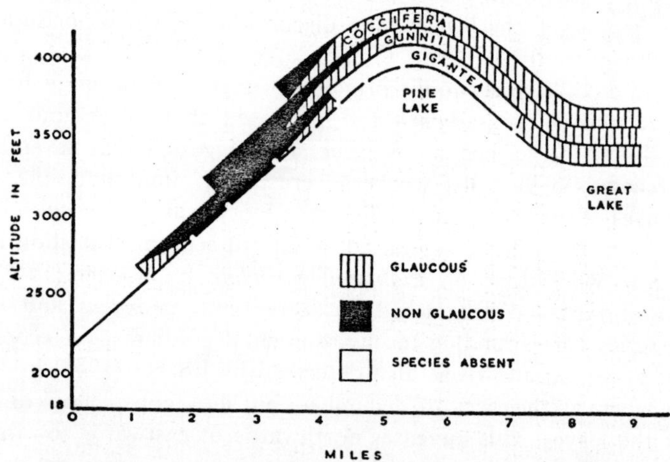

FIG. 32. Diagrammatic section to show altitudinal zonation of *Eucalyptus* species in N. Central Tasmania, and changes in proportion of glaucous to non-glaucous saplings in each species. Further explanation in text. (From Barber, 1955.)

Parallel clines have been recorded in 8 of the 25 species and are best correlated with increasing frost conditions: the more glaucous forms grow in regions with late spring frosts which can damage and kill the new spring leaves. Barber suggests that the incidence of the late spring frosts has been the main environmental factor of natural selection which has built up the clines. The pattern of clinal variation in three species, *E. gigantea*, *gunnii* and *coccifera*, falls into two types in a transect from 1200 ft. to 3400 ft. in the Central North of Tasmania (Fig. 32). In *E. coccifera* the altitudinal series starts with completely non-glaucous individuals with a rapid increase in the proportion of glaucous trees as the altitude increases, with mixed populations of green and glaucous individuals confined to a fairly narrow zone, succeeded by completely glaucous populations in the rest of the transect. Similar clinal patterns occur in other parts of its distribution area.

In *E. gunnii* there are two clines—one in glaucousness of the juvenile foliage and one in young adult foliage; while in *E. gigantea* the pattern is more complex. There the lowermost populations at 2500 ft. are uniformly glaucous and white-stemmed, but within one mile from them and a 100 ft. higher non-glaucous red-stemmed trees occur with a 50 per cent. frequency.

These clinal situations should be compared with those described by Clausen, Keck and Hiesey in their transect across Central California where the various populations of the species are more or less restricted to specialised habitats within the general climatic gradient and form a corresponding series of ecotypic races, most of which are sufficiently distinct phenotypically to be recognised as subspecies. In fact, within the various species studied by these workers (*Potentilla*, *Achillea*, *Viola*, *Penstemon*, etc.) a similar series of races was detected, corresponding to the sequence of climatic segments along the gradient.

When the distribution of a species population along a continuous environmental gradient is discontinuous, the ecocline will be broken up into units—ecoclinal ecotypes—which may be described if it is found convenient to do so. These are frequently referred to as *stepped clines*: the discontinuities and consequent "steps" or reference points may be quite arbitrary.

The choice is not, however, simply between discrete ecotypic populations and ecoclines (in their various forms). A third and probably as common situation is known in outbreeding species, as has been described by Bradshaw (1959) in *Agrostis tenuis*. The distribution of this species is essentially continuous in the areas in Central Wales studied, but made up of a series of local populations between which there appears to be effective isolation permitting differentiation to occur in response to local variations in the environment. The environment forms a graded patchwork of different conditions with a corresponding pattern of ecotypic differentiation which does not lend itself to taxonomic recognition. Not only does the absence of discontinuities make any delimitation of units quite arbitrary, but the overall picture does not form any coherent clinal pattern. There are indeed clines, but they are many and complex, as would be expected from the habitat factors which make up a complicated set of gradients. Moreover, Bradshaw points out that individual environmental factors do not necessarily form regular gradients, making the situation even more difficult. He suggests in such a case recording only differences between individual populations as they are found.

Similar patterns may be expected in other outbreeding species showing the same kind of pattern over a wide range of environmental conditions. Conspicuous, discrete, ecotypic variation cannot normally be expected in such cases except accidentally through inadequate sampling. Gregor & Watson (1961) comment that under the range of environmental conditions found in Britain discretely fractioned ecotypic variation seems to be the exception rather than the rule. Examples are, however, known such as lead-tolerant and lead-susceptible forms in *Agrostis tenuis* and *Festuca ovina* (cf. Wilkins, 1960a).

FREQUENCY OF ECOTYPIC DIFFERENTIATION

The above survey stresses the classic genecological investigations of Turesson, Clausen, Keck and Hiesey, Gregor and a few others, on a small number of plant groups. It is now widely accepted, however, that ecotypic differentiation occurs in many species. Gregor (1948) has emphasised that the ecotype concept embraces all intraspecific variation of ecological significance, the term ecotype being applicable to any population which is differentiated in respect of any feature which can be attributed to the selective action of environmental factors. We should remember, too, that adaptedness is much more likely to be expressed in physiological rather than morphological differences shown by an individual or population, as Turesson originally stressed (although he tended to emphasise morphologically recognisable ecotypes in his early work).

It is, as Daubenmire (1959) points out, possible for environmental selection to produce a series of races which are adapted to habitat variations even though they may not be recognisable morphologically. This would explain, as he says, the phenomenon frequently noted of the markedly different habitat preferences of the same species in different parts of its distribution area. It has been argued that some degree of ecotypic differentiation is a universal phenomenon in all species. Few instances are known of widespread species which do not show genecological differentiation when subjected to investigation. MacMillan (1956) suggests that this may be due to failure to report cases where it does not occur since they are not considered noteworthy. Cases of species which have what Daubenmire terms a wide *ecological amplitude*, due to the possession of a large reservoir of alleles not specialised into narrow ecotypes with complementary requirements, may be considered as comprising a *generalised ecodeme*.

From what we have said, not all ecotypic differentiation forms easily recognisable or describable patterns, nor forms discrete units or even trends, but a large number of cases has been published. Genecological differentiation has been found to be especially common in trees, possibly through extensive sylvicultural research into provenance trials; many cases have been investigated in agricultural crop plants (legumes, grasses), particularly by the Russian school (cf. Sinskaja, 1960). Other examples are known in shrubs, herbs and grasses; in inbreeders and outbreeders. An excellent survey of a wide range of ecological races, mainly from the viewpoint of their genetic structure, is given by Clausen & Hiesey (1958).

It is clear from this that it is rapidly becoming necessary for some system to be organised whereby the presence of ecotypic variation in species can be recorded and the data disseminated like other systematic information (Heywood, 1962).

The Russian School

An important school of genecological studies developed in Russia under Dr. Sinskaja of the Institute of Plant Industry at Leningrad. Sinskaja was responsible for much of the early work on ecotype recognition and classification, and the results of pre-war investigations of the school were published in her book *Dynamics of Species* (1948). The Russian approach differs in some respects from those of other schools and has paid great attention to geographical and zonal variability of populations in wild and cultivated plants. A survey of more recent work was published by Sinskaja in English in 1958 and an extended abstract in English of her researches on ecotypic differentiation of wild forage plants in the North Caucasus (1960) has been prepared by the Commonwealth Bureau of Plant Breeding, Cambridge.

The following account of work on zonal variability in the genus *Onobrychis* in the Northern Caucasus is taken from Sinskaja (1960).

Four vertical belts of *Onobrychis* populations may be distinguished:

(1) *The high mountain belt*. Here the plants are slow-growing, early-flowering, winter perennials.

(2) *Middle mountain belt*. Here only few plants come into flower during the first year, and their stems are considerably higher; in following years they begin to flower later.

(3) *Sub-mountain belt*. Here most plants flower in the first year but relatively late in the season (spring late-flowering perennials).

(4) *Adjacent Steppe belt*. Here all plants flower in the first year. They are drought persistent, spring perennials.

The morphological features of the populations show gradual changes in this series of belts: size of flowers and fruits diminishes from the upper to the lower belts; petal colour becomes less bright; leaflets become broader; etc. As a result it is very difficult to divide up the species according to breaks in morphological characters "so that the Caucasian *Onobrychis* species are really 'botanicorum crux et scandalum'. It is much easier to draw a distinguishing line between vertical belt-populations and to establish new species corresponding to these new belts of populations, on the basis mainly of biological and ecological characteristics."

This distribution pattern of *Onobrychis* biotypes follows Sinskaja's law of zonal population variability which refers to young species with more or less continuous distribution and without effective breeding barriers. This law states (Sinskaja, 1948): "The course of the successive variability in the population compositions in the process of zonal ecotype (and species) formation is expressed by the system of spiral intergrading series in which some variants are gradually eliminated, and others gradually change in the direction of the change in environmental conditions. At the same time, analogous variants (ecoelements) in neighbouring (zonal or vertical) series are not wholly identical and express successive stages of natural selection."

There is no doubt that situations described from other areas will be found to fit into this kind of ecological-geographical pattern.

Ecotypes in Agriculture and Forestry

With the realisation that species contain a large amount of ecological variation and that population samples from different habitats within a species area show different adaptive or physiological responses, a great amount of genecological research has been undertaken by agronomists and foresters as to the choice of the most suitable races for cultivation in particular areas and under particular conditions. Gregor and his school have in fact contributed significantly to this aspect.

Much of the research dealing with forest tree species (provenance trials) is published in specialist journals and is seldom reported on in botanical circles. This is unfortunate since the data add considerably to our knowledge of the intraspecific variation of species and often on a much vaster scale than can be contemplated by unsubsidised "pure" research. Frequently attempts at fitting the results into any taxonomic or biosystematic framework are soon abandoned and the different races recognised may be referred to by their provenances. This is especially true in sylviculture studies (cf. Röhrig, 1958). Svoboda (1953) has employed the term *climatype* for particular geographical provenance forms in Conifers.

Much of the variation in forest trees is clinal, and Böcher (1960a) suggests that the clear-cut ecotypes described in many wind-pollinated outbreeding tree species are probably parts of clines, their apparent distinctness being due to randomly chosen segments which are growing in different localities. From a practical point of view it is important for the forester to be able to distinguish geographical types so as to obtain seed which will produce trees that will be suitable for reafforestation in particular areas. That the geographical types form part of continuous variation ranges is not of prime importance to the forester.

The black pine, *Pinus nigra*, has already been mentioned as an example of a tree species showing clinal variation both locally and regionally, although there is an overall pattern of geographical subspeciation from North Africa and Spain across most of south Europe to the Caucasus. Certain features, such as needle length, vary clinally from east to west. The subspecies, however, contain ranges of variation in respect of these features but are characterised as well by other features which do not vary continuously. Many of them are, however, highly specialised physiological features or refer to mechanical properties of the wood and do not lend themselves to taxonomic recognition. Growth habit is also characteristic, but is a field character only and therefore of limited value. It is, furthermore, very liable to modification.

Pinus sylvestris, in which 140 "varieties" have been described (Carlisle, 1958), is reported to contain ecologically conditioned clinal variation (in leaf length, chlorophyll content, growth, etc.) in Scandinavia (Langlet, 1936), but it has been suggested that some of this variation is due to hybridisation between different races. In the Nearer East the ecotypic segregates cut across recognition of taxa on the basis of cone-scale shape, etc. The cone-scale varies within populations and it has not proved possible to recognise any infra-specific taxa in Anatolia on the basis of herbarium samples (Davis).

General Discussion

A number of points which have been brought out in this general survey will now be considered in more detail.

1. *Nature and Origin of Ecotypes*

The ecotype concept implies that some population fractions can be demonstrated to be adaptively related in respect of certain features to particular environmental factors within the species range. The basis of genecology, as Harberd (1958) says, is the observation that this intra-specific variability can be demonstrated to be of genetic origin and is not randomly dispersed throughout the species range, but is distributed in such a way that neighbouring plants tend to resemble one another. The adaptation can be demonstrated in morphological or physiological characters.

In ecotypic differentiation, adaptation refers to "any feature of an organism or its parts which is of definite value in allowing that organism to exist under the conditions of its habitat" (Daubenmire, 1959). It is recognised, however, that all the factors of the environment are interrelated and contribute to the selective forces operating on natural populations.

2. *Ecotypes and Ecads*

The ecotype is an experimental category so that its adaptive nature has to be proved by experiment. As we have seen, a single genotype may be able to grow successfully under a wide range of conditions, often assuming different morphological forms. Not all phenotypic variations which are apparently correlated with diverse habitats will therefore be genotypically controlled. Harberd (1957) has, for example, shown that in East Anglian fescues (*Festuca*) populations have a high genetic variability in certain characters which are widely assumed to have an adaptive significance (leaf-length, time of ear emergence) but there appears to be little genetic variation in overall ecological tolerance. Consequently it was not expected that selection would take place since the adaptability of the race lies in its ability to grow under a wide range of habitat conditions. It may well be, as Gregor & Watson (1961) suggest, that it is a cryptic feature of ecotypic selection to favour genotypes with a high plasticity potential so that the capacity for modification varies from plant to plant even within a population or ecotype, as Clausen, Keck & Hiesey (1940) have demonstrated in *Potentilla gracilis*.

Since not all genotypic differences between populations will be adaptive, we have to try in practice to distinguish between a number of different situations (Heywood, 1959a):

(1) Environmental selection of genotypes most fitted to a particular habitat, and selective elimination of unfitted genotypes, giving a genetically adapted, heritable ecological product, the *ecotype*.

(2) Environmental selection of genotypes which are able to grow in a range of habitats by consequence of their wide range of tolerance. The phenotypic product is termed an *ecad* or *modification*; although some botanists and zoologists especially have referred to it as an ecotype. It is the result of phenotypic plasticity: the genotypes develop the appropriate phenotype in the appropriate conditions, that is they have the capacity to function in a range of environments. This is what Mather (1955) has called "genetically determined adaptive plasticity of development".

(3) Fortuitous colonisation of habitats by different genotypes, especially of inbreeders, resulting in local habitat variants with slight morphological differences but with no adaptive significance.

(4) Fortuitous fragmentation into very small populations with the operation of "genetic drift", causing slight morphological variation in non-adaptive features.

The distinction to be made, therefore, is between the results of selection (the first two situations) and the results of drift, i.e. all variation in gene frequency which is random with respect to selection (the latter two situations). Random variations in gene frequency do not concern us here except in so far as they have to be guarded against as a possible source of error when sampling populations for genecological studies (see below).

Since the different results of environmental selection have both been referred to loosely as ecotypes, it has been suggested that the genetically adapted "fixed" situation be distinguished as a *genecotype*; the situation where it is the flexibility that is genetically determined and parallels the genecotype in its response to environmental conditions might usefully be termed a *phenecotype*. Thus the term ecotype could be used unqualified for both products of selection when the nature of the differentiation is not known.

In general, the occurrence of ecotypic differentiation (based on multiple gene inheritance) between populations occupying contrasting adjacent habitats will depend on the balance between the forces of selection and gene flow. The effect of gene flow from neighbouring populations is to prevent the building up of a complete response of the genotype complex to the local selection pressure, so that a specialised ecotype will not be formed.* This is more likely to happen in generalised habitats where alien gene infiltration will prevent the integration of local gene complexes adapted to the prevailing

* It should be noted that ecotypic differences based on simple gene inheritance may be built up rapidly in a population, as in industrial melanism in moths. No comparable plant examples are known.

selective pressures. Only a generalised ecotype will be formed in consequence (Gregor, 1956).

Not all apparent differences related to environmental differences between local habitats will be due to genecological differentiation but may be artefacts resulting from the kind of population sampling employed (Harberd, 1958). In perennial, vegetatively reproducing species, genotype reduplication in samples of populations is a serious potential source of error as it can make the statistical significance of the differences between populations spuriously high. Single genotypes can occupy a wide area and can in fact have the same range of ecological tolerance as the local population and indeed represent it. In *Festuca ovina* vegetative spread is weak and a single genotype may spread up to 2½ yards; *F. rubra*, on the other hand, has much greater powers of vegetative spread and Harberd (1958, 1961) has given convincing evidence that a single genotype can spread uninterrupted over 250 yards, producing a third of the tillers of the species in a 9-acre area of grassland. In such a case it is unlikely that genotype reduplication can be avoided in sampling. Another consequence of such a situation is that it may be exceedingly difficult to demonstrate how the populations may be ecotypically adapted since their characteristics will correspond largely to selective pressures of the past rather than those operating today. Harberd (1961) estimates that the Scottish genotype of *F. rubra* studied was anything from 400 to over 1000 years old!

3. *Hybridisation and the Origin of Ecotypes*

Genecological differentiation normally arises *de novo*, but it could be a consequence of hybridisation between taxa giving the initial variability necessary for selection to act upon. Curtis (1959) suggests as a model the centripetal migration of two related taxa across recently deglaciated terrain. Hybrids between them may be produced on contact and these would have a great variety of habitats in which to become established (the substrate might not be suitable for either parent). Continued backcrossing from such hybrid populations might introduce adaptive genes from one species into the other and further migrations might then lead to ecotypic differentiation.

Harberd (1961) suggests that ecotypic differentiation in *Galium saxatile* may have been a result of hybridisation with *G. pumilum*, although the process is normally distinguished as introgression: "If, however, the parental taxa are completely submerged by hybridisation, and so not describable, a situation might arise having many of the characteristics of differentiation by selection." He further argues that if all species show genecological differentiation, then not all would be the result of hybridisation, but if only a few species do so, then we should be suspicious of selection as an adequate interpretation. Instances where an environmental factor was responsible for the differentiation of one species and not another, when the species are generally similar, are not likely to be the result of selection alone.

On the other hand, Barber & Jackson (1957) argue against introgression as the cause of many variation patterns which he considers it may be possible to show experimentally to be the consequence of intense selection. They say that in a region of great ecological change, simultaneous clinal variation can be expected in the frequencies of genes at a number of loci. Along an ecological gradient, variability controlled by selection could reach a peak and fall on either side of this peak, thus giving the "loose association of variables" which is considered by Anderson (1953) as a characteristic feature of introgression. Barber suggests that "if the environmental factors which are significant in selection vary mosaically, a complex pattern of interlocking clines could be built up by selection alone. It would be difficult if not impossible to decide whether 'introgression' or selection had built up the 'loose association of variables' without the experimental measurement of selective coefficients."

One of the stumbling-blocks in interpreting apparent introgressive patterns as the result of selection is the difficulty of interpreting the functional or adaptive significance of the apparently trivial changes in the phenotypic expressions involved.

It may be mentioned here that under conditions of geographical isolation, as on islands, so-called insular races may be built up and be considered as geo-ecotypes which have not been formed as the result of selection. How far selection has in fact been responsible for such differentiation and how far it may just be regarded as genetic "drift" is a moot point and much debated by population geneticists at the present day. This problem is to a large extent tied up with methods of sampling and analysis (see p. 406).

4. *Parallelism between Genecotypes and Phenecotypes*

The phenomenon of parallelism between phenecotypes and genecotypes has already been mentioned. Turesson (1922) recorded examples in *Solanum dulcamara* (shade form), *Dactylis glomerata* and *Lysimachia vulgaris*. As he explained it, the population in an extreme habitat may respond as a whole with a reaction-type (or ecophene, see p. 389) which is suggestive of a specialised ecotype although it may be partly or wholly due to the modificatory effect of particular habitat factors.

Many other cases have been reported in the literature and they are a source of trouble to the taxonomist, since in the absence of experiment it is seldom possible to determine whether an ecological variant is genetically fixed (a genecotype) and deserving of taxonomic recognition or simply an environmental modification (phenecotype).

Current practice is to disregard phenecotypes taxonomically, but as has been pointed out (Heywood, 1959a), although they may be "sunk" by reason of their non-genetic nature, allowance should be made for them in drawing up taxonomic descriptions. They should not be given formal Latin names (or if they have, they should not be employed) but their range of variation should be included in the description of the taxon which produces them and even specifically mentioned. While the modification itself may be of no taxonomic value, the fact that a certain population is capable of producing a particular phenotype in response to particular environmental conditions certainly is.

It has been suggested by Wilmott (1949) that such modifications should be indicated by elisory phrases indicating the conditions under which they may be produced. As an example, he cites *Plantago coronopus* var. *pygmaea* which could be referred to as "forma in arenosis siccis", although in this particular instance the named variety probably covers dwarf modifications of medium-sized races and genotypically fixed dwarf races (Böcher, Larsen & Rahn, 1955b). The suggestion is, however, to be recommended.

A taxonomically embarrassing situation is found in circumstances where the part of the total range of plasticity of a taxon actually expressed, varies significantly in different parts of its distribution area without corresponding genotypic differences. This has been found in the aquatic *Ranunculus circinatus* for example (Cook, *in verbis*), and may be quite common. The uniformly different plastic responses of populations of a taxon in different geographic areas might well deserve some form of taxonomic recognition.

The significance of ecological modifications in a micro-evolutionary context has probably been underestimated. Their role in the formation of genecotypes has tended to be overlooked since the early observations of Turesson, but evidence has been accumulating in recent years which suggests that they have a considerable importance. They seem to have a very direct relevance to the sequence of events known as the "Baldwin effect" which has been summarised by Simpson (1953b), Waddington (1953) and others. Briefly, the phenomenon is that a phenotypic response of adaptive significance may be accompanied or followed by one which is genotypically controlled. It has been explained by Waddington (1957) as "genetic assimilation" whereby "natural selection for the ability to develop adaptively in relation to the environment will build up an epigenetic landscape which in turn guides the phenotypic effects of the mutations available".

Although Waddington's work has been concerned with reactions in *Drosophila*, the probable significance for plant genecology has been summarised by Gregor & Watson (1961) as follows: "If a phenotype, valuable in a selective sense, happens to arise as a direct response to the prevailing environmental conditions, it may ultimately achieve genetic fixity as a result of the operation of selection acting in the same direction over a number of generations, but only in so far as the requisite genes already exist in the original population. Thus it is not strictly correct to regard the environment as a phenotypic sieve." Actual experimental examples in plants are not known but it seems a likely explanation of the co-existence in a particular habitat of phenecotypes and genecotypes.

This intensification in time of a certain developmental response once it has been initiated in a population or individual is termed "canalisation" by Waddington, and he explains it in terms of competition for the same biochemical substrate by various developmental processes. Canalisation is a homeostatic process which favours a particular developmental pattern. In other words, once a particular line of development has been initiated, certain sequences result through its following a particular channel. The whole phenomenon, which appears superficially neo-Lamarckian, is thoroughly discussed by Waddington (1957, 1962).

The question has also been discussed by Crosby (1956) from a theoretical point of view. It would appear, he says, that phenotypic plasticity may be the most efficient way in which a species can adapt itself rapidly to a new environment. In this way the appropriate environmental niche is occupied by the phenecotype while natural selection modifies the biotype in the required direction. Crosby suggests that selection pressure may favour genetic control rather than phenotypic plasticity. Phenecotypes may be adequate for a certain number of generations, but are not so efficient as the genetically controlled adaptation of the genecotype.

Gregor & Watson (1961) have described experiments with *Plantago* which suggest that when the differences between habitat conditions are only minor, the effect of phenotypic plasticity is to buffer genotypes against the full effect of the environmental pressures. A very considerable selection pressure may have to be exerted before the kind of effect usually associated with ecotypic differentiation is perceived.

5. *Sampling Procedures*

The methods of sampling have a very great bearing on the interpretation of results—whether the populations are significantly different or not, and what are the relative roles of isolation and ecotypic differentiation (Harberd, 1958). Much attention has been drawn recently to the problems of sampling populations for genecological studies (Clausen, 1960; Harberd, 1957, 1958; Wilkins, 1959).

It is essential that the sampling techniques and procedure be appropriate to the type of investigation in hand. Clausen (1960) emphasises the differences between sampling the variability within a population and testing its breeding structure. The latter requires progeny tests of various kinds which are best regarded as a second stage in a genecological study. If the aim is to show that two populations differ in genetic constitution, analysis of the "within population" variance will suffice (Wilkins, 1959). But in a genecological study, the aim is to relate the variation to the selective factors of the environment. Attempts should therefore be made to distinguish between differences caused by drift and those caused by genecological adaptation. The elimination of drift effects may be difficult and depends on the sampling techniques employed and the intensiveness of sampling.

The following sampling techniques are employed:

(1) Direct transplants (as employed by Bonnier, Hall, Turesson, etc.) to study the performance of plants that have actually grown in natural

populations. This has the disadvantage that these will give only poor representation of the variability of the plants that grow in the area over a period of time under slightly differing conditions.

(2) Clonal method whereby ramets of a single genotype can be tested in a variety of conditions. As Wilkins (1959) points out, it is of advantage to use plants that can be cloned when attempting to study the interaction of the genotype and environment in that the genotype remains constant. Seed clones are possible in apomictic plants such as *Poa* spp. where a large number of species reproduce asexually (Watson & Clausen, 1961). Apomictic lines are useful in that they show differences in phenotypic expression at different centres. In general, it is better to use propagated individuals rather than numerous individuals of varying heredity.

(3) Seed sampling. This is according to Clausen (1960) the best method to give an adequate impression of variability in a population. Both the Stanford and Gregor's group agree that 60-100 plants give a minimum seed sample. Details are given by Clausen (1960), who recommends that 300-600 seedlings per population should be raised from these samples and about 120 seedlings pricked out from these. When transferred to the experimental garden, the plants represent a fair sample of the potential variability occurring in the population. If perennial plants are used, they can themselves be cloned and studied in a variety of habitats.

Clausen observes that "sound sampling is more important than using refined statistical techniques on poorly sampled material." "Statistical methods are at best only crude tools for the determination of phenotypic variability. The variation curve is a better biometrical description of the variability of a character within a population than are statistical means, variance, and standard deviations that tend to submerge the pattern of variability" (1960: 69).

If the procedures outlined above are accepted, it follows that the limited samples on which much biosystematic work has been carried out in the past result in their being too restricted for consideration as valid genecological trials. Often the seed (spontaneous) has been obtained through Botanic Garden seed lists and no information about sampling data is normally available—the material is not collected systematically from populations—and few plants are raised. Such work is often more strictly part of, or incidental to, "experimental and cytological" studies as Böcher, one of the leading exponents, terms them. This is not to deny in any way their value in uncovering data of considerable value about the variability of plants—they demonstrate the probable existence of racial differences within species although not very precisely since small samples are cultured and are divorced from natural conditions by the very nature of the investigations. This makes it all the more difficult to explain the nature of the adaptations observed. The above methods are normally used with annual or pauciennial herbs (cf. Böcher, Larsen & Rahn, 1953, 1955a) where cloning is not possible.

Recognition and Classification of Ecotypes

In ecotypic differentiation adaptation refers to "any feature of an organism or its parts which is of definite value in allowing that organism to exist under the conditions of its habitat" (Daubenmire, 1959).

The ecotype status of a population depends primarily, therefore, on habitat recognition. Then a relationship has to be established between the population heredity and the prevailing habitat factors. This may be difficult to demonstrate, as it is recognised that all the habitat factors are interrelated and contribute to the selective forces operating on populations. Ecotypes are therefore recognised as the product of the selective action of the whole environment, but they are defined with reference to some particular environmental characteristic. As Gregor puts it, one is forced to select that part of the heredity which is related to certain, more conspicuous features of the habitat, although it must be remembered that factors other than these may have had a part in the pattern of variation observed.

Three main sets of ecotypes are normally recognised—climatic, edaphic and biotic—depending on which of the environmental factors has been the most dominant. Many other kinds of ecotypes have been described and are summarised in the accompanying table (modified from Lawrence, 1945).

1. Climatic ecotype (Turesson, 1925)
   Syn. Climatype (Sinskaja, 1928, 1931)
2. Edaphic ecotype (Sinskaja, 1929; Gregor, 1942)
3. Biotic ecotype (Sinskaja, 1931)
   (*a*) Synecotype (Sinskaja, 1931)
   Syn. Phytosocial ecotype (Gregor, 1946)
   (*b*) Agro-ecotype (Gregor, 1933)
4. Geographic ecotype (Lawrence, 1945)
   Syn. Reclusion-type (Turesson, 1927)
   Geo-ecotype (Gregor, 1931)

It must be remembered that selection operates not on single characters but on character complexes, even though one or a few characters may be found to show variation related to the chosen environmental variable. Overemphasis on single characters may, as Stebbins points out (1950), confuse rather than clarify our understanding in the variation pattern. This is particularly true when the variation follows a clinal pattern: a given plant or population may belong to as many ecoclines as genotypic and ecologically significant trends that cross its territory. Gregor & Watson (1961) describe such a situation in *Plantago maritima*, where they could show from garden data that "gamodeme samples collected from three recognisably different types of habitat on the Scottish mainland represented three ecoclinal geno-ecodemes on the basis of leaf and stem length and growth-habit variation. Samples collected on one of the Atlantic islands off the Scottish coast from types of habitat corresponding to those of the mainland could on the same criteria also be arranged in the same ecoclinal sequence. But because the stem and leaf lengths of the island topodeme were in general shorter, the two ecoclinal series of genoecodemes, though parallel, were not coincident; a pattern which reflected the influence of some other factor, perhaps climate, which had not been taken into account."

The general problem that has to be faced here is that the ecotypic fractionation of a species or population in response to one selected feature of the habitat (e.g. soil) may not coincide with its subdivision in response to another (e.g. climate). This is perhaps the major source of difficulty in the taxonomic treatment of genecological differentiation. As Gregor (1956) points out, the individuality of ecotypic populations which are visibly recognisable is one of their most characteristic attributes, yet if this individuality is analysed it will usually be found that only a small part of the variation which contributes to this individuality is of proved ecological significance. Taxonomic recognition will concentrate more on this individuality than on the demonstrable ecotypic picture, and in cases of ecoclinally distributed variation taxonomic treatment is bound to be unsatisfactory, as it will weight the characters which contribute to the individuality of the local differentiated population and in so doing cut across clinal patterns (Heywood, 1959). The difficulties of demonstrating adaptive variants have been considered recently by Bradshaw (1959) and by Wilkins (1960b). These are especially acute if one is concerned with physiological, not only morphological responses.

One of the problems of demonstrating the adaptation in garden trials by comparative cultivation is that there is always a slight uncertainty as to what is being measured—the important character may not be the one measured but another closely associated with or related to it; or it may be the morphological expression of an underlying physiological difference. Further, the plant may respond not to the habitat factor which is being measured but to one which is correlated with it.

Apart from such considerations there is the difficulty of extrapolation from the test habitat or from controlled conditions to the wild, which, as Wilkins says, must be extremely speculative.

Ecotypes and Taxonomy

When adaptive variants or ecotypes show correlated differences susceptible to morphological treatment they may be considered eligible for taxonomic consideration following normal taxonomic principles. In other words, it matters little the means by which morphologically different populations are detected: provided they are on a scale considered worth recognition this can be attempted.

Often ecotypes are found to coincide with recognised taxa— for example in *Orchis latifolia*, *Geranium robertianum*, *Silene dioica*, *Potentilla glandulosa*; or in other cases taxa have to be erected to accommodate them.

As we have already noted, most of Turesson's early ecotypes were taxonomically small groupings more or less equivalent to the variety, while Clausen, Keck and Hiesey originally considered the morphologically separable, climatic ecotype as the basis of the subspecies. Turrill (1946: 43) held the view that "the special genecological methods can be absorbed into orthodox taxonomy, its classification sometimes coincides, or can be made by slight modifications to coincide with the more general classification which is necessary if taxonomy is to fulfil its functions . . .". Likewise Böcher has found that in *Plantago* species the units of orthodox taxonomy correspond very well to certain genetical races (ecotypes) or to ranges on continuous character gradients (clines), although in some cases a revaluation was required.

There are, however, many difficulties of which the principal ones will now be considered:

(1) In several cases the adaptive features of the ecotype do not lend themselves to taxonomic treatment or are not correlated with features which do. Ecotypes are often distinguished by flowering time, seasonal growth rhythm, frost resistance or other physiological differences.

(2) Most morphological differences between ecotypes are quantitative, depending on polygenic (or multiple-gene) inheritance: they may require statistical treatment for their detection, and it may be found that the variation *within* the ecotypes obscures their racial differences.

(3) Genecological variation in populations may frequently occur at a lower level than formal nomenclatural classification can usefully do. At the micro-evolutionary level, orthodox classification undoubtedly has severe limitations and a problem that is seldom considered is the *need* for taxonomic recognition of ecotypic variation (see discussion in Heywood, 1959a: 96-102). The question that should be asked is what it is useful to recognise and for what purpose.

Daubenmire (1959) encourages the recognition of ecotypes taxonomically for ecological purposes—clearly it is useful for variants to have names so that they can be easily referred to. The significance of ecotypes in the description of vegetation is obvious in that if two communities of widely different ecological requirements are described floristically, the presence of many species in common to them will suggest a degree of similarity which may be highly misleading if the plants belong to different ecotypes (Curtis, 1959). This somewhat neglected aspect of the ecotype in community function has been studied especially by MacMillan (1956).

The question of recognition of micro-taxonomic units is bound up with a number of other considerations such as the disrepute into which formal taxonomy below the subspecies level has fallen. The subdivision of the species all the way down to the *forma* level as was practised in, for example, many *Pflanzenreich* monographs, and in Floras at the beginning of this century receives limited support today on the grounds that the lower units are artificial, not repeatable, are descriptions of individual specimens, or simply too small to occupy the serious attentions of the taxonomist. Many practising taxonomists do, however, find the need for occasional use of all the infra-specific units for dealing with the variation in well studied groups.

In more complex cases it will be found that as many genecological situations occur in nature as sets of criteria we employ to define them. One of the solutions proposed to deal with this problem is the deme *terminology* which avoids attaching formal names to these situations.

Deme terminology. A basic consideration that led to the introduction of the deme terminology was the strong belief that it is necessary to keep separate general-purpose, orthodox classifications which are morphologically expressible, from those concerned with micro-evolutionary units, both in aims and terminology. In consequence, the categories of formal taxonomy are rigidly excluded from the deme system, even as prefixes or suffixes in compound words. For example, terms such as eco-species and eco-type are not admitted.

The system was originally proposed by Gilmour & Gregor (1939) and developed more fully by Gilmour & Heslop-Harrison (1954). A succinct summary is given by Gilmour (1960) from whom the following extract is taken:

The essence of the terminology is the construction of a series of category-terms by the addition of one or more self-explanatory prefixes to the "neutral" suffix "-deme". The suffix -deme is defined as "a term, always used in this terminology with a prefix, denoting any group of individuals of a specified taxon". (The use of the phrase "specified taxon" indicates the way in which the "orthodox" categories can be used as a "framework" into which other classifications, for particular purposes, can be fitted.) An important point to note is that the suffix -deme is *neutral*; that is to say, it carries no implication that the individuals exhibit any relationship other than that they belong to a specified taxon. In particular, it carries no implication that they form a population, in either a topographical or interbreeding sense; these, and other implications, are indicated in the terminology by *prefixes*.

Ten basic terms were suggested as follows:

Denoting association with locality or habitat

topodeme: a deme occurring in a specified geographical area.
ecodeme: a deme occurring in a specified kind of habitat.

Denoting difference

phenodeme: a deme differing from others phenotypically.
genodeme: a deme differing from others genotypically.
plastodeme: a deme differing from others phenotypically but not genotypically.

Denoting reproductive behaviour

gamodeme: a deme composed of individuals which are so situated spatially and temporally that, within the limits of the breeding system, all can interbreed.

autodeme: a deme composed of predominantly autogamous individuals.

endodeme: a gamodeme composed of predominantly endogamous (i.e. closely inbreeding) dioecious plants (or bisexual animals).

agamodeme: a deme composed of predominantly apomictic plants (or asexual animals).

Denoting variational trends

clinodeme: one of a series of demes which collectively show a specified variational trend (i.e. which collectively form a cline).

It has to be admitted that this basic terminology has so far received only limited support. An additional term which is widely used is cytodeme which refers to demes differing in chromosomes from other demes; also used to some extent is polyplodeme to refer to demes which differ from other demes in level of polyploidy. What is more serious, however, is the growing misuse of the terms, especially by zoologists such as Huxley (1942), Carter (1951) and Simpson (1961), in a way which appears to show a complete misunderstanding of the system. Simpson (1961:176-7), for example, defines a deme as the minimal population unit, replacing the earlier expression "local population". Gilmour & Heslop-Harrison (1954: 152) stress that in the definition of a deme as any group of individuals of a specified taxon the word group replaces "assemblage" in the original Gilmour & Gregor 1939 definition "so as to indicate even more unequivocally that no idea of 'population' enters into the definition of the root, 'deme' . . . ". As Gilmour (1960) further comments, it is essential to keep the suffix -deme neutral so as to avoid the injection of the connotation of breeding population into all the derivative terms which would thereby destroy their validity. The compound "gamodeme" should be used for the kind of situation envisaged by Simpson: it is essentially synonymous with "breeding population".

Second-order derivatives denoting various differences in attributes can be constructed. The following are those associated with various genecological terms:

genoecodeme: an ecodeme differing from others genotypically.

hologamodeme: a deme composed of all those individuals which, within the limits of the breeding system, are believed to be able to interbreed with a high level of freedom under a specified set of conditions (cf. Turesson's "ecospecies").

coenogamodeme: all hologamodemes considered to be capable of exchanging genes through their members to some extent, but not with freedom, under a specified set of conditions (cf. Turesson's "coenospecies").

The first of these, genoecodeme, corresponds generally with Turesson's ecotype; the other two correspond to that part of the genecological terminology which is primarily concerned with the fertility relationships of populations (see pp. 389-90).

As has been pointed out, whatever the merits or demerits of the deme terminology, it is *not* intended to be a nomenclatural system of classification, and problems arise in connection with referring to and communicating about individual examples. If the terms are to be employed widely, there is a need for some link with formal taxonomy for purposes of reference, as Gregor has suggested. He proposes that genecological assemblages should be connected with the names of the taxa with which they coincide, overlap or most nearly approximate. Gilmour (1960) suggests that if, for example, a series of gamodemes in a particular area is being studied, each gamodeme could be referred to by a number or letter, or by a descriptive adjective. One could therefore talk about the Box Hill gamodeme of *Zerna erecta*, or the B12 gamodeme of *Ranunculus fluitans;* similarly, the prostrate plastodeme of *Salicornia ramosissima*. Such methods have grave disadvantages for reference purposes but it is the price that has to be paid for the flexibility of the system.

(4) Much ecotypic variation is continuous, following a clinal pattern, often of a complex nature. The principal factor involved in determining whether the genecological variation pattern will be broken up into discrete units (ecotypes) or be continuous (ecoclinal) is the breeding system operative in the populations. This has been discussed by Baker (1953b) and by Gregor (1956). In general, it seems true that the ecotypic differences recorded in outbreeders are clinal in nature and discrete ecotypes not encountered. There is a tendency on the other hand for inbreeders to produce discrete ecotypic units. The breeding system also affects gene-flow from neighbouring populations, as does the areal pattern of the populations. These also affect continuity versus discontinuity in the variation pattern as do the range and effectiveness of dispersal and pollination. Widespread distribution of pollen in the genus *Pinus*, for example, maintains a more or less continuous interbreeding between populations and this is reflected in the numerous clines that have been described in various species of the genus, such as *P. nigra* and *P. sylvestris* referred to elsewhere (p. 400). Sampling methods also have a bearing on whether the differentiation appears to be clinal or not. Inadequate sampling may well be one explanation of the number of discrete ecotypes recorded in early genecological literature.

The taxonomic treatment of clines presents problems since the cline, as Simpson (1961: 179) points out, is an arrangement of characters, not of organisms or populations. As we have already remarked, an individual population may participate in more than one cline, depending on the number of continuous variational trends that pass through its area; a population cannot, however, belong to more than one taxonomic unit. Moreover, taxonomists seek discontinuities in variation patterns so that clines are taxonomically an embarrassment, and while the cline concept is of great value in helping us *understand* or *arrange* the taxonomy of a group, it cannot by its very nature permit us to *describe* taxonomic units (Heywood, 1959a: 109; Simpson, 1961: 179-80).

It is true that some ecoclines are found to be broken up (stepped clines) by extrinsic factors and are therefore amenable to taxonomic treatment. This is only satisfactory if the populations divided off by the discontinuities are definable on characters other than the clinal features; otherwise there is the danger that the taxonomic units so delimited would be based on single *character* variation, not on group variation. From this it follows that clinal situations will have no relevance to taxonomic discrimination unless the groups are separable and recognisable on other considerations.

Bradshaw (1962) has drawn attention to the misleading conclusions that can be drawn through inadequate sampling of an ecocline. This is illustrated in Fig. 33 (a-d). If the samples are taken at widely scattered points, what are in fact arbitrary segments will appear to be distinct ecotypes (a). This will be aggravated if the cline is stepped (b). If, on the other hand, the distribution pattern of the species is discontinuous, recognition of the units into which it is broken will be easy (c). Recognition may well be difficult, however, if the ecological pattern varies in different regions. If the same irregular or fragmented ecocline occurs in two or more geographical areas, the units in comparable positions on the cline may not be the same, or the number of units judged to occur in the cline in each area may be different. This is because of the occurrence in the various areas of different gaps in the ecological gradient (c-d).

Since continuous ecotypic variation does not lend itself to formal taxonomic recognition, alternative methods of recording it have to be devised. Gregor & Watson (1961) suggest that "whatever entities the investigator is able to detect by the techniques he chooses to employ" should be recorded in terms of demes. The advantage of the deme terminology for such situations is its flexibility: the categories are not mutually exclusive. It is permissible for a given population to be referable to several different kinds of demes according to the characters or aspects which it is wished to stress.

Particular points on an ecocline can be referred to as ecoclinodemes or ecoclinal genoecodemes; or alternatively, they can be termed cline-forms (Pryor, 1956).

(5) A final series of problems arises from the different modes of formation and distributional pattern of ecotypic variants. Some ecotypes appear to have arisen polytopically in spatially separate but ecologically similar habitats where populations of a species have been subjected to similar selective forces. Many of Turesson's original ecotypes were of this nature and the system of formal nomenclature which he proposed was suited to the specification of this kind of situation. For example, the maritime ecotype of *Silene dioica* (described taxonomically as var. *crassifolius*) was termed *oecotypus salinus*, and

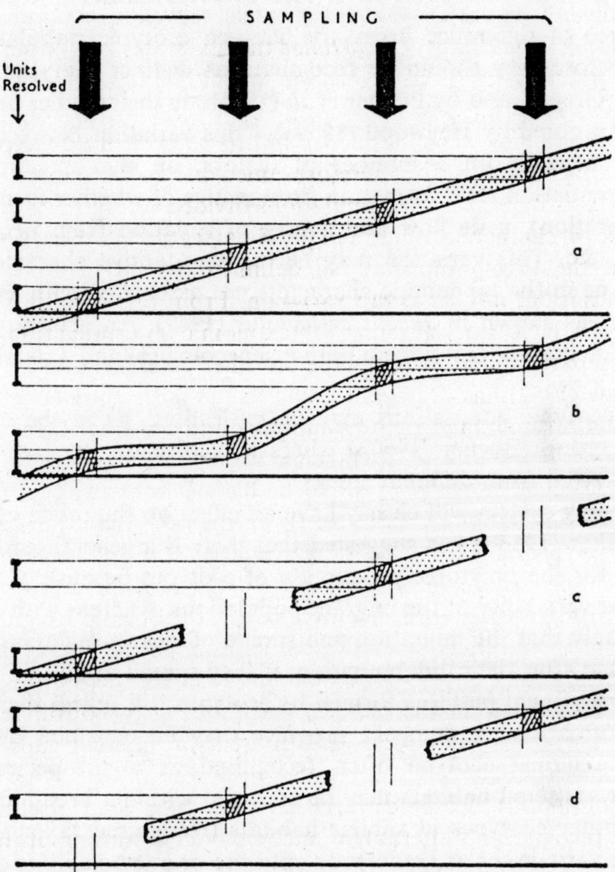

FIG. 33. The effects of limited sampling on various kinds of ecoclinal variation. (*a*) An ecocline resolved into distinct ecotypes. (*b*) A stepped ecocline resolved into very distinct ecotypes. (*c*) A discontinuous ecocline resolved into one set of ecotypes. (*d*) The same ecocline in another region resolved into another set of ecotypes. In each case the black base line delimits the same range of morphological variation. (From Bradshaw, 1962.)

similar *salinus* ecotypes were described in other species. As has been pointed out (Heywood, 1959a), this system when applied to polytopic ecotypes (where essentially the "same" ecotype occurs in different parts of the species' range and may correspond with different taxonomic varieties) would be a means of recognising *grades* of adaptation (see Chapter 2: 51). But whether or not there is a need to introduce a formal grade nomenclature, it does not seem necessary to add oecotypus to the categories of formal taxonomy: "var. (ec.)" or some such device would serve the purpose (cf. Wilmott, 1949; Clapham, 1949).

The degree of difference necessary between ecotypic populations that must exist before they should be recognised as distinct ecotypes has been discussed by Gregor, and by Böcher *et al.* (1955b) in their studies on *Plantago coronopus*. As noted by Heywood (1959a), "this variation between ecotypic components depends on a number of factors, on the breeding system, further differentiation after formation (irrespective of whether formed *in situ* or after migration), gene flow and degree of isolation from neighbouring populations, &c. This variation may be in the adaptive characters of the populations or in the taxonomic characters not associated with the habitat, as Clapham has shown in *Juncus articulatus* (1949). An arbitrary decision has to be made in most cases, from both a genecological and a purely taxonomic point of view."

Parallel ecotypic adaptations may be misleading, as in the *Campanula rotundifolia* group (Böcher, 1960b) where they obscure the fact that they have been derived from different stocks.

The breeding system will clearly have an effect on the mode of origin of ecotypes. Baker (1953b) has suggested that there is a general tendency with outbreeders for the polytopic production of ecotypes because of the actual and potential variability of the migrant populations, whereas with inbreeders it is more likely that the migration and spread of clearly differentiated races will take place after their differentiation, if they spread at all.

The distributional patterns formed by ecotypes will reflect their different modes of formation. A compact province may be occupied by the race (giving the regional ecotype often recognised as a subspecies); widely separated or scattered habitats may be occupied within a broad area, giving probably similar ecotypes in similar habitats (recognised taxonomically as one or more varieties or as grades); or only one or a few habitats of no great extent may be occupied (treated as varieties if recognised taxonomically at all). If, however, ecotypic variation forms a mosaic pattern, no taxonomic recognition is normally possible.

Chapter 13

EVOLUTION AND THE DIFFERENTIATION OF SPECIES

Introduction

At least in the higher plants and animals, evolution proceeds principally as a result of four indispensable processes: mutation, genetic recombination, natural selection and isolation (Stebbins, 1959b). None of these can be said to be more important than another, but the last process, reproductive isolation, is the one which marks the critical step in the evolution of species in the sense of the term adopted by many biosystematists. Dobzhansky (1950) considers that "it is recognised, except by a few conservatives, that the attainment of reproductive isolation between genetically diverging Mendelian populations is the essence of biological speciation. Reproductive isolation between populations is, accordingly, the criterion for the recognition of the specific entities which have biological meaning."

In the view of some taxonomists (e.g. Blackwelder, 1962) studies on the origins of species are not part of taxonomy although they should (and nowadays usually do) interest taxonomists. They are part of the new science of speciation. In other words, speciation is the concern of "speciationists" not of taxonomists. In fact, the application of speciation theories to the practice of taxonomy has been much less widely accepted than the literature of the "New Systematics" would lead one to believe. For our own part, we consider the study of speciation as coming within the province of taxonomy in the broad sense, but not as playing a necessary role in classification.

We have already discussed in Chapter 3 the various usages of the term species and have seen that it is currently employed in two very different senses or definitions—for groups of individuals possessing certain, mainly morphological, attributes in common, and for reproductively isolated Mendelian populations or gene-pools. We also gave reasons why the species should be retained in its traditional sense as a morphologically recognisable group of individuals, while gene-pool species should be denoted by some other terminology.

Speciation is applied to the origins and evolution of populations in terms of genetical processes, although the word itself suggests the evolution of morphological differences of the phenotype. Speciation is, in fact, concerned with the formation of reproductively isolated gene-pools. It would be idle to pretend today that such genetically defined units necessarily coincide with morphologically defined species (Walters, 1962).

This being so, the term speciation seems most unfortunate for a process leading to the formation of units which taxonomists may not consider species. Taxonomic species, in so far as they represent populations, have of course evolved through various genetic processes, but they do not all represent reproductively isolated, intrabreeding units. On the contrary, what taxonomists designate as species represent many different evolutionary situations (cf. Heywood, 1958 : 20). As Walters (1962) comments, it is surprisingly difficult to find an example of an "ideal" or uncomplicated species among the many cases that have been investigated genetically or biosystematically. Normally some complicating factor such as apomixis, inbreeding, polyploidy, hybridisation, or a combination of these, is operative.

While much of this and the following chapter is concerned with speciation in the sense of the usually gradual adoption by populations of different genetic patterns causing reproductive isolation, and with the ways in which this isolation may be broken down by hybridisation (all topics which may have little *direct* effect on taxonomic practice), one of the most interesting and absorbing lines of research for the taxonomist today is a study of the *relationship* between taxonomic species and the micro-evolutionary situations or units with which they most nearly coincide.

We feel this point needs considerable emphasis since any suggestion today that taxonomy is not vitally concerned with illuminating the evolution and genetic nature of taxa provokes strong protest. But to advocate the usage of the term species in a practical morpho-geographical sense in no way denies the importance of micro-evolutionary, population and genetic studies on the nature of taxa. It simply regards them as separate although complementary approaches. The genetic basis of species is not important for their recognition and naming by taxonomists; to suggest otherwise is demonstrably false, and to attempt a working classification with taxa based on gene-pools is simply to misunderstand the essentials of practical classification.

What then is the relevance of speciation to taxonomy? This is a far more difficult question to answer than might appear. An obvious answer is the one already suggested, namely that whenever possible the majority of taxonomists today wish to have an understanding of the evolutionary nature of the species and lower units that they describe in morphological-geographical terms, even though such knowledge may in no way alter their actual classificatory procedures or the units recognised in practice.

The following points have also to be considered:

(*a*) Since taxonomic species are not necessarily reproductively isolated populations, we cannot consider their origins as though they were.

(*b*) What we can do is to attempt to ascertain the evolutionary situations which our species represent.

(*c*) In so doing we may find that taxonomic species may cover more than one gene-pool, coincide approximately with one, or do not represent Mendelian populations (p. 353) at all, but much more complex and imprecise situations.

(*d*) As a result of such studies it may be found that a more natural and satisfactory means of delimiting the species may be possible, but this should only be employed if there are clearly definable morphological features supporting (i.e. correlated with) the new delimitation. In other words, species should make taxonomic sense and not be recast simply to meet gene-pool definitions irrespective of the possibility of their morphological recognition. Speciation studies may show that a taxonomic species comprises diploid and tetraploid cytodemes which cannot be morphologically separated; to recognise the different cytodemes as separate species simply because they represent different isolated gene-pools may satisfy the requirements of some geneticists but would make little taxonomic sense. Few taxonomists today, no matter what their persuasions, would advocate such a course of action (cf. Chapter 6).

(*e*) As we shall see below, it is a naïve assumption to believe that all plants are organised into clear cut, reproductively isolated gene-pools. The various factors causing reproductive isolation occur at different intensities, evolve at different rates, and operate in different parts of populations in different ways and with various degrees of effectiveness.

(*f*) It is widely overlooked that speciation studies presuppose a classification into taxonomic species. While evolutionary or population taxonomists are often ready to decry the outdatedness of orthodox taxonomic procedures and units, they are entirely dependent on the prior existence of a classification before they can themselves take their initial steps. If we could wipe the slate clean and start classification from scratch, it is debatable whether it would be possible for us to devise any more practicable method of procedure than attribute-grouping which is morphologically expressible (cf. Walters, 1962). However, we do not have such an opportunity and as a result there are repeated attempts to redefine species in genetic terms, attempts which are often doomed to failure. The fact that taxonomically and genetically defined species cannot necessarily be equated, clearly limits the role of speciation studies in classification, but it offers us a fascinating new field of relationships for study. It should also be remembered that knowledge of the origins (phylogeny) of species or of biosystematic units may give us little information about their present day relationships in respect of certain attributes, such as interfertility (cf. Jones, 1961).

Abrupt and Gradual Speciation

The two main processes by which reproductive isolation of populations is achieved have been distinguished as *abrupt* and *gradual* speciation. Valentine

(1949) introduced the concept of abrupt ecospecies (*a*-ecospecies) and gradual ecospecies (*g*-ecospecies) as biosystematic units with the following definitions:

Abrupt ecospecies: Groups differing in chromosome number between which there are well-defined ecological and geographical differences and which are capable of only limited gene-exchange.

Gradual ecospecies: Groups with the same chromosome number between which there are well-defined morphological, ecological and geographical differences and which, under artificial or natural conditions, are capable of only limited gene-exchange.

Abrupt speciation, a phrase derived from the above terminology, is normally regarded as the result of a sudden change in chromosome number or constitution, producing instantly an almost irreversible barrier between populations and thus effectively isolating them. This abrupt change is generally the result of polyploidy. Speciation of this sort is initially sympatric in that the new ecospecies and its progenitor(s) have coincidental or at least overlapping distributions. Polyploids may, however, if successful, come to occupy distinct eco-geographical areas.

Gradual speciation, as the name suggests, is a step-by-step process, through the accumulation of differences caused by mutation, recombination, selection and isolation. Internal isolating barriers normally arise slowly and not necessarily in step with other genetic changes or with morphological differentiation, which may outstrip the development of reproductive barriers. Geographical or spatial isolation is normally considered as a necessary part of this process, so that gradual speciation is usually allopatric, at least in most cross-breeding organisms; some possible exceptions are given on page 430. Löve (1960b) suggests that minor chromosomal rearrangements, "by restricting the amount of gene recombination possible in crossings between the original and modified forms, may serve to break up the species into groups with different genetic properties, while not immediately lowering the interfertility to any extent". This view (unsubstantiated by adequate experiments) "would imply that almost all gradual speciation starts sympatrically and later is intensified by allopatric isolation".

The distinction between abrupt and gradual speciation and their respective biosystematic units is not clear cut. There is always a gradual element in speciation. In the case of polyploidy, for example, success in nature depends not so much on the first formation of polyploid individuals as on the factors which allow them to propagate, increase, build up a new population and acquire new genetic variability (Stebbins, 1959b).

Considering the evolutionary significance of abrupt versus gradual speciation, Löve (1960b) points out that all successful new species seem to have developed tolerances somewhat different from those of their parent species; this is, however, more pronounced in polyploids than diploids. He considers this to be probably the reason why the general incidence of polyploidy increases with latitude and altitude, and to be certainly the cause of the frequency of polyploidy being much higher in old groups than in young ones. This can be exemplified by the horsetails and ferns, both of which are high polyploid groups that are but a fraction of their former size. The wider tolerance of polyploids has enabled them to survive when unfavourable conditions caused the extinction of their diploid progenitors. Ultimately, however, the polyploids will succumb, since they cannot return to the diploid condition and their genetic evolution appears to be hampered by their high chromosome number. Löve concludes that "gradual speciation based on mutation, recombination and selection is the essential source of variation; abrupt speciation produces combinations which are more tolerant and thus more resistant to the extremes . . . All speciation inevitably must come to an impasse, but although polyploidy has been termed an evolutionary dead-end, these species will resist the destructive forces of selection considerably longer than their diploid relatives."

Abrupt speciation can therefore be conceived as superimposed on the processes of gradual speciation. As we noted in Chapter 6, species evolution in polyploid complexes normally takes place at the diploid level with the polyploid levels appearing as an embroidery.

The part that abrupt speciation is likely to have played in the evolution of the higher categories is discussed by Stebbins (1950) and Swanson (1958: 514). Although the general belief is that most generic and family differentiation has taken place at the diploid level, there is indirect evidence to suggest that this view requires modification. Many woody groups (e.g. *Rosaceae-Pomoideae*), which now behave as diploids, must have had a polyploid ancestry.

Swanson visualises the sequence of evolution, in which abrupt speciation has played an important part, as follows: "Gene mutation, aberrations, recombination, aneuploidy and selection will provide the principle means for the production of favourable types at the diploid level, and out of this welter of ecotypes will come most of the divergent lines that lead to the formation of the higher categories. Autoploidy will have relatively little influence on this complex, for there is little evidence that autopolypoids will be appreciably different from the diploids giving rise to them. They may, however, facilitate hybridisation and thus permit the intermingling of diverse genomes, which might otherwise be prevented by barriers at the diploid level. . . . At the tetraploid level, diploidisation may set the stage for another cycle of evolution comparable to that which took place at the diploid level. The limitations, however, are more strict here, and will be increasingly so as the degree of polyploidy increases. Consequently the diploid and to a lesser extent the tetraploid levels appear to be the major sources of new genera and families; the higher levels contribute only to an increase in the species." Stebbins (1950), however, finds evidence that such progressive floral trends as sympetaly or epigyny have not been caused or promoted by polyploidy.

Highly evolved families (e.g. *Compositae*, *Gramineae*) still have numerous species at the diploid level.

Races, Species and Isolating Mechanisms

Since the evolution and operation of various isolating mechanisms obviously affect the degree of phenotypic distinction between populations from which taxonomists construct their taxa, it is important for us to consider examples of these mechanisms and their bearing on the species and infra-specific units.

The various factors causing isolation between populations normally build up gradually (except in the case of polyploidy), and we must therefore consider briefly the distinction that is made between ecospecies and races. We have already discussed in Chapter 12 the genecological differentiation of populations giving rise in many instances to ecotypes, and the geographical differentiation of populations giving rise to subspecies. These are infra-specific products, and are often referred to as races—ecological or geographical. They may be distinguished taxonomically as subspecies or varieties, or the differentiation may take on a continuous (clinal) pattern.

In the genetic sense, races of the same species are "open systems" capable of exchanging genes (Keck, 1957). It follows that they must normally be restricted to different geographical or eco-geographical areas, since they might otherwise merge and lose their identities through hybridisation or introgression (cf. Fig. 34). Overlapping subspecies are expected to show some degree of intermediacy and to this extent taxonomic and genetic ideas of races may coincide.

Isolation, however caused, is a key factor in speciation in allowing new evolutionary lines to be built up and maintained. Detailed discussions of isolating mechanisms, largely from an evolutionary point of view, have been published by Stebbins (1950), Clausen (1951) and others. We will consider only some examples here which have a particular significance for taxonomy.

The impression must not be given, however, that there is a universal tendency towards the crystallising out of clearly delimitable species. In certain groups, especially in those which are predominantly outbreeding and occupy diverse open habitats, there may be a selective advantage during considerable periods of time for maintaining or increasing variability. This appears to be the case in the genus *Isatis* in which even the fruit form (usually so constant in the species of *Cruciferae*) varies to such an extent in the Near East that the recognition of species has to be somewhat arbitrary. Many wind-pollinated trees (e.g. *Quercus*, *Salix*) are excessively variable, much of this variability being undoubtedly attributable to hybridisation.

1. *Geographical and Ecological Isolation*

Spatial or geographical isolation is on an entirely different level from other mechanisms since its cause may be independent of the plant. This in no way detracts from its taxonomic value; in fact it is the kind of isolation about which the taxonomist most frequently possesses information affecting his decisions. Although long-distance dispersal may account for some striking cases of geographical isolation (as on oceanic islands), in the majority of cases fragmentation of a once more or less continuous distribution due to historical changes in climate or topography is probably much more common. Whatever the cause of geographical isolation, it may well be, in cases where

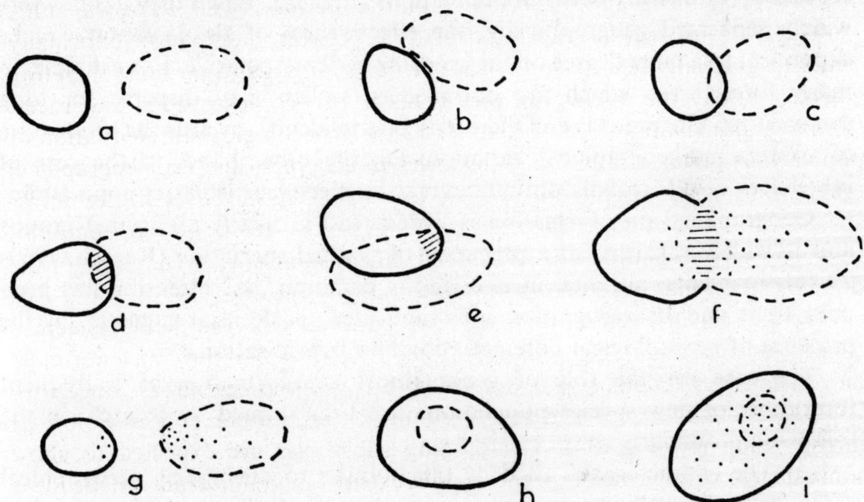

FIG. 34. Present day distributional relationships of species. (*a*) Allopatric. (*b*) Contiguous. (*c*) Overlapping (partially sympatric) *without* hybridisation. (*d*) Overlapping (partially sympatric) *with* hybridisation. (*e*) Partially sympatric with *local hybridisation* (a situation commonly found in "semi-species"). (*f*) Allopatric introgression, the "pure" species being separated by a zone of intermediates (cf. p. 468). (*g*) Allopatric introgression as a result of previous contact or of present pollen transfer (cf. p. 468). (*h*) Fully sympatric *without* hybridisation. (*i*) Fully sympatric *with* introgression.—For simplicity of presentation, only two species are shown in the diagram. Hatching indicates hybridisation, stippling denotes introgression.

the discontinuities are well-marked, more effective in preventing gene-exchange between populations than other kinds of barriers.

The degree of effectiveness of geographical isolation will depend to a very large extent on the areal distribution of the populations concerned. At one extreme we may have the situation whereby a population or series of connected populations covers a very wide area, so that members at extreme ends may be effectively isolated from each other despite the theoretical possibility of gene-exchange between them. At the other extreme, separation of populations may be trans-oceanic or trans-continental and gene-exchange will be impossible except through chance dispersal across the gap. This is the situation in disjunct populations of the same species: numerous cases have been reported where no taxonomic separation between the widely separated segments appears possible, even though there is evidence that they are genetically or physiologically differentiated. (In many other cases of vicariads there are marked morphological, genetic, physiological or chromosomal differences between them.) Between these extremes there are all degrees of separation or overlap between populations (Fig. 34). When they are not very widely separated geographically, the effectiveness of the isolation will be dependent to a large degree on the breeding system operative. For example, in many forest trees which are outbreeders, pollen may disperse for long distances (cf. Chapter 11) and there may be a tendency towards the production of continuous geographical variation. On the other hand, in the case of inbreeders, small areal discontinuities may be effective in isolating populations.

Geographical race-formation is widespread in nearly all animal groups and is widely regarded as a precursor of gradual speciation (Rensch, 1959: 22-46). In plants, although it is certainly common, less attention has been paid to it and its recognition is complicated, as Rensch suggests, by the presence of genecological differentiation and hybridisation.

The case for the role of geographical race-formation as a frequent forerunner of new species populations has been argued by Rensch on the following grounds:

(*a*) The characteristics used by taxonomists to distinguish geographical races have generally been shown in those cases which have been studied experimentally to have a genotypic background. Most geographical races differ from each other in several genes, and "by the formation of various geographic races, each of which is a more or less harmonious unit, a species becomes a very complicated genetic system capable of adapting in various directions, and thus possessing an enormous evolutionary plasticity".

(*b*) There are numerous intermediate stages between geographical races (subspecies) and species.

As regards the first point, it is to be expected that different geographical populations of a species will show some degree of genetic differentiation, and experimental and collateral cultivation experiments, such as those of Böcher and his school referred to in the previous chapter, have confirmed this in several cases. Such genetic variation may be related to ecological factors or no such relationship may be detectable, in which case it is regarded as simply geographical variation (although not necessarily non-adaptive). It should be remembered, too, that the major (ecotypic) geographical races recognised by Clausen *et al.* were regarded as subspecies.

A clear distinction between purely geographical and purely ecological variation is not possible, since different areas have different climatic and ecological conditions. Stebbins (1950) suggests that in the evolution of geographically isolated subspecies or species, initial separation and divergence has been caused by the response of the ancestral population to different selection pressures; this results from ecological differentiation in various parts of its originally more uniform environment. Most geographical races may therefore have arisen primarily as ecotypes; and their separate distributions may be an incidental secondary result of ecotypic differentiation. This concept is in accordance with the fact that species tend to occur in more restricted, and often more extreme, habitats at the edge of their range than in the centre of it. If these peripheral populations become spatially isolated they will often be ecologically differentiated from the rest of the species. Many examples are known of wide-ranging species with a series of satellite microspecies mostly scattered around the margin of their area, and they may have arisen through such a process (*Teucrium polium* complex, *Cytisus* Sect. *Tubocytisus*, etc.). Often there is a difference in breeding system, too, between the satellites and the main body—the former tending to inbreeding, as in *Epipactis* in N.W. Europe. Taxonomic treatment vacillates between treating the satellites as subspecies (as one can conveniently do in *Arabis caucasica* and *Teucrium chamaedrys*) or by treating them as species, as is usually done in *Euphrasia* and *Phlomis* Sect. *Gymnophlomis*. In very large complexes like *Teucrium polium* it is obviously convenient to reduce the polymorphic bulk of the species by separating off distinct marginal units as species and attempting to divide the main body into various subspecies. When the latter can be achieved, it is often found that the marginal subspecies are much less variable than the central mass of the species. Subspecies may therefore be of very unequal value, even within the same species*. In some cases local populations are so numerous (as in *Capsella bursa-pastoris*) or variation so poorly correlated with distinct patterns of distribution (as in *Teucrium orientale*) that it is very difficult to recognise any satisfactory subspecific units. In large critical groups of species the difficulty of deciding to which species various subspecies should be subordinated encourages the taxonomist (often against his inclinations) to maintain numerous microspecies. In such cases more thorough comparative studies (including, in the case of *Euphrasia*, knowledge of chromosome numbers) may show how a natural subgrouping can be achieved.

Ecological differentiation can be very subtle: very slight differences in climate or soil, or even in the flora with which the species is competing, may be sufficient to initiate divergence of the population.

On the other hand, spatial isolation of populations may persist for very long periods without necessarily producing any marked divergence between them (Stebbins, 1942; S. A. Cain, 1944). Caution should be exercised,

* The Code of nomenclature (Art. 26) lays down that when a subspecies is described, another (including the type of the species) is automatically created. Although the "type" subspecies is often more variable than the marginal subspecies, this will probably not be the case if the type of a species happens to have been collected from a marginal population.

however, in assuming that populations of species showing disjunct distributions are in fact the same, since in only a few cases have detailed comparative studies been made. Plant geographers must rely to a large extent on published information on distributions and there is wide disagreement concerning the identity (or non-identity) of many examples of, say, trans-atlantic disjunct "populations". The risks of citing extra-territorial distributions of species in Floras tend to be underestimated (see discussion in Heywood, 1960a: 174). Several discussions have been published on the problems of interpreting plant distributions from this point of view (cf. Walters, 1951), and reference should be made to Hultén's major study on amphi-atlantic plants (1958).

The tools of cytology and comparative cultivation are now available for the study of vicariant populations and in many cases chromosomal differences have been detected (cf. Löve, 1955; Á. & D. Löve, 1958a; Favarger & Contandriopoulos, 1961; Hara, 1962).

So far, we have seen that the facts of distribution do not necessarily indicate whether geographical isolation is operative in speciation or not, since other factors may be involved. A further difficulty arises when considerable differences are shown to exist between geographically separated populations which are nevertheless interfertile when tested experimentally. In other words, there is no necessary correlation between the development of genetic isolating mechanisms and phenotypic differentiation of populations. This has considerable repercussions on the species problem and will be discussed later.

There is some reason to believe that spatial isolation, divorced from other factors, does not lead to the formation of differentiated geographical races ("subspeciation"), still less to the evolution of ecospecies. In very small populations, however (frequent in the early stages of colonisation), rapid divergence in non-adaptive characters may occur by means of genetic drift (i.e. the "Sewall Wright effect"—cf. Mayr, 1942). Wright showed that in small populations the chance fixation and elimination of genes may be more important than selection. Furthermore, in such populations recessive mutations (many of them non-adaptive) have a much better chance to become homozygous than in large panmictic populations. Populations have to be very small, however, before the effect of drift can outweigh selection—how small is a very much debated topic. The breeding system and the life history of the species (e.g. whether annual herb or tree) are bound to affect the amount of drift and the rate at which it proceeds.

Populations that have been fragmented and reduced by the climatic changes of the Pleistocene may show aberrant characters as a result of "drift". In the case of colonisation of a new area, however, a similar state of affairs may arise from a different cause. If the original colonisers are unrepresentative samples of their population, the developing colony will also be unrepresentative. This situation, resulting from the atypical genetic complement of the original founders, has been termed by Mayr (1942: 237) the "founder principle". If the original founders are one or few (as they are likely to be in the case of long-distance dispersal), genetic drift will probably operate subsequent to the establishment of the founders. The likelihood of a divergent, relatively uniform population being rapidly built up is thereby enhanced (cf. Mayr, 1954; Stebbins, 1959b). Waddington (1957: 87), however, describes a probable "drift-reinforcing mechanism". The selective value of any particular gene in a population will depend to some extent on the other genes with which it is associated. As the "founders" represent only a fragment of the population from which they came, the selective value of the remaining genes may be materially changed by the altered chance of co-adaptive genes becoming associated. "Selection will tend to pull the gene frequencies in the new population towards an equilibrium" different from that of the ancestral population. A temporary escape from the control of natural selection (as will occur when the population is extremely small) may therefore have far-reaching consequences. Many isolated populations (on islands, in gorges or on mountain tops), classified as vicarious species or subspecies, may owe their distinctive characters to the interaction of drift and selection.

Acute practical problems arise in the taxonomic assessment of disjunct races when there is no evidence available other than morphology and distribution. The phenomenon of micro-endemics in mountain floras, as in the Mediterranean, is well known. Populations are often fragmented into local, more or less discrete units, and in complex cases each mountain range or even mountain peak in an area may have its endemic representative of a species group. The morphological differences between them may be small but fairly constant, as in the genus *Petrocoptis* (*Caryophyllaceae*) monographed by Rothmaler (1941) in which six of the seven species form a chain across the north of the Iberian Peninsula from the Pyrenees to León.

The question of the taxonomic rank to be accorded to such disjunct units has been discussed by Hultén (1949), Heywood (1960c) and van Steenis (1957). A practical approach has been suggested by Hedberg (1958b) in his studies on Afro-alpine vascular plants. He has proposed minimum requirements in numbers of character differences for the assessment of specific and subspecific rank in allopatric populations. These requirements are detected more easily by simple biometric techniques that are easily applied if adequate material is available.

Closely allied, fully allopatric taxa have often been termed *vicariads* (or vicariants). Löve (1954b) advocates the use of the term for closely allied allopatric taxa that are believed to have penetrated into separate areas prior to their morphological and genetic differentiation. He distinguishes those taxa which have differentiated prior to their occupation of different territories as *false vicariads* or *substitution taxa.* The distinction between these two types is related to their mode of speciation. The majority of those which have arisen by gradual speciation (the different taxa being homoploid) are true vicariads. Allopatric taxa that have arisen by abrupt speciation (mainly through polyploidy) are necessarily false vicariads. Löve suggests using the term *corresponding taxa* when information is insufficient for a distinction to be drawn.

These concepts have been used by Favarger & Contandriopoulos (1961) and Contandriopoulos (1962) in their recognition of four types of endemic taxa in the Corsican flora: (1) *palaeoendemics*, which have no closely related species in the modern flora; (2) *schizoendemics*, representing true vicariads with the same chromosome number; (3) *patroendemics*, which are false vicariads in which the endemic species (usually diploid) has given rise by polyploidisation to a more widespread (usually allopolyploid) species elsewhere; (4) *apoendemics*, representing the opposite kind of false vicariism—where a more widespread species (usually diploid) has given rise to a polyploid endemic. Their method of analysing endemism is most suitable for relatively small areas of endemism and of course cannot be used above the species level.

The second of Rensch's considerations, that intermediate stages between geographical races and species are of common occurrence, is also true of plants. As Stebbins (1950) has observed, the differences between closely related species are often parallelled in races or subspecies of the same species. In other words, any feature separating species may be subject to geographical variation. The morphological differences between species may be more pronounced than between subspecies but are not necessarily so; much will depend, of course, on the species concept that has been used in the group's classification.

In species which show geographical replacement, it is common to find that the more widespread species are subdivided into allopatric races. This is the case in *Primula* Sect. *Primula* (Wright Smith & Fletcher, 1947; Valentine, 1953) in which the attainment of specific status appears to have arisen as a continuation of race-formation. Inconsistencies in taxonomic treatment can readily obscure such relationships. Plant populations that are often classified as belonging to different species would, if they had been birds, have been classified as different subspecies. Even within the Angiosperms, treatments have often been so inconsistent (in terms of splitting and lumping, etc.) that a phytogeographer comparing say, the number of species recorded from the Alps with those recorded from the Caucasus, would obtain figures that would not be biologically comparable.

The commonest kind of isolation in the higher plants is that which results from the existence of related types in different geographical areas which differ in their climatic or edaphic conditions. As eco-geographical isolation is already well documented, we shall only stress a few points here.

For the taxonomist, one of the major difficulties in the delimitation of species is the fact that morphological divergence may or may not be correlated with the establishment of sterility barriers. It is, of course, a general tendency that the more unlike two plants are in their morphology, the more unlikely they are to be able to interbreed if given the opportunity to do so. But exceptions are so numerous that sterility criteria frequently conflict with morphological ones in the delimitation of species.

Two classic examples will serve to illustrate the point. The plane trees, *Platanus orientalis* (E. Mediterranean) and *P. occidentalis* (Eastern United States) have been isolated for millions of years but, despite being morphologically readily separated, are still potentially interfertile, the London Plane (*P.* × *hybrida*) being their vigorous and fertile artificial hybrid. Because of this potential interfertility, Clausen & Hiesey (1958a) have preferred to treat them as subspecies—a conclusion which we are disinclined to accept. The same applies to *Catalpa ovata* (China) and *C. bignonioides* (Eastern United States), but whereas in *Platanus* the geographically isolated species grow under very different climatic conditions, the *Catalpa* species occupy climatically similar areas and grow in similar habitats. From this we may draw the implication that, though *Catalpa* might interbreed in the future if climatic and topographical changes resulted in their migration bringing them together, this could never happen to *Platanus* because their different ecological requirements would prevent their growing together. Such arguments are often used to suggest that vicarious species which are ecologically differentiated are somehow "better" species than those which grow under conditions which are ecologically equivalent. This is true in so far as ecological preference is an additional taxonomic character, but classification is surely difficult enough on the present-day distribution of plants without having to take into account what might happen in the future. If any taxonomists are alive in the distant future, let them reclassify the flora on its behaviour *then* as we should do on its behaviour *now*. If we accept the principle that plants should be classified as they are at the present day, then the vicarious species of *Platanus* and *Catalpa* are as effectively isolated from one another as they would be if they possessed inherent isolating mechanisms. Genetic and ecological isolation are ineffective when populations are completely isolated by geographical barriers, and in allopatric populations can only be treated as additional taxonomic characters. Although it is desirable that the morphological species concept should coincide with the biological one wherever it can conveniently do so, in cases where the concepts conflict, preference must be given to the former.

The ecological separation of closely allied sympatric species appears to be commoner in plants than in animals. Examples are well known and will not be discussed here (cf. Stebbins, 1950: Chapter 6). In the majority of cases, however, ecological isolation is only one of several mechanisms which help to keep sympatric populations separate; it is usually reinforced by seasonal isolation or by partial hybrid sterility often associated with polyploidy.

The fact that many sympatric species may retain their identity as a result of ecological isolation by no means implies that they evolved as a result of sympatric ecological divergence. Mayr, in company with many evolutionists, has claimed that gradual speciation has come about while the populations were geographically isolated, their ecological and morphological differentiation being established or accentuated during spatial isolation. If this were not so, cross-breeding would prevent the establishment of the combination of genetic differences which distinguish the majority of species. When once the specific pattern has been built up, however, ecological isolation can prove fairly effective in maintaining it if migration brings the two species in contact.

In a discussion of the factors which enable related species to grow in the same habitat (and thus to be in direct competition with one another), Harper *et al.* (1961) have recently drawn attention to the fact that the mechanisms by which reproductive isolation is achieved constitute only a part of the process by which systematic diversity is developed. "The presence of two or more systematically closely allied species living in the same habitat must imply . . . that in the course of their evolutionary divergence these species have become subject to different controlling factors. Differences in site requirements for seedling establishment, differences in susceptibility to parasites or predators,* differences in germination time or requirement for breaking dormancy seem to offer the kinds of differences between closely related species which might permit cohabitation and which might relatively easily be acquired during the course of species differentiation." Harper's work shows how unfounded is the claim that reproductive isolation alone should define even "biological" species, let alone taxonomic ones.

Mayr's explanation for the origin of sympatric species by way of an earlier allopatric isolation (a view adopted by Harper *et al.*) is no doubt valid in cross-breeding animals, but gradual sympatric speciation in the Angiosperm seems at least theoretically possible in the following cases:

(1) In self-pollinated groups, since cross-pollination, if reduced to a minimum, would prevent the breakdown of genetic differences built up in response to different edaphic conditions. Speciation by this method may follow the occasional crossing of allied sympatric species. Löve (1960b) suggests that hundreds of species of autogamous plants may have arisen in this way. Such species are, of course, reproductively isolated by their pollination mechanism rather than by inherent intersterility.

(2) In parasitic groups, like *Orobanche*. Many of these grow on various hosts, whereas others (like *O. hederae*) are confined to one. A mutation changing the host/parasite relationship would effectively isolate the mutant form from its ancestral population because any new hybrids between them would probably not be adapted to the host of either parent.

(3) By seasonal or temporal isolation, involving differences in flowering time (see below). The effectiveness of this method would be inversely proportional to the degree of overlap in flowering time, and would be more

* J. B. Gillett (1962) has recently discussed the important part which "pest pressure" may have played in accelerating the rate of plant evolution.

likely to result in speciation if reinforced by some self-pollination.

(4) By a change in the form, colour or scent of the flower which would lead to the adoption of a different pollinator. This may have been the case, for instance, in *Ophrys* (p. 434) and *Pentstemon* (p. 474).

(5) By "catastrophic selection" in ecologically marginal populations. In *Clarkia* (*Onagraceae*) H. Lewis (1962) has shown how exceptional drought, by eliminating all but one or a few deviant individuals, has led to rapid speciation at the diploid level. This is an extreme example of aridity as an evolutionary stimulus (cf. Stebbins, 1952).

(6) By "disruptive selection" (Mather, 1953).

The use of geographical distribution in classification itself (in contrast to its interpretive role) is where possible best avoided. If allopatric distribution is stressed in classification, there is always a danger that species (or even genera) will be credited with morphological differences which they do not in fact possess. This can happen if the number of specimens available is small and the differences between them are not confirmed when more material becomes available. It is also possible that the very fact that populations are solated may lead taxonomists to imagine, or exaggerate, differences between them—such is the power of suggestion.

The use of geographical distribution as a primary lead in keys must be avoided. It encourages bad taxonomy. On the other hand, mention of distribution as a subsidiary character (placed at the end of a lead) is justifiable when critical subspecies (or even species) are being keyed out. Such information, however, is an aid in identification, not a basis for the recognition of taxa. The same applies to differences in ecology (such as a preference for acid versus basic soils) which can provide helpful subsidiary characters in the key of a field Flora. Just how useful such characters are will depend on the ecological constancy of the taxa and the magnitude of the difference between the habitat of each.

2. *Seasonal and Temporal Isolation*

Seasonal isolation, unaccompanied by any other form of isolation, is apparently rare in plants. An example is found in *Lactuca canadensis* and *L. graminifolia* which occur together as roadside weeds in the south-eastern United States (Whitaker, 1944). Having the same chromosome number and being fully interfertile, they are normally kept apart by the fact that *L. graminifolia* flowers in early spring and *L. canadensis* in summer.

Far more often seasonal isolation reinforces ecological, geographical or cytogenetic isolation. For example, *Genista hystrix* and *G. polyanthos* are weakly differentiated morphologically yet the former, which occurs in N.W. Spain and Portugal, flowers in June and July, while the latter, which grows in S. and S.W. Spain and Portugal, flowers from February to April. On the Monterey peninsula (California) *Pinus radiata* and *P. muricata* are isolated by a combination of ecological preferences and a difference in flowering time; nevertheless, a few hybrids occur, apparently less vigorous and productive than their parents (Stebbins, 1950: 209). In British *Juncaceae*, differences in flowering time (allowing for a considerable overlap) are reinforced by ecological differences (*Juncus effusus* and *J. conglomeratus*) and often by cytological differences as well (*J. acutiflorus* (n=40) and *J. articulatus* (n=80)); in the last example sterile triploid hybrids, spreading by vigorous rhizomes, often form large stands.

Differences in flowering time have been used a good deal in classification and in the construction of keys. Apart from smell and taste, flowering time is the only intrinsic non-morphological feature of the plant that has been used in a general purpose classification. Its value as a character increases with the degree of seasonal difference. In several genera with underground storage organs, autumn- and spring-flowering habit may separate (biologically and taxonomically) closely allied species or even subspecies. In *Galanthus nivalis*, for instance, subsp. *reginae-olgae* flowers in autumn, subsp. *cilicicus* in midwinter and subsp. *nivalis* in early spring (Stern, 1956). In this case the subspecies are allopatric and small morphological differences are of course also present. In *Cyclamen*, *C. cilicium* flowers in autumn and the closely allied *C. coum* (*orbiculatum*) in early spring. These two species are the only members of their series, have similar karyotypes and, in south Anatolia, are partially sympatric (Schwarz, 1955).

More often, however, groups of related species are either spring-flowering or autumn-flowering, as they are in *Crocus* and *Colchicum*. In these genera, species which flower in early autumn (in the Mediterranean before the autumn rains) flower before the leaves (i.e. their leaves are hysteranthous); those that flower later in the autumn (during the rainy season) produce leaves and flowers together (their leaves being synanthous). The relationship between flowering time and leaf production can therefore be a useful character (maintained in cultivation) for keying out species of *Crocus* and *Colchicum*. Even in these genera, however, the value of differences in flowering time for identification is somewhat lessened by the existence of a few winter-flowering species. In other genera found in the Mediterranean, autumn-flowering species are the rare exception (e.g. *Narcissus serotinus*, *Allium autumnale*, *Muscari parviflorum*, *Euphorbia dimorphocaulon*). At the generic level, the Mediterranean genus *Biarum* is little more than an autumn flowering *Arum* with entire hysteranthous foliage and uniovulate ovaries. In most if not all of these cases the autumn flowering habit is probably a derived condition and therefore of value for phylogenetic interpretation.

In cultivation, of course, flowering time can differ considerably from that in the field. At high altitudes and latitudes species usually flower later than they do at lower ones, and only collateral cultivation can show whether such differences are genetically fixed. Some high altitude species (or races of them) may even flower earlier when cultivated at low altitudes than their counterparts growing lower down—a result of being "genetically early". These possibilities should be borne in mind when deciding if flowering time is to be used in classification and in the construction of keys.

Temporal isolation, in which pollen is shed (or stigmas are receptive) at different times of day, is probably a more widespread mechanism than is realised at present. Species of *Agrostis*, for instance, shed their pollen at different times of day (Philipson, 1937). Isolation by this means, however, is not necessarily very effective. Although Philipson found that *Agrostis stolonifera* shed its pollen from 10 to 11.30 a.m. and *A. tenuis* from 1 to 2.30 p.m. in the experimental plots at Kew, Bradshaw (1958) records extensive hybridisation between these species, even in closed communities.

3. *Mechanical and Ethological Isolation*

Mechanical isolation is based on structural contrivances of the flower and is particularly effective in plants with elaborate structures, such as we find in *Asclepiadaceae* and *Orchidaceae*.

In the United States several species of *Asclepias* are prevented from interbreeding by their "lock-and-key" pollination mechanisms. Cross-pollination is effected by the accidental insertion of the pollinia (which stick to the pollinating insect's legs) into the stigmatic slits between the anthers. This operation is difficult enough within the same species, but virtually impossible between different ones due to specific differences in the shape of the pollinia and stigmatic slits. Inter-specific hybridisation is therefore extremely rare (Woodson, 1941), and even in the same species seed-set is often low.

As Stebbins (1950) points out, mechanical isolation, although it can greatly reduce the reproductive potential of the individual, maintains the integrity and thereby the adaptive peak represented by the species. In *Asclepias* the disadvantages of mechanical isolation are probably offset by the longevity of the individual plants and by the large number of seeds produced.

We can now consider *ethological isolation*, a phenomenon named and studied by Grant (1949, 1950a)*. Here cross-pollination between two species is mechanically possible, but does not occur because of the constancy of the pollinator. Like mechanical isolation, ethological isolation presupposes a certain level of complexity both of the flower and its pollinator. These two types of isolation, however, are so intimately connected that, in our present state of knowledge, it is often impossible to decide to which a particular case should be referred.

No aspect of ethological isolation has received so much attention as the flower-constancy of bees, particularly of the honeybee, probably introduced from Indo-Malaya very early in man's history and now world wide.†

* The term has been widely employed in animal behaviour and zoological literature (e.g. Mayr, 1942) from which Grant adopted it.

† Butler (1954) suggests that it may be native in parts of Europe.

Constancy is not only maintained throughout each flight from the hive, but often extends over a period of days. The constancy of bumble bees is somewhat lower, and of butterflies probably considerably lower. Different patterns of bee activity are discussed by Linsley (1958) in his review of the ecology of solitary bees.

Mather's classic experiment (1947) on *Antirrhinum majus* and *A. glutinosum* showed that pollination between these two interfertile species was opposed by a barrier more potent than micro-geographical separation. That barrier was the flower constancy of the pollinating honeybee, founded on the perception of likeness and differences in the corollas, or a difference in scent.

Since Mather's experiment, Grant (1949) has shown that bees can sometimes distinguish between the flowers of one subspecies and another. Grant grew three subspecies of *Gilia capitata* (*Polemoniaceae*); these were planted at random and pollinated by honeybees. He found that the bees visited either subsp. *chamissonis* or subspp. *capitata* and *tomentosa*, flying over intervening plants to reach them. This fidelity became less extreme as summer advanced and *Gilia* flowers became more scarce. The bees' behaviour suggested that they were guided by the form of the corolla and/or flower scent —not by its colour, in this case different shades of violet.

The remarkable phenomenon known as *pseudo-copulation* in *Ophrys*, a predominantly Mediterranean genus of Orchids, is now well known. Many of these flowers are pollinated by male bees of the genera *Andrena* and *Scolia*. The differences between the flowers of *Ophrys* species consist partly of superficial resemblances to the females of different species of bee, so that each species of *Ophrys* usually attracts only one species of bee*. The males hatch out before the females and are first visually attracted to the *Ophrys* flower. Stimulated by contact with the hairs on the labellum, the insect attempts to copulate with the flower and, in the process, carries off the pollinia and cross-pollinates the flower of the next plant it visits. It is common to find several species of *Ophrys* growing together in the Mediterranean, but hybrids between them are scarce. Considering that striking variations in labellum shape and markings are known to occur in several species of *Ophrys*, sympatric speciation may possibly have been brought about in this way by means of the "adoption" of a different pollinator (Heslop-Harrison, 1958b). Stebbins & Ferlan (1956) consider the North African *Ophrys murbeckii* to be a hybrid derivative of *O. lutea* × *O. fusca*, an example of sympatric speciation made possible by the strong and peculiar selective pressure exerted by the unusual method of cross-pollination. Internal reproductive barriers are weak or absent in the *Orchidaceae* (even between many genera) and are largely replaced by mechanical or ethological barriers (cf. Adams & Anderson, 1958).

* The Australian genus *Cryptostylis* shows a similar relationship with the males of species of Ichneumon wasp.

Field observations on the bee pollination of tropical American Orchids are given by Dodson & Frymire (1961) and Dodson (1962). Though pseudo-copulation was not observed, pollination involves other forms of "deception" on the plant's part, and such behaviour as aggressive defence of territory on the part of the male bees. Indeed, much of the floral diversity in the family appears to be based on various ways of deceiving the pollinator.

From the work of von Frisch (1914, 1919, 1950), Knoll (1926), Kühn (1927) and Manning (1956a, b), etc., we now know a great deal about the bee's "appreciation" of colour and form, and something about its keen perception of scent (Ribbands, 1955). Butler (1954) concludes that honey-bees can almost certainly recognise four distinct qualities of colour, which we may call yellow, blue-green, blue and ultraviolet (the latter invisible to man). Many red flowers, however, reflect much ultraviolet light and therefore probably appear bluish to bees. So far as the bee is concerned, white flowers are of two types—those which reflect ultraviolet and probably appear as a kind of white, and those which absorb ultraviolet and appear as a shade of blue-green. Bees cannot distinguish between fine shades. If blues and purples predominate in bee flowers, it may well be because these colours contrast most strongly with green foliage. Nectar-guides (or "honey-guides", those markings on the corolla which mark the entrance to the nectar) are frequently based upon simple colour contrasts. Many bicoloured flowers are blue and yellow (*Myosotis*), blue and white (*Anchusa*) or white and yellow (*Limnanthes*); in such cases the white or yellow zone is in the middle of the flower, indicating the nectar-guide function of the pattern. In other flowers the nectar-guides take the form of dark lines or rows of dots on a contrasting colour. Nectar-guides are probably much commoner in flowers than was at first believed, since many (invisible to us) rest upon the bee's appreciation of ultraviolet. Colour patterns found in "pollen flowers" (e.g. *Anemone*, *Papaver*) seem to have received less attention from investigators. The contrast is often between the colour of the petals and the stamens—a feature which is also found in some "nectar flowers", e.g. *Saintpaulia*.

Von Frisch (1914) and others have claimed that bees have a keen sense of form and a good memory for simple geometrical figures; furthermore, they can distinguish third dimensions and forms of bilateral symmetry. Bees, however, do not perceive form in the same way as man. Wolf & Zerrahn (1935) have shown that what they really respond to are margins of colour contrast whose image moves from one ommatidium of the bee's eye to the next. Other things being equal, the more "flicker" produced in this way the better the bees like it. Manning (*in litt.*) does not consider the bee's memory for simple shapes to be very effective unless they differ widely in the amount of outline. Leppik (1956, 1957), however, has accredited insects with a greater power of recognition and "literacy", claiming that they can learn to distinguish a series of "figure numerals" found in an evolutionary sequence of flower types. He recognises a series of six evolutionary grades (amorphic, haplomorphic, actinomorphic, pleomorphic, stereomorphic and zygomorphic flowers) corresponding to the stages of sensory development of the pollinating insects. His explanation, particularly with regard to bees learning a "floral alphabet", seems unnecessarily complicated. Van der Pijl (1960) suggests that Leppik may have confused innate preferences with temporary flower-constancy.

Manning (1956a), using large artificial models, found that bumble bees reacted first to the edge of plain shapes, where there is a line of colour contrast with the background. Even if a nectar-guide pattern is present, the bee still reacts initially to the outline of the model; having approached close to the model, it then reacts to the colour margins of the nectar-guides*. In real flowers, however, it is the scent of the flower, not the presence of nectar-guides, which stimulates the bee to alight. It therefore appears that the bee is first attracted by the colour of the flower; it then reacts to the corolla margin by hovering over it, but does not alight until it has recognised the scent; the pattern of the nectar-guide (if present) serves to direct the insect to the nectar. It has been shown that bees can detect and react to the scent of flowers which we should usually describe as scentless (Kugler, 1932; Ribbands, 1955). Manning (1956a) considers that "apart from its function as a characteristic for the identification of the flower and as an important stimulus to alighting, scent may actually help a bee to find the nectaries, and this effect must interact with that of the honey-guide. Normally, one supposes, the two features, scent and honey-guide, would 'co-operate' in causing a bee to move towards the flower's centre as it alights, and to probe in the right place." There is some evidence to suggest that bees are able to orientate by scent alone. Lex (1954) has shown that scent distribution in flowers is "patterned".

Grant suggests that ethological isolation may play two roles in the evolution of bee plants. Firstly, it may initiate a primary evolutionary divergence, the result of assortative pollination of mutant floral types (cf. Heslop-Harrison, 1958b). This, however, remains an unproven possibility. Secondly, it prevents or reduces interbreeding between sympatric populations which have acquired different floral characters during previous allopatric isolation. This last role—the augmentation of geographical isolation—is probably of very considerable evolutionary importance. Manning (1957) maintains that scent changes will lead to better discrimination by bees than small (or even quite large) visual changes.

It would be a mistake, however, to overestimate the flower constancy of bees. Many hybrids (and also hybrid swarms) have been brought about by the agency of bees, although it is the conditions of the habitat which usually control the survival of hybrids. In European marsh orchids, hybridisation often occurs when one species is abundant and the other scarce, the implication being that the insect working the rare species moves on to the commoner

* As Manning used such large models (up to 5 in. across), how far his results are valid for much smaller flowers is open to question.

one in its search for food. In New Zealand (where only oligotropic*, solitary bees are native) the introduction of the honeybee may be partly responsible for the high proportion of hybrids in that country's flora. It is evident that bees play a part in the breakdown of species as well as in their evolution and establishment.

The role played by flower constancy in the evolution of bees themselves has been discussed by Linsley & MacSwain (1958).

Since bee plants possess a method of pollination not available to promiscuous plants, they can undergo more rapid evolution than less specialised groups of Angiosperms. We should therefore expect a greater morphological diversity in bee plants than in promiscuously pollinated plants. Grant found that this was indeed the case in Southern California, where there are an average of 5·9 species per genus in bee plants, as opposed to 3·4 species per genus in promiscuously pollinated entomophilous Angiosperms. In such calculations the greater attention often given by taxonomists to the *Gamopetaleae* than to other Dicotyledons (and consequently the tendency for more species to be described in gamopetalous genera) should be taken into account.

There can be no doubt that flower constancy has played an important part in the evolution of the Angiosperms. In the Mesozoic, none of the groups of specialised pollinators was probably present, so that insect pollination must have been promiscuous and inefficient†. As pointed out by Grant (1950a), the Mesozoic prototypes of *Magnolia* were probably pollinated by beetles of little skill and no constancy, such as pollinate some species at the present day. Natural selection would have favoured mutations that tended to confine insects to one kind of flower at a time—less pollen would be wasted on interspecific visits, resulting in an increase in seed-set. Mutations causing gamopetaly, especially when tubular round the stamens, would increase flower constancy: fewer types of insect would find access to the flower. The results of the screening process exercised by tubular perianths are apparent today. In the Alps, unspecialised pollinators (beetles, wasps and short-tongued flies, etc.) constitute 86 per cent. of the insects on flowers with open nectar, but form only 37 per cent. of the visitors to flowers with concealed nectar (Müller, 1881: 490).

Flowers pollinated by wind, water or unspecialised insects will probably be incapable of developing barriers (other than temporal) to interspecific pollination, being open to all comers. Floral isolating mechanisms are therefore confined to those groups of Angiosperms with flowers sufficiently elaborate to ensure non-promiscuous pollination by more specialised insects

* *Oligotropic*: confined to a few allied species of plants throughout the lifetime of the insect. Honey bees are *polytropic* (being capable of visiting a wide range of flower types), but become steadfast to a particular species for a period of time after a previous choice. *Monotropic* insects visit only a single species of plant during their lifetime.

† Corner (1961) however, claims a Mesozoic origin for *Ficus*, a large genus now pollinated by monotropic gall-wasps.

with longer tongues. We should therefore expect "non-promiscuous" Angiosperms to be separated taxonomically by floral characters that serve as barriers to interspecific pollination. These floral differences should be less marked in "promiscuous" Angiosperms. This situation is well exemplified in *Ranunculaceae*: *Delphinium* and *Aconitum* (with hidden nectar) show great diversity in floral structure (particularly in the shape of the nectariferous

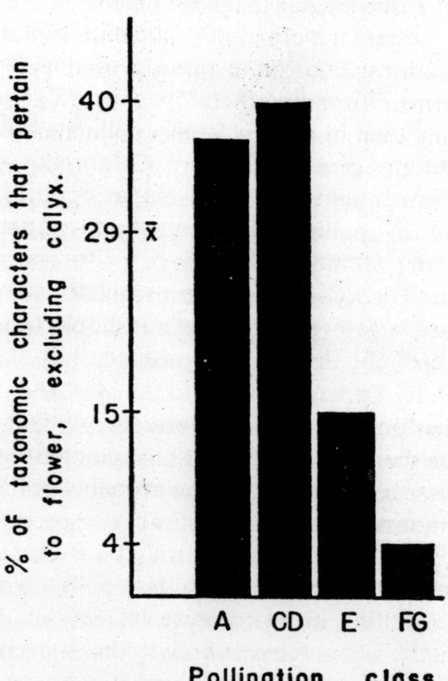

FIG. 35. The relative taxonomic importance of floral characters (excluding those of calyx, fruit and seeds), as correlated with mode of pollination in Angiosperms. *A*, Bird-pollinated plants. *CD*, Bee and long-tongued fly plants. *E*, Plants pollinated promiscuously by unspecialised insects. *FG*, Wind and water-pollinated plants. $\bar{x}$, Mean for all classes. (From Grant, 1949.)

petals), whereas in *Ranunculus* (with readily accessible nectar) most of the specific differences are in the achenes (specialised for dispersal) or vegetative parts. Grant (1949) found this type of relationship to be widespread for the Angiosperms, as may be seen from his bar diagram reproduced in Fig. 35. One cannot help feeling, however, that many floral differences which separate species and genera (even when calyx, fruit and seed are excluded) cannot be adequately explained on the basis of adaptation to specialised pollinators (e.g. differences in style number in the *Caryophyllaceae-Silenoideae*).

It has been shown that a significant correlation exists between the type of ovule protection and the kind of pollinator (Grant, 1950b). This relationship can be considered under the following headings, given in the order of the evolution of their pollinators.

(1) Flowers visited by *beetles*, which chew up the inner parts of the flower, usually have their ovules protected by elevation of the floral chamber above them; such flowers are perigynous (*Rosa*) or epigynous (*Umbelliferae*); or the flowers may be so closely aggregated that they form a surface on which the beetles walk about. Eighty per cent. of beetle-pollinated flowers show one or other of these adaptations. This may seem surprising when it is considered that primitive Angiosperms are generally believed to have been hypogynous. However, the dense imbrication of the numerous stamens and carpels of *Magnolia*, borne on an elongated receptacle, may itself provide protection against marauding beetles, since the inner sporophylls are protected by the outer ones. It is interesting that Sporne (1949), in his statistical approach to the nature of the primitive flower, concluded that "the convex receptacle is neither more nor less primitive than the inferior ovary, but that both began to evolve at a very early stage from some intermediate form with a flat or saucer-shaped receptacle".

(2) Flowers pollinated by adult *bees, flies, wasps, butterflies and moths*, which suck nectar or collect pollen and do not injure the inner parts of the flower, mostly lack the special modifications characteristic of bird and beetle flowers; or, if they possess them, may have retained them from a previous period of beetle pollination. Grant suggests that the closed carpel itself may be a protective adaptation to beetle pollination.

(3) Flowers pollinated by *birds* (sunbirds and humming-birds). The majority (58 per cent.) of known bird flowers show one or other of the following features: an inferior ovary (*Fuchsia*), gynophore (*Embothrium*), or staminal tube between the ovary and the nectar (*Abutilon*). These specialisations, which may often be pre-adaptations, separate the ovary from the nectary and thus serve to protect the ovules from the probing beak of the bird. A big exception is found in the *Personales* (25 per cent.), most of which have a superior, sessile ovary; in the American ornithophilous *Gesneriaceae-Gesnerioideae*, however, the ovary is often inferior. Bird-pollinated flowers are frequently bright red—a colour (contrasting with green foliage) distinguished by birds but not by bees—of sturdy construction and usually scentless; nectar production is copious. Many are also visited by insects (cf. Baker, 1963)

Although information on bat-pollination is still rather sparse, bat flowers occur in many genera that have produced bird flowers, and the total frequency of special ovule protection in the two groups is considered by Grant to be much the same. Chiropterophily (bat-pollination) has been a subject of recent research in tropical floral biology (cf. van der Pijl, 1936, 1961; Baker & Harris, 1957). The syndrome includes nocturnal anthesis, large strong inflorescences, abundant nectar and pollen, and a peculiar position of the flowers outside the foliage, often associated with caulifiory or flagelliflory.

Many genera owe their success (in terms of numbers of species) to adaptation not only to different species of related insects but to completely different animal groups. *Aquilegia*, for instance, has species pollinated by bumble bees, hawk-moths, humming-birds and probably beetles: both *Salvia* and *Delphinium* have blue, yellow and white flowered species pollinated by bees, and red flowered species pollinated by birds. Here ornithophily appears to have been superimposed on an entomophilous construction. As such genera are usually well-diversified and of wide distribution, they are probably relatively old. Dodson (1962) has observed bird-pollination in some tropical American *Orchidaceae*. Several unrelated genera show convergence in floral morphology in relation to humming-bird pollination. He considers *Oncidium* to be a polyphyletic genus composed of sections of separate origin; these are held together generically by common characteristics which in many cases are merely adaptations to a common type of pollinator.

Whether the barriers to interbreeding are mechanical or ethological, they have certainly played an important part, if not in the initial divergence of populations then at least as an accessory means of isolation in the evolution of species. Much work remains to be done in this fascinating field of research. The recent papers of van der Pijl (1960, 1961) are to be welcomed for the information they give of the association of characters (syndrome) found in flowers adapted to different types of pollinator, especially in tropical families for which factual observations were all too scarce. Van der Pijl recognises a series of zoophilous flower classes ("classes with bad boundaries, but with a clear centre") according to their visitors—beetles, carrion and dung flies, bees, hawkmoths, butterflies, ants, birds and bats. To these we may add the remarkable gall-wasp inflorescences of *Ficus* (Corner, 1961). As van der Pijl points out, the study of flower classes does not lead to a delimitation of groups, but supplies the ecological background for evolutionary trends in such groups.

We have devoted considerable space to the relationship between flowers and pollinators, however, not only because of its importance as an isolating mechanism, but because it clearly underlies the evolution of the Angiosperms and thus their classification. This is true at every level of the hierarchy—species, genus and family. As Stebbins has shown, adaptations to animal and wind pollination, together with adaptations to seed dispersal, underly most of the floral diversity found in the Angiosperms, as discussed on p. 121. The combinations of adaptive floral characters found in the largest families of Angiosperms (*Compositae*, *Leguminosae*, *Gramineae*, *Orchidaceae*) occur in few or no other families. However, Stebbins' suggestion (1951) that successful exploitation of these combinations by other groups has been retarded by competition with the large families which first acquired them is open to question. This would largely depend on the vegetative adaptations and tolerance of the groups concerned.

So much attention has been focused on the fascinating relationship between insect and floral evolution, that the evolution of wind pollinated groups has tended to be neglected. Yet the success of wind pollination is all too obvious: the dominance of the *Coniferae*, temperate *Amentiferae* and *Gramineae* over a large area of the earth emphasises the selective advantage of wind pollination (probably derived from insect pollination) under certain conditions and in certain groups. The very large number of species found in the *Cyperaceae* and *Gramineae* must have evolved within the framework of wind pollination. Here floral structure is relatively uniform, specific and generic diversity being largely based on different modifications of the spikelet connected with seed dispersal.

4. *Internal barriers*

All the methods of isolation discussed above are *external* barriers which prevent or reduce the frequency of cross-pollination between different populations in nature. The barriers discussed below are *internal* ones in that their principal action is within the tissues of the plant; they prevent, after cross-pollination, the formation of hybrids or reduce the fertility of those formed.

Internal isolating barriers may be arranged in a progressional series in which the barrier appears later and later in the plant's life history. Only the first in the series—prevention of hybridisation—occurs between the parental species; the remainder occur within the hybrid progeny. We shall not here go into genetic details, which are fully discussed by Stebbins (1950, Chapter 6), but briefly review them and discuss their taxonomic implications.

Prevention of fertilisation is most frequently caused by the failure of pollen to germinate on the stigma or to grow down the style. Though this must often be chemically controlled, in other cases it results directly from differences in style length, the pollen tube being unable to reach the ovules; in the latter event, seed is more likely to be set when the shorter styled species is the maternal parent. In such cases fertile seed may be obtained by stylar grafting (a very delicate operation) or even by introducing pollen directly into the ovary—a sort of artificial insemination. The latter method has resulted in some viable seeds being obtained in a wide cross between *Helleborus* and *Paeonia* (Bosio, 1940); this result is particularly interesting because *Helleborus* is the genus in *Ranunculaceae* which is morphologically closest to *Paeonia* (now usually placed in a family of its own). The relatively thick pollen tubes of polyploids may have difficulty in penetrating the more slender styles of related diploids. It is unusual, however, for prevention of fertilisation from such internal causes to be the primary, or at least the only, cause of isolation between closely allied species in nature; internal barriers of hybrid inviability or sterility are usually operative as well.

Stebbins (1950: 215) considers that the most widespread of the barriers that separate plant species and probably the most fundamental are those that prevent growth or reproduction of F_1 hybrids. While this may be true, the claim that such barriers are more widespread than those of geographical isolation (which he implies) would be hard to prove.

Weak or inviable hybrids may result from disharmony between parental genomes, between the developing embryo and its surrounding endosperm, or between the hybrid endosperm and the maternal tissue which surrounds it. In the first case, normal mature hybrids cannot be obtained under any conditions; in the other cases vigorous hybrids may be obtained by dissecting out the young embryo and culturing it in a nutrient solution. Crosses have been achieved by this means between cherry, plum and peach. Similarly vigorous hybrids have been raised between *Hordeum jubatum* and *Secale cereale* by embryo culture. Such hybrids may not be fertile, but the fact that they can be raised at all shows that the intergeneric cross is not lethal (Maheshwari, 1950: Chapter 5). Such circumventions, however, can play no part in the origin of species in nature.

In some cases, a hybrid embryo may produce a plant which is too weak to reach maturity or to flower normally. McNaughton & Harper (1960) have recently recorded hybrids of this nature between *Papaver rhoeas* and *P. dubium* which simulate plants infected by a virus.

The fact that hybrids are not found in nature between two sympatric species should not be taken as evidence that they cannot occur. Their establishment, particularly in closed communities, may be prevented by competition instead of by any internal barrier.

Hybrid sterility has been extensively reviewed by Dobzhansky (1941: Ch. 9) and summarised by Stebbins (1950: Ch. 6). It is well established that the failure of a hybrid to set viable seed is not connected with any constitutional weakness of the hybrid—the latter may often be more vigorous than the parents, the condition known as *heterosis*. Dobzhansky recognises two types of hybrid sterility: genic and chromosomal. *Genic sterility* covers those cases in which the sex organs fail to develop to the point at which meiosis can occur, or in which genically controlled abnormalities appear at meiotic division (e.g. abnormalities in spindle formation, etc.). *Chromosomal sterility* covers those cases in which sterility results from a lack of homology between the parental genomes, as when the parents differ in chromosome number or are distantly related to one another. Chromosomal sterility, however, may be difficult to detect when there is complete bivalent formation but heterozygosity for very small structural differences.

There are numerous examples in the Angiosperms in which hybrids between closely allied species having the same chromosome number show apparently normal bivalent pairing at meiosis despite the fact that they are partly or completely sterile. Stebbins (1950: 221-22) suggests that in the great majority of such hybrids the decision as to whether sterility is genic or chromosomal can only be taken after allopolyploids have been obtained. Chromosomal sterility of this type, in which structural differences are too small to materially affect chromosome pairing, has been termed *cryptic structural hybridity* (cf. Stebbins, 1950: 222). Genic and chromosomal sterility factors are not mutually exclusive—both can occur in the same hybrid.

A common way in which interspecific hybridisation is prevented is through what has been termed *seed incompatibility*, a phenomenon that has been most intensively studied by Valentine in *Primula* Sect. *Primula* (*Vernales* —Valentine, 1961, and earlier papers). Here fertilisation occurs "but there is always some abnormality in the subsequent development of the seeds (e.g. partial or complete failure of the endosperm), so that though the number of seeds produced is not affected, their quality is below normal. Germination of hybrid seed is always less than that of comparable non-hybrid seed. This phenomenon is quite general in interspecific crosses in flowering plants and probably represents the main barrier to crossing between allied species" (Valentine, 1961). Seed incompatibility in the Sect. *Primula* is of a special kind; the types of seed produced from reciprocal crosses differ in a constant and easily recognised way, both in regard to their morphology and development. This has enabled Valentine to draw up a "crossability series" in which the species are arranged in a definite order according to the amount of "good" seed they produce of each type. Valentine claims that in this group the crossability series gives a clear lead on species relationships, and that it should therefore be used for exploring phylogenetic problems. In *Primula*, however, crossability often cuts across phenetic relationships. How can one decide what *is* phylogenetic relationship if only one line of evidence is used? Unless it can be shown that crossability correlates with other evidence, or unless the discrepancies can be satisfactorily explained, any claims to have unravelled phylogeny should be treated with reserve.

It is a widely held belief that interfertility is a reliable measure of genetic affinity, whereas intersterility cannot be taken as indicative of a lack of relationship unless its cause is understood. These are acceptable generalisations. The claim that a sterile hybrid, just because it is sterile, indicates a lack of close relationship between the parents, is surely unjustified. A vigorous hybrid which is sterile may be just as much a result of close genetic relationship as would be a fertile one. To claim otherwise is to confuse the evolutionary importance of a sterility barrier in speciation with genetic and taxonomic relationship—two (and often three) completely different things.

It must on no account be assumed that because an F_1 hybrid is fully fertile, its progeny will necessarily be so. Several cases are described in which the F_2 progeny are too weak to survive, so that the presence or absence of an isolating mechanism cannot be confirmed until the hybrid has been followed through to the F_2 generation. On the other hand, it would be wrong to assume that because fertility is low it is not necessarily of evolutionary importance. Under the natural conditions in which hybridisation occurs, backcrosses are often more likely to survive than F_2 progeny, and by this means the fertility of the hybrid plants is likely to be increased. It has also been demonstrated that introgression can occur against quite strong internal barriers (Heiser,

1951). Moore (1959) records (on biometrical evidence) extensive introgression in S.W. England between *Viola lactea* and *V. riviniana*, despite the low fertility of the F_1 hybrid and the fact that artificial backcrosses could not be obtained with either parent. Nature has more time and more material for her experiments than man.

The degree of hybridisation possible under experimental conditions is often expressed in the form of a "crossing polygon". In its simplest form

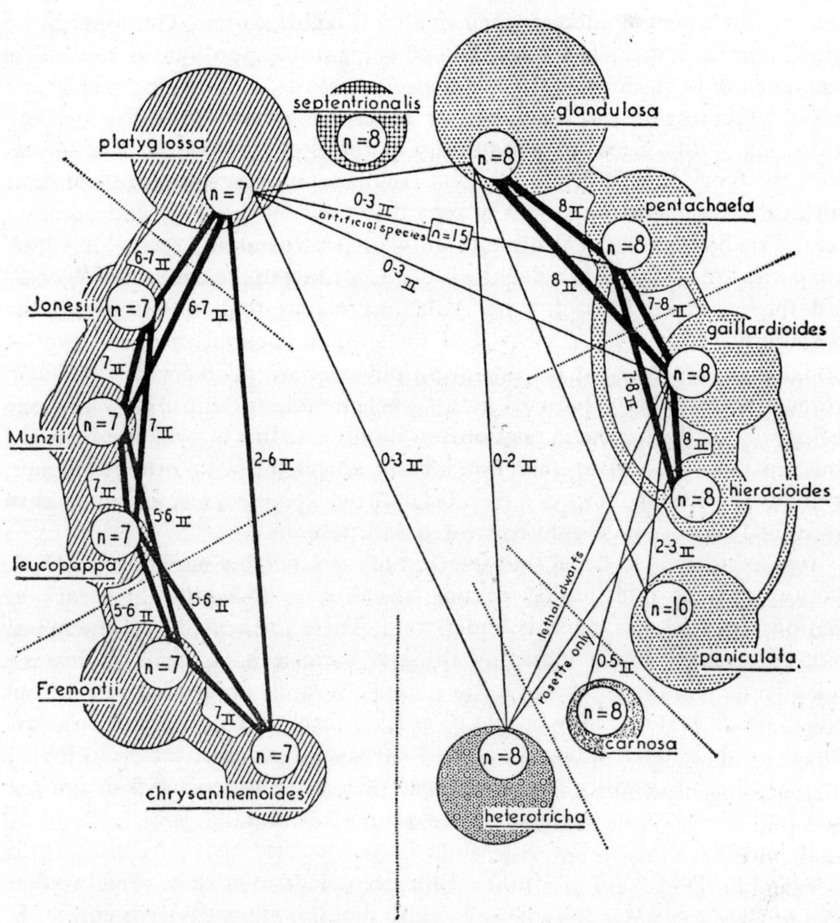

FIG. 36. Crossing polygon of *Layia*, showing cytogenetic relationships. The black lines connecting species indicate successful hybridisations, and the width of the lines the approximate degree of pairing between the parental chromosomes in the hybrid. The widths of shaded connections between the species indicate degrees of gene interchange possible. Dotted radials denote the major morphological discontinuities within the genus. (From Clausen, Keck & Hiesey, 1941.)

such a diagram shows only which taxa can be connected by viable hybrids; in more elaborate types, chromosome numbers, degree of pairing at meiosis and taxonomic grouping are shown. An elegant example for *Layia* (*Compositae*) from Clausen et al. (1941) is reproduced in Fig. 36, summarising the evidence obtained from crossing experiments, the results of studies on chromosome behaviour, and the analysis of morphological differences.

Crossability and Species Definitions

Discussion on this inflammatory topic will be confined here to two major aspects which most affect classification:

(a) *Morphology at variance with crossability*. The occurrence of this phenomenon varies very much from group to group. In groups with highly specialised flowers (e.g. *Aquilegia*, *Asclepias* and *Ophrys*), in which mechanical or ethological isolation is highly effective in the separation of sympatric species, the occurrence of a high degree of artificial interfertility would seem to be of little significance either in evolutionary or in taxonomic terms; there can have been little selection for the build-up of internal sterility barriers. The taxonomist would be justified in ignoring the crossability relationships if they conflicted with other evidence of relationship.

Few, if any, moderately large genera can have been more intensively studied cytogenetically than *Geum* (Gajewski, 1957, 1959). This natural genus of 56 species has been divided into several subgenera on the basis, particularly, of stylar morphology, one series of subgenera being adapted to animal dispersal, another to wind dispersal. Gajewski considers *Geum* to be an old polyploid complex, and finds that rather strong sterility barriers exist between most of the subgenera, whereas inter-specific hybridisation is very common within them. A notable exception is *G. montanum* in Subgenus *Oreogeum* which hybridises freely (despite a difference in chromosome number) with species of Subgenus *Geum* (*Eugeum*), although the floral morphology of the two groups is very different. Gajewski has had no hesitation in maintaining *G. montanum*, and the species with which it hybridises, in different subgenera, nor has he attempted to reduce the specific status of the hexaploid species in Sect. *Geum* (including the notorious *G. urbanum* and *G. rivale*) which are fully interfertile.

Valentine (1961) on the other hand, has suggested that some taxa in *Primula* Sect. *Primula*, treated by their monographers (Wright Smith & Fletcher, 1947) as subspecies, might be recognised as species (notably *P. lofthousei*) on the grounds of intersterility in the experimental plot. The reverse procedure—the reduction of species to subspecies—has been frequently advocated on the grounds of artificial interfertility, a famous example being the reduction of *Rodigia commutata* (*Compositae*) to a subspecies of *Crepis foetida*—a decision justified in this case by close morphological similarity, *Rodigia* differing from *Crepis foetida* mainly in the presence of receptacular scales; this character is known to have a simple genetic basis (Babcock & Cave, 1938).

The taxonomist's position in such cases should not be in doubt. Species and subspecies should be defined on the basis of the variation found in nature and with due regard to their morphological discontinuity. Despite the undoubted importance of intersterility in the evolution of species, crossability criteria must be assessed, for purposes of a general classification, like any other taxonomic character and not be unduly weighted. Having heard it claimed so often that sterility criteria are "fundamental" in the delimitation of species, there is a danger that an increasing number of taxonomists, wooed by these siren voices, may come to regard it as true.* It is only in cases of doubt, in which evidence from other sources conflicts, that evidence from crossability can justifiably be used to tip the scales.

(b) *Different fertility relations in different parts of a species' area, and between related species.* These situations, which are reflected in the morphological variation of the populations concerned, are taxonomically embarrassing, and becoming known in an increasingly large number of taxa. Differences in fertility relations both *within* and *between* taxonomic species are very often associated with cryptic polyploidy (p. 217), but this is by no means necessarily so. They may also result from very subtle chromosomal rearrangements that are not readily detectable and have little, if any, effect on the phenotype. Within species, intersterile populations having the same chromosome number have sometimes been reported. Snyder (1950) found that in the very variable *Elymus glaucus*, only 6 out of 20 strains tested from a transect in the Sierra Nevada (California) showed a high degree of interfertility. In this case sterility barriers appear to have spread into the species through introgression with *Sitanion*, hybridisation occurring despite the rather strong internal sterility barriers between the species (Snyder, 1951). Anderson (1953) points out that these barriers continue to segregate within the populations of *Elymus glaucus*, so that morphologically similar plants of *E. glaucus* may be separated by sterility barriers almost as strong as those which separate *Elymus* and *Sitanion*. On morphological grounds these populations have been included within *E. glaucus*, although the twenty tested by Snyder were classified in 15 different ecospecies of a single coenospecies!

The situation in which two species will cross in one part of their area and not in another may be strongly suspected in the field and even in the herbarium, although it cannot be confirmed without experiment.

Praeger (1951) records frequent hybridisation between *Senecio aquaticus* and *S. jacobaea* in Ireland, whereas it is much less common in England. In California *Quercus douglasii* and *Q. lobata* are sympatric over a large part of their range but hybridise in only one small part of this area (Stebbins, 1950). This type of case often results from differences in selection pressure instead

* It is ironic how many taxonomists today declare their acceptance of this principle but do not (indeed cannot!) put it into practice.

of from the existence of internal barriers, so that hybrids are at an advantage in one area and not in another. When herbarium material is available from all parts of a species' range, it is not uncommon to find that it remains distinct from its allies in one country and not in another. This is the case, for instance, with *Veronica anagallis-aquatica* and *V. catenata*, which are readily separable (and are known to form sterile hybrids) in Britain but show morphological intergradation in continental Europe.

This phenomenon may often be related to the fact that species (and lower units) are usually more variable in the centre of their range than they are at their periphery (cf. p. 425). Turrill (1934) has shown how *Ajuga chamaepitys* is uniform in the western part of its European range, but very variable in S.E. Europe where it meets the more eastern and very polymorphic *A. chia.* In western Europe, where only *A. chamaepitys* occurs, we have no difficulty in recognising it, whereas large intermediate populations occupy much of S.E. Europe where the distribution of the two taxa is contiguous. The uniformity of *A. chamaepitys* in western Europe was interpreted by Turrill as due to "purification" (implying decrease in genetic variability) associated with its westward migration. In the light of Anderson's work, however, the same situation might be due to extensive introgression between two initially allopatric taxa. Whichever explanation is correct, the point to make is that the difficulties of separating closely related taxa are often greater when they occur in the same area than when they do not.

Morphological intergradation may be due to homologous variation or to hybridisation—or to both; it may also result from direct modifications of the phenotype. In the majority of cases the causes of this intermediacy remain undetected, but there is no doubt that, whether the causes are understood or not, the phenomenon presents taxonomists with one of their most difficult problems.

We have stressed that any taxon, if it is being revised and reclassified, should whenever possible be studied throughout its entire range. In a monograph only one general classification is possible, and, if the monographer is honest with himself, if a taxon behaves like a subspecies in one country and a species in another, then it must be classified as a subspecies (as the lowest common denominator) throughout its range. This is by no means the case in Floras; the same entity being given varietal, subspecific or specific status according not only to the taxonomic concepts of the author but also (one hopes) because of the performance of the taxon in the country whose Flora is being written. In a Flora covering a small part of a species' range, the subspecies is not usually indicated if the only member present is the type subspecies; or, if the only subspecies is not the type subspecies, it is often accorded specific rank since it is clearly distinct from all other species present in the area covered by the Flora. *Crepis taraxacifolia*, for instance, was treated in Babcock's monograph (1947b) as a subspecies of *C. vesicaria*, a decision taken on its continental variability. In Britain, however, where *taraxacifolia* is the only representative of the complex, it has often been maintained as a separate species. Such treatments may be the most convenient for identification in a circumscribed area, but they tell us less about the plant's biological relationships than the recognition of subspecific units would do; if the latter are not used, a note should be given on the performance of the plant outside the area whose Flora is being written.

The inconsistency of treatment found in monographs and Floras is often deplored, and so far as it is based on inequalities of judgement it is certainly to be regretted. From the biological point of view, however, a case might be made out for such inequalities of status if they are based on the performance of the taxon itself. If a taxon behaves like a species in Spain, and like a subspecies in Greece, the reality of the situation is frankly acknowledged by treating it as a species in a Flora of Spain and a subspecies in a Flora of Greece. This cannot be done, of course, in a monograph or in a Flora of Europe where a decision is forced upon the author. In such situations the use of the *semi-species* concept, whatever the technical difficulties of applying it, has something to recommend it for plants which behave in the manner described above. "*Silene vulgaris* semisp. *maritima*", for instance, would combine the nomenclatural use of *S. maritima* as a binomial with an informal attachment (not a formal trinomial combination) to the closely allied *S. vulgaris*, the varied performance of the taxa thereby being reflected in the classification (cf. Heywood, 1963: 34).

POLYTOPIC ORIGIN

When a species (or any other taxonomic group) occurs in two or more separate areas, it is said to be polytopic. The fact of polytopy leads us to the question of its causes (fully discussed by S. A. Cain, 1944) and taxonomic treatment. The most frequent explanations of polytopy are that disjunct areas have been bridged in the past either by long-distance dispersal or by slow migration, the latter explanation being known as the climatic migration hypothesis. In either case, the polytopic forms are genetically and immediately related to one another.

Other explanations suppose a *polytopic origin* of the disjunct populations; that is, they do not stand in a direct parent-descendant relationship. This theory has two explanations: (1) The intervening area was, or still is, occupied by an original interbreeding population (hologamodeme) out of which the modern disjunct populations have independently evolved; this is termed by S. A. Cain the polytopic differentiation hypothesis. (2) The disjunct populations have not had an immediate common ancestry but have converged from originally diverse genetic stocks—the phenomenon of polyphylesis.

Although the climatic migration hypothesis is probably valid in the majority of cases of polytopy, in recent years evidence has accumulated for the origin of certain species and taxa of lower rank by polytopic differentiation. The problem, however, is inevitably tied up with the different delimitations of species and the view of many taxonomists that, if a taxon can be shown to be of polytopic origin, it should be split into two or more taxa.

The evolution of similar ecotypes (or ecoclines) in different parts of a species' range is a well-established phenomenon. As these ecotypes are selected out in response to certain environmental conditions, there is no reason why they should not be selected out "wherever the general species population migrates into an area where the differential selective habitat factors operate" (Heslop-Harrison, 1953b). This is the case in the sexual *Hieracium umbellatum*, in which a sand dune gene-ecotype has been selected out in different areas from the local inland population. As Turresson (1922b) showed, this can be demonstrated by differences in minor, not obviously adaptive leaf characters which are shared with neighbouring inland populations. As the evolution of discrete ecotypes is a basic step in gradual speciation, it follows that the polytopic origin of species by this means is a possibility to be seriously considered. Because isolated populations, however, are unlikely to have been subject to identical selection pressure, and were probably genetically somewhat different to start with, they will gradually tend to diverge both genetically and morphologically. The likelihood of a taxon originating by polytopic differentiation will therefore decrease the higher the taxonomic level of the group. In fact, independent origin of closely allied vicarious species seems far more probable than the polytopic origin of the same species.

The situation, however, is obscured by the different species definitions employed by taxonomists. Dwarf mountain forms appear to have frequently arisen polytopically from more widespread lowland populations. Though differing slightly in not obviously adaptive characters, these are often for convenience included within a single variety or subspecies: *Solidago virgaurea* var. *cambrica* and *Scutellaria orientalis* subsp. *alpina* appear to represent such cases. But so does the variable *Veronica kurdica* from alpine regions in the Levant. This appears to have arisen polytopically from *V. orientalis*, a species widespread and polymorphic at lower elevations. *Fraxinus pallisae*, differing from *F. oxycarpa* mainly in its pubescence, is believed by Turrill (1938b) to have arisen independently from the latter in two isolated river deltas in S.E. Europe. In such cases (in which, it must be admitted, we are often in ignorance of the cytological facts), specific rank seems usually unjustified. As the morphological differences are slight and intermediates not infrequent, infra-specific rank would in most cases seem more appropriate.

A striking case of what appears to be multiple polytopic differentiation has been described by Hedge (1961a) in the *Ziziphora clinopodioides* complex (*Labiatae*) in the Near East. Here numerous local microspecies have been described by Austrian and Russian taxonomists. Many of these, however, so closely parallel one another in different areas that one cannot tell to which "species" a gathering should be referred unless its provenance is known. In order to avoid describing still more microspecies to accommodate the variation covered by recent collections from Anatolia, Hedge has preferred to adopt an informal classification, submerging the taxa into four major groups on the basis of growth form, leaf shape and calyx indumentum. These are, in fact, evolutionary grades which appear to have been selected out repeatedly in different parts of the species' range. In some areas both the major groups, and locally differentiated populations within them, are linked by intermediates. Hedge's treatment is admittedly provisional, pending more extensive studies of the group.

Simple genetic variants may, of course, turn up sporadically in a species range. Albinos, tenuisect leaf variants and glabrous variants may often be in this category, and have often been treated collectively as varieties or forms, whether they form local populations or not (cf. p. 383, and polytopic pseudo-vivipary mentioned in Ch. 12).

There seems no reason why nearly identical obligate apomicts should not become established in different areas—sufficiently similar, that is, to be classified as belonging to the same taxon. S. M. Walters (*in litt.*) suggests that this is very probably the case in *Alchemilla faeroensis*, evidently an apomict of hybrid origin between the *alpina* and *vulgaris* groups, and in which the Icelandic and Faroese plants are slightly different. On the other hand, in *Sorbus* very similar apomicts of allopolyploid origin (especially those derived from hybridisation between *S. torminalis* and *S. aria* sensu lato) appear to have arisen independently in different parts of Europe. Here taxonomists, influenced by the unlikelihood of polytopy, or possibly as a result of failing to study the group throughout its range, have applied different binomials to plants which are for practical purposes almost identical.

S. A. Cain (1944: 305), while admitting the polyphyletic origin of species through hybridisation, knew of none that was proved to be polytopic. Recent literature, however, cites several examples where it seems to be the most likely explanation. Tutin (1957) has shown that not only is the ubiquitous *Poa annua* almost certainly an allotetraploid derived from hybridisation between the diploid *P. infirma* and *P. supina*, but that it probably arose on several different occasions—an interpretation supported by Chrtek & Jirásek (1962). One of the most convincing examples of polytopism of this type is given for *Tragopogon* by Ownbey (1950). During the present century, three diploid species of this Old World genus have become rather widely naturalised in N. America, and wherever two or three of them meet, sterile F_1 hybrids are formed. But Ownbey was able to detect four small amphidiploid populations that were of recent and apparently independent origin—two similar ones derived from *T. dubium* × *T. porrifolius*, and another pair from *T. dubium* × *T. pratensis*. Each of these pairs of true-breeding entities (which Ownbey described as new species) resembled the diploid hybrids except for their fertility and conspicuous "gigas" features. As Stebbins (1959a) remarks, these are the first spontaneously arising species which have received scientific attention almost from their birth. Other examples of allopolyploids that are probably of polytopic origin have been described in *Pentstemon* (Keck, 1957), *Phacelia* (Heckard, 1960) and *Dactylis* (Borrill, 1961a).

The sporadic occurrence of autopolyploidy has already been referred to (p. 215). The building-up of independent and extensive autopolyploid populations, however, is probably considerably less common, and its occurrence could only be established by careful mapping of chromosome numbers and the detailed morphological comparison of neighbouring diploid and autopolyploid populations.

Apparent polytopism resulting from polyphyly, except in the case of allopolyploids and possible hybrid swarms that pass into literature as species, applies to the higher categories. In fact, the likelihood that a taxonomic group is of polyphyletic origin increases with the taxonomic status of the group (S. A. Cain, 1944). It is, however, usually a pseudo-polytopism which results from faulty classification based on unjustified character weighting. If the group is reclassified on overall resemblances, distributional anomalies will often disappear.

To conclude, polytopic differentiation of species and units of lower rank almost certainly occurs, but whether it is recognised or not will very much depend on the species definition of the taxonomist and how carefully the group has been studied. There seems every justification for classifying cases of suspected polytopism on the basis of their overall resemblance. Inferred polytopism (like polyphylesis) may lead the taxonomist to look for additional differences between the populations concerned, but should not in itself dictate the form the classification should take.

Taxonomy and Biosystematics

1. *The Relationships of Taxonomy and Biosystematics*

As we have seen in this and previous chapters, a consequence of the great developments in cytology, genetics, ecology and statistics during the past few decades has been that biologists have been able to comprehend to an increasing extent some of the ways in which the small-scale evolution of plants and animals proceeds.

The effects of these developments on taxonomy and classification are still being worked out. Some botanists see them as largely *replacing* orthodox taxonomic procedures and denote this by referring to the revitalised discipline as the New Systematics, Biological or Evolutionary Taxonomy, Biosystematics, etc. Others, perhaps a minority, regard the newer approaches as distinct from orthodox taxonomy and would recommend keeping the two separate both in aims and in terminology. And many biologists move around between the two camps.

Biosystematics is, amongst Angiosperm taxonomists at least, the most widely used of the new descriptive terms to cover these newer approaches and its aims were stated by Camp & Gilly (1943), who introduced it (as "Biosystematy") : "(1) to delimit the natural biotic units and (2) to apply to these units a system of nomenclature adequate to the task of conveying precise information regarding their defined limits, relationships, variability and dynamic structure".

We may usefully compare this with the definition given by Clausen *et al.* (1945) who regard genetics, cytology, comparative morphology and ecology as furnishing the critical data which together, when applied to the study of organic evolution, make up biosystematics. There is, as Valentine (1961) points out, a difference in these definitions which reflects different methods of approaching what are essentially the same problems of variation.

In the approach used by Camp & Gilly, classification or categorisation of the variation studied plays an important role. This is indicated by the series of terms which these authors proposed to define the "kinds" of species units, an aspect which we discuss further below.

Clausen and his associates appear to emphasise much more the *evolutionary* aspects of the variation and are, as Valentine says, more concerned with processes than with end-products. Much of the debate, in fact, centres around whether it is the aim of taxonomy to study products or processes.

A pertinent discussion is given by Camp (1961), who notes that some equate the study of population genetics with experimental systematics. "The experimental systematist usually begins with the classical interpretation of species and works backwards so as to understand the genetic mechanics of his material. The population geneticist begins with the raw population, discarding in his own mind any classical concept of species, and works from there into a series of group concepts which may or may not be comparable to the taxonomist's concept of species."

We have already discussed in Chapter 11 the role of the population and its various definitions in the different approaches to present day taxonomy and biosystematics. From what we said there, it is clear that the contrast suggested by Camp is not so marked with regard to the use of the orthodox taxonomic framework as a starting-off point. In fact, the experimentalist cannot divest himself of the orthodox system since he must go into the field to look for his populations armed with the accumulated knowledge of generations of taxonomists; without this knowledge he would not know where to start or how to relate his new knowledge to an established frame of reference (cf. Walters in discussion following Camp, 1961).

In broad terms, therefore, it appears that biosystematics is taken to cover, on the one hand, a revivified systematics which takes into account the genetic and evolutionary nature of groups and may or may not redefine them in such terms, and on the other hand a separate discipline concerned with microevolutionary phenomena and not with the formal naming and description of formal taxonomic units, although using group concepts based on such criteria as reproductive isolation and genetic structure.

In both cases, the aim is to find out as much as possible about the biological nature of populations and to organise this information in the best way possible. The practical difference, however, arises from the fact that biosystematics in the first sense is essentially an expansion of classical taxonomy. It still retains the purpose of employing comparative morphology as a primary means of describing, delimiting and defining taxa, even though other kinds of evidence may be incorporated. This means that a general classification can be produced which can be utilised both by the amateur and professional botanist and other biologists, as well as by specialist groups of workers interested in certain aspects of relationship. As we have repeatedly pointed out in this book, the only successful method of achieving this aim so far is that of employing the methods and procedures of orthodox taxonomy based on overall relationship and not weighting excessively any one line of evidence. It is not denied that morphological characteristics are weighted, for practical considerations, since unless groups can be characterised phenotypically the problems of description and communication of results are almost insuperable. As Keck (1957) has said, "If a classification should have to depend upon the results of experiments in an experiment garden, it would prove to be intolerably cumbersome and would be summarily rejected".

Clearly then we are faced with a situation whereby the utilization of certain kinds of information for the production of a classification is incompatible with the general utility of that classification. It is for this reason that we advocate the production of one classification for general purposes, and others to convey different kinds of experimental, cytogenetic and other relationship which result in groups which do not necessarily correlate with those required for a general classification.

It would be wrong, however, to suggest that such views are held by all taxonomists and biosystematists. On the contrary, many botanists are convinced that a combined classification is possible. They believe that purposes of biology are best served if the existing taxonomic units are redefined in such a way as to stress their population-genetic nature. This is most clearly indicated by the many attempts to replace the taxonomic species by the "biological" species as the basic unit of classification (see Chapter 3).

The species group lies at the centre of the problem. It is because the species is invested with such a special importance that biosystematists often wish to retain it as a basic unit even though they redefine it in such a way as to detract from its general utility. The dilemma is illustrated by Keck's recent review of taxonomy (1957) where he recognises, as pointed out above, that a species definition relying on experimental evidence would be impracticable. Yet he cites the case of the genus *Layia* (*Compositae*) which has been most thoroughly investigated by the Carnegie Institution team of experimental taxonomists and draws a number of conclusions, of which the following are pertinent:

1. The genus is represented in the different environments it occupies by groups of populations physiologically adapted to these environments.

2. These groups are arranged into species, which are separated by

genetic barriers of various distinctness. Such species are ecospecies (or in a few cases, monotypic coenospecies) usually containing but one ecotype. (A few species contain two, but none more than two, ecotypes in *Layia*.)

3. The species can be *recognised* morphologically, but these characters *merely* serve as markers for the boundaries determined by genetic methods. *The genetic basis of the species is of more importance than the particular combination of morphological characters it brings forth.* (Our emphasis.)

When he talks of the importance of the genetically defined species we must assume he means for evolutionary studies, but this is not the sole purpose of taxonomy, albeit a significant one.

We have already discussed (p. 98) the validity of the dualistic approach whereby morphological evidence is used to *recognise* species which are *defined* in genetic terms. Such a procedure, in which the basis for the definition of a unit is quite different from the evidence available and employed for deciding whether a particular sample meets the definition, is perfectly valid in itself (cf. C. Wright, 1950). Unfortunately the correlation between the morphological markers and the genetically defined units breaks down in a large number of cases.

In those instances where species defined both morphologically and genetically agree in circumscription, the results will be highly satisfactory to all parties and the species may then be regarded as *natural* in the best sense—showing a high correlation between morphological, ecological, cytological and genetical evidence. And clearly the more evidence of all kinds that there is available, the better position we shall be in to detect these natural species, and for them a more stable classification will result. If, however, we regard such natural species as an ideal to achieve in all cases, we will certainly be striving for an unattainable goal. It is being increasingly found that species and other units have to be delimited with respect to the morphological-geographical evidence *despite* knowledge that breeding between the populations so circumscribed occurs, and in other cases no practicable morphological separation can be made between populations which fulfil the requirements of genetical species definitions.

It seems clear that a single classification with species defined both genetically and morphologically may not be at the end of the road, but rather a series of different classifications for different purposes, each stressing certain biological aspects.

Not every piece or kind of information can be accommodated within the strait-jacket of a binomial system and its subunits, and no amount of redefinition of the species will enable this to be achieved. That is precisely why we advocate the use of broad species or species aggregates as a framework in groups such as *Erophila verna* where no overall correlation between morphological, geographical, ecological, cytological and genetical evidence appears possible. This is preferable to the creation of a series of "experimental" binomials into which plants or populations cannot often be satisfactorily accommodated on phenetic evidence. When Mayr (1959b) says that the subspecies category is not suitable for describing the population structure of the species (which is the aim of many biosystematists), he does not suggest that it should be abandoned as a classificatory unit. On the contrary, he notes that although the subspecies is not a unit of evolution but conceals much of the inter- and intra-population variability of species, it is precisely this oversimplification that makes it such a useful tool for the taxonomist. In other words, it is adequate for one purpose but not for another. This is equally true for the species.

There is no doubt that if a workable experimental taxonomic system of readily identifiable units conveying information on population structure and genetics could be devised, it would be widely employed and would in time replace the existing orthodox system. Nature and the evolutionary processes have seen to it that such a system is unattainable. A basically static formal system is needed for everyday use, and a series of dynamic, largely *ad hoc*, systems are needed for micro-evolutionary information.

It is becoming increasingly recognised that much of the variation being brought to light by experimental studies is such that it cannot be simply categorised within the special experimental units (let alone the formal taxonomic categories), thus raising a whole new series of classificatory problems for biosystematics. Much experimental work has been of necessity undertaken on limited samples in circumscribed areas. The effect of extending such work has been to show that many of the earlier pictures obtained have been oversimplifications and are no longer tenable, and also that the categories proposed for dealing with such variation are no longer adequate. We will consider these problems in the next section.

As a general procedure we envisage biosystematic knowledge of species as an extension of taxonomy, some of which can be incorporated into the formal nomenclatural system, but much of which will be found to be best accommodated in a separate system of special categories which can then be attached to the traditional framework.

This is why we advocate the closest possible co-operation between taxonomists and experimentalists, evolutionists, biosystematists and others. As we said in the introduction to this book, a symbiotic relationship should exist between the two: the better the taxonomic classification, the more use it is for the evolutionist. The more experimental, cytological, genetical and other evidence there is for the taxonomist to work with, the more likely it is that species may be shown to have been overlooked or misinterpreted taxonomically. It is quite as important to reduce "bad" species to synonymy as it is to describe new ones.

The ways in which biosystematic evidence may be employed within the orthodox framework of taxonomy has been little studied. We have given numerous examples in previous chapters of how it may be done in particular cases, and a general discussion has been published by Valentine & Löve (1958).

A list of the kinds of biosystematic data which monographers should become informed about as far as possible was worked out at the Paris Botanical Congress in 1954 and put by Stebbins to a meeting of taxonomists, cytologists and geneticists in the form of a series of proposals. These proposals, which have not been as widely publicised as they deserve, may be conveniently termed *Stebbins' Ten Points* and are listed below.

1. Data on variability in terms of populations.
2. Chromosome numbers adequately obtained and recorded.
3. Reproductive biology.
4. Apomixis.
5. Structural heterozygozity.
6. Presence and type of polyploidy.
7. Degree to which species can be crossed.
8. Presence and extent of introgression.
9. Detection of homoploid populations or species of hybrid origin.
10. Habitat.

It was further suggested that Floras and monographs include indications where possible by word or symbols of the condition of every taxonomic group possible in respect of these characteristics and should include key references to the literature. And, further, the terminology of conventional taxonomy and of the newer disciplines should be related to each other as closely as possible, and new terms should be adopted only when a general agreement is reached that they represent a real advance in our understanding of natural phenomena.

One of the best explicit comparisons between orthodox and experimental taxonomy is that given by Heslop-Harrison in Chapter 8 of his monograph, *New Concepts in Flowering Plant Taxonomy* (1953b). Although we are in agreement with his general thesis of the separation of the two approaches, his table of contrast between the aims and methods of "experimental" and "classical" taxonomy is somewhat misleading, since even though he states that he is referring to classical taxonomy as practised in museums and herbaria up to the beginning of the present century, there is an implication that it largely holds true today. He does, however, go on to consider the special field of "experimental" taxonomy and also the expansion of orthodox taxonomy. We will comment on a few of the more important points compared.

1. *Unit of study.* For *experimental* taxonomy, Heslop-Harrison gives: in *non-sexual organisms, the biotype*; *in sexual organisms, the breeding population or large representative samples of it. Materials preferably living.* For *classical*: *individuals or small numbers of individuals, usually dead.* On this we would comment that one of the features of herbarium taxonomy is that it has for the last 60-70 years the great advantage over experimental studies of utilising large amounts of material from throughout the range of species, and as we have pointed earlier in this book a much more balanced classification has resulted. It is simply not true of much experimental taxonomy that large

numbers of individuals (population samples, etc.) have been employed. All too often only very limited numbers of plants are used and grown in experimental plots for hybridisation and cytological purposes. In wide-ranging species detailed population studies are nearly always confined to small parts of the total distribution and comparisons are frequently made on this basis with other species known only from orthodox field and herbarium studies. A further point is that classical taxonomy is based on countless checkings and re-checkings of conclusions by taxonomists in the field, laboratory and herbarium, and frequently in the garden. It is this much neglected aspect which makes taxonomic classifications so valuable. Species are described, re-collected time and time again, redescribed in further Floras and taxonomic revisions and become known intimately by orthodox taxonomists as variable populations. It is misreading the record to suggest that biological knowledge of species is the prerogative of experimental taxonomists: much of what is termed Biosystematics is what used to be called Natural History. Such knowledge may be legitimately regarded as an expansion of taxonomy: that it is not available for all species is, if anything, a reflection of the lack of workers in the field of taxonomy. It is not due to a lack of interest. A vast amount of extremely valuable biological information is hidden away in thousands of formal taxonomic papers and in the field books of plant collectors. The problem is largely one of presentation of data.

2. *Source of data.* However true it may be that the data of the herbarium taxonomist are derived largely from dead material, it must not be forgotten that the plants were alive when collected and observed in the field! Too strict a contrast between the original collection and description of a new taxon, and a biosystematic study of it, is misleading. Field observations on hybridisation (or its absence) with plants of related species are frequently made by collectors and in turn assessed by taxonomists. In fact the absence of hybridisation between sympatric species, and the existence of genuine mixed gatherings in the herbarium, provide most useful evidence to confirm specific status. Similarly, field observations on variability and habitat preferences recorded on herbarium labels aid the taxonomist in his assessment of rank.

3. *Tests of characters.* From what we have said above it is evident that neo-classical taxonomy relies on more than "intuition (based on accumulated experience) and trial and error" (Heslop-Harrison). Observation in the field and in cultivation on the constancy of characters has long been practised and observational methods using statistical techniques are widely employed today. Trial and error, in the sense of the constant checking and re-checking of characters, is one of the strengths of classical taxonomy and one which experimentalists would do well to emulate.

4. *Methods of description.* In classical taxonomy descriptions are, as Heslop-Harrison says, based on individuals, but these individuals are recognised as being population samples (however inadequate) and in well-

studied groups are taken from a wide range of the taxon concerned and often cover a greater range of its variation than is possible with limited experimental samples. Nomenclature is an important aspect as a consequence of the need for general communication; informal nomenclature is also necessary for experimental taxonomy, although there is not the same need for the use of Latin epithets. Some experimentalists do, however, regard formal nomenclature as important (cf. Sinskaja, 1950, on the naming of ecotypes).

5. *Concept of natural variation.* In classical taxonomy the system is by its very nature static although it is, of course, recognised that present-day taxa are the result of evolutionary processes. It is perhaps clearer to say that it is concerned with the past and the present and not with the future. Experimental taxonomy is concerned with evolutionary dynamics and hence is preoccupied with the future evolution of populations. It is not true to say, as Heslop-Harrison suggests, that units above the species level are based exclusively upon degrees of morphological difference. They may be described and differentiated on such a basis but in decisions on circumscription other kinds of comparative evidence are taken into account (see Chapter 4). Only comparative data can be used for defining and describing the higher units since experimentation is virtually limited to the coenospecies-comparium level, so that no general classification of the plant kingdom can be attempted by the experimentalist.

It follows from this reassessment that much of what characterises experimental taxonomy has been assimilated by classical taxonomy. What separates them still is largely their aims and particularly the fact that in the classical approach classification for general purposes and for communication occupies an essential and primary role, whereas in the experimental approach only limited attempts are made to categorise information and then usually of a special kind and for a limited purpose.

We have gone into the relationships between orthodox taxonomy and the newer disciplines at some length since there are widespread misunderstandings as to their relative roles—an unhealthy situation for the progress of our science. Perhaps the greatest misunderstanding stems from the idea of *separation* between the two approaches. The separation is not between taxonomists and experimentalists but between some of their activities. Often the same botanists are engaged in both.

To summarise, much of what is included under the general title of biosystematics is simply part of the modern approach to taxonomy and represents usually an extended state of knowledge of taxa. The term biosystematics itself is in fact largely superfluous and serves mainly to describe one of the stages through which taxonomy passes. The non-classificatory aspects of biosystematics which are concerned with the evolutionary and genetic nature of populations might be better termed *evolutionary dynamics* and considered as a complementary approach to the general taxonomic study of plants.

2. *The Units of Micro-evolutionary Dynamics*

No generally agreed terminology has been devised for the micro-evolutionary units. There is a great multiplicity of often overlapping terms, many of which are defined in different ways by different workers (Gilmour & Heslop-Harrison, 1954).

One of the main reasons for this diversity of terminology is that what are chosen as important evolutionary units depend on the interest and approach of the investigator. As Keck (1957) points out, the geneticist may start with the individual and then turn to the biotype-ecotype-ecospecies-coenospecies-comparium hierarchy based largely on relative capacity of gene-exchange; the cytologist, on the other hand, may start off with the gene and build up his hierarchy with the chromosome segment, chromosome and genome. In such a complex field as evolutionary dynamics, no one classification can be expected to satisfy the needs of all workers.

(*a*) *Kinds of species*. Many attempts have been made during the past two or three decades to enhance the value of the term species by distinguishing different *kinds* of species according to their breeding systems, origins, etc. Darlington (1933) distinguished six kinds according to their genetic-reproductive mechanism:

1. The habitually self-fertilised diploid.
2. The habitually cross-fertilised diploid.
3. The sexually reproducing fertile polyploid.
4. The mixed species containing both diploid and polyploid forms.
5. The complex-heterozygote species (balanced lethal type), as in *Oenothera*.
6. The clonal species not reproducing sexually at all.

To these Huxley (1942) added (for Angiosperms):

7. The subsexual species like *Rosa canina*. A similar, although distinct, breeding system has been found in *Leucopogon juniperinus* by Smith-White (1955b).

The most ambitious system is that proposed by Camp & Gilly (1943), who decided that there were various kinds of species depending on (1) the type of genetic system operative within the group concerned, and (2) the stage of evolution at which it is found. In describing these kinds of species they also took into account the appearance of species populations in the field. The names they applied were: homogeneon, phenon, parageneon, dysploidion, euploidion, alloploidion, micton, rheogameon, cleistogameon, heterogameon, apogameon, agameon. We are not concerned with the details of their definition (many of them are self-explanatory) but with the principle behind Camp & Gilly's proposals. They stress the responsibility of the taxonomists to list not only the genetic systems operative in species populations but to organise the available information relating to these systems so that it is expressed in a functional nomenclature. And not only must the species of biosystematics be recognisable but "organised about a reasonably repetitive genetical system".

This taxonomists' approach to the species problem, as Heslop-Harrison terms it, is in essence no more than a sophisticated way of saying that species represent different evolutionary situations. Doubtless there was a need for this to be stressed when Camp & Gilly wrote their introduction to biosystematy, but very few taxonomists would agree that the information needs to be built into the categories, at least for general purposes. The obvious drawback of the system is that while many species may be fitted into these types, many others cannot because they combine the characteristics of several types. Even for those species about which a decision can justifiably be made, there would appear to be little value in designating them by special terms or by prefixes except for limited purposes.

The impracticability of the system is evidenced by the fact that virtually no use has been made of it. Similar considerations apply to other systems, such as that distinguishing between *allo-species*, *auto-species* and *apo-species*, *genospecies*, *plerospecies*, etc. (cf. Wanscher, 1948).

Another related approach stems from Shull's ideas of the basic genotypic "core" of "good" species (Shull, 1929; cf. Clausen & Hiesey, 1958a). Lamprecht, for example, finds the use of the species (*Art*) for the species of systematics intolerable. He proposes, therefore, to use the following modifications in papers employing formal nomenclature (1959):

1. *Primary species* (*Art*)—bearing a binomial, e.g. *Phaseolus coccineus* L.
2. *Addospecies*—also bearing a binomial to which are added the "foundation" species, e.g. *Galeopsis tetrahit* L. adsp. *pubescens* Bess + *speciosa* Mill.
3. *Superspecies*—e.g. *Empetrum nigrum* L. ssp. *hermaphroditum* (Lge.) Hagerup (4x). This is to indicate that the superspecies is the product of a doubling of the genomes of *E. nigrum*. (The use of the abbreviation "ssp.", which is often employed for subspecies, is unfortunate, quite apart from any other considerations.)
4. *Mixtospecies*—e.g. *Pisum arvense* L. mixsp. *abyssinicum* Braun. This is intended to show that *P. abyssinicum* has "developed" from *arvense*. (In this case one allele causes an interspecific barrier through a small change in chromosome structure. This type of species is also intended to cover other structural changes, translocations, etc.)

The emphasis on "ideal" or "primary" species emerges in another guise in Alvarez López's (1957) establishment of three classes or models of species, namely, the *euspecies* in which morphological and genetic differentiation have reached specific level; *gonospecies* in which genetic differentiation occurs but morphological differentiation is slight or virtually absent (sibling species); and *morphospecies* in which morphological differentiation is advanced but is associated with weak genetic isolation or none at all.

It is widely agreed that these various attempts at distinguishing kinds of species according to different criteria do little to solve any of the major classificatory problems of taxonomy or evolutionary dynamics. There are two main reasons for this. Firstly, specification of particular kinds of genetic or evolutionary evidence is inconsistent with the functions and aims of practical classification, even if it were possible to fit variational units into such categories. Secondly, the kinds of genetic and evolutionary situation with which we are concerned are too many and various to be accommodated within any one such excessively simplified framework. In other words, if we are going to continue to employ the formal taxonomic hierarchy which is admittedly an oversimplification of the "facts of variation", the less we attempt to redefine its categories in evolutionary and genetic terms the better. To do so is to enter such a quaking bog of conflicting aims and interests that practical classification would be paralysed by disputes as to what factor or aspect should be stressed and with which of many conflicting situations any sample should be equated. From a micro-evolutionary or genetic viewpoint, the various terms and categories proposed are no more than partial solutions. Much more flexibility is needed and certainly no one classification is going to serve all special classificatory purposes.*

(b) *The genecological hierarchy*. We have already considered in Chapter 12 the ecotype as an adaptational product of environment selection; but it is also, as we pointed out, the basic unit in the genecological hierarchy proposed by Turesson: ecotype-ecospecies-coenospecies. Of all the experimental terminologies, this is the one to have found most favour and the components have been subjected to many redefinitions (cf. Ch. 12).

Although now recognised by some leading genecologists as too restrictive a system, based as it is on the single character of relative capacity to exchange genes (and one which is virtually impossible to apply satisfactorily in most cases—Gregor, *in verbis*), the ecotype-ecospecies-coenospecies classification is still widely favoured by many experimental taxonomists for conveying crossing relationships.

(c) *Other categories*. The only radically new system of terminology for micro-evolutionary units is that provided by the deme system already outlined in Chapter 12. This can be extended to cover virtually any micro-evolutionary situation and, when intelligently used, together with informal descriptive phrases, offers the greatest likelihood of a solution to these problems (cf. Gregor, 1963). Being intentionally non-hierarchical, it is an auxiliary system of units, not a classification.

* A good example of the consistent usage of both taxonomic and auxiliary units will be found in the *Flora of Cultivated Plants of the USSR* (ed. Sinskaja, 1950, 1961).

Chapter 14

HYBRIDISATION AND TAXONOMY

Introduction

To the taxonomist, hybridisation normally implies the known or assumed crossing between plants belonging to different taxa, usually species, subspecies or, in rare cases, genera. The most appropriate definition of hybridisation for evolutionary purposes is considered by Stebbins (1959a) to be as follows: *crossing between individuals belonging to separate populations which have different adaptive norms*. This definition includes secondary intergradation as defined by Mayr (1942)*, as well as hybridisation between reproductively isolated species. In so far as the hybridising populations can be recognised as different taxa, this broad definition is taxonomically useful.

Although the most obvious effect of hybridisation is to break down and blur the distinctions between species, this is only one aspect of hybridisation; as discussed below, it plays a constructive role, not only in speciation but in macro-evolution as well. The evolutionary role of hybridisation is often a cyclical one, alternating phases of hybridisation and isolation making up a "hybridisation cycle" (cf. Ehrendorfer, 1959). Polyploidy and (more rarely) apomixis are often associated with it.

Stebbins (1959a), in his important review of the role of hybridisation in evolution, classifies the evolutionary processes involved under three main headings: (1) stabilisation of hybrid derivatives by means of amphidiploidy, (2) stabilisation by introgression, (3) stabilisation of segregates from hybridisation; these are discussed below. It must be pointed out, however, that introgression and segregation may result in non-stabilisation, instead of in stabilisation—and in every transition between them. Consideration of non-stabilisation must therefore be included here.

Stabilisation of Hybrids

1. *Amphidiploidy*

As a method of speciation, this has already been considered in Chapter 6. Some of the classical examples of amphidiploids (cf. p. 222; Clausen *et al.*,

* "*Secondary intergradation* refers to cases in which two units, now connected by a steeply sloping character gradient, were separated completely at one time and have now come into contact again after a number of differences have evolved" (Mayr 1942: 99). *Primary intergradation* exists if the steepening of the slope develops gradually, while all the populations involved were in continuous contact.

1945; Darlington, 1956) provide the most clear-cut evidence for the ancestry of the stabilised hybrid. One of the most famous examples is provided by the estuarine grass, *Spartina* "*townsendii*". This name was first applied to a male-sterile F_1 hybrid between the native *S. maritima* ($2n=60$) and an introduced American species, *S. alterniflora* ($2n=62$), the hybrid being discovered in 1870 at Southampton Water (S. England) where it still persists. A vigorous fully fertile amphidiploid ($2n=120$, 122, 124) arose from this hybrid and quickly spread round the coast of Britain, largely replacing the parent species (Huskins, 1931).

An example of an amphidiploid species of very recent origin is provided by *Senecio cambrensis* ($2n=60$), first discovered by a roadside in Wales in 1948 (Rosser, 1955). This evidently arose by allopolyploidy following hybridisation between the native *S. vulgaris* ($2n=40$) and the S. Italian alien, *S. squalidus* ($2n=20$). In this case both the F_1 hybrid and the amphidiploid have been synthesised, the latter differing from the former mainly in its fertility.

Allopolyploidy has been most studied in groups of economic importance, especially in cultivated plants. The latter include several allopolyploids that have arisen from crosses between different genera: *Triticale* ($n=\pm 28$) from *Triticum aestivum* ($n=21$) × *Secale cereale* ($n=7$); *Raphanobrassica* ($n=18$) from *Raphanus sativus* ($n=9$) × *Brassica oleracea* ($n=9$)—cf. Clausen *et al.*, 1945, for references. The most complex situation of this type, however, is provided by the cultivated bread wheats (*Triticum aestivum*) whose tangled history since Neolithic times has been largely unravelled by cytogeneticists. *Triticum aestivum* is an allohexaploid group found to contain genomes (AABBDD) derived through chance hybridisation between two genera, *Triticum* and *Aegilops*; this probably involved crossing between the tetraploid (AABB) wheats and the diploid DD genome species (*Aegilops squarrosa*), cf. Bowden, 1960.*

In recent years, detailed studies of karyotype morphology, in conjunction with studies of gross morphology and geographical distribution, have enabled the ancestry of certain allopolyploid species to be traced in several groups which would not otherwise have given up their secrets. As in the bread wheats, the complexity of some of these cases has often lead to more than one interpretation of an amphidiploid's phylogeny, quite apart from the question of time and place of origin. We will take examples from *Aegilops* and *Gossypium* in which the full story is still to be unravelled.

The annual grass, *Aegilops triuncialis*, is a common weed in the Mediter-

* At one time *Agropyron* was thought to have contributed the B genome (Anderson, 1954a), but a recent view is that *Aegilops* has contributed both the B and D genomes—cf. Bell (1961) who summarises the evolution of the bread wheats. For an account of the taxonomy and nomenclature of the wheats, barleys and ryes, see Bowden (1959). Hybridisation between amphidiploids in the wheat (*Aegilops-Triticum*) group is discussed by Zohary & Feldman (1962).

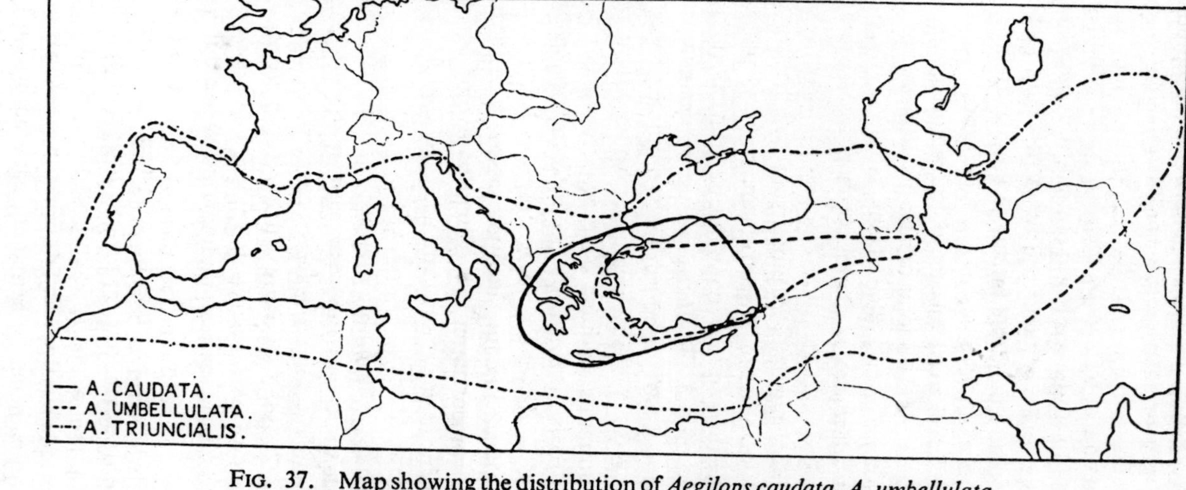

FIG. 37. Map showing the distribution of *Aegilops caudata*, *A. umbellulata* and their amphidiploid derivative, *A. triuncialis*. (From Stebbins, 1959a.)

A. caudata CC   A. triuncialis CCC^uC^u A. umbellulata C^uC^u

FIG. 38. Karyograms of the three species of *Aegilops* mapped in Fig. 37; the karyogram of *A. triuncialis* is of var. *persica*. (From Stebbins, 1959a, after Kihara, 1954.)

ranean. Kihara (1954) has shown that this species, with 28 chromosomes, is an amphidiploid of two ancestral diploids, *A. caudata* and *A. umbellulata*; as shown in Fig. 38, it combines the genomes of both of them. These two diploid species are confined to the Eastern Mediterranean where their ranges overlap in Anatolia (Fig. 37); hybridisation between them and the renewed formation of the amphiploid may still be occurring (Stebbins, 1959a). Being exclusively a plant of weedy habitats, *A. triuncialis* has spread from Morocco to Central Asia, and is even becoming a troublesome weed in parts of California. *A triuncialis* has in fact been synthesised from the artificial diploid hybrid, *A. caudata* × *A. umbellulata* (Kihara & Kondo, 1943), but further analysis of the wild amphidiploid shows that at least in some races of *A. triuncialis* the "*caudata*" genome has undergone modification. Zohary & Feldman (1962) show that this variable amphidiploid species possesses variants (belonging to subsp. *orientalis*) whose spikelets disarticulate at maturity—a character of the tetraploid *A. cylindrica* which appears to have been introduced into *A. triuncialis* by introgressive hybridisation. Although the diploid species of *Aegilops* are well distinguished from one another, specific boundaries between the tetraploids are often blurred, indicating gene interchange between tetraploids sharing a common genome—a supposition confirmed by field observations of natural hybridisation. On the basis of their findings, Zohary and Feldman disbelieve Kihara's hypothesis that the polyploid species of the *Aegilops-Triticum* group evolved independently as simple amphidiploids. To account for the cytological peculiarities and variation patterns found in the group, they maintain that the formation of initial amphidiploids was followed in most cases by extensive hybridisation between the raw polyploids, i.e. at the tetraploid level. Most of the polyploids are considered to be hybrid derivatives. They are believed to have evolved from only a few initial amphidiploids which shared one of their two genomes; this common genome may well belong to a diploid species which exists today The karyotype and interspecific relationships in *Aegilops* are fully reviewed by Chennaveeraiah (1960).

Cytogenetic evidence has been almost conclusive in unravelling the ancestry of the cultivated New World cottons (J. B. Hutchinson *et. al*, 1947). The cotton species of the Old World are all diploid ($n = 13$) with the genome we can distinguish as AA. On the other hand, the cultivated (linted) New World cottons are all tetraploid ($n = 26$) and contain two genomes, AA, DD. The D genome is that of the wild American lintless cottons, whereas the A genome is that of the Old World linted cottons in Sect. *Herbacea*. This remarkable fact can only mean that the cultivated New World cottons arose by amphidiploidy between diploids Old World cottons and wild American species. The original hypothesis of Hutchinson and co-workers has had to be modified in the light of more recent research (Gerstel, 1953; J. B. Hutchinson, 1954). The New World cottons are evidently amphidiploids derived from crossing between the Peruvian *G. raimondii* and *Gossypium herbaceum* (or its ancestor) which still exists as a wild plant in Africa.* Stebbins (1959b) discusses the probable distributional history which enabled the tetraploid American cottons to evolve.

2. *Introgression*: *Sympatric and Allopatric*

The process of *introgressive hybridisation* (*introgression*) was first extensively studied and named by Anderson (1949, 1953). Introgression involves the gradual infiltration of the germplasm of one species into that of another, and usually occurs between species which show some measure of reproductive (physiological) isolation. Introgression is achieved in three phases: (1) the initial formation of F_1 hybrids, (2) their back-crossing to one or other of the parental species, (3) natural selection of certain favourable recombinant types. The process is illustrated in Fig. 39. As a result of the particulate

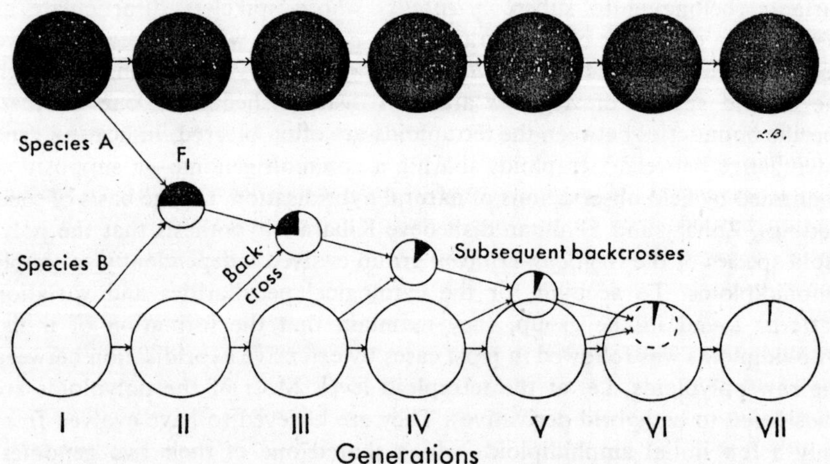

FIG. 39. Diagram illustrating introgression—the interbreeding of species, followed by back-crossing to one parent and ultimate absorption of some genes from one parent into at least some members of the population of the other. (From Benson, 1962.)

nature cf heredity, the products of introgression may be regerded as permanent additions to the gene pool of the recurrent species (Stebbins, 1959a). Anderson believes that it is when hybridisation is least obvious that it plays the most important role in evolution. This is achieved by keeping up the variability of the introgressed species, and by introducing new gene combinations. If natural selection favours introgression, it can take place against strong (but of course not absolute) reproductive barriers, as between diploid and tetraploid species of *Tradescantia*.

* To support the wild status of *G. herbaceum* var. *africanum*, Hutchinson points out that it is parasitised by a distinct species of Bollworm which is found on no other cotton.

The term introgression has often been used in a rather wide sense, to include examples of hybridisation where there is no evidence for the spread of genes away from the area of hybridisation. As Heiser (1959) points out, such hybrid swarms hardly deserve to be treated as examples of introgression.*

Two types of introgression are generally recognised: sympatric and allopatric, according to whether the parental species grow in the same area or in different areas. As with introgression itself, the terms are relative and therefore difficult to define unambiguously.

(a) *Sympatric introgression.* Introgression is sympatric when it occurs between species which share the same general geographical distribution but are usually separated by ecological barriers. Both the "pure" species are therefore found in the area of hybridisation (Fig. 34i, p. 423).

The phenomenon has been studied in numerous North American genera, from the obvious cases like introgression between *Iris fulva* and *I. hexagona* var. *giganti-caerulea* (Anderson, 1949), to examples which are much harder to analyse due to the more subtle differences between the species, as in stemless white violets (Anderson, 1954b). In California the introduced *Helianthus annuus* has introgressed with the native sunflower of serpentine soils, *H. bolanderi.* Though the hybrids are partially sterile, enough introgression has occurred to produce a vigorous weedy variant of the native species which is now spreading into irrigated areas (Anderson, 1949). One such hybrid *Helianthus* population has been studied by Stebbins and Daly (1960) over an eight-year period. During this time the size of the population has fluctuated widely, due both to climatic fluctuations and to changes in the associated flora. A sod grass (*Distichlis spicata* var. *stricta*) invaded the population and divided it into two halves. These halves, although separated by only about 120 m., nevertheless behaved like separate populations. The western half maintained a high frequency of *bolanderi*-like plants, whereas the eastern half, which at the beginning of the observations had a modal frequency of intermediate plants, first changed in the direction of *H. bolanderi* and then reversed in the direction of *H. annuus* due to the elimination of *bolanderi*-like plants. The pollen fertility of plants with intermediate frequency increased significantly throughout the period studied.

Hybrid swarms are usually found in disturbed habitats, particularly in ones which are intermediate between the habitats of the parents and to which neither parent is therefore as well adapted as the hybrids. The closer the habitat is to one of the parents, the more likely are the back-crosses to the

* The term *miscibility* has been used by Dansereau and van Steenis to denote extensive hybridisation between interfertile species or subspecies whereby *a complete series of transitions is formed.* Two species may thereby blend into one (at least locally) without the spread of genes away from the initial area of hybridisation that is implied by "introgression". *Geum urbanum* and *G. rivale* can be considered as locally miscible. The terms, however, are relative and "miscibility" seems to have been little used.

parent to become established. The population may eventually come to look like one variable species, the parent having been "contaminated" by genes from the other. This is the case with *Silene* (*Melandrium*) *alba* in parts of England. *Silene alba* is spreading in weedy habitats, whereas the area of *S. dioica* (which grows in woods) is contracting. In many areas all that is left is a variable population of "*S. alba*" containing genes from *S. dioica* which has otherwise disappeared from the area (Baker, 1948a). This does not appear to have happened to such an extent in Scotland, where the more humid climate allows *S. dioica* to flourish outside woodland habitats, particularly on hedge banks and on cliff ledges; in the latter habitat it may remain uncontaminated by *S. alba*.

Provided the F_1 hybrids show some degree of fertility, introgression may take place. But the outcome of hybridisation depends very strongly upon the environment in which it occurs. As summarised by Stebbins (1959a), the sequence of events will be as follows: in a stable habitat, where two or more related species occur which are well adapted to the conditions there, and when no other suitable ecological niches are available, the amount of hybridisation will be restricted and any hybrids formed will have little or no effect on the evolution of subsequent populations. However, should the same species meet in a marginal environment, and especially one that is changing so that new niches are being made available, the initial hybridisation will not only be more frequent in many cases but new populations, strongly influenced by hybridisation, will tend to be built up as a result of the selective action of the new environment in favouring some of these derivatives. Moreover, this selective action may persist in some cases for many generations after the original hybridisation, and may allow the new derivative population to spread far from the place where hybridisation first occurred.

(b) *Allopatric introgression* is more difficult to define. If it occurs between species which are now fully allopatric, we must assume that there has been contact between them in the past (i.e. that they were contiguous or partially sympatric), or that pollen is transferred between them at the present time (Fig. 34g). In practice, however, the term is widely used for cases in which the species are centred in different areas but share an intermediate region largely occupied by the products of hybridisation, so that the "pure" species do not occur together in the same area (Fig. 34f, p. 423).

One of the most detailed studies of allopatric introgression has been made by Hall (1952) on the hybridisation between *Juniperus ashei* and *J. virginiana* (cf. Anderson, 1953). *J. virginiana* is a characteristic tree of the mesophytic flora of eastern North America, and *J. ashei* a low bushy xerophyte from dolomitic outcrops in Texas and Okhlahoma. The area of intergradation between these two species has been demonstrated by Hall to be one of introgression. Throughout that area variation is exclusively confined to partial recombinations of characters between these two well-differentiated species.

Hall went on to study the variation of *J. virginiana* as a whole. As shown in Fig. 40, the central core of "pure" *J. virginiana* is surrounded by a series of geographical subspecies. In addition to the subspecies formed by introgression with *J. ashei*, there are similar areas of introgression from *J. scopulorum* of the Rocky Mountains, *J. horizontalis* in the north and *J. barbadense* in the south. These areas of introgression have become sufficiently stabilised to be recognised as subspecies of *J. virginiana*.

In a well reasoned paper Muller (1961) traces the origin of *Quercus fusiformis* in North America. This oak species, centred in Texas and adjacent Mexico, is morphologically intermediate between *Q. brandegei* (now confined to the Cape Region of Baja California) and *Q. virginiana* from the S.E. coastal plain of the U.S.A. The fossil history of the area's vegetation suggests that *Q. brandegei* extended much further to the west in Tertiary times, coming in contact with *Q. virginiana*. As the two species are ecologically differentiated, it seems probable that the previous zone of contact was a narrow one, confined to the Edwards Plateau escarpment area. Allopatric introgression occurred, but genes spread slowly because of the limited contact of the parental species and the effect of rhizomatous propagation which must have reduced the frequency of reproduction by seed (Muller, 1951). The introgressed population (now in contact only with *Q. virginiana*) is sufficiently stable to be classified as a distinct species. This appears to be a case akin to the fossil hybrid complex described for *Galium* on p. 321, though in *Quercus* introgression, rather than hybridisation in the narrow sense, seems to have been the cause. To distinguish, however, between past introgression and past hybridisation seems over-optimistic. The analysis of introgression in *Typha* (Fasset & Calhoun, 1952) shows every transition between introgression and hybridisation *sensu stricto*, apparently depending on the habitat, the local frequency of the parents, and the length of time they have been in contact.

A remarkable case of introgression between two well-marked woody genera in the *Rosaceae* is described by Stebbins (1959a). In western North America sympatric and allopatric introgression occurs between *Purshia tridentata* and *Cowania stansburiana*. The "species" recognised as *Purshia glandulosa* is simply *P. tridentata* plus some characteristics of *Cowania stansburiana* obtained through introgression, and extends somewhat beyond the area of overlap between the ancestral species. It appears to have become more or less stabilised and to have replaced *P. tridentata* in the S.W. part of its range, no doubt due to gene combinations for greater tolerance of heat and aridity "borrowed" from *Cowania*. Since the limits between *Purshia tridentata* and *P. glandulosa* are blurred by extensive introgression, Stebbins regards them as subspecies of a single species. The generic status of *Cowania* and *Purshia*, however (based on conspicuous differences in the number of carpels and in fruit morphology), is not called in question.

As Anderson (1953) points out, "There is no very evident association

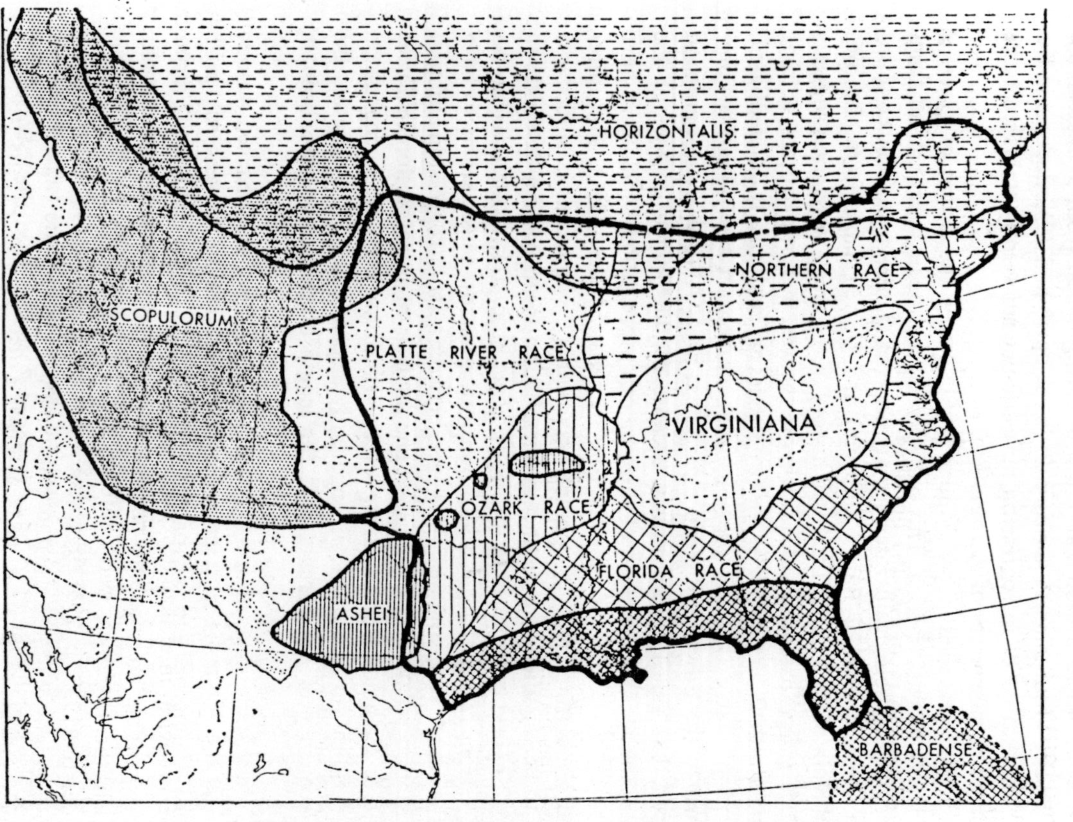

FIG. 40. The effect of introgression in forming geographical subspecies within *Juniperus virginiana* in U.S.A. The broadest line indicates the limits of *J. virginiana*. (From Grant, 1958, modified from Hall.)

between the strength of the internal barriers between species and the degrees to which the species introgress in nature. . . . With weak internal barriers, as in the genus *Aquilegia*, introgression produces subspecies or varieties which intergrade imperceptibly and smoothly into the parental species. With strong internal barriers, introgression tends to bring the barriers into the populations of the recurrent parent" (cf. hybridisation between *Elymus* and *Sitanion* on p. 446).

Introgression has been so well studied in North America that Anderson (1953) has been able to detect the areas in which the ancestral species are centred and to recognise the floristically younger region of the Mississippi basin as the chief area of introgression. The phenomenon has been much less studied in Eurasia, and may indeed be less common, at least in Europe, than it is in N. America. The classical examples (e.g. in *Silene* (*Melandrium*) described on p. 468) are mainly cases of sympatric introgression, or of hybridisation rather than introgression. Whether the apparent scarcity of introgression is the result of different climatic-historical causes, or only of inadequate investigation, remains to be seen. Several of the groups in which introgression has been detected in America (*Delphinium*, *Quercus*, *Juniperus*) are certainly common enough in Europe and the Near East, and their taxonomical difficulty there suggests that the phenomenon deserves more attention than it has hitherto received in the Old World.

Bobrov (1961) states that the Baikal part of Siberia has gone through several successions of its vegetational cover during the Quaternary and has become a favourable area for the extensive distribution of hybrid populations. Within this area the frequent occurrence of introgressive hybridisation is evident in such genera as *Betula*, *Adenophora*, *Larix*, *Oxytropis*, *Polygonum*, *Sanguisorba*, etc. In this part of Siberia introgression is so widespread that Bobrov speaks of the "introgression of plant formations". It hardly needs stressing that such a phenomenon (in this case not substantiated by population analyses or breeding experiments) makes classification particularly difficult.

More or less uniform species are often separated by areas occupied by morphological intermediates. It has frequently been assumed in the past that these variable intermediate populations represent the ancestral stock from which the extremes were differentiated—the "genetic pool" hypothesis. Anderson (1953, p. 296) suggests several tests which may be applied to such intermediate populations, and claims that in every case published and for which sufficient information is available, the "genetic pool" hypothesis is incorrect, and that the intermediate populations have arisen by introgression between ancestral populations which initially evolved in isolation from one another. This interpretation has been challenged by Barber & Jackson working on clinal variation in *Eucalyptus*. They suggest that many of the examples of introgression claimed by Anderson (even *Juniperus* and *Oxytropis*!) are due to selection on an ecological gradient—i.e. they should be interpreted as clines (cf. discussion see p. 403). There is no reason to believe that introgressive hybridisation *between* species and clinal variation *within* them are not both common ways in which intermediate populations arise. There is plenty of room for both processes.

The taxonomic treatment of the results of introgression will depend on how far it has gone. If very slight, it may pass unrecognised; or the population may be classified as a subspecies of one of the parents (as in *Juniperus virginiana*); or as a separate species (as in *Quercus fusiformis*). In other cases, when the character gradient is still well marked, it may be recognised as a cline, a phenomenon which may also arise, without interspecific introgression, from selection of a variable population over an ecological or topographical gradient (cf. p. 55). In many cases the complex results of extensive introgression or hybridisation are bound to be difficult to classify. G. W. Gillett (1961), battling with these phenomena in *Phacelia* (*Hydrophyllaceae*), puts the problem well: "When genetic materials undergo the shuffling and recombining characteristic of hybrid populations, and this is followed by the selection of many recombinations over a mosaic of habitats, the resultant productions stand in open defiance to a precise classification scheme." In any solution, convenience is bound to be an important consideration.

3. *Hybrid Segregates*

Although amphidiploidy and introgression are probably the commonest ways in which hybrid progeny can become stabilised, this result can arise without either of these processes being involved (Stebbins, 1959a). It is most likely to occur when two interfertile species or subspecies come into contact, often as a result of climatic changes inducing migration. As Stebbins puts it, segregation in such hybrid progeny "is very complex, owing to the large number of gene pairs which are segregating simultaneously. As a result, genotypes exactly like the original parent will be recovered rarely, and a tremendous number of new gene combinations will be presented to the environment, on which natural selection can act."

It has been experimentally demonstrated in the Californian *Potentilla glandulosa* (Clausen & Hiesey, 1958a) that many of these new combinations may have quite new adaptive properties. Some individuals show "transgressive variation"—i.e. in some characters the limits of variation found in either parental species are exceeded. This applies to both morphological and physiological features. In *Potentilla glandulosa,* among the progeny of a cross between subsp. *glandulosa* (*typica*, adapted to the coastal area) and subsp. *nevadensis* (adapted to alpine conditions in the Sierra Nevada), there occur genotypes that are *better* adapted to middle altitudes in the Sierra Nevada than are those of either parent, as well as other genotypes that do equally well in coastal, mid-altitude and sub-alpine habitats.

Hybridisation between *Nicotiana langsdorfii* and *N. alata* produces some types which have flowers larger than those of either parent, though the flowers of *N. alata* are the largest in the genus (H. H. Smith, 1953).

Natural hybridisation between *Violas* in Subsect. *Boreali-Americanae* produces an extraordinary range of leaf shapes, many of which are quite different from any found in the parental species (Brainerd, 1924). Stebbins (1950), commenting on the similarity of form between several species of north American violets and the hybrid segregates between *Viola pedatifida* and *V. sagittata*, suggests that the species may well be derived from stabilisation of such hybrid derivatives. Camp (1961), describing natural hybridisation in this polyploid group of stemless blue violets, records a remarkably rapid shift in the frequency of certain leaf shapes and in the segregation of new forms; these changes are associated with seral developments in the woodland communities in which they grow. In New Zealand, hybridisation between *Ranunculus lyalii* (with simple leaves) and *R. buchananii* (with multisect leaves) produces the whole gamut of variation between these two extremes (du Rietz, 1958, f. 3). This variation is as great as we should expect to find in half a dozen species of buttercup, but the segregates have not (at least for the time being) been stabilised by natural selection.

The gradual evolution of new species and subspecies from segregates of hybrids between interfertile taxa is often difficult to recognise because it requires considerable knowledge of the group's evolutionary history (Stebbins, 1959a). In the Sierra Nevada *Potentilla glandulosa* subsp. *hanseni* appears to have arisen by hybridisation between subsp. *reflexa* and subsp. *nevadensis*, between which it is both morphologically and ecologically intermediate (Clausen *et al*., 1940). A Balkan fir, *Abies borisii-regis*, is often interpreted as the stabilised product of hybridisation between *Abies alba* and *A. cephalonica*. In Greece the variability is being increased by an influx of genes from the adjacent parents, but on Mt. Athos and on the Rhodope mountains the intermediate race has become relatively constant and true-breeding so that recognition as a species seemed justified to Mattfield (1930). *Abies equi-trojana*, endemic to Mt. Ida in Anatolia, is interpreted by Aytug (1959), on the basis of morphology and pollen studies, as having arisen in a similar way from hybridisation between *A. cephalonica* and *A. bornmuelleriana*, although *A. cephalonica* does not extend into Anatolia at the present day. Ehrendorfer (1958, 1959) records that in the Aegean islands *Galium graecum* and *G. canum* hardly cross at all, although they occur sympatrically and are in direct competition with one another. But in Anatolia crossing barriers have broken down, which has lead to the formation of the highly polymorphic *Galium dumosum* which represents a stabilised hybrid complex. As these must be of considerable antiquity, Ehrendorfer refers to them as "fossil hybrid complexes".

A complex case has been recorded in *Amaranthus* in California. Here five diploid species are involved, only one of which is native to the Sacromenta-San Joaquin Delta where hybridisation between native and introduced species has produced various giant weedy Amaranths which are confined to the damp organic soils of the area. Here the evolutionary situation is still in a state of flux. It is interesting to note that although the parental species all have dehiscent utricles, in several of the hybrids they are indehiscent. The latter feature is a character of several other sections of the genus, and appears to have been derived by breakdown of the mechanism controlling circumcision of the utricle in some hybrid genotypes (Tucker & Sauer, 1958).

Lewis & Epling (1959) claim that the taxonomic difficulties in the genus *Delphinium* largely stem from hybridisation at the diploid level. The Californian *D. gypsophilum* closely resembles artificial hybrids between *D. hesperium* and *D. recurvatum* which occupy different areas and environments but which overlap slightly with *D. gypsophilum*. Lewis and Epling suggest that *D. gypsophilum* had its origin by a rapid sequence of hybridisations and fixations of a limited number of hybrid genotypes, without change of chromosome number. The rapidity of the adaptive spread of a population of hybrid origin and its establishment and genetic integration as a species would depend on the extent and prevailing rates of environmental change.

Since *D. gypsophilum* occupies its own area and characteristic habitat, as well as being morphologically readily separated from its allies, it is justifiably treated as a species despite its interfertility with other members of its group. In this case artificial F_1 hybrids raised between *D. recurvatum* and *D. hesperium* proved remarkably variable, the variation exceeding that of either parent. This variability seems to stem from previous hybridisation between *D. hesperium* and another species, *D. variegatum*, which is known to hybridise with it in areas of contact. In addition to the diploid *D. gypsophilum*, the species includes a tetraploid race which is apparently a segmental allopolyploid. Lewis and Epling claim that "*Delphinium* would appear to be a long continued anastomosing phylad within which species [in the "biological" sense] are approached but seldom realised." Having some acquaintance with the taxonomic complexities of the genus in the Old World, we can well believe it.

Evidence to suggest that some good taxonomic species have been evolved from segregating hybrid progeny is provided by the work on *Pentstemon* by Straw (1955, 1956b). The Southern Californian *Pentstemon spectabilis*, with somewhat bilabiate, purplish blue corollas, is apparently derived by hybridisation between *P. grinnellii* that has widely bilabiate, bright blue corollas and *P. centranthifolius* with narrowly tubular, red corollas. The parents are normally kept apart by the fact that one is pollinated by carpenter bees and the other by humming-birds. Hybrids occur rarely in nature, the F_1 normally having no pollinator to which it is well adapted although it is fertile when artificially pollinated. *P. spectabilis* strongly resembles the F_1 hybrid and has probably been derived from it by segregation and selection; it is pollinated by a genus of wasps. A new pollinating agent has apparently "adopted" the F_1 hybrid and, while ensuring its increase in an intermediate ecological niche, has caused it to be reproductively isolated from the two parental species.

As described on p. 434, a similar method of speciation has probably occurred in *Ophrys*, and Stebbins (1959a) suggests that it may have been important in the evolution of other genera in the *Orchidaceae*. Selection and isolation from hybrid swarms by specialised pollinators is necessarily confined to plants which have flowers highly adapted to cross-pollination by insects or birds, and between which the hybrids are fertile.

In the above discussion on the role of hybridisation in evolution, emphasis has naturally been on the long-term results of hybridisation, particularly on the breakdown of specific patterns of variation and the build-up of new ones. A great deal of hybridisation, however, seldom if ever gets past the F_1 generation which is often the taxonomist's main concern, as discussed on p. 484. F_1 hybrids and their derivatives can sometimes be very successful, despite a high degree of sterility. An unusual situation has been described by Raven (1963) in *Circaea* (*Onagraceae*). The hybrid (*C.* × *intermedia*) between *C. alpina* and *C. lutetiana* is sterile and better suited at the present day to the environmental conditions of certain parts of Britain than either parent, often replacing them there. Raven suggests that this situation results from the ability of hybridisation to produce a large number of relatively uniform genotypes which are suited to a particular environment, apparently linked with a loss of evolutionary flexibility which is evidently advantageous under present conditions. *C. alpina* had previously been accorded a spuriously extensive distribution in Britain, through confusion with the hybrid which is now much more extensive than *C. alpina*. Similar examples of successful F_1 populations occur in *Nuphar* × *intermedia*—a sterile diploid (Y. Heslop-Harrison, 1953), and *Rorippa* × *sterilis*—a sterile triploid.

A somewhat different situation has been described by Stebbins & M. S. Walters (1949) among Californian grasses, where the area of *Elymus condensatus* overlaps with that of the more widespread *E. triticoides*. These two species have the same chromosome number and hybridise readily to form a vigorous but extremely sterile F_1. The F_1 hybrids, however, must occasionally set a few seeds, since intermediate clones are found which cover the whole spectrum of variability between the morphologically very distinct parental species. Furthermore, these intermediate clones, though common in the area of the parents, extend considerably beyond the range of *E. condensatus*. This is explained by the great longevity and vigorous propagation of the clones.

Sterile F_1 hybrids may not appear to play any part in evolutionary change; they are, however, the raw material on which amphidiploidy may be superimposed, and hence, further speciation. In other cases fertility in the F_1 may lead to hybrid swarms being formed without any noticeable introgression occurring. Such populations may not, at present, be leading to further speciation, but in the future a change in environment may give the hybrid segregates a selective advantage so that they may come to play a more dynamic role in evolution.

In most apomictic complexes, hybridisation seems to have initiated the diversity of the group. As an example we may take the European *Ranunculus auricomus* complex, though here the issue has been complicated by both pseudogamy and polyploidy. "The whole group gives the impression of an enormous hybrid swarm, ranging from *auricomus*-like types to *cassubicus*-like types and displaying a wealth of intermediate forms" (Marklund & Rousi, 1961). Nearly all the persisting segregates are now obligate apomicts and therefore come true from seed.

Methods of Analysis

Anderson has devised several methods for the analysis of populations suspected of hybridisation. All these depend on a general characteristic of hybridity: characters associated together in the parents tend to be associated together in the hybrids, even after generations of back-crossing, the association becoming progressively weaker. Most of the taxonomic characters separating species are controlled by several (even many) genes—i.e. they are multifactorial. These genes tend to be linked together in the hybrids in the same way as they were in the parents, certain combinations being favoured by natural selection.

Anderson's most powerful analytical tool is the *pictorialised scatter diagram*. We can best exemplify this by describing the analysis of 50 plants of a variable population of *Oxytropis* (Anderson, 1953) as illustrated in Fig. 41. *O. albiflorus* is a common species of the Rocky Mountains; while ordinarily white-flowered, in some populations pale blue- and pink-flowered plants outnumber the white ones. When a preliminary sample of such a population was taken it was soon apparent that the racemes varied from condensed to open, the peduncles from being longer than the leaves to much shorter, the outer pubescences from adpressed to woolly, etc. A random sample of 50 plants was then collected, a flowering peduncle with a subtending leaf being taken from each plant. Two of the variables proved easy to score: the openness of the raceme and the extent to which the peduncle did or did not extend beyond the leaves. A scatter diagram was then prepared to show the relation between density of raceme and exsertion of peduncle—the correlation was slight and negative,* i.e. the more open the racemes, on the average, the more they tended to be exserted beyond the leaves. Re-examination of the population showed that the correlation, though slight, still held, and that it did so in adjacent populations. The two extreme types, low-dense and tall-open, were examined for the other variables in the manner shown in Fig. 41. Coloured calyx, coloured corolla, appressed pubescence, and unrestricted microscopic hairs were all represented by long rays on the scatter diagram,

* If laxity of raceme had been scored, instead of density, the correlation would have been slight and positive.

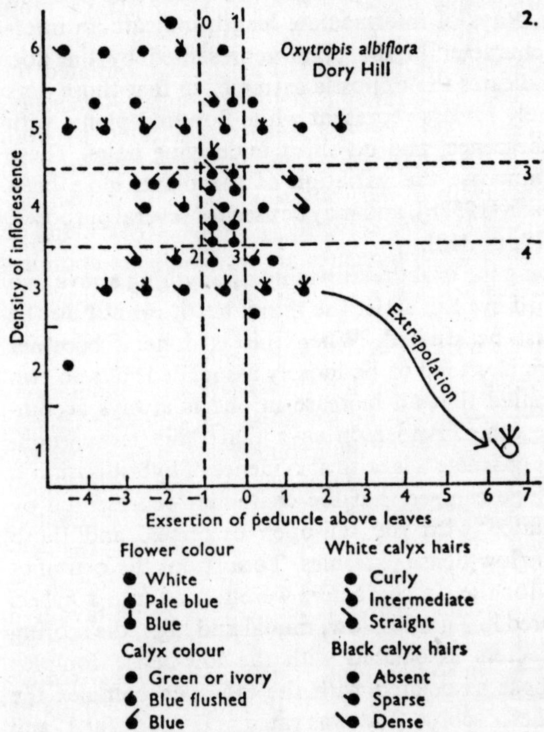

FIG. 41. Pictorial scatter diagram showing interrelationships of six variables in a population of fifty plants of *Oxytropis*. Explanation in text. (From Anderson, 1953.)

FIG. 42. Frequency distribution of the fifty plants of Fig. 41 on a hybrid index with values of 0 for "pure" *Oxytropis albiflora* and of 12 for the opposite extreme. (From Anderson, 1953.)

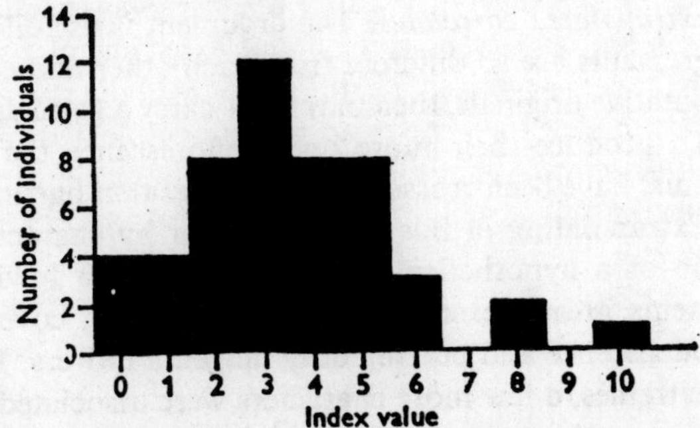

each of these four additional characters being scored at a particular position on each dot on the diagram. Rays of intermediate length indicate an intermediate expression of this character in the plant represented by the dot. Complete absence of a ray indicates the opposite extreme, so that those dots in Fig. 41 which are completely rayless represent white flowered plants with a green calyx, very curly pubescence, and no short underlying hairs. These pictorialised dots, which summarise the variation of the plant, have been termed *metroglyphs* by Anderson (1957b), and may be used for several purposes, including mapping (cf. Fig. 19, p. 312).

The mind is poor at grasping the exact relationship of anything above two variables. In Fig. 41 the trained eye can assist the mind, but it cannot do the whole job; the diagram must be studied. When this is done it becomes apparent that all 6 of the variables tend to be loosely associated, but no two of them are so closely associated that an increase in one is always accompanied by an increase in the other. Anderson says that "this loose, non-random association of various characters is critical evidence of hybridisation".

By studying the diagram one can see that the characters represented by long rays tend to be associated with the tall-open extremes, and those represented by no rays with the low-dense extremes. To sort out the extremes in the population, the scores for the six characters were turned into a *hybrid index*. Each character was scored in 3 grades, low, modal and high, the scoring being adjusted so that characters associated with the low-dense complex scores had the lowest and those associated with the tall-open complex the highest scores. For each character scored, low was rated at 0, modal at 1, and high at 2. The frequency distribution of the sample for this index is shown in Fig. 42. The low-dense extremes proved to be very much alike, but the 3 tall-open extremes were unlike one another. It was therefore assumed that the low-dense extremes represented more or less the original population before it had undergone introgression, and the tall-open extremes were the plants out of a potentially large number of hybrids which might result from introgression from another species. Had they been due to independent mutations, it would be difficult (if not impossible) to suggest any mechanism which would account for the character correlation shown.

The probable introgressing species was then detected by the method of *extrapolated correlates*. The argument is as follows: if these three tall introgressants are so different from each other and on the whole so similar to the putative originals, then they must carry a good deal of the original germplasm. To produce their more open inflorescence the original dense inflorescence must have been crossed with a species that had an extremely lax inflorescence. Extrapolating in this way, character by character, a description was drawn up of a hypothetical species of *Oxytropis* having a long lax inflorescence, stems greatly exserted beyond the leaves, covered with a highly adpressed pubescence and bearing deep magenta flowers. By comparing the two sets of extremes, a few more characters were associated with the two complexes and added to the description. This description enabled the species to be run down in the local Flora and identified as *Oxytropis lambertii*, a common plant at slightly lower altitudes.

To test the validity of his method, Anderson repeatedly analysed *Oxytropis* populations with his students who, using the same method but making their own choice of characters, succeeded in drawing up effective descriptions of *O. lambertii*. "When two different people can in this way make predictions of a species unknown to either of them and then, out of all the conceivable species that might exist, predict exactly that one, correct even to technical details, then the method of extrapolated correlates is as valid as any in biology." To clinch the matter, artificial hybrids have been raised which resemble intermediate plants in the hybrid swarm.

The *Hybrid Index* referred to above is a very simple, even crude, method of analysis for rapid field analysis. If six characters are scored (each as 0, 1 or 2), then one "pure" species will have an index value approaching 0, the other of approaching 12; hybrids and back-crosses, combining the characters of both, will have an intermediate value. Anderson suggests that it is advisable to give different score values to certain characters, "either because they can be more accurately measured and therefore deserve more consideration as criteria, or because they are known to rest upon a wider genic basis". To weight the characters in this way, however, seems a hazardous procedure, particularly when insufficient is known about the variation pattern.

The Hybrid Index can be safely used only for plant by plant comparisons; if the mean of the index scores of a population is used alone it cannot take their spread into account. The use of a mean Hybrid Index alone could give a very false idea of the true picture: for instance, a change in the proportion of two species in a non-hybridising population could give the same mean as a population where hybridisation is taking place. Bimodal and unimodal population curves could receive the same value. To overcome this limitation, Gay (1960) has devised a method which combines the hybrid index with the mean of another factor which is simply related to it but which is not limited by considerations of normality. This additional factor, which is calculated for each specimen of the sample, is termed the *Hybrid Number* and represents the deviation of the Hybrid Index of the specimen from the nearer of the two extremes values of the Hybrid Index scale. Reference should be made to Gay's paper for details of the method which was used to compare different hybridising populations of *Erica ciliaris* and *E. tetralix*. It eliminates some of the subjective element inherent in the Hybrid Index technique and facilitates the comparison and summary description of complex populations containing hybrids. There is no means of discovering, however, if the differences between populations with different values are statistically significant.

Hatheway (1962) has recently described a method for constructing a *Weighted Hybrid Index*, the weight given to each character being proportional to its usefulness in demonstrating a known or suspected relationship. "As degree of association is conveniently measured by the coefficient of correlation, it seems natural to seek weights which maximize the correlation between the two complexes of characters." Hatheway met these requirements by using the technique of canonical analysis described by Hotelling (1936); this resulted in scatter diagrams which suggested genetic relationships among members of the sample. Hatheway emphasises that multivariate statistical methods (including canonical analysis) should be applied only after more general methods, particularly metroglyphic analysis, have thrown light on the nature of the problem.

Hybrid Complexes

Groups of species in which hybridisation has obscured the basic diploid types have been termed collectively *hybrid complexes* by V. Grant (1953). He recognises five types which we may cite in order of increasing taxonomic complexity:

(a) *Homogamic complex*, in which the hybrid derivatives are sexual (with a normal meiotic cycle) and predominantly or entirely diploid. Structural hybridity, if present in F_1 hybrids, is eliminated in later progeny by selection for sexual fertility. Examples are found in *Gilia* Sect. *Gilia*, *Quercus* and *Ceanothus*.

(b) *Clonal complex*, in which the hybrids reproduce mainly or entirely by vegetative division and therefore remain more or less localised, as in *Opuntia* and *Mentha*.

(c) *Heterogamic complex*, where the products of hybridisation are permanent structural heterozygotes, as in *Oenothera* Sect. *Oenothera* (Cleland, 1950).

(d) *Polyploid complex* (Babcock & Stebbins, 1938), in which the derivatives of hybridisation are sexual polyploids. Examples are discussed in Chapter 6.

(e) *Agamic complex* (Babcock & Stebbins, 1938), in which the hybrids or their derivatives reproduce entirely or partially by unfertilised seeds or by bulbils, thus often achieving a wide dispersal. Examples include certain groups in *Taraxacum* and *Hieracium*, *Crepis* Sect. *Psilochaenia*, and parts of *Sorbus* and *Alchemilla*, being in fact particularly common in *Compositae* and *Rosaceae*. When no fertilisation is involved, this group is basically little different from the Grant's clonal complex, the main difference being in the wider distribution of the disseminules.

In their historical development, the last four of these complexes may go through a period of expansion and a period of decline. Complexes in which sexual recombination is inhibited can only expand until they reach a climax of variability; from this point their fate is gradually to die out. The *homogamic complex* stands apart from the rest in possessing an open system of evolution. The derived forms are as fully capable of new gene combinations and variations as the ancestral species. There would seem to be no limit to the complex's evolutionary potentialities. Unlike the agamic, polyploid, heterogamic and clonal complexes, which can be recognised by their cytological and genetic behaviour when the ancestral species are extinct, we can only recognise the homogamic complex if both ancestral and derived species are in existence. Impressed by his study of the leafy-stemmed *Gilias*, Grant puts up a strong case for the very important role played by hybridisation in the evolution of the Angiosperms. He interprets the reticulate pattern of variation which is characteristic of relationships between Angiosperm families (so different from the divergent pattern dominant in mammals) as being the result of widespread hybridisation in the early stages of Angiosperm evolution. This hypothesis is consistent with the widespread occurrence of allopolyploidy in the Angiosperms (cf. p. 421) and with the lack of clear discontinuities between phyletic lines. Grant suggests that the Angiosperms have evolved as a succession of hybrid complexes. During their evolution, hybridisation at the diploid (or diploidised) level has produced the basic interwoven pattern of variation, though the complexity of the net has no doubt been increased by allopolyploidy.

Taxonomic Treatment of Hybrids

Although Appendix I of the *International Code of Botanical Nomenclature* (1961) is explicit about the methods that can be used to designate hybrids, the fact that there are (quite rightly) alternatives has lead to different usages (cf. Little, 1960; Rowley, 1961). These inconsistencies in treatment are inevitable and reflect the different nature and occurrence of hybrids.

The Code states the "hybrids or putative hybrids between two species of the same genus are designated by a formula and, whenever it seems useful or necessary, by a name", thus putting the onus on the botanist to decide when a name is desirable. The formula consists of the names of the two parents connected by the multiplication sign (×), or the name of the genus followed by the two specific epithets connected by the same sign.* When a name is used, it must be validated by a Latin description, citation of type, etc., in the same way as for any other binomial name; it is distinguished from a species by placing × before the second epithet, e.g. *Salix* ×*capreola*=*Salix aurita* ×*S. caprea*, or *Salix aurita*×*caprea* (the latter is the commoner practice). Wild ternary hybrids (as in *Salix*) are more helpfully given formulae than binomials. If binomials are given to hybrids between two species belonging to different genera, the × should precede the binomial. It is common practice for the generic name of the hybrid to combine euphoniously the name of the parent genera, e.g. ×*Festulolium loliaceum* (*Festuca pratensis*×*Lolium perenne*). Hybrids at the generic level are named according to parentage, not morphology.

Whenever binary Latin names are used for hybrids, they cover all obvious manifestations of hybridity between the parents—i.e. F_1 and later generations

* When the parentage is known, the formula may be made more precise by inserting ♀ after the specific epithet of the female parent and ♂ after that of the male.

as well as back-crosses (e.g. *Geum* × *intermedium*, covering the full range of hybridity between *Geum urbanum* and *G. rivale*). It therefore stands to reason that a binomial is more useful and biologically more meaningful for a more or less sterile F_1 (pauciform hybrid) than it is for a hybrid swarm (multiform hybrid).

Since most cross-breeding organisms are probably hybrids in the narrow genetic sense, the hybrid concept is a relative one. Part of the variability of a species may in fact be due to introgression from another species (whether recognised or not), but for convenience will be covered by the name of one of the species, not by a hybrid formula or name. We are faced with the practical difficulty of recognising a hybrid when we see one. The disagreement, for instance, on the extent of hybridisation in Britain between *Calystegia sepium* and *C. sylvatica* partly depends on the range of variation which the botanist considers to be characteristic of the "pure" species, quite apart from the fact that the two species appear to hybridise more readily in the southern part of the country than in the northern part. An example of considerable economic importance is found in *Quercus* where there is much debate as to the limits of the species *Q. robur* and *Q. petraea*, and about the extent to which they hybridise. In the light of reports that the two species when crossed on the Continent of Europe set an average of only 2 per cent. fertile seed, Jones (1959) amplified the circumscription of the species and claimed that hybrids were not likely to exceed 5 per cent. However, Cousens (1963) in a careful and extensive study of Scottish oaks finds that the majority of trees had to be considered as intermediate between the two species. Intermediate values for the diagnostic features he scored predominated; moreover, "extreme" *robur* and *petraea* characters often occurred on the same tree. He concluded that most Scottish oak populations consist largely of hybrids. In *Quercus*, as in *Salix*, it is largely hybridisation (associated with feeble internal sterility barriers) which makes the genus a *bête noire* for all but the devoted monographer.

As failure to recognise hybrids, or their misinterpretation, has been responsible for much faulty taxonomy, we will briefly mention ways in which misunderstandings have arisen, drawing examples from both Angiosperms and ferns. (1) The hybrid may be mistaken for one of its parent species, or pass unnoticed. Hybrids between *Digitalis purpurea* and *D. thapsi* may appear so similar to the latter species that they can be readily overlooked unless one has seen artificial hybrids and knows what to look for in the field. McNaughton & Harper (1960) have described weak hybrids (readily passed over in the field) between *Papaver rhoeas* and *P. dubium*; these resemble stunted, virus-infected plants and are not intermediate between the parents. The main danger of the failure to recognise hybrids (as with *Calystegia* and *Quercus* cited above) is that the circumscription of the species will be amplified to include interspecific hybrids, quite apart from the justifiable inclusion of a

* The term *nothomorph* can be used to distinguish different derivatives of hybridisation between the same species (Article H. 5 of the Code). This excludes allopolyploids.

limited amount of variation that may be brought into a species through introgression. It should be borne in mind that the more closely species resemble one another, the harder hybridisation will be to detect but the more likely it may be to occur. (2) Hybrids may be described as subspecies or varieties of one of the parental species. *Dryopteris spinulosa* var. *fructuosa* is in fact the sterile hybrid between *D. spinulosa* and *D. intermedia* (Manton & Walker, 1953). (3) Hybrids may frequently be mistaken for distinct species. In N. America *Typha glauca*, for instance, appears to be not a specific entity but a series of individuals belonging to hybrid swarms between *T. latifolia* and *T. angustifolia* (Fasset & Calhoun, 1952). T. G. Walker (1958) has demonstrated experimentally that the fern from Ceylon described as *Pteris otaria* is in fact a particularly common segregate of hybrid swarms between *P. multiaurita* and *P. quadriaurita*. Many "species" described in the New Zealand flora have turned out to be hybrids (e.g. *Myrtus ralphii* which is really *M. bullata* × *M. obcordata*), and the same applies to several European "paper species" described in *Quercus* and *Salix*. (4) The so-called hybrids may be imaginary. Putative hybrids have often been reported between *Spergularia media* (*marginata*) and *S. marina* (*Caryophyllaceae*); these are actually apterous-seeded variants of *S. media* and form a race of that species (Ratter, 1959). *Drosera intermedia*, a "good" species, was at one time considered a hybrid between *D. rotundifolia* and *D. anglica*, between which a true hybrid, does, in fact, occur. (5) Allopolyploids may be mistaken for hybrids. *Dryopteris clintoniana* has been interpreted as a hybrid between *D. goldiana* and *D. cristata*, but is really an allopolyploid of that cross and classifiable as a separate species (S. Walker, 1962).

The likelihood that an intermediate is indeed a hybrid is increased by collecting it together with both its putative parents, but we cannot assume that it is not a hybrid if we do not find it with them. Hybrids often grow with only one of the parents, or in "neutral" ground between them; they may even become established and form large swarms many miles from either parent, as described by Rollins (1954) for *Lesquerella* (*Cruciferae*). In the ferns, Blasdell (1963) claims that plants of hybrid origin in the *Cystopteris fragilis* complex occur thousands of miles from one of the parental species, a phenomenon made possible by the minute spores being carried great distances by wind currents. In anemophilous Angiosperms (especially trees), wind-borne pollen might possibly produce a somewhat similar, though less extreme, effect.*

It is evident that every line of evidence should be considered for the recognition of hybrids: distribution, character by character comparisons, evidence of intersterility (which of course need not be present in hybrids), chromosome number (only helpful if the chromosome numbers of the parents are different), irregularities at meiosis, synthesis of the artificial hybrid, and progeny tests. If these are insufficient, other trails may be followed: Randolph *et al.* (1961) have described the use of marker chromosomes for detecting hybrids

* For recent references to introgression in forest trees, see J. W. Wright (1962).

in the Louisiana irises, and chromatographic techniques have been adopted for similar purposes in *Baptisia* (*Leguminosae*; Alston & Turner, 1963).

Although the majority of F_1 hybrids are probably more or less intermediate between their parents, this need not apply to hybrids between species with very different chromosome numbers, or in cases of cytoplasmic inheritance. In a recent paper on artificial intergeneric hybrids between *Fragaria* (seed parent) and *Potentilla* (pollinator), Ellis (1962) shows how much closer the hybrids are in habit and leaf to *Fragaria* than to *Potentilla*.

Whether we use names or formulae to designate hybrids will depend on how common the hybrid is (or its economic importance) and the way in which it behaves in nature. When the hybrid occurs frequently and reproduces vegetatively so vigorously as to be readily mistaken in the field for a species, it should be given a binomial: e.g. *Stachys* ×*ambigua* (*S. palustris* ×*sylvatica*), *Circaea* ×*intermedia* (*C. alpina*×*lutetiana*) and *Mentha* ×*piperita* (*M. aquatica*×*spicata*). In such cases it is usually the sterile F_1 that is found. Vegetative reproduction, however, often enables these pauciform hybrids to spread (or survive) beyond the present range of their parents (cf. p. 475). In such cases an argument might be advanced for dropping the × before the binomial and treating the plant as a species. Against this procedure, however, is the fact that the parentage of such populations is known, and that they are not self-reproducing (a criterion of species) except by vegetative spread. Binomials may also be useful for pauciform hybrids that are of common occurrence but do not spread vegetatively or give rise to hybrid swarms, e.g. *Primula* ×*variabilis* (*P. veris*×*vulgaris*). Cultivated hybrids may be given formulae plus a cultivar name, according to the *International Code of Nomenclature for Cultivated Plants* (1958).

The above examples may be said to cover "true" hybrids in the taxonomic sense (i.e. between different taxa). In the case of hybrid swarms that have become so stabilised that they behave as species—and have in fact become species—a specific epithet (without ×) should be applied (e.g. *Abies borisii-regis*, *Delphinium gypsophilum*). The same applies to amphidiploids which, breeding true and being distinguishable from the parent species, should be given specific names.

The taxonomist will often find himself required to name specimens which are intermediate between two taxa without being able to decide if hybridisation between them is involved. In such cases he may write (for instance): "*Fibigia* inter *clypeatam* (L.) Medicus et *eriocarpam* (DC.) Boiss." or simply "*F. clypeata* (L.) Medicus $\gtrless$ (or $\rightleftarrows$) *eriocarpa* (DC.) Boiss." If a specimen approaches a known species but one is not sure of its identity, a phrase like "*Fibigia* ad *clypeatam* (L.) Medic. vergens" will serve. (See also Terrell, 1963.)

Although we know a great deal more about the detailed mechanics and patterns of hybridisation than Linnaeus did, it must be admitted that little significant advance has been made in the taxonomic treatment of it. Much remains to be learnt about the extent of its occurrence and its role in evolution.

CONCLUSIONS

By way of conclusion and at the risk of oversimplification, the major principles and issues discussed in the preceding chapters will be summarised.

1. Classification is a fundamental activity and is undertaken primarily for practical purposes.

2. Biological classification is an extension of general classification and is governed by the same basic principles.

3. Different classifications are needed for different purposes; no one classification can serve all purposes.

4. As far as the Angiosperms are concerned, the primary need today is the completion of a workable phenetic classification based on correlations of the maximum number of attributes possible. For practical reasons this needs to be *expressed* in terms of directly appreciable characters, particularly morphological ones (although this need not be true of non-Angiosperm groups). Such a classification is basic for all users of biological material and serves as an information retrieval system. It allows the maximum number of inferences to be drawn from it.

5. The classification of a large part of the world's flora is still in its early or even pioneer stages. Consequently a vast amount of basic work remains to be achieved. At a more advanced stage a great increase in monographic work, especially of larger groups, is a major desideratum.

* * *

6. In the Angiosperms classifications based on phylogeny are unattainable except for certain special cases and then mostly at or near the species level. Because of the complexity and multidimensional nature of evolutionary relationships, what pass for such classifications vary from unconsciously phenetic classifications to distorted special classifications produced by unjustifiable weighting of characters.

7. Phylogenetic classifications cannot be generally attained for the following reasons:

a. Lack of factual evidence, especially concerning the emergence of taxa in time and the direction of evolutionary trends.

b. Different rates at which both organs, and groups, evolve.

c. Difficulty (or inability) of detecting convergence.

d. Inability to detect which characters are functionally, or otherwise necessarily, correlated.

e. Inability to provide a justifiable basis for weighting characters of phylogenetic importance.

8. The search for evolutionary relationships is what inspires the majority of taxonomists today to pursue their studies. Apart from its intrinsic interest, this has the advantage for classification that it leads to a search for new data, and allows phenetic groups to be arranged in sequences according to their possible or probable evolution. It is essential, however, that phenetic classification should be clearly distinguished from evolutionary interpretation, and that speculations about evolutionary relationships should not be allowed to conflict with the construction of general-purpose classifications. It is significant that the application of neo-Adansonian classification (based on the maximum correlation of unweighted characters) in Numerical Taxonomy is forcing taxonomists to appreciate this distinction.

* * *

9. The main causes of unsatisfactory classification include the following:

a. Failure to distinguish between what is possible and what is desirable in classification, and confusion between the different purposes of taxonomy.

b. Insufficient material, mainly due to inadequate collecting or sampling.

c. Lack of information, especially morphological, anatomical and cytological, or the overlooking of previously published data.

d. Failure to study groups throughout their range or to consult sufficient herbaria.

e. Misobservation and bad descriptive work.

f. Lack of information from the field and the experimental garden, concerning the breeding system, genetic basis of variation, and behaviour in nature of populations.

g. Unjustifiable weighting of characters.

h. Lack of balance between synthesis and analysis; inconsistency of treatment.

An otherwise good classification can be marred by bad presentation and incorrect or unorthodox nomenclature.

10. The incompleteness and inadequacy of existing documentation and abstracting services for taxonomy, the lack of indexes to subspecies and lectotypes, for example, all contribute to faulty taxonomy and ensure that a disproportionate amount of the taxonomist's time has to be spent in bibliographic searching instead of in studying his material.

11. Although the handling of large numbers of characters in the production of general classifications is facilitated by numerical techniques and computers, the initial selection of characters and their definition, the statistical techniques used, and the interpretation of the results are entirely dependent on the skill, judgment and fallibility of the taxonomist.

12. Non-morphological characters may or may not correlate with the morphological attributes on which the basic classification of a group was founded. Evidence from non-morphological disciplines is, in the vast majority

of cases, very incomplete and (unlike morphology) has seldom been tested for phenotypic, genetic, ontogenetic and ecological variation. Because of this, such evidence must be used with caution; on no account should it be over emphasised in classification.

* * *

13. The ranks or categories of the taxonomic hierarchy are abstract and can be only defined in terms of their relative position in the hierarchy. The groups or taxa for which the categories are employed may be considered objective to the extent that they comprise plants which occur in nature, and subjective in so far as they form generalisations or concepts in the taxonomist's mind. Unlike the categories, they can be defined.

14. Species and other taxonomic groups in the Angiosperms are traditionally defined in such a way that they can be morphologically recognised. Species are groups of individuals sharing characters in common and separable from each other by highly correlated discontinuities in two or more features, excluding geographical distribution.

* * *

15. Attempts to define taxonomic species in terms of intrabreeding populations separated from each other by genetic barriers to gene-exchange are now being recognised as special classifications based on one or two criteria only. Such attempts served a useful purpose in focusing attention on the dynamic nature of species populations but have probably outlived their usefulness. Units based on such criteria as restriction of gene-exchange (genetic species, ecospecies, etc.) do have a useful purpose in specialised fields of taxonomy and should be designated by special terminologies.

16. Taxonomic species represent different evolutionary situations. No assumptions about their breeding relationships are implicit in their definition, and it must not be assumed that sexual species comprise only interbreeding individuals, unless this can be demonstrated. Species and infra-specific units are not suitable for describing population variation: formal classification should not be pushed too far. Special terms, such as those of the deme terminology, should be employed for such purposes, supplemented by descriptive phrases when necessary. In many cases, however, variation can only be adequately portrayed by graphical or statistical methods or models. Attempts to classify variation or fit it into special terminologies may have the effect of distracting biologists from the more important task of studying evolution at work.

* * *

Taxonomy may be considered as a way of arranging and interpreting information. Even with the information at present available, we have no doubt that classifications can be greatly improved if the general principles of phenetic classification are followed. Such improvements are, in fact, steadily going on in many centres, and most of them result from the adoption of phenetic classification based on multiple correlations and its corollary, de-weighting of characters. In addition to this welcome development, the advances that can be made in taxonomy are largely dependent on those that are possible in the basic fields from which it draws its evidence. Systematists are becoming more and more responsible for the developments in some of these basic fields (e.g. anatomy, embryology, pollen morphology, karyology) as they are discarded by the rapidly advancing spearheads of biology, as well as for the use that is made of them. As the person responsible for synthesis, the taxonomist might well be expected to direct the research of collaborators in these fields basic to classification.

The practice of taxonomy involves making decisions on the material and resources available. All a taxonomist's decisions are subject to revision in time. In no case is an ideal or final classification possible, since in even the best worked groups, unstudied. (and even unsuspected) attributes may become available at some future date which will cause judgments to be reconsidered and probably modified. Taxonomy is a series of progressive stages of knowledge.

BIBLIOGRAPHY AND AUTHOR INDEX

This bibliography differs from the usual kind by: (1) giving in square brackets the page numbers on which the references it lists are referred to in our text (including figure legends); (2) listing the names of those authors who are referred to in our text without a dated reference, the appropriate page being again given in square brackets but at the end of each author's entry.

Works cited by full title are not included unless referred to by date elsewhere in the text. A few works of great historical importance are included in the *Index to Subjects and Organisms.*

ADAMS, H. & ANDERSON, E. 1958. A conspectus of hybridisation in the *Orchidaceae. Evol.*, **12**: 512–518. [434]

ADANSON, M. [23, 24, 29, 111]

AFZELIUS, B. M. 1954. La structure du sporoderme vue au microscope électronique. *8me Int. Bot. Congr. Paris Rapp. & Comm.* sect. 6: 241. [185]

1956. Electron microscopic investigations into exine stratification. *Grana Palynologica*, **12**: 22–37. [185]

AIRY SHAW, H. J. 1950. Post-Darwinian development of taxonomy (botany). *Lectures on the Development of Taxonomy*: 60-79. The Linnean Society, London. [57]

ALBERTUS MAGNUS. [13]

ALEX, J. F. [284]

ALMQUIST, E. 1921. Studien über *Capsella*, II. *Acta Hort. Berg.*, **7**: 41–93. [379]

ALSTON, R. E. & IRWIN, H. S. 1961. The comparative extent of variation of free amino acids and certain "secondary" substance among *Cassia* species. *Amer. J. Bot.*, **48**: 35–39. [253]

ALSTON, R. E. & TURNER, B. L. 1959. Application of paper chromatography to systematics: Recombination of parental biochemical components in a *Baptisia* hybrid population. *Nature*, **184**: 285–6. [253, 254]

1963. Natural hybridisation among species of *Baptisia* (*Leguminosae*). *Amer J. Bot.*, **50**: 159–173. [484]

ALSTON, R. E. [284]

ALVAREZ LÓPEZ, E. 1957. *Especificación, subespecificación y Biogeografía.* Editorial Summa, Madrid. [460]

ANDERSON, E. 1936. The species problem in *Iris*. *Ann. Missouri Bot. Gard.*, **23**: 457–509. [138]

1940. The concept of the genus, II: A survey of modern opinion. *Bull. Torrey Bot. Club*, **67**: 363–369. [107]

1941. The technique and use of mass collections in plant taxonomy. *Ann. Missouri Bot. Gard.*, **28**: 287–292. [265]

1949. *Introgressive Hybridisation.* New York. [55, 329, 466, 467]

1953. Introgressive hybridisation. *Biol. Rev.*, **28**: 280–307. [329, 404, 446, 466, 468, 469, 471, 476–479]

1954a. *Plants, Man and Life.* London. [463 f.n.]

1954b. An analysis of introgression in a population of stemless white violets. *Ann. Missouri Bot. Gard.*, **41**: 263–269. [467]

1957a. An experimental investigation of judgements concerning genera and species. *Evol.*, **11**: 260–261. [94, 105]

ANDERSON, E. 1957b. A semigraphical method for the analysis of complex problems. *Proc. Nat. Acad. Sci.*, **43**: 923–927. [477]

——— [447]

ANDERSON, E. & SAX, K. 1936. A cytological monograph of the American species of *Tradescantia*. *Bot. Gaz.*, **97**: 433–476. [202]

ANDREWS, H. [56]

ARBER, A. 1938. *Herbals, their Origin and Evolution*. Cambridge Univ. Press. [13]

1953. From medieval herbalism to the birth of modern botany. In: E. A. Underwood (ed.), *Science, Medicine and History; essays on the evolution of scientific thought and medical practice written in honour of Charles Singer*, **2**: 317–336. London. [13, 14]

ARBER, E. A. N. & PARKIN, J. 1907. On the origin of Angiosperms. *J. Linn. Soc. (Bot.)*, **38**: 28–80. [60, 61]

ARCANGELI, G. [99]

ARENDT, G. [161]

ARISTOTLE. [12, 14]

ARTSCHWAGER, E. & SMILEY, E. M. 1925. *Dictionary of Botanical Equivalents: German-English, Dutch-English, Italian-English, French-English*. Baltimore. [271]

ASCHERSON, P. [58, 59]

ASHBY, E. 1948. Studies in the morphogenesis of leaves. I. An essay on leaf shape. *New Phytol.*, **47**: 153–176. [340]

ASTIÉ, M. 1963. Tératologie spontanée et expérimentale. Exemples. Application à l'étude du quelques problèmes de biologie végétale. *Ann. Sci. Nat.*, sér. 12, **3**: 619–844. [358 f.n.]

AVDULOV, N. P. 1931. Karyo-systematische Untersuchungen der Familie Gramineen. *Bull. Appl. Bot. Genet. Pl. Breed.*, suppl., **43**: 1–428. [208, 210]

AXELROD, D. [55]

AYTUG, B. 1959. *Abies equi-trojani* Asch. & Sint. est une espèce d'origine hybride d'après l'étude des pollens. *Pollen et Spores*, **1**: 273–278. [473]

BABCOCK, E. B. 1947a. *The Genus Crepis. Part One: The Taxonomy, Phylogeny and Evolution of Crepis*. Univ. Calif. Publ. Bot., **21**. [67, 68, 99, 101, 199, 204, 208, 209, 316]

1947b. *The Genus Crepis. Part Two: Systematic Treatment*. Univ. Calif. Publ. Bot., **22**. [67, 101, 294, 316, 321, 447]

BABCOCK, E. B. & CAVE, M. S. 1938. A study of intra- and inter-specific relations of *Crepis foetida* L. *Zeitschr. Ind. Abst. u. Vererbungsl.*, **75**: 124–160. [384, 446]

BABCOCK, E. B. & STEBBINS, G. L. 1938. *The American Species of Crepis: their Relationships and Distribution as affected by Polyploidy and Apomixis*. Carnegie Inst. Washington Publ. No. 504. [227, 363, 385, 480]

1942. Genetic and evolutionary processes in *Crepis*. *Amer. Nat.*, 76: 337–363. [203, 204]

BADER, R. S. 1958. Similarity and recency of common origin. *Syst. Zool.*, **7**: 184–187. [66]

BAILEY, I. W. 1944. The development of vessels in Angiosperms and its significance in morphological research. *Amer. J. Bot.*, **31**: 421–428. [175]

1951. The use and abuse of anatomical data in the study of phylogeny and classification. *Phytomorphology*, **1**: 67–69. [144, 167]

1953. Evolution of tracheary tissue of land plants. *Amer. J. Bot.*, **40**: 4–8 [175]

1954. *Contributions to Plant Anatomy*. Waltham, Mass. [169, 171]

1957. The potentialities and limitations of wood anatomy in the study of phylogeny and classification of Angiosperms. *J. Arn. Arb.*, **38**: 243–254. [172–175]

——— [172]

BAILEY, I. W. & NAST, C. G. 1945a. The comparative morphology of the *Winteraceae*. VII. Summary and conclusions. *J. Arn. Arb.*, **26**: 37–47. [171]

1945b. Morphology and relationships of *Trochodendron* and *Tetracentron*. I. Stem, root and leaf. *J. Arn. Arb.*, **26**: 143–154. [171]

BAILEY, I. W. & SWAMY, B. G. L. 1948. *Amborella trichopoda* Baill.: A new morphological type of vesselless Dicotyledon. *J. Arn. Arb.*, **29**: 245–254. [171, 180]

BAILEY, I. W. & SWAMY, B. G. L. 1949. The morphology and relationships of *Austrobaileya*. *J. Arn. Arb.*, **30**: 211–226. [171]

1951. The conduplicate carpel of the Dicotyledons and its initial trends of specialisation. *Amer. J. Bot.*, **38**: 373–379. [180]

BAILEY, L. H. [267, 269]

BAILEY, P. C. 1952. Differential reactivity in six species of *Trillium*. *Bull. Torrey Bot. Club*, **79**: 451–458. [206]

1954. Differential reactivity in chromosomes as an indicator of species in *Trillium*. *Bot. Gaz.*, **115**: 241–248. [206]

BAILLON, H. E. [322]

BAKER, H. G. 1948a. Stages in invasion and replacement demonstrated by species of *Melandrium*. *J. Ecol.*, **36**; 96–119. [468]

1948b. Corolla size in gynodioecious and gynomonoecious species of flowering plants, Part II. *Proc. Leeds Phil. Lit. Soc.*, **5**; 136 [369]

1953a. Dimorphism and monomorphism in the *Plumbaginaceae*, III. Correlation of geographical distribution patterns with dimorphism and monomorphism in *Limonium*. *Ann. Bot.* (N.S.), **17**: 615–627. [372]

1953b. Race-formation and reproductive methods in flowering plants. *Symp. Soc. Exp. Biol.*, **7**: 114–143. [372, 413, 416]

1954a. Dimorphism and incompatibility in the *Plumbaginaceae*. *8me Congr. Int. Bot. Paris Rapp. & Comm.* sects. 9 & 10: 133-134. [372]

1954b. [Abstract]. *J. Ecol.*, **42**: 571. [381]

1955. Self-compatibility and establishment after "long-distance" dispersal. *Evol.*, **9**: 347–348. [376]

1956. Pollen dimorphism in the *Rubiaceae*. *Evol.*, **10**: 23–31. [187]

1959a. Reproductive methods as factors in speciation in flowering plants. *Cold Spring Harb. Symp. Quant. Biol.*, **24**: 177–191. [369, 370, 371, 373, 377, 379, 380, 381]

1959b. The contribution of autecological and genecological studies to our knowledge of the past migrations of plants. *Amer. Nat.*, **93**: 255–272 [376]

1961. Rapid speciation in relation to changes in the breeding system of plants. *Recent Advances in Botany*, 881–885. Toronto Univ. Press. [373, 381]

1963. Evolutionary mechanisms in pollination biology. *Science*, **139**: 877–883. [439]

BAKER, H. G. & HARRIS, B. J. 1957. The pollination of *Parkia* by bats and its attendant evolutionary problems. *Evol.*, **11**: 449–460. [440]

BAKER, R. T. & SMITH, H. G. 1920. *A Research on the Eucalypts and their Essential Oils*. Sydney. [251]

BAKER, W. & OLLIS, W. D. 1961. Biflavonyls. In: W. D. Ollis (ed.), *Recent Developments in the Chemistry of Natural phenolic Compounds*. Oxford. [246]

BAKSAY, L. 1956. Cytotaxonomical studies on the Flora of Hungary. *Ann. Hist.-Nat. Mus. Nat. Hung.*, ser. nov., **7**: 321–334. [212]

1957. The chromosome numbers and cytotaxonomical relations of some European plant species. *Ann. Hist.-Nat. Mus. Nat. Hung.*, ser. nov., **8**: 169–174. [212]

1958. The chromosome numbers of Ponto-Mediterranean plant species. *Ann. Hist.-Nat. Mus. Nat. Hung*, ser. nov., **9**: 121–125. [212]

BALDWIN, J. T. 1941. *Galax:* the genus and its chromosomes. *J. Heredity*, **32**: 249–254. [216]

BALL, P. W. & HEYWOOD, V. H. 1962. The taxonomic separation of cytological races of *Kohlrauschia prolifera* (L.) Kunth sensu lato. *Watsonia*, **5**: 113–116. [165, 195, 224]

BARBER, H. N. 1955. Adaptive gene substitutions in Tasmanian Eucalypts: I. Genes controlling the development of glaucousness. *Evol.*, **9**: 1–14. [123, 395, 396]

1956. The natural history of natural selection. *Austral. J. Sci.*, 1956: 148–159. [123, 238, 252, 360, 395]

BARBER, H. N. & JACKSON, W. D. 1957. Natural selection in action in *Eucalyptus*. *Nature*, **179**: 1267–1269. [395, 403, 471]

BARKOUDAH, Y. I. 1962. *A revision of Gypsophila, Bolanthus, Ankyropetalum and Phryna*. Amsterdam. [291]

BARLOW, B. A. 1959. Polyploidy and apomixis in the *Casuarina distyla* species group. *Austral. J. Bot.*, **7**: 239–251. [228, 231]

BARNARD, C. 1957. Floral histogenesis in the Monocotyledons. I. The *Gramineae*. *Austral. J. Bot.*, **5**: 1–20. [181]

1961. The interpretation of the Angiosperm flower. *Australian J. Sci.*, **24**: 64–72. [58, 125]

BARNETT, J. A. 1960. Comparative studies of yeasts. *Nature*, **186**: 449–451. [233]

BARROS NEVES, J. DE. 1944. *Contribuição para o estudo cáriosistemático das espécies portuguesas do género Ranunculus L.* Dissert. Univ. Coimbra. [204]

BARTLETT, H. H. 1940. The concept of the genus. *Bull. Torrey Bot. Club*, **67**: 349–362. [103]

BATEMAN, A. J. 1952. Self-incompatibility systems in Angiosperms. *Heredity*, **6**: 283–310. [372]

1954. The diversity of incompatibility systems in flowering plants. *8me Congr. Int. Bot. Paris Rapp. & Comm.*, sects. 9 & 10: 138–145. [372]

BATE-SMITH, E. C. 1954. Leuco-anthocyanins. I. Detection and identification of anthocyanidins formed from leucoanthocyanins in plant tissues. *Biochem. J.*, **58**: 122–125. [241]

1956. The commoner phenolic constituents of plants and their systematic distribution. *Sci. Proc. Roy. Dublin Soc.*, **27**: 165–176. [241]

1958. Plant phenolics as taxonomic guides. *Proc. Linn. Soc. Lond.*, **169**: 198–211. [241, 243–245]

1959. A chemical approach to plant taxonomy. *Low. Temperature Research Station Record Memorandum* No. 344. [235, 241]

1961. Chromatography and taxonomy in the *Rosaceae*, with special reference to *Potentilla* and *Prunus*. *J. Linn. Soc.* (*Bot.*), **58**: 39–54 [241, 245]

1962. The phenolic constituents of plants and their taxonomic significance. *J. Linn. Soc.* (*Bot.*), **58**: 95–173. [241, 243, 244]

——— [232, 242]

BATE-SMITH, E. C. & LERNER, N. H. 1954. Leuco-anthocyanins. 2. Systematic distribution of leuco-anthocyanins in leaves. *Biochem. J.*, **58**: 126–132 [241]

BATE-SMITH, E. C. & METCALFE, C. R. 1957. Leuco-anthocyanins. 3. The nature and systematic distribution of tannins in dicotyledonous plants. *J. Linn. Soc.* (*Bot.*), **55**: 669–705. [241, 244]

BATE-SMITH, E. C. & RIBEREAU-GAYON, P. 1959. Leuco-anthocyanins in seeds. *Qualitas Plant. et Mater. Veget.*, **5**: 189–198. [241, 247]

BATTAGLIA, E. 1957. *Allium ascalonicum* L., *A. fistulosum* L., *A. Cepa* L.: analisi cariotipica. *Caryologia*, **10**: 1–28. [204]

BAUHIN, C. [15, 282, 284]

BAUMANN, M. G. 1946. *Mydocarpus* und die Phylogenie der Umbelliferen-Frucht. I. *Ber. Schw. Bot. Ges.*, **55**: 13–112. [167]

BAUR, E. 1914. *Einführung in die experimentelle Vererbungslehre*. Ed. 2. Berlin. [349]

BEAUDRY, J. R. 1960. The species concept: its evolution and present history. *Rev. Canad. Biol.*, **19**: 219–240. [91, 95, 101]

BECKNER, M. 1959. *The Biological Way of Thought*. Columbia Univ. Press. [76]

BEER, G. R. DE. 1951. *Embryos and Ancestors* (revised edition). Oxford Univ. Press. [41, 44, 156]

——— [157 f.n.]

BEER, G. R. DE & STEARN, W. T. [284]

BEERS, R. J. & LOCKHART, W. R. 1962. Experimental methods in computer taxonomy. *J. Gen. Microbiol.*, **28**: 633–640. [134, 137 f.n.]

BELL, C. R. 1954. The *Sanicula crassicaulis* complex (*Umbelliferae*). A study of variation and polyploidy. *Univ. Calif. Publ. Bot.*, **27**: 133–230. [217, 227]

BELL, G. D. H. 1961. Phylogeny, genetic variation and plant breeding. In: P. J. Wanstall (ed.), *A Darwin Centenary* (B.S.B.I. Conference Report No. 6): 121–129. [463]

BENSON, L. B. 1957. *Plant Classification*. Boston. [57, 65, fig. facing 66; 70, 108]

1962. *Plant Taxonomy: Methods and Principles*. New York. [55, 56, 466]

BENTHAM, G. 1866. *Flora Australiensis*, **3**. London. [161]

1874. On the recent progress and present state of knowledge of systematic botany. *Rep. Brit. Ass. Adv. Sci. for* 1874, 27–54. [293, 305]

——— [28, 29, 272, 296, 301]

BENTHAM, G. & HOOKER, J. D.

——— [4, 22, 26, 28–30, 59–62, 67, 76 f.n., 84, 176, 190, 259, 267, 270, 287, 303]

BERG, R. L. [35]

BERGGREN, G. 1962. Reviews on the taxonomy of some species of the genus *Brassica*, based on their seeds. *Sv. Bot. Tidskr.*, **56**: 65–134. [165]

BESSEY, C. E. 1915. The phylogenetic taxonomy of flowering plants. *Ann. Missouri Bot. Gard.*, **2**: 109–164. [61]

——— [35, 60, 62–64]

BIAŁOBRZESKA, M. & TRUCHANOWICZOWNA, J. 1960. The variability of shape of fruits and scales of the European Birches (*Betula* L.) and their determination in fossil materials. *Monogr. Bot.*, **9**: 1–93. [146 f.n.]

BIGELOW, R. S. 1958. Classification and phylogeny. *Syst. Zool.*, **7**: 49–59. [31, 46, 50, 66]

1959. Similarity, ancestry and scientific principles. *Syst. Zool.*, **8**: 165–168. [66]

——— [33, 66]

BINZ, A. & THOMMEN, E. [296]

BJÖRKMAN, S. O. 1960. Studies in *Agrostis* and related genera. *Symb. Bot. Upsal.*, **17** (1). [153]

BLACKWELDER, R. E. 1962. Animal taxonomy and the new systematics. In: B. Glass (ed.), *Survey of Biological Progress*, **4**: 1–57. [74, 76 f.n., 90, 266, 417]

BLAKE, S. F. & ATWOOD, A. C. 1942–61. *Geographical Guide to the Floras of the World. Part* **1** (1942): *Africa, Australia, N. & S. America & Islands of the Atlantic, Pacific and Indian Oceans* (Miscell. Publ. No. 401). *Part* **2** (1961): *Western Europe* (Miscell. Publ. No. 797). U.S. Dept. of Agriculture, Washington. [267, 299]

BLAKELY, W. F. 1934. *A Key to the Eucalypts*. Sydney. [161]

BLAKESLEE, A. F. 1922. Variations in *Datura* due to changes in chromosome number. *Amer. Nat.*, **56**: 16–31. [199]

1930. Extra chromosomes, a source of variations in the Jimson weed. *Smithsonian Report*, **1930**: 431–450. [199]

1934. New Jimson weeds from old chromosomes. *J. Heredity*, **25**: 80–108. [199, 200]

BLASDELL, R. F. 1963. A monographic study of the fern genus *Cystopteris*. *Mem. Torrey Bot. Club*, **21**: 1–102. [483]

BLUNT, W. 1950. *The Art of Botanical Illustration*. London. [13, 322]

BLUNT, W. & SITWELL, S. [308]

BOBROV, E. G. 1959. The development of the concept of race and of the series method of the Russian school of taxonomy. *Bot. Zhurn*, **44**: 1565–1566 (English summary). [102, 103 f.n.

1961. Introgressive hybridisation in the flora of the Baikal part of Siberia. *Bot. Zhurn.*, **46**: 313–327. [471]

BÖCHER, T. W. 1941. On the origin of *Saxifraga Nathorstii* (Dusén) v. Hayek. *Medd. Grönl.*, **131** (2): 1–14. [223]

1943. Studies on variation and biology in *Plantago lanceolata* L. *Dansk. Bot. Arkiv.*, **11**, No. 3: 1–18. [395]

1944. The leaf size of *Veronica officinalis* in relation to genetic and environmental factors. *Dansk. Bot. Arkiv.*, **11**, No. 7: 1–20. [395]

1954. Natural populations of *Pulsatilla* in Zealand. *Bot. Tidsskr.*, **51**: 33–47. [215]

1959. Tetraploid and hexaploid *Trisetum spicatum* coll. A cytotaxonomical study. *Bot. Tidsskr.*, **55**: 23–29. [219]

1960a. Infraspecific differentiation, Coreferate. *Botanica Medica*, 8 Jahrg.: 224–225. [400]

1960b. Experimental and cytological studies on plant species. V. The *Campanula rotundifolia* complex. *Biol. Skr. Dan.*, **11**, No. 4. [215, 218, 228, 416]

1961a. The development of cytotaxonomy since Darwin's time. In: P. J. Wanstall (ed.), *A Darwin Centenary* (B.S.B.I. Conference Report No. 6): 26–43. [215, 216, 218, 225]

BÖCHER, T. W. 1961b. Experimental and cytological studies on plant species. VI. *Dactylis glomerata* and *Anthoxanthum odoratum*. *Bot. Tidsskr.*, **56**: 314–335. [228]

1963. The study of ecotypical variation in relation to experimental morphology. *Regn. Veg.*, **27**: 10–16. [146]

BÖCHER, T. W. & LARSEN, K. 1958. Secondary polyploidy and ecotypical differentiation in *Sarothamnus scoparius*. *New Phytol.*, **57**: 311–317.

BÖCHER, T. W., LARSEN, K., & RAHN, K. 1953. Experimental and cytological studies on plant species. I. *Kohlrauschia prolifera* and *Plantago coronopus*. *Hereditas*, **39**: 289–302. [310, 407]

1955a. Experimental and cytological studies on plant species. II. *Trifolium arvense* and some other pauciennial herbs. *Biol. Skr. Dan.*, **8**: No. 3. [215, 407]

1955b. Experimental and cytological studies on plant species. III. *Plantago coronopus* and allied species. *Hereditas*, **41**: 423–453.

BOCK, H. [395, 405, 416] [14]

BOCQUET, G. 1958. Structure de l'ovule-campylotrope. *Actes Soc.-Helv. Sci. Nat. Glaris*, **138**: 135. [189]

1959a. The structure of the placental column in the genus *Melandrium* (*Caryophyllaceae*). *Phytomorphology*, **9**: 217–221. [181]

1959b. The campylotropous ovule. *Phytomorphology*, **9**: 222–227. [189]

BOCQUET, G. & BAEHNI, C. 1961. Le Caryopyllacées-Silenoideés de la Flore Suisse. *Candollea*, **17**: 191–202. [181]

BOCQUET, G. & BERSIER, J. D. 1960a. La valeur systématique de l'ovule: développements teratologiques. *Archiv. Sci.* (*Genève*), **13**: 475–496. [189]

1960b. A propos de la vascularisation florale dans le genre *Silene* sect. *Gastrolychnis* (Fenzl) Chowdhuri. *Actes Soc. Helv. Sci. Nat. Argovie*, **1960**: 113–114. [181]

BODMER, W. F. 1958. Natural crossing between homostyle plants of *Primula vulgaris*. *Heredity*, **12**: 363–370.

BOISSIER, E. [373 f.n.]

BOIVIN, B. 1962. Persoon and the subspecies. *Brittonia*, **14**: 327–331. [33, 109, 303, 387]

BONNIER, G. [98 f.n.]

BORCHT, P. VAN DER. [387, 388, 406] [13]

BOREAU, E. (ed.). 1956–62. Rapport sur la Paleobotanique dans le Monde. *Regn. Veg.*, **7, 11, 19, 24.** [56]

BORGAONKAR, D. S. & DE WET, J. M. J. 1960. A cytogenetical study of hybrids between *Dichanthium annulatum* and *Dichanthium fecundum*. *Фyton*, **12**: 137–144.

BORGSTRÖM, G. [376 f.n.] [367]

BORRILL, M. 1956. A biosystematic study of *Glyceria* in Britain. *Watsonia*, **3**: 291–298. [326]

1961a. *Dactylis marina* Borrill, sp. nov., a natural group of related tetraploid forms. *J. Linn. Soc.* (*Bot.*), **56**: 431–440. [152, 228, 451]

1961b. Epidermal characteristics in the diploid subspecies of *Dactylis glomerata* L. *J. Linn. Soc.* (*Bot.*), **56**: 453–458. [153, 228]

1961c. The pattern of morphological variation in diploid and tetraploid *Dactylis*. *J. Linn. Soc.* (*Bot.*), **56**: 441–452. [123, 126, 228]

BOSEMARK, N. O. 1956a. On the accessory chromosomes in *Festuca pratensis*. III. Frequency and geographical distribution of plants with accessory chromosomes. *Hereditas*, **42**: 189–210. [201]

1956b. Cytogenetics of accessory chromosomes in *Phleum phleoides*. *Hereditas*, **42**: 443–466. [201]

1957. On accessory chromosomes in *Festuca pratensis*. V. Influence of accessory chromosomes on fertility and vegetative development. *Hereditas*, **43**: 211–235. [201]

BOSIO, M. G. 1940. Ricerche sulla fecondazione intraovarica in *Helleborus* & *Paeonia*. *Nuovo Giorn. Bot. Ital.*, **47**: 591–598. [441]

BOUREAU, E. 1954–57. *Anatomie végétale*. Vol. 1 (1954), Vol. 2 (1956), Vol. 3 (1957). Paris. [168]

BOWDEN, W. M. 1959. The taxonomy and nomenclature of the wheats, barleys and ryes and their wild relatives. *Canad. J. Bot.*, **37**: 657–684. [463 f.n.]

1960. An experimental taxonomist examines the classification of the grasses. *Rev. Canad. Biol.*, **19**: 279–292. [463]

BRADLEY, D. E. 1958. The study of pollen grains in the electron microscope. *New Phytol.*, **57**: 226–229. [186]

BRADSHAW, A. D. 1958. Natural hybridisation of *Agrostis tenuis* Sibth. and *A. stolonifera* L. *New Phytol.*, **57**: 68–84. [126, 433]

1959. Population differentiation in *Agrostis tenuis*. I. Morphological differentiation. *New Phytol.*, **58**: 208–227. [397, 409]

1962. The taxonomic problems of local geographical variation in plant species. *Systematics Assoc. Publ.*, No. 4: 7–16. London. [414, 415]

BRADSHAW, M. E. 1963. Studies on *Alchemilla filicaulis* Bus., *sensu lato*, and *A. minima* Walters. Introduction and I. Morphological variation in *A. filicaulis* Bus. *sensu lato*. *Watsonia*, **5**: 304–320. [366]

BRAINERD, E. 1924. Some natural Violet hybrids of North America. *Vermont. Agr. Exp. Sta. Bull.*, No. 239. [473]

BRAINERD, E. & PIETERSON, A. K. 1920. Blackberries of New England—their classification. *Vermont Agr. Exp. Stat. Bull.*, No. 217. [339]

BRANDT, J.-P. 1961. Cytotaxinomie et cytogéographie de *Veronica prostrata* L. *Bull. Soc. Neuchât. Sc. Nat.*, **84**: 35–88. [154, 219, 329]

BRAUN, A. [58, 59]

BRAUN-BLANQUET, J. & RUBEL, E. [300]

BREMEKAMP, C. E. 1939. Phylogenetic interpretations and genetic concepts in taxonomy. *Chronica Botanica*, **5**: 390. [111]

1953. A re-examination of Caesalpino's classification. *Acta Bot. Neerl.*, **1** (4): 580–593. [14]

BRIQUET, J. 1916. In: E. Burnat, *Flore des Alpes Maritimes*, **6**: 71 seq. Lyon. [167]

BRITTON, N. L. [190]

BROCK, R. D. & BROWN, J. A. M. 1961. Cytotaxonomy of Australian *Danthonia*. *Austral. J. Bot.*, **9**: 62–91. [228]

BRONGNIART, A. 1843. *Enumération des Generes des Plantes, &c.* Paris [59]

——— [58]

BROWN, R. [27, 288 f.n.]

BROWN, W. H. 1938. The bearing of nectaries on the phylogeny of flowering plants. *Proc. Amer. Phil. Soc.*, **79**: 549–594. [36 f.n.]

BROWN, W. V. 1958. Leaf anatomy in grass systematics. *Bot. Gaz.*, **119**: 170–178. [169]

BROWN, W. V., HEMISCH, C. & EMERY, W. H. P. 1957. The organisation of the grass shoot apex and systematics. *Amer. J. Bot.*, **44**: 590–595. [191]

BRUMMITT, R. K. & HEYWOOD, V. H. 1960. Pink-flowered *Calystegiae* of the *Calystegia sepium* complex in the British Isles. *Proc. Bot. Soc. Brit. Isles*, **3**: 384–388. [363]

BRUNFELS, O. [13]

BUNGE, A. VON. 1872. Die Gattung *Acantholimon*. *Mém. Acad. Pét.*, sér. 7, **18** (2): 1–72. [315]

——— [92, 99]

BURMAN, J. [284]

BURROWS, C. J. 1960. Studies in *Pimelea*. I. The breeding system. *Trans. Roy. Soc. New Zeal.*, **88**: 29–45. [370]

BURSER, J. [281, 284]

BURTT, B. L. 1951. Dealing with the raw material (plants). *Lectures on the Practice of Botanical and Zoological Classification* (1949–50). Linn. Soc. London. [101, 265]

1961a. *Compositae* and the study of functional evolution. *Trans. Bot. Soc. Edinburgh*, **39**: 216–232. [124, 162]

1961b. Interpretive morphology (Review). *Notes Roy. Bot. Gard. Edinburgh*, **23**: 569–572. [182]

BUTLER, C. G. 1954. *The World of the Honey Bee.* London. [433, 435]

CAESALPINUS, A. [5 f.n., 14, 25, 126]

CAIN, A. J. 1954. Subdivisions of the genus *Ptilinopus* (*Aves, Columbae*). *Bull. Brit. Mus. Nat. Hist. Zoology*, **2**: 265–284. [127]

1956. The genus in evolutionary taxonomy. *Syst. Zool.*, **5**: 97–109 [4]

1958. Logic and memory in Linnaeus' system of taxonomy. *Proc. Linn. Soc. Lond.*, **169**: 144–163. [17, 24, 121]

1959a. Deductive and inductive methods in post-Linnaean taxonomy. *Proc. Linn. Soc. Lond.*, **170**: 185–217. [21, 23, 25, 49, 123]

1959b. The post-Linnaean development of taxonomy. *Proc. Linn. Soc. Lond.*, **170**: 233–244. [17, 21, 33 f.n., 128, 140]

1959c. Function and taxonomic importance. *Systematics Assoc. Publ.*, No. 3: 5–19. London. [25, 35, 39, 121, 126]

1959d. Taxonomic concepts. *Ibis*, **101**: 302–318. [112]

——— [33, 66, 110]

CAIN, A. J. & HARRISON, G. A. 1958. An analysis of the taxonomist's judgement of affinity. *Proc. Zool. Soc. Lond.*, **131**: 85–98. [113]

1960. Phyletic weighting. *Proc. Zool. Soc. Lond.*, **135**: 1–31. [8, 42, 44, 127, 130, 131]

CAIN, S. A. 1944. *Foundations of Plant Geography.* New York & London. [315, 317, 425, 448, 450, 451]

CAMERARIUS, R. J. [16]

CAMP, W. H. 1951. Biosystematy. *Brittonia*, **7**: 113-127. [92, 94]

1961. The pattern of variability and evolution in plants. In: P. J. Wanstall (ed.), *A Darwin Centenary* (B.S.B.I. Conference Report No. 6): 44–69 [452, 473]

CAMP, W. H. & GILLY, C. L. 1943. The structure and origin of species. *Brittonia*, **4**: 323–385. [136 f.n., 451, 459, 460]

CANDOLLE, ALPHONSE L. P. P. DE. 1880. *La Phytographie, ou L'Art de décrire les Végétaux.* Paris. [281]

——— [25]

CANDOLLE, A. P. DE. 1813. *Théorie élémentaire de la Botanique.* Paris. [8, 29]

——— [4, 12, 20, 22–30, 33, 67, 72, 84, 104, 166, 281, 288 f.n., 387]

CANNON, H. G. [322]

CARLISLE, A. 1958. A guide to the named variants of the Scots Pine (*Pinus silvestris* Linnaeus). *Forestry*, **31**: 203–224. [400]

CARLQUIST, S. 1959. Handling the anomalous genus. *Proc. 9th Int. Bot. Congr. Montreal*, **2**: 59. [171]

1961. *Comparative Plant Anatomy. A guide to taxonomic and evolutionary application of anatomical data in Angiosperms.* New York. [151, 156 f.n., 167 f.n., 173]

CAROLIN, R. C. 1960. The structures involved in the presentation of pollen to visiting insects in the order *Campanales*. *Proc. Linn. Soc. N.S. Wales*, **85**: 197–207. [161]

CARTER, G. S. 1951. *Animal Evolution, a Study of recent Views of its Causes.* London. [412]

CAVE, M. S. 1948. Sporogenesis and embryo-sac development of *Hesperocallis* and *Leucocrinum* in relation to their systematic position. *Amer. J. Bot.*, **35**: 343–349. [191]

1953. Cytology and embryology in the delimitation of genera. *Chronica Botanica*, **14**: 140–153. [191]

1958–(ed.). *Index to plant Chromosome Numbers* (in progress). Univ. North Carolina Press. [195]

1959. Embryological characters of taxonomic value. *Proc. 9th. Int. Bot. Congr. Montreal*, **2**: 62. [188]

CAZIER, M. A. & BACON, A. L. 1949. Introduction to quantitative systematics. *Bull. Amer. Mus. Nat. Hist.*, 93, Article 5: 343–388. [328]

CHANDLER, M. E. J. 1923. Geological history of the genus *Stratiotes*. *Quart. J. Geol. Soc.*, **79**: 117–138. [56]

CHANEY, R. V. [55]

CHASSAGNE, M. [300]

CHAYTOR, D. A. & TURRILL, W. B. 1935. The genus *Clypeola* and its intraspecific variation. *Kew Bull.*, **1935**: 1–24. [325]

CHEADLE, V. I. 1953. Independent origin of vessels in the Monocotyledons and Dicotyledons. *Phytomorphology*, **3**: 23–44. [175]

CHENNAVEERAIAH, M. S. 1958. Karyomorphologic and cytotaxonomic studies in *Aegilops*. *Acta Hort. Gotoburg.*, **23** (4): 85–178. [465]

CHESTER, K. S. 1937. A critique of plant serology: a review in English of the serodiagnostic work of Karl Mez. *Quart. Rev. Biol.*, **12**: 19–46, 165–190, 294–321. [254]

CHITTENDEN, F. J. [269]

CHOUARD, P. 1931. Types de développement de l' appareil végétatif chez les Scillées. *Ann. Sci. Nat.*, sér. 10, **13**: 131–323. [105, 148]

CHOWDHURI, P. K. 1957. Studies in the genus *Silene*. *Notes Roy. Bot. Gard. Edinburgh*, **22**: 221–278. [163, 181]

CHRTEK, J. & JIRÁSEK, V. 1962. Contribution to the systematics of the *Poa* genus, section *Ochlopoa* (A. et Gr.) V. Jirás. *Preslia*, **34**: 40–68 [450]

CHURCH, G. L. 1949. A cytotaxonomic study of *Glyceria* and *Puccinellia*. *Amer. J. Bot.*, **36**: 155–165. [152]

CLAPHAM, A. R. 1949. Taxonomic problems in *Galium* and *Juncus*. In: A. J. Wilmott (ed.), *British flowering Plants and modern Systematic Methods*: 72–74. London: Bot. Soc. Brit. Is. [415]

CLAPHAM, A. R., TUTIN, T. G. & WARBURG, E. F. 1952. *Flora of the British Isles*. Cambridge Univ. Press. [102, 301]

1962. *Flora of the British Isles*, ed. 2. Cambridge Univ. Press. [102, 296]

——— [297]

CLARKE, C. B. [286]

CLAUSEN, J. 1951. *Stages in the Evolution of Plant Species*. Cornell Univ. Press. [32, 96, 120, 390, 394, 422]

1954. Partial apomixis as an equilibrium system in evolution. Proc. 9th Int. Congr. Genetics, Bellagio, 1953. *Caryologia*, suppl. vol. **6**: 469–470. [365]

1960. A simple method for the sampling of natural populations. *Report, Scottish Pl. Breeding Station*, **1960**: 69–75. [382, 406, 407]

CLAUSEN, J. & HIESEY, W. M. 1958a. *Experimental Studies on Nature of Species, IV. Genetic structure of ecological races.* Carnegie Inst. Washington Publ. No. 615. [120, 349, 358, 359, 379, 390, 392, 398, 429, 460, 472]

1958b. Phenotypic expression of genotypes in contrasting environments. *Report, Scot. Pl. Breeding Station*, **1958**: 41–51. [349, 365]

CLAUSEN, J., KECK, D. D. & HIESEY, W. M. 1939. The concept of species based on experiment. *Amer. J. Bot.*, **26**: 103–106. [389]

1940. *Experimental Studies on the Nature of Species, I. The effect of varied environments on western North American plants.* Carnegie Inst. Wash. Publ. No. 520. [228, 338, 390, 393, 401, 473]

1941. *Experimental Taxonomy.* Carnegie Inst. Washington Year Book No. 40: 160–170. [444, 445]

1945. *Experimental Studies on the Nature of Species, II. Plant evolution through amphidiploidy and autoploidy, with examples from the Madiinae.* Carnegie Inst. Washington Publ. No. 564. [223, 389, 390, 452, 462, 463]

1948. *Experimental Studies on the Nature of Species, III. Environmental responses of climatic races of Achillea.* Carnegie Inst. Washington Publ. No. 581. [228, 390]

——— [54, 391, 397, 398, 410, 424]

CLAUSEN, R. T. 1941. On the terms "subspecies" and "variety". *Rhodora*, **43**: 157–167. [100]

CLELAND, R. E. 1950. Studies in *Oenothera* cytogenetics and phylogeny. *Indiana Univ. Publ. Sci. ser.*, No. 16. [480]

1951. Extra, diminutive chromosomes in *Oenothera*. *Evol.*, **5**: 165–176. [201]

CLIFFORD, G. [283]

CLUSIUS, C. [13, 284]

COCHRAN, W. G. 1938. Some difficulties in the statistical analysis of replicated experiments. *Emp. J. Expt. Agric.*, **6**: 157–175. [357]

1953. *Sampling Techniques.* New York. [357]

COGNIAUX, A. [170]

COLE, M. J. 1961. Interspecific relationships and intraspecific variation of *Chenopodium album* L. in Britain. I. The taxonomic delimitation of the species. *Watsonia*, **5**: 47–58. [164]

CONSTANCE, L. 1951. The versatile taxonomist. *Brittonia*, **7**: 225–231. [294]

1955. The systematics of the Angiosperms. In: *A Century of Progress in the Natural Sciences, 1853–1953*: 405–484. California Academy of Sciences, San Francisco. [51, 55]

1957. Plant taxonomy in an age of experiment. *Amer. J. Bot.*, **44**: 88–92. [3]

CONTANDRIOPOULOS, J. 1962. *Recherches sur le Flore endemique de la Corse et sur ses origines.* Thèse, Faculté des Sciences, Montpellier. [428]

COOK, C. D. K. [405]

COOPER, J. 1951. Studies on growth and development in *Lolium.* II. Pattern of bud development of the shoot apex and its ecological significance. *J. Ecol.*, **39**: 228–270. [378]

COPELAND, H. F. 1940. The phylogeny of the Angiosperms. *Madroño*, **5**: 209–218. [57]

1957. Forecast of a system of the Dicotyledons. *Madroño*, **14**: 1–9. [65]

CORDUS, V. [14]

CORNER, E. J. H. 1949a. The Durian theory on the origin of the modern tree. *Ann. Bot.* (N.S.), **13**: 367–414. [38, 147]

1949b. The Annonaceous seed and its four integuments. *New Phytol.*, **48**: 332–364. [41]

1951a. Review of P. Maheshwari, *An Introduction to the Embryology of Angiosperms. Phytomorphology*, **1**: 243. [188]

1951b. The Leguminous seed. *Phytomorphology*, **1**: 117–150. [166]

1958. Transference of function. *J. Linn. Soc.* (*Bot.*), **56**: 33–40. [151]

1961. Evolution. In: A. M. Macleod & L. S. Cobley (ed.), *Contemporary Botanical Thought*: 95–114. Edinburgh. [437 f.n., 440]

1962. The classification of *Moraceae. Gardens' Bull. Singapore*, **19**: 187–252. [10 f.n., 189 f.n.]

——— [147]

COUSENS, J. E. 1963. Variation in some diagnostic characters of the sessile and pedunculate oaks and their hybrids in Scotland. *Watsonia*, **5**: 273–286. [482]

COWAN, J. M. 1950. *The Rhododendron Leaf. A Study of the epidermal Appendages.* Edinburgh. [155, 170]

COWAN, S. T. 1962. The microbial species—a macromyth? In: G. C. Ainsworth & P. H. A. Sneath (ed.), *Microbial Classification*: 433–455 (12th Sympos. Soc. Gen. Microbiol. 1962). [91, 93]

CRAWSHAY-WILLIAMS, R. 1961. *In report on:* The mathematical assessment of taxonomic similarity including the use of computers. *Taxon*, **10**: 97–101. [111]

CRITCHFIELD, W. B. 1957. Geographical variation in *Pinus contorta. Maria Moors Cabot Foundation Publ.* No. 3 (Harvard). [160]

CROCKER, W. & BARTON, L. V. 1953. *The Physiology of Seeds.* Waltham, Mass. [165]

CROIZAT, L. 1960. *Principia Botanica.* Caracas. [151]

CROMBIE, A. C. 1950. The notion of species in mediaeval philosophy and science. *Actes VI Congr. Int. d'Hist. Sci. Amsterdam*, **1**: 261–269. [89]

CRONQUIST, A. 1957. Outline of a new system of families and orders of Dicotyledons. *Bull. Jard. Bot. État. Bruxelles*, **27**: 13–40. [65, 303]

——— [63]

CROSBY, J. L. 1949. Selection of an unfavourable gene-complex. *Evol.*, **3**: 212–230. [373]

1954. Population and evolution: the significance of incompatibility. *8me Congr. Int. Bot. Paris Rapp. & Comm.*, sects. 9 & 10: 163–169. [356]

1956. A suggestion concerning the possible role of plasmagenes in the inheritance of acquired adaptations. *J. Genetics*, **54**: 1–8. [406]

1959. Outcrossing of homostyle primroses. *Heredity*, **13**: 127–131. [373 f.n.]

CULLEN J. 1962. A Taxonomic Revision of *Anthyllis* sect. *Vulneraria* DC. in Europe. (Ph.D. thesis, Liverpool University, unpublished.) [140, 161]

——— [150, 312, 320]

CURTIS, J. H. 1959. *The Vegetation of Wisconsin.* Univ. Wisconsin Press: Madison. [403, 410]

CUVIER, G. L. C. F. D. [25]

DADAY, H. 1954a. Gene frequencies in wild populations of *Trifolium repens.* I. Distribution by latitude. *Heredity,* **8**: 61–78. [237]

1954b. Gene frequencies in wild populations of *Trifolium repens.* II. Distribution by altitude. *Heredity,* **8**: 377-384. [237]

1958. Gene frequencies in wild populations of *Trifolium repens.* III. World distribution. *Heredity,* **12**: 169–184. [237]

DALLA TORRE, K. W. V. & HARMS, H. [259]

DANSER, B. H. 1950. A theory of systematics. *Bibl. Biotheor.,* ser. D, **4**: 117–180. [9, 10]

DANSEREAU, P. 1945. Essai de corrélation sociologique entre les plantes supérieures et les poissons de la beine du lac Saint-Louis. *Rev. Canad. Biol.,* **4**: 369–417 [146]

1957. *Biogeography. An ecological Perspective.* New York. [146]

——— [167 f.n.]

DANSEREAU, P. & LEMS, K. 1957. The grading of dispersal types in plant communities and their ecological significance. *Contr. Inst. Bot. Univ. Montréal,* No. 71. [124, 165 f.n.]

DARLINGTON, C. D. 1933. Chromosome study and the genetic analysis of species. *Ann. Bot.,* **47**: 811–814. [459]

1937. *Recent Advances in Cytology.* London. [197, 206, 218]

1939. Misdivision and genetics of the centromere. *J. Genetics,* **37**: 341–364. [199]

1956. *Chromosome Botany.* London. [193, 197–204, 207, 209, 215, 231, 463]

DARLINGTON, C. D. & LA COUR, L. F. 1950. Hybridity selection in *Campanula. Heredity,* **4**: 217–248. [199]

DARLINGTON, C. D. & MATHER, K. 1949. *The Elements of Genetics.* London. [197, 209, 372]

DARLINGTON, C. D. & SHAW, G. W. 1958. Parallel polymorphism in the heterochromatin of *Trillium* species. *Heredity,* **13**: 89–121. [206]

DARLINGTON, C. D. & WYLIE, A. P. 1955. *Chromosome Atlas of Flowering Plants.* London. [195, 213]

DARRAH, W. C. [56]

DARWIN, C. 1859. *The Origin of Species by means of Natural Selection.* London. [29, 31, 41, 51, 67]

1876. *The Effects of Cross and Self Fertilization in the Vegetable Kingdom.* London. [373]

——— [25, 49, 53, 89, 366, 387]

DAUBENMIRE, R. F. 1959. *Plants and Environment. A Textbook of Plant Autecology.* New York & London. [339 f.n., 347, 348, 388, 398, 401, 408, 410]

DAVEY, V. MCM. & LANG, J. M. S. 1939. Experimental taxonomy. III. Correlation of characters within a population. *New Phytol.,* **38**: 32–61. [394]

DAVIDOF, N. N. & BAXTEEVA, F. K. [271]

DAVIDSON, J. F. 1947. The polygonal graph for simultaneous portrayal of several variables in population analysis. *Madroño,* **9**: 105–110. [329]

DAVIS, M. T. 1957. A guide and an analysis of Engler's "Das Pflanzenreich". *Taxon,* **6**: 161–182. [59]

DAVIS, P. H. 1953. Notes on the summer flora of the Aegean. *Notes Roy. Bot. Gard. Edinburgh,* **21**: 101–142. [288]

1960. Materials for a Flora of Turkey: IV. *Ranunculaceae,* II. *Notes Roy. Bot. Gard. Edinburgh,* **23**: 103–161. [148, 159, 187]

1961. Hints for hard-pressed collectors. *Watsonia,* **4**: 283–289. [259, 264, 265]

——— [401]

DAVIS, P. H. & KHAN, M. S. 1961. *Aristolochia* in the Near East. *Notes Roy. Bot. Gard. Edinburgh,* **23**: 515–546. [148]

DELISLE, A. L. 1937. The influence of auxin on secondary branching in two species of *Aster*. *Amer. J. Bot.*, **24**: 159–167. [338]

DESFONTAINES, R. L. [190]

DE WOLF, G. P. Jr. 1957. Chromosome numbers in the higher plants. *Rhodora*, **59**: 241–244. [213]

DEYL, M. 1950. Diminishing and repeated variations in the evolution of plants. *Studia Bot. Čecksl.*, **11** (4): 245–261. [63]

DIELS, L. [122]

DILLEMANN, G. 1960. La differentiation chimique infraspécifique. *Planta Medica*, 8 Jahrg.: 263–274. [238, 252]

DIOSCORIDES, P. A. [13, 14]

DOBZHANSKY, T. 1941. *Genetics and the Origin of Species*, ed. 2. Columbia Univ. Press. [142]

1950. Mendelian populations and their evolution. *Amer. Nat.*, **84**: 401–418. [351, 417]

——— [32, 353, 355, 359]

DOBZHANSKY, T. & EPLING, C. 1942. Genetics of natural populations. VI. Microgeographic races in *Linanthus parryae*. *Genetics*, **27**: 317–332. [359]

DODSON, C. H. 1962. The importance of pollination in the evolution of the orchids of tropical America. *Amer. Orchid Soc. Bull.*, **31**: 525–534, 641–649, 731–735. [435, 440]

DODSON, C. H. & FRYMIRE, G. P. 1961. Natural pollination of orchids. *Missouri Bot. Gard. Bull.*, **49**: 133–152. [435]

DONY, J. G. [300]

DORMER, K. J. 1946. Vegetative morphology as a guide to the classification of the *Papilionatae*. *New Phytol.*, **45**: 145–161. [244]

1961. The crystals in the ovaries of certain *Compositae*. *Ann. Bot.* (N.S.), **25**: 241–254. [236]

1962. The taxonomic significance of crystal forms in *Centaurea*. *New Phytol.*, **61**: 32–35. [237]

——— [151]

DOSTÁL, J. [296]

DOUGLAS, G. E. 1957. The inferior ovary. II. *Bot. Rev.*, **23**: 1–46. [179, 182]

DREIDING, A. 1961. The betacyanins, a class of red pigments in the *Centrospermae*. In: W. D. Ollis (ed.), *Recent Developments in the Chemistry of natural Compounds*. Oxford. [242]

DRUCE, G. C. [309]

DUFOU, O. 1961. Structure cytologique du pollen de quelques Angiospermes. *Rev. Gén. Bot.*, **68**: 73–110. [186]

DUKE, J. A. 1961. Preliminary revision of the genus *Drymaria*. *Ann. Missouri Bot. Gard.*, **48**: 173–268. [164]

DÜRER, A. [13]

DU RIETZ, G. E. 1930. The fundamental units of biological taxonomy. *Svensk. Bot. Tidskr.* **24**: 333–428 [92, 99, 100]

1931. Life forms of terrestrial flowering plants. I. *Acta Phytogeogr. Suecica*, **3**: 1–95. [146]

1958. The hydrid concept. *Uppsala Univ. Årsskr.*, **6**: 216–223. [473]

DUVAL-JOUVE, M. J. 1875. Histotaxie des feuilles de Graminées. *Ann. Sci. Nat. Bot.*, **8**: 227–346. [169]

DYER, A. F. 1963. Allocyclic segments of chromosomes and the structural heterozygosity that they reveal. *Chromosoma*, **13**: 545–576. [205, 206]

EAMES, A. J. 1951. Again: "The new morphology". *New Phytol.*, **50**: 17–35. [179, 181]

1953. Floral anatomy as an aid in generic limitation. *Chronica Botanica*, **14**: 126–132. [179, 182]

1961. *Morphology of the Angiosperms*. New York, Toronto & London. [36, 37, 58, 151, 163, 169, 173, 174, 177, 179–181, 187, 190, 191]

ECKARDT, T. 1957. Vergleichende Studie über die morphologischen Beziehungen zwischen Fruchtblatt, Samenanlage und Blütenachse bei einigen Angiospermen. *Neue Hefte zur Morphologie*, No. 3: 1–91 [111]

EHRENDORFER, F. 1951. Rassengleiderung, Variabilitätszentren und geographisch Merkmalsprogression als Ausdruck der raumzeitlichen Entfaltung des Formenkreises *Galium incanum* S.S. *Öst. Bot. Zeitschr.*, **98**: 428–490. [325]

1952. Cytotaxonomic studies in *Achillea*. *Carnegie Inst. Wash. Yearbook*, **51**: 125–131. [228]

1953. Systematische und zytogenetische Untersuchungen an europäischen Rassen des *Achillea millefolium*-Komplexes. *Öst. Bot. Zeitschr.*, **100**: 583–592. [228]

1958. Zur Phylogenie der Gattung *Galium*, VI. Ein Variabilitätszentrum als "fossiler" Hybrid-Komplex: Der Ostmediterrane *Galium graecum* L.—*G. canum* Req. Formenkreis. *Öst. Bot. Zeitschr.*, **105**: 229–279. [318–321, 473]

1959. Differentiation-hybridisation cycles and polyploidy in *Achillea*. *Cold. Spring Harb. Symp. Quant. Biol.*, **24**: 141–151. [228, 462, 473]

1961. Neufassung der Sektion *Lepto-Galium* Lange und Beschreibung neuer Arten und Kombinationen. (Zur Phylogenie der Gattung *Galium*, VII.) *Sitzungber. Österr. Akad. Wiss., math.-naturw. Kl.*, Abt. I, **169**: 407–421. [228]

1962. Cytotaxonomische Beiträge zur Genese der mitteleuropäischen Flora und Vegetation. *Ber. Deutsch. Bot. Gesellsch.*, **75**: 137–152. [228]

EHRLICH, H. G. & HALL, J. W. 1959. The ultrastructure of Eocene pollen. *Grana Palynologica*, **21**: 32–35. [186]

EICHLER, A. W. [52, 58, 59]

ELLIS, J. R. 1962. *Fragaria-Potentilla* intergeneric hybridization and evolution in *Fragaria*. *Proc. Linn. Soc. Lond.*, **173**: 99–105. [484]

EMBERGER, L. 1960. *Les Végétaux Vasculaires*. Vol. 2 of M. Chadefaud & L. Emberger's *Traité de Botanique Systématique*. Paris. [64]

ENDLICHER, S. L. 1836–40. *Genera Plantarum*. Vienna. [27–29, 58]

ENGLER, A. [52, 58, 60, 62, 65, 98, 101, 268]

ENGLER, A. & GILG, E. [266]

ENGLER, A. & PRANTL, K. [58–61, 63, 176, 191, 267]

ERDTMANN, G. 1943. *An Introduction to Pollen Analysis*. Waltham, Mass. [183]

1952. *Pollen Morphology and Plant Taxonomy. Angiosperms.* Stockholm. [183]

1954. Palynology and plant taxonomy. *8ᵐᵉ Congr. Int. Bot. Paris Rapp. & Comm.*, sects. 2, 4, 5 & 6: 26–36. [185]

1956. Current trends in palynological research work. *Grana Palynologica*, **1**: 127–139. [185]

1959. UV micrographs and photomicrographs from the palynological Laboratory, Stockholm-Solna. *Grana Palynologica*, **21**: 36–39. [185]

1960. Pollen walls and Angiosperm phylogeny. *Bot. Notis.*, **113**: 41–45. [186]

1963. Palynology. In: R. D. Preston (ed.), *Advances in Botanical Research*, **1**: 149–208. [183, 187]

—— [183]

ERDTMANN, G. & STRAKA, H. 1961. Cormophyte spore classification. *Geol. Fören. Förhandl.*, **83**: 65–78. [187]

ERDTMAN, H. 1956. Flavonoid heartwood constituents of conifers. *Sci. Proc. Roy. Dublin*, **27**: 129–138. [234, 246]

1959. Conifer chemistry and taxonomy of conifers. In: K. Kratzl & G. Billek (ed.), *Biochemistry of Wood*, **2**: 1–28. London. [246]

ERNST, A. 1953. "Basic numbers" und polyploidy und ihre Bedeutung für das Heterostylie-problem. *Archiv. Julius Klaus-Stiftung*, **28**: 1–159. [373]

ESAU, K. 1953. *Plant Anatomy*. New York. [168]

1960. *Anatomy of Seed Plants*. New York & London. [168]

ESAU, K., CHEADLE, V. & GIFFORD, E. 1953. Comparative structure and possible trends of specialisation of the phloem. *Amer. J. Bot.*, **40**: 9–19. [175]

EVENSON, J. P., SPILSBURY, J. F. & HEYWOOD, V. H. [155]

FAEGRI, K. 1956. Recent trends in palynology. *Bot. Rev.*, **22**: 639–664. [183]

FAGERLIND, F. 1937. Embryologische, zytologische und bestäubungs-experimentelle Studien in der Familie *Rubiaceae* nebst Bemerkungen über einige polyploiditätsprobleme. *Acta Horti Berg.*, **11**: 195–470. [217]

FAHN, A. 1953. The topography of the nectary in the flower and its phylogenetical trend. *Phytomorphology*, **3**: 424–426. [161]

FAIRBROTHERS, D. E. & JOHNSTON, M. A. 1959. The precipitation reaction as an indicator of relationship in the subfamily *Festucoideae* of the family *Poaceae* (*Gramineae*). *Proc. 9th Int. Bot. Congr. Montreal*, **2**: 110–111. [256]

FARNSWORTH, N. R. 1961. The pharmacognosy of the periwinkles: *Vinca* and *Catharanthus*. *Lloydia*, **24**: 105–138. [240]

FASSET, N. C. & CALHOUN, B. 1952. Introgression between *Typha latifolia* and *Typha angustifolia*. *Evol.*, **6**: 367–379. [469, 483]

FAVARGER, C. 1949. Notes de caryologie alpine. *Bull. Soc. Neuchât. Sci. Nat.*, **72**: 15–22. [212]

1953. Notes de caryologie alpine II. *Bull. Soc. Bot. Neuchât. Sci. Nat.*, **76**: 133–169. [212]

1956. A propos des "races chromosomiques". *Arch. Jul. Klaus. Stift.*, **31**: 277–285. [219]

1959. Notes de caryologie alpine III. *Bull. Soc. Bot. Neuchât. Sci. Nat.*, **82**: 255–285. [212]

1960a. Recherches cytotaxonomiques sur les populations alpines d'*Arenaria ciliata* L. (sens. lat.). *Bull. Soc. Bot. Suisse*, **70**: 126–140. [212, 218]

1960b. Étude cytologique du *Cicendia filiformis* et du *Microcala pusilla* (Gentianacées). *Bull. Soc. Bot. Fr.*, **107**: 94–98. [211]

1962. Contributions a l'étude cytologiques des genres *Minuartia* et *Arenaria*. *Bull. Soc. Neuchât. Sci. Nat.*, **85**: 53–81. [212]

FAVARGER, C. & CONTANDRIOPOULOS, J. 1961. Essai sur l'endémisme. *Bull. Soc. Bot. Suisse*, **71**: 384–408. [231, 426, 428]

FAWCETT, W. & RENDLE, A. B. [296]

FEDDE, F. 1909. *Papaveraceae: Hypecoideae-Papaveroideae*. *Pflanzenreich*, **40** (IV. 104). Leipzig. [101]

FENZL, E. 1840. *Alsineae*, in S. L. Endlicher, *Genera Plantarum*, 963–970. Vienna. [116

FERNALD, M. L. [296]

FERNANDES, A. 1951. Sur la phylogénie des espèces du genre *Narcissus*. *Bol. Soc. Brot.*, sér. 2, **25**: 113–192. [204, 212, 218]

1953. Sur l'origine de *Narcissus taitii* Henriq. *Bol. Soc. Brot.*, sér. 2, **27**: 163–174. [204]

1959a. On the origin of *Narcissus romieuxii* Br.-Bl. & Maire. *Bol. Soc. Brot.*, sér. 2, **33**: 103–117. [204]

1959b. On the origin of *Narcissus cantabricus* DC. *Bol. Soc. Brot.*, sér. 2, **33**: 47–60. [204]

FERNANDEZ-MORAN, H. & DAHL, A. O. 1952. Electron microscopy of ultra-thin frozen sections of pollen grains. *Science*, **116**: 465–467. [185]

FERRIS, G. F. 1928. The principles of systematic entomology. *Stanford Univ. Publ. Biol. Sci.*, **5**: 103–269. [293, 305]

FIRBAS, F. 1962. In: R. Harder, F. Firbas, W. Schumacher & D. von Deuffer, *Lehrbuch der Botanik für Hochschulen*, ed. 28. [64]

FISHER, R. A. 1941. *Statistical Methods for Research Workers*, ed. 8. Edinburgh & London. [138]

1954. *Statistical Methods for Research Workers*, ed. 12. Edinburgh & London. [357]

FITCH, W. H. [322]

FOSBERG, F. R. [285 f.n.]

FOWDEN, L. & STEWARD, F. C. 1957. Nitrogenous compounds and nitrogen metabolism in the *Liliaceae*, I. The occurrence of soluble nitrogenous compounds. *Ann. Bot.* (N.S.), **21**: 53–67. [253]

FREI, E. 1955. Die Innervierung der floralen Necktarien dikotyler Pflanzenfamilien. *Ber. Schweiz. Bot. Ges.*, **65**: 60–114. [161]

FRIES, E. [287 f.n.]

FRISCH, K. VON. 1914. Der Farbensinn und Formensinn der Biene. *Zool. Jb. Physiol.*, **35**: 1–188. [435]

1919. Über den Geruchsinn der Biene und seine Blütenbiologische Bedeutung. *Zool. Jb. Physiol.*, **37**: 1–238. [435]

1950. *Bees, their Vision, Chemical Senses and Language*. Ithaca, N.Y., Cornell Univ. Press. [435]

FRITZSCHE, K. J. [183]

FRÖST, S. 1956. The cytological behaviour of accessory chromosomes in *Centaurea scabiosa*. *Hereditas*, **42**: 415–431. [201]

1958a. The geographical distribution of accessory chromosomes in *Centaurea scabiosa*. *Hereditas*, **44**: 75–111. [201]

1958b. Studies of the genetical effects of accessory chromosomes in *Centaura scabiosa*. *Hereditas*, **44**: 112–122. [201]

FRYXELL, P. A. 1956. Effect of varietal mass on percentage of outcrossing in *Gossypium hirsutum* in New Mexico. *J. Heredity*, **47**: 299–301. [375]

1957. Mode of reproduction of higher plants. *Bot. Rev.*, **23**: 135–233. [370 f.n., 374, 376]

1961. The evolutionary position of inbreeding systems. In: *Recent Advances in Botany*, 887–891. Toronto Univ. Press. [374]

FUCHS, H. P. 1958. Historische Bemerkungen zum Begriff der Subspezies. *Taxon*, **7**: 44–52. [98 f.n.]

FUCHS, L. [13]

FUKAREK, P. 1960. Prilog Poznavanju Crnog Bora (*Pinus nigra* Arn. s. lat.). *Radowa Poljoprivredno-šumarskog fakulteta Sarajevo*, **3** (2): 1–92. [380

FUKUDA, I. & KOZUKA, Y. 1958. Evolution and variation in *Trillium*, V. A list of chromosome composition in natural populations of *Trillium kamschaticum* Pall. *J. Fac. Sci. Hokkaido Univ.*, ser. 5, **6**: 273–319. [206]

GAERTNER, J. [163]

GAJEWSKI, W. 1957. A cytogenetic study on the genus *Geum*. *Monogr. Bot.*, **4**. [67, 96, 197, 359, 445]

1959. Evolution in the genus *Geum*. *Evol.*, **13**: 378–388. [445]

GARBER, E. D. 1958. The genus *Collinsia*. III. The significance of chiasmata frequencies as a cytotaxonomic tool. *Madroño*, **14**: 172–176. [206]

GATES, R. R. 1938. The species concept in the light of genetics and cytology. *Amer. Nat.*, **72**: 340–349. [96]

GAUSE, G. F. 1934. *The Struggle for Existence*. Baltimore. [374]

GAUSSEN, H. 1942–60. Travaux du Laboratoire Forestier de Toulouse. II. Études Dendrologiques. [169]

1960. *Les Gymnosperms actuelles et fossiles*. Fasc. VI, Ch. XI. *Generalités, genre Pinus*. Toulouse. [186]

GAY, A. 1960. A new method for the comparison of populations that contain hybrids. *New Phytol.*, **59**: 218–226. [321, 479]

GEIGER, R. 1950. *The Climate near the Ground*. Harvard Univ. Press. [348]

GEISSMAN, T. A. 1962. The Chemistry of the Flavanoid Compounds. Oxford. [241]

GELL, P. G. H., HAWKES, J. G. & WRIGHT, S. T. C. 1959. The application of immunological methods to the taxonomy of species within the genus *Solanum*. *Proc. Roy. Soc. Ser. B*, **151**: 364–383. [255]

GERSTEL, D. U. 1953. Chromosomal translocations in interspecific hybrids of the genus *Gossypium*. *Evol.*, **7**: 234–244. [465]

GIBBS, R. D. 1945. Comparative chemistry as an aid to the solution of problems in systematic botany. *Trans. Roy. Soc. Canada* (3), Sect. V, **39**: 71–103. [253]

1954. Comparative chemistry and phylogeny of flowering plants. *Trans Roy. Soc. Canada* (3), Sect. V, **48**: 1. [239, 241, 244, 253]

1958a. Chemical evolution in plants. *J. Linn. Soc. Lond. Bot.* **56**: 49-57. [236]

1958b. Biochemistry as an aid to establishing the relationship of some families of Dicotyledonous plants. *Proc. Linn. Soc. Lond.* **169**: 216-230. [236, 244]

——— [232]

GILBERT, G. & LÉONARD, J. 1954. Importance des plantules pour la délimitation des genres. *8me Congrès Int. Bot. Paris Rapp. & Comm.*, sects. 2, 4, 5 & 6: 49 [160]

GILES, N. H. 1942. Autopolyploidy and geographical distribution in *Cuthbertia graminea* Small. *Amer. J. Bot.*, **29**: 637–645. [216]

GILG, E. [253]

GILLETT, G. W. 1961. An experimental study of variation in the *Phacelia sericea* complex. *Amer. J. Bot.*, **48**: 1–7. [472]

GILLETT, J. B. 1960. *Indigofera hirsuta* L. and *I. astragalina* DC. *Kew Bull.*, **14**: 290–295. [326, 327]

1962. Pest pressure, an underestimated factor in evolution. *Systematics Assoc. Publ.*, No. 4: 37–46. London. [430 f.n.]

GILMOUR, J. S. L. 1940. Taxonomy and philosophy. In: J. S. Huxley (ed.), *The New Systematics:* 461–474. Oxford Univ. Press [74]

1958. The species: yesterday and tomorrow. *Nature*, **181**: 379–380. [94, 98]

1960. The deme terminology. *Report, Scottish Pl. Breeding Station*, **1960**: 99–105. [411, 412, 413]

1961. Taxonomy. In: A. M. MacLeod & L. S. Cobley (ed.), *Contemporary Botanical Thought:* 27–45. Edinburgh. [74]

——— [33, 66, 87]

GILMOUR, J. S. L. & GREGOR, J. W. 1939. Demes: a suggested new terminology. *Nature*, **144**: 333–334. [411, 412]

GILMOUR, J. S. L. & HESLOP-HARRISON, J. 1954. The deme terminology and the units of micro-evolutionary change. *Genetica*, **27**: 147–161. [411, 412, 459]

GILMOUR, J. S. L. & TURRILL, W. B. 1941. The aim and scope of taxonomy. *Chronica Botanica*, **6**: 217–219. [111]

GLÜCK, H. 1905–1924. *Biologische und morphologische Untersuchungen über Wasser-und Sumpfgewächse.* I (1905), II (1906), III (1911), IV (1924). Jena. [343]

GOEBEL, K. 1897. *Ueber Jugendformen von Pflanzen und deren künstliche Wiederhervorrufung. Sitzungsber. d. Math.-Phys. Class d. k. b. Akad. Wissensch. München*, **26**. [342]

——— [125]

GOOD, R. D'O. [125]

GOODSPEED, T. H. 1954. *The Genus Nicotiana.* Waltham, Mass. [67, 202, 204, 294]

GOODWIN, R. H. 1937a. Notes on the distribution and hybrid origin of × *Solidago asperula. Rhodora*, **39**: 22–28. [96]

1937b. The cyto-genetics of two species of *Solidago* and its bearing on their polymorphy in nature. *Amer. J. Bot.*, **24**: 425–432. [96]

GOULDEN, C. H. 1952. *Methods of statistical Analysis.* Ed. 2. New York & London. [357]

GRAHAM, A. 1962. Palaeobotany and the university herbarium. *Taxon*, **11**: 218–220. [260]

GRANT, S. A. & HUNTER, P. F. 1962. Ecotypic differentiation of *Calluna vulgaris* L. in relation to altitude. *New Phytol.*, **61** (1): 44–55. [147]

GRANT, V. E. 1949. Pollination systems as isolating mechanisms in the Angiosperms. *Evol.*, 3: 82–97. [433, 434, 438]

1950a. The flower constancy of bees. *Bot. Rev.*, **16**: 379–398 [433, 437]

1950b. The protection of ovules in flowering plants. *Evol.*, **4**: 179–201. [439]

1953. The role of hybridization in the evolution of the leafy stemmed *Gilias*. *Evol.*, **7**: 51–64. [108 f.n., 480, 481]

1954. Genetic and taxonomic studies in *Gilia*. V. *Gilia clivorum. El Aliso*, **3**: 19–34. [223]

1957. The plant species in theory and practice. In: E. Mayr (ed.), *The Species Problem:* 39–80. Amer. Assoc. Adv. Sci. Washington, D.C. [95, 224]

1958. The regulation of recombination in plants. *Cold Spring Harb. Symp. Quant. Biol.*, **23**: 337–363. [96, 470]

——— [436]

GRANT, W. F. 1960. [illegible] categories of classical and experimental taxonomy and the species concept *Rev. Canad. Biol.*, **19**: 241–262. [98]

GRAVIS, A. 1943. Observations anatomiques sur les embryons et les plantules. *Lejeunia*, **3**: 180–196. [160 f.n.]

GRAZI, F., UNAERUS, M. & ÅKERBERG, E. 1961. Observations in the mode of reproduction and the embryology of *Poa pratensis*. *Hereditas*, **47**: 489–541. [365]

GRAY, A. [29, 104, 270]

GREEN, H. E. [164]

GREGG, J. R. 1954. *The Language of Taxonomy*. Columbia Univ. Press. [76]

GRÉGOIRE, V. 1938. La morphogénèse et l'anatomie morphologique de l'appareil floral. 1. Le carpelle. *Cellule*, **47**: 287–452. [178 f.n.]

GREGOR, J. W. 1930. Experiments on the genetics of wild populations. I. *Plantago maritima*. *J. Genetics*, **32**: 15–25. [394]

1938. Experimental taxonomy. II. Initial population differentiation in *Plantago maritima* L. of Britain. *New Phytol.*, **37**: 1–49. [147, 359, 394]

1948. Some reflections on intra-specific ecological variation and its classification. *Trans. Bot. Soc. Edinburgh*, **34**: 377–391. [398]

1939. Experimental taxonomy. IV. Population differentiation in North American and South American sea plantains allied to *Plantago maritima* L. *New Phytol.*, **38**: 293–322. [55, 389, 394]

1956. Adaptation and ecotypic components. *Proc. Roy. Soc. Ser. B*, **145**: 333–337. [403, 409, 413]

1963. Genecological (Biosystematic) classification: the case for special categories. *Regn. Veg.* **27**: 24-26. [461]

GREGOR, J. W., DAVEY, V. MCM. & LANG, J. M. S. 1936. Experimental taxonomy. I. Experimental garden techniques in relation to the recognition of the small taxonomic units. *New Phytol.*, **35**: 323–350. [390, 394]

GREGOR, J. W. & SANSOME, F. W. 1930. Experiments on the genetics of wild populations. II. *Phleum pratense* L. and the hybrid *P. pratense* × *P. alpinum* L. *J. Genetics*, **22**: 373–387. [223]

GREGOR, J. W. & WATSON, P. J. 1961. Ecotypic differentiation: observations and reflections. *Evol.*, **15**: 166–173. [397, 401, 405, 406, 414]

GREGORY, W. C. 1941. Phylogenetic and cytological studies in the *Ranunculaceae*. *Trans. Amer. Phil. Soc.* (N.S.), **31**: 443–521. [81, 210]

GREGUSS, P. 1955. *Identification of living Gymnosperms on the Basis of Xylotomy*. Budapest. [171]

GRESHOFF, M. 1909. Phytochemical investigations at Kew. *Kew Bull.*, **1909**: 397. [241]

GUINEA, E. 1960. Some considerations on the future of the spontaneous vegetation of Europe. *Feddes Rep.*, **63** (2): 222–228. [3]

GUINEA, E. & HEYWOOD, V. H. [228]

GULLIVER, G. 1866. On raphides as natural characters in the British flora. *Quart. J. Micr. Sci.* (N.S.), **6**: 1. [236]

——— [236]

GUNDERSEN, A. 1950. *Families of Dicotyledons*. Waltham, Mass. [173]

GUSTAFSSON, Å. 1943. The genesis of the European Blackberry flora. *Lund. Kgl. Fysiogr. Sallsk. Handlingar* (N.S.), **54** (6): 1–200. [385]

1946. Apomixis in higher plants. Part I. The mechanism of apomixis. *Lunds Univ. Årsskr.*, **42** (3): 1–67. [362, 384]

1947a. Apomixis in higher plants. Part II. The causal aspect of apomixis. *Lunds Univ. Årsskr.*, **43** (2): 69–179. [362, 384]

1947b. Apomixis in higher plants. Part III. Biotype and species formation. *Lunds Univ. Årsskr.*, **43** (12): 183–370. [362, 384]

HAAS, O. & SIMPSON, G. G. 1946. Analysis of some phylogenetic terms, with attempts at redefinition. *Proc. Amer. Phil. Soc.*, **90**, No. 5. [34]

HAECKEL, E. H. P. A. [156]

HAGERUP, E. & HAGERUP, O. 1953. Thrips pollination of *Erica tetralix*. *New Phytol.*, **52**: 1–7. [368]

HAGSTRÖM, J. O. 1916. Critical researches on the Potamogetons. *Svenska Vet. Akad. Handl.*, **55**, No. 5. [148]

HAIR, J. B. & BEUZENBERG, E. J. 1961. High polyploidy in a New Zealand *Poa*. *Nature*, **189**: 160. [196]

HALL, A. D. 1940. *The Genus Tulipa*. London. [241]

HALL, B. A. 1951. The floral anatomy of the genus *Acer*. *Amer. J. Bot.*, **38**: 793–799. [247]

1954. Variability in the floral anatomy of *Acer negundo*. *Amer. J. Bot.*, **41**: 529–532. [182]

HALL, H. M. [406]

HALL, M. T. 1952. A hybrid swarm in *Juniperus*. *Evol.*, **6**: 347–366. [468]

HALLIER, H. [64]

HARA, H. 1954. Morphological and chromosome variations in widespread Eurasiatic species. *8me Congr. Int. Bot. Paris Rapp. & Comm.*, sects, 9 & 10: 71–72. [212]

1956. Contributions to the study of variations in the Japanese plants closely related to those of Europe or North America. Part 2. *J. Fac. Sci. Univ. Tokyo*, sect. 3. **6**: 343–391. [212]

1957. Synopsis of the genus *Chrysosplenium* L. *J. Fac. Sci. Univ. Tokyo*, sect. 3, **7**: 1–90. [212, 294]

1962. Racial differences in widespread species, with special reference to those common to Japan and North America. *Amer. J. Bot.*, **49**: 647–652. [212, 426]

HARBERD, D. J. 1957. The within population variance in genecological trials. *New Phytol.*, **56**: 269–280. [401, 406]

1958. Progress and prospects in genecology. *Report, Scottish Pl. Breeding Station*, **1958**: 52–60. [401, 403, 406]

1961. The case for extensive rather than intensive sampling in genecology. *New Phytol.*, **60**: 325–338. [403]

HARLING, G. 1950. Embryological studies in the *Compositae*. Part I. *Anthemideae—Anthemidinae*. *Acta Hort. Berg.*, **15**: 135–168. [190]

1951. Embryological studies in the *Compositae*. Part II. *Anthemideae—Chrysanthеminae*. *Acta Hort. Berg.*, **16**: 1–56. Part III. *Astereae*. *Ibid.*, 73–120. [190]

1954. The embryo sac development of *Vittadinia triloba* (Gaud.) DC. *Sv. Bot. Tidskr.*, **48**: 489–496. [191]

1958. Monograph of the *Cyclanthaceae*. *Acta Hort. Berg.*, **18**: 1–428. [294]

1960. Further embryological and taxonomical studies in *Anthemis* L. and some related genera. *Sv. Bot. Tidskr.*, **54**: 571–590. [190]

HARPER, J. L., CLATWORTHY, J. N., MCNAUGHTON, I. H. & SAGAR, G. R. 1961. The evolution and ecology of closely related species living in the same area. *Evol.*, **15**: 209–227. [430]

HARTL, D. 1956. Die Beziehungen zwischen den Plazenten der Lentibulariaceen und Scrophulariaceen nebst einem Excurs über Spezialisationsrichtungen der Plazentation. *Beitr. Biol. Pflanzen*, **32**: 471–490. [181]

HARZ, C. D. 1885. *Handwirthschaftliche Samenkunde*. Berlin. [164]

HASKELL, G. 1949. Variation in the number of stamens of the common Chickweed. *J. Genetics*, **49**: 291–301. [162]

1961. Seedling morphology in applied genetics and plant breeding. *Bot. Rev.*, **27**: 382–421. [160]

HATHEWAY, W. H. 1962. A weighted hybrid index. *Evol.*, **16**: 1–10. [479]

HAYATA, B. 1921. The natural classification of plants according to the dynamic system. *Icon. Pl. Formos.*, **10**: 75–95. [65]

HAYEK, A. VON. [268]

HAYES, S. F., KEENAN, J. & COWAN, J. M. 1951. A survey of the anatomy of the *Rhododendron* leaf in relation to the taxonomy of the genus. *Notes Roy. Bot. Gard. Edinburgh*, **21**: 1–34. [170]

HECKARD, L. R. 1954. Cytotaxonomy of a polyploid complex. *8me Congr. Int. Bot. Paris 1954 Rapp. & Comm.*, sects. 9 & 10: 72–73. [228]

1960. Taxonomic studies in the *Phacelia magellanica* polyploid complex: with special reference to Californian members. *Univ. Calif. Publ. Bot.*, **32** (1): 1–126. [228, 230, 451]

HEDBERG, I. 1961. Cytotaxonomic studies in *Anthoxanthum odoratum* s. lat. I. Morphologic analysis of herbarium specimens. *Sv. Bot. Tidskr.*, **55**: 118–128. [220]

HEDBERG, O. 1946. Pollen morphology in the genus *Polygonum* s. lat. and its taxonomic significance. *Sv. Bot. Tidskr.*, **40**: 371–404. [186]

1955. A taxonomic revision of the genus *Sibthorpia* L. *Bot. Notiser*, **108**: 161–183. [294]

1957. Afroalpine vascular plants. A taxonomic revision. *Symb. Bot. Upsal.*, **15** (1). [113, 316, 325, 328]

1958a. Cyto-taxonomic studies in Scottish mountain plants, notably *Deschampsia caespitosa* (L.) P.B. s. lat. *Sv. Bot. Tidskr.*, **52**: 37–46. [215, 383]

1958b. The taxonomic treatment of vicarious taxa. *Uppsala Univ. Årsskr.*, **6**: 186–195. [82, 94, 99, 427]

1961. Monograph of the genus *Canarina* L. (*Campanulaceae*). *Sv. Bot. Tidskr.*, **55**: 17–62. [323, 325]

HEDGE, I. C. 1960. Studies in East Mediterranean species of *Salvia*, III. *Notes Roy. Bot. Gard. Edinburgh*, **23**: 167–171. [310]

1961a. Some remarks on the perennial species of *Ziziphora* with special reference to those in Turkey. *Notes Roy. Bot. Gard. Edinburgh*, **23**: 209–221. [163, 370, 449]

1961b. Materials for a Flora of Turkey: V. *Notes Roy. Bot. Gard. Edinburgh*, **23**: 547–558. [314]

——— [286]

HEDWIG, J. [287 f.n.]

HEGENAUER, R. [232 f.n.]

HEGI, G. [297 f.n.]

HEILBORN, O. 1939. Chromosome studies in *Cyperaceae*. III-IV. *Hereditas*, **25**: 224–240. [198]

HEISER, C. B. 1951. Hybridisation in the annual sunflowers. *Helianthus annuus* × *H. debilis* var. *cucumerifolius*. *Evol.*, **5**: 42–51. [443]

HEISER, C. B. & SMITH, D. M. 1954. New chromosome numbers in *Helianthus* and related genera (*Compositae*). *Proc. Indiana Acad. Sci.*, **64**: 250-253. [228]

1959. Natural Hybridisation and Introgression. Natural hybridisation and introgression with particular reference to *Helianthus*. *Recent Advances in Botany*, **1**: 874–877. Toronto Univ. Press. [467]

HELLER, J. L. & STEARN, W. T. [286]

HEMENWAY, A. F. 1913. Studies on the phloem of the Dicotyledons. II. The evolution of the sieve-tube. *Bot. Gaz.*, **55**: 236–243. [175]

HEMINGWAY, J. S. 1961. Volatile mustard oils of *Brassica juncea* seeds. *Nature*, **192**: 993. [253]

HESLOP-HARRISON, J. 1948. Field studies in *Orchis* L. I. The structure of Dactylorchid populations on certain islands in the Inner and Outer Hebrides. *Trans. Bot. Soc. Edinburgh*, **35**: 26–66. [118]

1951. A comparison of some Swedish and British forms of *Orchis maculata* L. s. lat. *Sv. Bot. Tidskr.*, **45**: 608–635. [118, 323]

1952a. Notes on the distribution of the Irish Dactylorchids. *Veröff. Geobot. Inst. Rübel Zürich*, **25**: 100–113. [118]

1952b. Statistical methods in plant taxonomy. *Taxon*, **1**: 53–59, 73–78. [116, 118, 351, 352]

1952c. A reconsideration of plant teratology. *Phyton*, **4**: 19–34. [358 f.n.]

1953a. Studies in *Orchis* L. II. *Orchis Traunsteineri* Sant. in the British Isles. *Watsonia*, **2**: 371–391. [118]

1953b. *New Concepts in Flowering Plant Taxonomy*. London. [32, 53, 326, 339, 343, 449, 457, 458]

1954. A synopsis of the Dactylorchids of the British Isles. In: E. Rübel & W. Lüdi, *Bericht über Geobotanische Forschungsinstitut Rübel in Zürich für das Jahr 1953*. Zürich. [322]

1955. The conflict of categories. In: J. E. Lousley (ed.), *Species Studies in the British Flora*: 160–172. London: Bot. Soc. Brit. Isles. [97]

HESLOP-HARRISON, J. 1957a. The experimental modification of sex expression in flowering plants. *Biol. Rev.*, **32**: 38–90. [60]

1957b. *vide* MEIKLE, R. D. 1957. [100]

1958a. The unisexual flower—a reply to criticism. *Phytomorphology*, **8**: 177–184. [45, 46, 60]

1958b. Darwin as a botanist. In: S. A. Barnett (ed.), *A Century of Darwin*: 267–295. London. [373, 434, 436]

1959. Growth substances and flower morphogenesis. *J. Linn. Soc.* (*Bot.*), **65**: 269–281. [338]

1960. Infraspecific differentiation. *Planta Medica*, 8 Jahrg.: 208–218. [237]

1961. The experimental control of sexuality and inflorescence structure in *Zea mays* L. *Proc. Linn. Soc. Lond.*, **172**: 108–123. [338, 376]

1962. Purposes and procedures in the taxonomic treatment of higher organisms. In: G. C. Ainsworth & P. H. A. Sneath (ed.), *Microbial Classification*, 14–36. Cambridge Univ. Press. [90, 132, 137]

——— [145, 237, 354, 460]

HESLOP-HARRISON, Y. 1953. *Nuphar intermedia* Ledeb., a presumed relict hybrid, in Britain. *Watsonia*, **3**: 7–25. [475]

HEYWOOD, V. H. 1954a. A revision of the Spanish species of *Tanacetum* subsect. *Leucanthemopsis* Giroux. *Anal. Inst. Bot. Cavanilles*, **12** (2): 313–377. [167, 190]

1954b. Notulae criticae ad floram Hispaniae pertinentes. *Bull. Brit. Mus. Nat. Hist. Bot.*, **1** (4). [324]

1957. A proposed Flora of Europe. *Taxon*, **6**: 33–42. [94]

1958. *The Presentation of Taxonomic Information: a short guide for contributors to Flora Europaea.* Leicester Univ. Press. [92, 99, 100, 268, 296, 298, 304, 418]

1959a. The taxonomic treatment of ecotypic variation. *Systematics Assoc. Publ.*, **3**: 87–112. London. [100, 121, 123, 125, 389, 390, 402, 404, 409, 410, 414, 415, 416]

1959b. A check list of the Portuguese *Compositae-Chrysanthemineae*. *Agron. Lusit.*, **20**: 205–216. [167]

1960a. (ed.) Problems of taxonomy and distribution in the European flora. *Feddes Rep.*, **63**: 107–228. [115, 129, 382, 386, 426]

1960b. The taxonomy of polyploids in Flora Europaea. *Feddes Rep.*, **63**: 179–192. [215, 226]

1960c. Problems of geographical distribution and taxonomy in the Iberian peninsula. *Feddes Rep.*, **63**: 160–168. [94, 427]

1960d. *The presentation of taxonomic information: supplement.* Alcobaça (Portugal) [268, 296]

1961. On the identity and relationships of *Ionopsidium heterospermum* and *I. abulense*. *Feddes Rep.*, **64**: 9–11. [166]

1962. The information problem in biosystematics. *Nature*, **193**: 935–936. [398]

1963. The "species aggregate" in theory and practice. *Regn. Veg.*, **27**: 26–37. [102, 225, 448]

——— [286]

HEYWOOD, V. H. & PICHI-SERMOLLI, R. E. G. [268]

HEYWOOD, V. H. & WALKER, S. 1961. Morphological separation of cytological races in *Ranunculus ficaria* L. *Nature*, **189**: 604. [162, 220]

HIESEY, W. M. 1953. Growth and development of species and hybrids of *Poa* under controlled temperatures. *Amer. J. Bot.*, **40**: 205–221. [341]

HILDITCH, T. P. 1952. The seed and fruit fats of plants. *Endeavour*, **11**: 173–182. [252]

HILLIS, W. E. & ORMAN, H. R. 1962. The extractives of New Zealand *Nothofagus* species. *J. Linn. Soc.* (*Bot.*), **58**: 175–184. [254]

HIRMER, M. [56]

HITCHCOCK, A. S. 1925. Methods of descriptive systematic Botany. New York. [293, 301]

——— [281]

HOFMEISTER, W. [27]

HOLMBOE, J. [99]

HOOKER, J. D. 1855. Introductory essay to J. D. Hooker & T. Thomson, *Flora Indica*: 1–44. London. [293, 295]

——— [28, 29, 167, 289, 306]

HOPWOOD, A. T. 1959. The development of pre-Linnaean taxonomy. *Proc. Linn. Soc. Lond.*, **170**: 230–234. [1 f.n., 17]

HORT, A. (transl.). 1916. *Theophrastus, Enquiry into Plants.* [13]

HOTELLING, H. 1936. Relations between two sets of variables. *Biometrika*, **28**: 321–377. [480]

HOUARD, C. 1908–1913. *Les Zoocécidiés des Plantes d'Europe et du Bassin de la Mediterranée.* Paris. [347]

HOWARD, A. W. & MANTON, I. 1940. Alloploid nature of the wild tetraploid watercress. *Nature*, **146**: 303–304. [223]

HUDSON, J. P. 1957 (ed.). *Control of the Plant Environment.* London. [348]

HULTÉN, E. 1949. On the races in the Scandinavian flora. *Sv. Bot. Tidskr.*, **43**: 383–406. [427]

1958. The amphi-atlantic plants and their phytogeographical connections. *Handlingar, Kungl. Svenska Vetenskapsakademiens*, IV, **7**: 1–340. [426]

——— [310]

HUMBOLDT, F. H. A. VON. [27]

HUSKINS, C. L. 1931. The origin of *Spartina townsendii*. *Genetica*, **12**: 531–538. [223, 463]

HUTCHINSON, A. H. 1936. The polygonal presentation of polyphase phenomena. *Trans. Roy. Soc. Can.* Ser. 3. Sect. V, **30**: 19–26. [329]

HUTCHINSON, J. 1924. Contributions towards a phylogenetic classification of flowering plants: IV. *Kew Bull.*, **1924**: 115–134. [62]

1946. Evolution and classification of Rhododendrons. *Rhododendron Year Book*, **1946**: 42–47. London. [155]

1959. *The Families of Flowering Plants*, ed. 2. Oxford Univ. Press. [35, 62–64, 266, 267]

——— [39, 46, 60, 62, 84, 108, 176, 205, 239, 245]

HUTCHINSON, J. B. 1954. New evidence on the origin of the Old World cottons. *Heredity*, **8**: 225–241. [465, 466 f.n.]

HUTCHINSON, J. B., SILOW, R. A. & STEPHENS, S. G. 1947. *The Evolution of Gossypium and the Differentiation of the cultivated Cottons.* Oxford Univ. Press. [53, 65, 67, 465]

HUXLEY, J. S. 1938. Clines: an auxiliary taxonomic principle. *Nature:* **142**: 219–220. [55]

1942. *Evolution: the modern Synthesis.* London. [412, 459]

1955. Morphism and evolution. *Heredity*, **9**: 1–52. [165, 359]

1958. Evolutionary processes and taxonomy with special reference to grades. *Uppsala Univ. Årsskr.*, **6**: 21–39. [6, 50, 51]

——— [32, 121, 129, 318, 367, 395]

HUXLEY, T. H. 1853. On the morphology of cephalous mollusca. *Phil. Trans. Roy. Soc.*, **143**: 29–65. [10]

HUZIWARA, Y. 1959. Chromosomal evolution in the subtribe *Asterinae*. *Evol.*, **13**: 188–193. [202]

HYDE, H. A. & WILLIAMS, D. A. [183 f.n.]

HYLANDER, N. [129, 382, 386]

ILTIS, H. H. 1959. Studies in the *Capparidaceae*. IV. *Cleome* sect. *Physostemon*. *Brittonia*, **11**: 123–162. [38, 39, 70]

INGER, R. F. 1958. Comments on the definition of genera. *Evol.*, **12**: 370–384. [105]

ISELY, D. 1947. Investigations in seed classification by family characteristics. *Iowa Agric. Exp. Sta, Res. Bull.*, 351. [165]

JACKSON, B. D. [270]

JACQUES-FÉLIX, M. 1958. Evolution des idées sur l'embryon des Graminées. *83me Congr Soc. Sav.*, 323–337. [191]

JAMINET, F. 1960. La différentiation chimique infraspécifique, Corapport. *Planta Medica*, 8 Jahrg.: 275–281. [252]

JANCHEN, E. 1932. Entwurf eines Stammbaumes der Blutenpflanzen nach Richard Wettstein. *Öst. Bot. Zeitschr.*, **81**: 161–166. [64]

——— [268]

JANSSONIUS, H. H. 1950a. Wood anatomy and relationship. *Blumea*, **6**: 407–461. [172]

1950b. The variability of wood-anatomy in large and small genera. *Blumea*, **6**: 462–464. [172]

JARDENI, D. 1938. The dimorphism of *Ononis leiosperma* Boiss. var. *tamarae* var. nov. *Pal. J. Bot.*, *Jerns. ser.*, **1** (2): 235. [340]

JENTYS-SZAFEROWA, J. 1937. Z badán biometrycznych nad zbiorowym gatunkiem *Betula alba* L. I. Wiepostaciowość liści brzóz. *Rozpr. Inst. Bad. Las. Państw.*, Seria A, Nr. 40. [148]

1950a. Analysis of the collective species *Betula alba* L. on the basis of leaf measurements. Part I. Aim and method of the work on the example of *Betula verrucosa* Ehrh. *Bull. Acad. Polon. Sci. Lett. B* (*1*), **1949**: 175–214. [148]

1950b. Analysis of the collective species *Betula alba* L. Part II. *Betula pubescens* Ehrh., *B. tortuosa* Ledeb., *B. carpatica* Waldst. et Kit. *Bull. Acad. Polon. Sci. Lett. B* (*1*), **1950**: 1–63. [148]

1952. Analysis of the collective species *Betula alba* L. Part III. *Betula oycoviensis* Bess. and *B. obscura* Kotula. Determination on the basis of a single leaf. *Bull. Acad. Polon. Sci. Lett. B* (*1*), **1951**: 1–40. [148]

1959. Problems connected with *Betula obscura* Kotula. *Rocznik Dendrologiczny*, **13**: 1–49. [146 f.n.]

JEPSON, G. L. 1944. Phylogenetic trees. *Trans. New York Acad. Sci.*, ser. **6**: 81–92. [43]

JOHANSEN, D. A. 1950. *Plant Embryology. Embryogeny of the Spermatophyta.* Waltham, Mass. [188, 192]

JONES, E. W. 1959. Biological Flora of the British Isles: *Quercus* L. *J. Ecol.*, **47**: 169–222. [462]

JONES, K. 1954. A consideration of certain aspects of cytotaxonomic investigation in the light of results obtained in *Holcus*. *8me Congr. Int. Bot. Paris 1954, Rapp. & Comm.*, sects. 9 & 10: 75–77. [194, 218, 225]

1955. The role of population cytology in evolutionary studies. *Heredity*, **9**: 418. [194]

1961. The status and development of abrupt-ecospecies. *Recent Advances in Botany*, 862–866, Toronto Univ. Press. [218, 419]

——— [285, 286]

JONES, K., CARROLL, C. P. & BORRILL, M. 1961. A chromosome atlas of the genus *Dactylis*. *Cytologia*, **26**: 333–343. [310]

JORDAN, A. [378, 387]

JØRGENSEN, C. A., SØRENSEN, T. & WESTERGAARD, M. 1958. The flowering plants of Greenland, a taxonomical and cytological survey. *Biol. Skr. Dan. Vid. Selsk.*, **9**: 1–172. [195, 212]

JUEL, H. O. [284]

JUNIPER, B. 1959. The surfaces of plants. *Endeavour*, **18**, No. 20–25. [344]

JUSSIEU, A. L. DE. [23–26, 30, 59, 76, 84]

JUSSIEU, B. DE. [24]

JUST, T. 1946. The relative value of taxonomic characters. *Amer. Midl. Nat.*, **36**: 291–297. [115]

JUZEPČZUK, S. V. 1958. The Komarov conception of the species, its historical development and expression in the flora U.R.S.S. *Akad. Nauk S.S.S.R., Bot. Inst. V.L. Komarova*, 1958, **1**: 130–204. (In Russian—extended summary in translation by P. S. Hudson, Commonwealth Bureau of Plant Breeding and Genetics, Cambridge, England—Roneo copy.) [86, 99 f.n.,102]

KAPPERT, H. 1956. Experimentelle Untersuchungen über die Variabilität eines Totalapomikten (*Taraxacum officinale* Web.). *Ber. Deutsch. Bot. Ges.*, **67**: 325–333. [384]

KARPECHENKO, G. D. 1928. Polyploid hybrids of *Raphanus sativa* L. × *Brassica oleracea* L. *Zeitschr. ind. Abstamm. Vererbungsl.*, **39**: 1–7. [222]

KAWANO, S. 1963. Cytogeography and the evolution of the *Deschampsia caespitosa* complex. *Canad. J. Bot.* **41**: 719–742. [383]

KECK, D. D. 1957. Trends in systematic botany. *Survey of Biological Progress*, **3**: 47–107. [193, 215, 216, 223, 225, 422, 451, 453, 459]

KERNER, VON MARILAUN, A. 1869. *Die Abhängigkeit der Pflanzengestalt von Klima und Boden. Ein Beitrag zur Lehre von der Entstehung und Verbreitung der Arten, gestützt auf die Verwandtschaftsverhältnisse, geographische Verbreitung und Geschichte der* Cytisus *arten aus dem Stamme* Tubocytisus *DC.* Innsbruck. [99]

——— [92, 315]

KERNER, A. VON M. & OLIVER, R. W. [151 f.n.]

KIHARA, H. 1954. Considerations of the evolution and distribution of *Aegilops* species based on the analyser-method. *Cytologia*, **19**: 336–357. [228, 464, 465]

KIHARA, H. & KONDO, N. 1943. Studies on amphidiploids of *Aegilops caudata* × *A. umbellulata* induced by colchicine. *Seiken Ziho*, **2**: 24–42. [465]

KIMURA, Y. 1956. Système et phylogénie des Monocotyledones. *Mus. Nat. Hist. Paris Notul. Syst.*, **15**: 137–159. [236]

KINGES, H. 1961. Merkmale des Gramineenembryos. Ein Beitrag zur Systematik der Gräser. *Bo . Jahrb.*, **81**: 50–93. [191]

KLAN, Z. 1947. *Srovnávacl anatomie Plodu Rostlin Okoličnatych* (Comparative anatomy of the fruits of Umbelliferous plants in the area of the Czechoslovak Republic). Prague. [167]

KLEINSCHMIDT, O. 1900. Arten oder Formenkreise? *Journ. f. Ornithol.*, **48**: 134–139. [102]

KOBEL, F. [375]

KORSCHINSKY, S. [315]

KOSTOFF, D. 1938. Studies on polyploid plants. XVIII. Cytogenetic studies on *Nicotiana sylvestris* × *tomentosiformis* hybrids and amphidiploids and their bearing on the problem of the origin of *N. tabacum*. *C. R.* (*Doklady*) *Acad. Sci. U.R.S.S.*, **18**: 459–462. [223]

KOWAL, T. & KUŹNIEWSKI, E. 1959. The generalisation of the dendritic method and its application to the systematics of plants as shown on the sample of *Chenopodium* L. and *Atriplex* L. *Act. Soc. Bot. Polon.*, **28**: 249–262. [133]

KREMP, G. O. W. & AMES, H. T. 56]

KRUSCHKE, E. P. 1955. *The Hawthorns of Wisconsin*, I. Milwaukee Publications in Botany, no. 2. Milwaukee, Wisconsin. [264]

KRUSSMANN, G. [269 f.n.]

KUGLER, H. 1932. Blütenökologische Untersuchungen mit Hummeln, IV. *Planta*, **16**: 534–553. [436]

KUHL, W. [322]

KÜHN, A. 1927. Über der Farbensinn der Bienen. *Zeitschr. Vergl. Physiol.*, **5**: 762–800. [435]

KUNTH, K. S. [22]

KUPCHAN, S. M., ZIMMERMAN, J. H. & AFONSO, A. 1961. The alkaloids and taxonomy of *Veratrum* and related genera. *Lloydia*, **24**: 1–26. [240]

KUPRIANOVA, L. A. 1948. Morphology of the pollen of the Monocotyledons. *Trudi Bot. Inst. Komarov Acad. Sci. U.S.S.R.*, ser. 1, **7**: 163–262. (In Russian.) [187]

KURABAYASHI, M. 1958. Evolution and variation in Japanese species of *Trillium*. *Evol.*, **12**: 286–310. [206]

KURITA, M. 1958a. Chromosome studies in *Ranunculaceae*. VIII. Karyotype and phylogeny. *Rep. Biol. Inst. Ehime Univ.*, **5**: 1–14. [202, 204, 210]

1958b. Chromosome studies in *Ranunculaceae*. IX. Comparison of chromosome volume between a 14- and a 16-chromosome species in *Anemone* and in *Ranunculus*. *Rep. Biol. Inst. Ehime Univ.*, **6**: 1–7. [202, 204]

1961. Chromosome studies in *Ranunculaceae*. XIX. Chromosome size in *Ranunculus* species in the eight-chromosome series. *Mem. Ehime Univ.*, Sect. II (Science), Ser. B (Biology), **4** (2): 263–268. [202]

LAM, H. J. 1936. Phylogenetic symbols, past and present. *Acta Biotheor.*, **2**: 153–194. [43]

1947. A new system of the *Cormophyta*. *Blumea*, **6**: 282–289. [179]

LAM, H. J. 1948. Classification and the new morphology. *Acta Biotheor.*, **8**: 107–154. [179]

1950. Stachyspory and phyllospory as factors in the natural system of the *Cormophyta*. *Sv. Bot. Tidskr.*, **44**: 517–534. [179]

1955. Phylogeny of the *Cormophyta*. *Acta Bot. Neerl.*, **4**: 410–428. [179]

1959. Taxonomy: general principles and Angiosperms. In: W. B. Turrill (ed.), *Vistas in Botany*: 3–75. London. [9, 110, 111, 122]

LAMARCK, J. B. A. P. DE. [24, 281]

LAMPRECHT, H. 1959. Die Artbegriff. *Agri. Hort. Gen.*, **17**: 105–264. [460]

LANGLET, O. F. J. 1932. Über chromosomenverhältnisse und Systematik der *Ranunculaceae*. *Sv. Bot. Tidskr.*, **26**: 381–400. [204, 210]

1936. Studienüber die physiologi sche Variabilität der Kiefer und deren Zusammenhang mit dem Klima. Beiträge zur Kentniss der Öcotypen von *Pinus silvestris* L. *Meddel. Statens. Skogsförsöksanstalt*, **29**: 219–470. [400]

LANJOUW, J. 1958. On the nomenclature of chemical strains. *Taxon*, **7**: 43–44. [252]

LANJOUW, J. & STAFLEU, F. A. (ed.). 1954–1959. *Index Herbariorum.* I: The herbaria of the world (ed. 4), *Regn. Veg.*, **15** (1959). II (1): Collectors A-D. *Regn. Veg.*, **2** (1954). II (2): Collectors E-H. *Regn. Veg.*, **9** (1957). [308]

LARSEN, K. 1955. Cytotaxonomical studies on the Mediterranean flora. *Bot. Not.*, **108**: 263–275. [212, 215]

1956a. Chromosome studies in some Mediterranean and South European flowering plants. *Bot. Not.*, **109**: 293–307. [212, 215]

1956b. Cyto-taxonomical studies in *Lotus*. III. *Bot. Tidsskr.*, **53**: 49–56. [215]

1957. Cryptospecies in *Lathyrus pratensis*. *Bot. Tidsskr.*, **53**: 291–294. [218]

1960a. Infraspecific cytological differentiation. *Botanica Medica*, 8 Jahrg.: 226–239. [194, 195]

1960b. Cytological and experimental studies on the flowering plants of the Canary Islands. *Biol. Skr. Dan. Vid. Selsk.*, **11**, No. 3. [212]

LARSON, D. A. & LEWIS, C. W. 1961. Fine structure of *Parkinsonia aculeata* pollen. I. The pollen wall. *Amer. J. Bot.*, **48**: 934–943. [186]

LAWALRÉE, A. 1960. Indicateur des principaux ouvrages de floristique européenne. *Natura Mosana*, **13**: 29–68. [268]

LAWRENCE, G. H. M. 1940. Armerias, native and cultivated. *Gentes Herbar.*, **4**: 391–418. [381]

1951. *Taxonomy of Vascular Plants*. New York. [57, 160, 181, 265, 268, 270, 271, 281, 301, 303, 307]

1953. Plant genera: their nature and definition. *Chronica Botanica*, **14**: 117–120. [83]

LAWRENCE, W. E. 1945. Some ecotypic relations of *Deschampsia caespitosa*. *Amer. J. Bot.*, **32**: 298–314. [349, 392, 408]

LAWRENCE, W. J. C. & PRICE, J. R. 1940. The genetics and chemistry of flower variation. *Biol. Rev.*, **15**: 35–58. [241]

LAWRENCE, W. J. C. & SCOTT-MONCRIEFF, R. 1935. The genetics and chemistry of flower colour in *Dahlia*: a new theory of specific pigmentation. *J. Genetics*, **30**: 155–226. [241]

LECOMTE, H. 1918. *Atlas des Bois d'Indochine*. Paris. [168]

1922. *Les bois de la forêt d'Analamazaotra* (*Madagascar*). Paris. [168]

LEE, A. T. 1948. The genus *Swainsonia*. *Contrib. New South Wales Nat. Herb.*, **1**, No. 4. [294]

LEENHOUTS, P. W. 1959. Revision of the *Burseraceae* of the Malaysian area in a wider sense. Xa: *Canarium*. *Blumea*, **9**: 275–475. [294]

LE MAOUT, E. & DECAISNE, J. [270]

LEMÉE, A. [267]

LEMESLE, R. 1948. Trachéides aréolées du type cycadéen dans le genre *Paeonia*, leur intérêt au point de vue systématique et phylogenetique. *C.R. Acad. Sci. Paris*, **226**: 2172–2173. [171]

LEMS, K. 1960. Botanical notes on the Canary Islands. II. The evolution of plant forms in the islands: *Aeonium*. *Ecology*, **41**: 1–17. [171]

LEPPIK, E. E. 1953. Some viewpoints on the phylogeny of rust fungi. I. Coniferous rusts. *Mycologia*, **45**: 46–74. [257]

LEPPIK, E. E. 1956. The form and function of numeral patterns in flowers. *Amer. J. Bot.*, **43**: 445–455. [435]

1957. Evolutionary relationship between entomophilous plants and anthomophilous insects. *Evol.*, **11**: 466–481. [435]

LEROY, J. F. 1955. Étude sur les *Juglandaceae*. A la recherche d'une conception morphologique de la fleur, feuille et du fruit. *Mém. Mus. Nat. Hist. Nat.*, sér. B, Bot., **6**: 1–246. [179]

1960. Structure du bois et classification. *Mém. Soc. Bot. France*, 1960: 20–29. [171]

LEVITSKY, G. A. 1931a. The morphology of chromosomes. *Bull. Appl. Bot. Genet. Pl. Breed.*, **27**: 1–74. [203]

1931b. The "karyotype" in systematics. *Bull. Appl. Bot. Genet. Pl. Breed.*, **27**: 187–239. [203, 204]

——— [211]

LEVITSKY, G. A. & TRON, E. J. 1930. Zur Frage der karyotypischen Evolution der Gattung *Muscari*. *Planta*, **9**: 760–775. [203]

LEWIS, D. 1954a. Incompatibility in relation to physiology, genetics and evolutionary taxonomy. *8me Congr. Int. Bot. Paris Rapp. & Comm.*, sects. 9 & 10: 124–132. [371, 372]

1954b. Comparative incompatibility in angiosperms and fungi. *Advances in Genetics*, 6: 235–285. [372]

LEWIS, D. & CROWE, L. K. 1956. The genetics and evolution of gynodioecy. *Evol.*, **10**: 115–125. [370, 372]

1958. Unilateral interspecific incompatibility in flowering plants. *Heredity*, **12**: 232–256. [374]

LEWIS, H. 1953. Quantitative variation in wild genotypes of *Clarkia*. *I.U.B.S. Symposium on Genetics of Population Structure*, Pavia, 114–125. [201]

1957. Genetics and cytology in relation to taxonomy. *Taxon*, **6**: 42–46. [97, 217]

1962. Catastrophic selection as a factor in speciation. *Evol.*, **16**: 257–271. [431]

1963. The taxonomic problem of inbreeders, or how to solve any taxonomic problem. *Regn. Veg.*, **27**: 37–44. [379 f.n.]

LEWIS, H. & EPLING, C. 1959. *Delphinium gypsophilum*, a diploid species of hybrid origin. *Evol.*, **13**: 511–525. [474]

LEX, T. 1954. Duftmale an Blüten. *Z. Vergl. Physiol.*, **36**: 212–234. [436]

LINDLEY, J. 1830. *Introduction to the Natural System of Botany*. London. [149]

——— [27, 62, 183, 270]

LINDSEY, A. A. 1938. Anatomical evidence for the *Menyanthaceae*. *Amer. J. Bot.*, **25**: 480–485. [182]

LINNAEUS, C. [3, 5, 9, 13, 15–23, 25, 26, 29, 46, 50, 76, 84, 86, 89, 91, 98, 104, 105, 108 f.n., 126, 149, 281–286, 291]

LINSBAUER, K. 1930. Die Epidermis. In: K. Linsbauer (ed.), *Handbuch der Pflanzenanatomie*, **4**. [152 f.n.]

LINSLEY, E. G. 1958. The ecology of solitary bees. *Hilgardia*, **27**: 543–599. [434]

LINSLEY, E. G. & MACSWAIN, J. W. 1958. The significance of floral constancy among bees of the genus *Diadasia* (*Hymenoptera*, *Anthophoridae*). *Evol.*, **12**: 219–223. [437]

LITTLE, E. L. 1960. Designating hybrid forest trees. *Taxon*, **9**: 225–231. [481]

L'OBEL, M. DE. [13, 284]

LOCKE, J. [91]

LONGLEY, A. E. 1927. Supernumary chromosomes in *Zea mays*. *J. Agr. Res.*, **35**: 769–784. [200]

LOTSY, J. P. [20]

LÖVE, Á. 1951. Taxonomical evaluation of polyploids. *Caryologia*, **3**: 263–284. [196, 213]

1954a. The foundation of cytotaxonomy. *8me Congr. Int. Bot. Paris Rapp. & Comm.*, sects. 9 & 10: 59–71. [193]

1954b. Cytotaxonomical evaluation of corresponding taxa. *Vegetatio*, **8**: 212–220. [231, 427, 428]

1955. Biosystematic remarks on vicariism. *Acta Soc. Fl. Fenn.*, **72**: 1–14. [426]

LÖVE, Á. 1960a. Taxonomy and chromosomes—a reiteration. *Feddes Rep.*, **62**: 192–202. [207]

1960b. Biosystematics and the process of speciation. In: T. W. M. Cameron (ed.), *Evolution, its Science and Doctrine*, 115–122. Toronto. [420, 430]

1960c. Biosystematics and classification of apomicts. *Feddes Rep.*, **62**: 136–148. [364, 382, 384, 385]

——— [386]

LÖVE, Á. & LÖVE, D. 1943. The significance of differences in distribution of diploids and polyploids. *Hereditas*, **29**: 145–163. [213]

1944. Cytotaxonomic studies on boreal plants. II. Some notes on the chromosome numbers of *Juncaceae*. *Arkiv. Bot.*, **31B** (1): 1–6. [226]

1948. Chromosome numbers of northern plant species. *Icel. Univ. Inst. Appl. Sci. Dept. Agric. Rep.* B, **3**: 1–131. [195]

1949. The geobotanical significance of polyploidy. I. Polyploidy and latitude. *Portug. Acta Biol.* (A), R. B. Goldschmidt vol.: 273–252. [213]

1956. Cytotaxonomical conspectus of the Icelandic flora. *Acta Horti Gotoburg.*, **20**: 65–290. [212, 383]

1957. Arctic polyploidy. *Proc. Genet. Soc. Canada*, **2**: 23–27. [212]

1958a. The American element in the flora of the British Isles. *Bot. Notis.*, 1958: 376–388. [426]

1958b. Biosystematics of *Triglochin maritimum* agg. *Naturaliste Canad.*, **85**: 156–165. [219]

1961. Chromosome numbers of central and northwest European plant species. *Opera Botanica Lund*, **5**. [194–196, 198, 386]

——— [220]

LÖVE, Á., LÖVE, D. & RAYMOND, M. 1957. Cytotaxonomy of *Carex* section *Capillares*. *Canad. J. Bot.*, **35**: 715–761. [198, 226, 227]

LÖVE, D. & DANSEREAU, P. 1959. Biosystematic studies on *Xanthium:* taxonomic appraisal and ecological status. *Canad. J. Bot.*, **37**: 173–208. [380]

LÖVE, D. & NADEAU, L. 1961. The Hutchinson polygraph, a method of simultaneous expression of multiple and variable characters. *Canad. J. Gen. Cytol.*, **3**: 289–294. [330, 331]

LÖVKVIST, B. 1956. The *Cardamine pratensis* complex. Outlines of its cytogenetics and taxonomy. *Symb. Bot. Upsal.*, **14** (2). [225]

1962. Chromosome and differentiation studies in flowering plants of Skåne, South Sweden. I. General aspects. Type species with coastal differentiation. *Bot. Notis.*, **115**: 261–287. [388 f.n.]

LOWE, J. 1961. The phylogeny of Monocotyledons. *New Phytol.*, **60**: 355–387. [62, 68]

LUBBOCK, J. 1892. *A Contribution to our Knowledge of Seedlings*. London. [160 f.n.]

LVOV, S. D. 1954. Signification physiologique du procès de la formation des essences pour la plante. *Acad. Sci. Bot. U.S.S.R. Essays Bot. Moscow*, **2**: 561–676. [248]

LYSENKO, O. & SNEATH, P. H. A. 1959. The use of models in bacterial classification. *J. Gen. Microbiol.*, **20**: 284–289. [137]

MACHATSCHKI-LAURICH, B. 1926. Die Arten der Gattung *Biscutella* L. sectio *Thlaspidium* (Med.) DC. *Botan. Archiv. Koenigsberg*, **13**: 1–115. [139]

MACLEOD, A. M. & MCCORQUODALE, H. 1958. Water-soluble carbohydrates of seeds of the *Gramineae*. *New Phytol.*, **57**: 168–182. [248]

MCCLURE, D. S. 1957. Seed characters of selected plant families. *Iowa State Coll. J. Sci.*, **31**: 649–682. [165]

MCKELVEY, S. D. & SAX, K. 1933. Taxonomic and cytological relationship of *Yucca* and *Agave*. *J. Arn. Arb.*, **14**: 76–81. [205]

MCMILLAN, C. 1956. Ecotypes and community function. *Amer. Nat.*, **94**: 245–255. [398, 410]

MCNAUGHTON, I. H. & HARPER, J. L. 1960. The comparative biology of closely related species living in the same area. II. Aberrant morphology and a virus-like syndrome in hybrids between *Papaver rhoeas* L. and *P. dubium* L. *New Phytol.*, **59**: 27–41. [442, 482]

McNeill, J. 1962. Taxonomic studies in the *Alsinoideae*. I. Generic and infra-generic groups. *Notes Roy. Bot. Gard. Edinburgh*, **24**: 79–155. [116]

——— [325 f.n.]

Madueño Box, M. & Palau Medina, F. 1941. Contribución a la distinción de semillas del género "*Brassica*". *Inst. Nacional Investig. Agron.*, Cuaderno No. 11. [165]

Maheshwari, J. K. 1962. Cleistogamy in Angiosperms. In: P. Maheshwari, B. M. Johri & I. K. Vasil (eds.), *Proc. Summer School Botany, Darjeeling*: 145–155. New Delhi. [367]

Maheshwari, P. 1950. *An Introduction to the Embryology of Angiosperms*. New York, Toronto & London. [188, 190, 191, 442]

Maheshwari, S. C. & Baldev, B. 1958. A contribution to the morphology and embryology of *Commelina forskalaei* Vahl. *Phytomorphology*, **8**: 277–298. [367]

Manning, A. 1956a. The effect of honey guides. *Behaviour*, **9**: 114–139. [435, 436]

1956b. Some aspects of the foraging behaviour of bumble bees. *Behaviour*, **9**: 164–201. [435]

1957. Some evolutionary aspects of the flower constancy of bees. *Proc. Roy. Phys. Soc.*, **25** (3): 67–71. [436]

——— [435]

Mansfeld, R. 1958. Zur Frage der Behandlung nur physiologisch, aber nicht morphologisch verschiedener Sippen in der botanischen Systematik. *Taxon*, **7**: 41–43. [252]

Manske, R. H. F. 1944. The alkaloids. *Ann. Rev. Biochem.*, **13**: 533–548. [239]

Manton, I. 1932. An introduction to the general cytology of the *Cruciferae*. *Ann. Bot.*, **46**: 509–556. [167, 210]

1934. The problem of *Biscutella laevigata*. *Z. indukt. Abstamm. u. Vererblehre*, **67**: 41–57. [219, 228]

1950. *Problems of Cytology and Evolution in the Pteridophyta*. Cambridge Univ. Press. [213]

1958. The concept of aggregate species. *Upps. Univ. Årsskr.*, **6**: 104–112. [102, 225]

Manton, I. & Walker, S. 1953. Cytology of the *Dryopteris spinulosa* complex in E.N. America. *Nature*, **171**: 1116–1118. [483]

Maritain, J. [18]

Marklund, G. 1939. Einige Sippen der *Ranunculus auricomus*-Gruppe. *Mem. Soc. Flora Fenn.*, **16**: 45–53. [385]

1954. Fröbildning utan bekruktning och därmed sammanhängande mångformighet hos växter. *Soc. Sci. Fenn. Årsbok*, **31B** (2): 1–23. [385]

Marklund, G. & Rousi, A. 1961. Outlines of evolution in the pseudogamous *Ranunculus auricomus* group in Finland. *Evol.*, **15**: 510–522. [366, 476]

Marmey, F. 1958. Contribution à l'étude morphologique et anatomique du genre *Marrubium* L. an Maroc. *Trav. Inst. Scient. Chérif.*, sér. Bot. No. 14. [171]

Marsden-Jones, E. M. 1935. *Ranunculus ficaria*: life history and pollination. *J. Linn. Soc. (Bot.)*, **50**: 39–55. [220]

Marsden-Jones, E. M. & Turrill, W. B. 1930-1945. Reports on the transplant experiments of the British Ecological Society at Potterne, Wilts. *J. Ecol.*, **18**: 352–378 (1930); **21**: 268–293 (1933); **23**: 443–469 (1935); **25**: 189–212 (1937); **26**: 359–389 (1938); **33**: 57–81 (1945). [345]

Martin, A. C. 1947. The comparative internal morphology of seeds. *Amer. Midl. Nat.*, **36**: 513–660. [165, 166]

Martinez, M. 1931. Contribución al estudio de las Digitales (4ª nota). *Bol. Soc. Esp. Hist. Nat.*, **31**: 509–520. [155]

Martius, K. F. P. von. [33]

Mason, H. L. 1950. Taxonomy, systematic botany and biosystematics. *Madroño*, **10**: 193–208. [8, 74, 78, 88]

1959. Formal instruments of the taxonomic system. *Proc. 9th Int. Bot. Congr. Montreal*, **2**: 254. [90]

Massart, J. 1902. L'accomodation individuelle chez le *Polygonum amphibium*. *Bull. Jard. Bot. État Bruxelles*, **1**: 73–95. [343]

——— [388]

MASTERS, M. T. 1869. *Vegetable Teratology: an Account of the principal Deviations from the usual Construction of Plants*. London. [197]

MATHER, K. 1946. *Statistical Analysis in Biology* (ed. 2). London. [354]

1947. Species crosses in *Antirrhinum*. I. Genetic isolation of the species *majus*, *glutinosum* and *orontium*. *Heredity*, **1**: 175–186. [434]

1953. The genetical structure of populations. *S.E.B. Symp. VIII: Evolution*: 66–95. [358, 361]

1955. Response to selection. *Cold Spring Harb. Symp. Quant. Biol.*, **20**: 158–165. [335, 336 f.n., 402]

MATHIEU, A. [168]

MATTFELD, J. 1930. Über hybridogene Sippen der Tannen. *Biblioth. Bot.*, **25**, No. 100: 1–84. [473]

MATTHEWS, B. 1962. [Broadcast discussion. B.B.C. Network Three] see: M. PYKE, Out of your senses. *Listener*, **67**: 250–251. [113]

MATTIOLI, P. A. [14]

MAUDE, P. F. 1939. The Merton catalogue. A list of the chromosome numerals of species of British flowering plants. *New Phytol.*, **38**: 1–31. [195]

MAUGINI, E. 1953. Ricerche cito-sistematiche su *Scilla Hughii* Tin. (*Liliaceae*). *Caryologia*, **5**: 167–177. [204]

1956. Ricerche cito-sistematiche su *Scilla cupani* Guss. (*Liliaceae*). *Caryologia*, **8**: 370–378. [204]

MAYR, E. 1931. Notes on *Halcyon chloris* and some of its subspecies. *Amer. Mus. Novitates* No. 469. [102]

1942. *Systematics and the Origin of Species*. Columbia Univ. Press. [32, 122, 358, 426, 427, 433 f.n., 462]

1954. Change of genetic environment and evolution. In: J. Huxley, A. C. Hardy, E. B. Ford (ed.), *Evolution as a Process*, 157–180. London. [427]

1957. *The Species Problem*. Amer. Assoc. Adv. Sci. Pub. No. 50. [2, 89, 91, 96, 97]

1959a. Where are we? *Cold Spring Harb. Symp. Quant. Biol.*, **24**: 1–14. [352, 354]

1959b. Trends in Avian systematics. *Ibis*, **101**: 293–302. [275]

——— [54, 353, 429, 430]

MAYR, E., LINSLEY, E. G. & USINGER, R. L. 1953. *Methods and Principles of Systematic Zoology*. New York. [84, 100, 266, 293, 301, 303, 304, 322]

MEDAWAR, P. B. 1955. Review of J. Huxley (ed.), *Evolution as a Process*. *Heredity*, **9**: 274–277. [156]

MEDICUS, F. C. [19]

MEEUSE, A. D. J. 1961. The *Pentoxylales* and the origin of the Monocotyledons. *Proc. Konikl. Nederl. Akadamie van Weltenschappen*, ser. C, **64**: 543–559. [65, 107, 182]

MEIKLE, R. D. 1957. What is the subspecies? *Taxon*, **6**: 102–105. [100]

MELCHIOR, H. 1943. Entwicklungsgeschichte der Primulaceen-Gattung *Dionysia*. *Mitt. Thür. Bot. Ver.* N.F., **50**: 156–174. [150]

MELVILLE, R. 1937. The accurate definition of leaf shapes by rectangular co-ordinates. *Ann. Bot.* (N.S.), **1**: 673–679. [331]

1939. Contribution to the study of British elms. II. *J. Bot. Lond.*, **77**: 138. [148]

1951. On the application of biometrical methods in plant taxonomy. *Proc. Linn. Soc. Lond.*, **162**: 153–159. [148, 332]

1953. Growth and plant systematics. *Proc. Linn. Soc. Lond.*, **164**: 173–181. [159, 332, 333]

1955. Morphological characters in the discrimination of species and hybrids. In: J. E. Lousley (ed.), *Species Studies in the British Flora*: 55–64. Oxford: Bot. Soc. Brit. Is. [148]

1960a. A metrical study of leaf shapes in hybrids. *Kew Bull.*, **14**: 88–109, 161–177. [331, 332]

1960b. A new theory of the Angiosperm flower. *Nature*, **188**: 14–18. [179]

1962. A new theory of the Angiosperm flower, I. The gynoecium. *Kew Bull.*, **16**: 1–50. [65, 179 f.n.]

MENDEL, G. [53, 387]
MERRILL, E. D. & WALKER, E. H. [268]
MERXMÜLLER, H. [386]
MESQUITA RODRIGUES, J. E. DE. 1953. *Contribuiçao para o conhecimento cariológica das halófitas e psamófitas literais.* Dissert. Univ. Coimbra. [213]
METCALFE, C. R. 1954. Recent advances in taxonomy as seen by an anatomist. *8me Congr. Int. Bot. Paris Rapp. & Comm.* sects. 2, 4, 5 & 6: 37–43. [167, 169, 176]
1961. The anatomical approach to systematics. General introduction with special reference to recent work on Monocotyledons. In: *Recent Advances in Botany*: 146–150. Toronto Univ. Press. [153]
——— [177]
METCALFE, C. R. & CHALK, L. 1950. *Anatomy of the Dicotyledons.* Oxford. [168, 243]
METCALFE, Z. P. 1954. The construction of keys. *Syst. Zool.*, **3**: 38. [301]
MEUSEL, H. 1952. Über Wuchsformen, Verbreitung und Phylogenie einiger mediterran-mitteleuropäischer Angiospermen-Gattungen. *Flora*, **139**: 333–393. [146]
MEYER, A. [13]
MEZ, K. [254]
MICHENER, C. D. & SOKAL, R. R. 1957. A quantitative approach to a problem in classification. *Evol.*, **11**: 130–160. [135, 137]
MILLINER, L. H. 1961. Day length as related to vegetative development in *Ulex europaeus.* I. The experimental approach. *New Phytol.*, **60**: 339–354. [340]
MIROV, N. T. 1938. Phylogenetic relations of *Pinus jeffreyi* and *P. ponderosa. Madroño*, **4**: 169–171. [249]
1954. Chemical composition of gum turpentines of pines of the United States and Canada. *J. Forest Prod. Res. Soc.*, **4**: 1–7. [249]
1956. Composition of turpentines of lodgepole × jack pine hybrids. *Canad. J. Bot.*, **34**: 443–457. [248, 250]
1958. Distribution of turpentine components among species of the genus *Pinus.* In: K. V. Thimann, (ed.), *The Physiology of Forest Trees.* New York. [248, 249, 250]
1959. Biochemical geography of the genus *Pinus. Proc. 9th Int. Bot. Congr. Montreal*, **2**: 265. [251]
1961. *Composition of Gum Turpentines of Pines.* U.S. Dept. Agric. Tech. Bull. 1239. [248, 250]
——— [232]
MOELIONO, B. M. 1959. A preliminary note on the placenta of *Stellaria media* (L.) Vill. and *Stellaria graminea* L. A possible axile origin of ovules? *Acta Bot. Neerl.*, **8**: 292–303. [181]
MOHAN RAM, H. Y. 1959. Embryological characters in the classification of the *Acanthaceae. Proc. 9th Int. Bot. Congr. Montreal*, **2**: 267. [191]
MOHL, H. VON. [183]
MOLL, J. W. [149, 150 f.n., 168]
MOMOTANI, Y. 1961. Taxonomic studies of the genus *Acer* with special reference to the seed proteins. I. Taxonomic characters. *Mem. Coll. Sci. Univ. Kyoto* ser. B., **28**: 455–470. [183 f.n.]
1962a. Taxonomic studies of the genus *Acer* with special reference to the seed proteins. II. Analysis of protein. *Mem. Coll. Sci. Univ. Kyoto* ser. B., **29**: 81–102. [183 f.n.]
1962b. Taxonomic studies of the genus *Acer* with special reference to the seed proteins. III. System of *Aceraceae. Mem. Coll. Sci. Univ. Kyoto* ser. B., **29**: 177–189. [183 f.n.]
MOORE, M. 1959. Population studies on *Viola lactea* Sm. and its wild hybrids. *Evol.*, **13**: 318–332. [444]
MOREAU, R. & BAILLAUD, L. 1960. Les variations de la couleur chez les fleurs. II. Variations de la couleur des fleurs dans le Jura et certaines régions limitrophes. *Ann. Scient. Univ. Besançon*, sér. 2, Medecine, fasc., 4: 95–116. [163]
MOREAU, R., BAILLAUD, L. & SCHAEFER, R. 1960. Les variations de la couleur chez lés fleurs. I. Idées sur le determinisme de la pigmentation chez les végétaux supérieurs. *Ann. Scient. Univ. Besançon*, sér. 2, Medecine, fasc. 4: 85–93. [163]
MORLEY, F. H. W. 1959. Natural selection and variation in plants. *Cold Spring Harb. Symp. Quant. Biol.*, **24**: 47–56. [237, 342]

MORLEY, T. 1953. The genus *Mouriri* (*Melastomaceae*). A sectional revision based on anatomy and morphology. *Univ. Calif. Publ. Bot.*, **26**: 223–312. [170]

MUHLETHALER, K. 1953. Untersuchungen über die Struktur der Pollenmembran. *Mikroskopie* (*Wien*), **8**: 103–110. [185]

MULLER, C. H. 1951. The significance of vegetative reproduction in *Quercus*. *Madroño*, **11**: 129–137. [469]

1961. The origin of *Quercus usiformis* Small. *J. Linn. Soc.* (*Bot.*), **58**: 1–12. [469]

MÜLLER, H. 1881. *Alpenblumen, ihre Befruchtung durch Insekten und ihre Anpassungen an dieselben*. Leipzig. [437]

MÜNTZING, A. 1932. Cyto-genetic investigations on synthetic *Galeopsis tetrahit*. *Hereditas*, **16**: 105–154. [223]

1954. The cytological basis of polymorphism in *Poa alpina*. *Hereditas*. **40**: 459–516. [226, 384]

1957. Frequency of accessory chromosomes in rye strains from Iran and Korea, *Hereditas*, **43**: 682–685. [201]

1959. A new category of chromosomes. *Proc. 10th Congr. Genet.*, **1**: 453–467. [200, 201]

MÜNTZING, A. & G. [365]

MURBECK, S. V. 1920. Beiträge zur Biologie der Wüstenpflanzen. II. Die Synaptospermie. *Lunds Univ. Årsskr.* N.F., Avd. 2, **17** (1): 1–52. [39]

1933. Monographie der Gattung *Verbascum*. *Lunds Univ. Årsskr.* N.F., Avd. 2, **29** (2). [161]

NAIRN, A. E. M. (ed.). 1961. *Descriptive Palaeoclimatology*. New York. [57]

NATHO, G. 1959. Variationsbreite und Bastardbildung bei mitteleuropäischen Birkensippen. *Feddes Rep.*, **61**: 211–273. [148]

NAVASHIN, M. 1925. Morphologische Kernstudien der *Crepis*—Arten in Bezug auf die Artbildung. *Zeitschr. Zellf. Mikrosk. Anat.*, **2**: 98–111. [204]

1926. Variabilität des Zellkerns bei *Crepis*—Arten in Bezug auf die Artbildung. *Zeitschr. Zellf. Mikrosk. Anat.*, **4**: 171–215. [204]

NÈGRE, R. 1957. Les *Frankenia* du Maroc. *Trav. Inst. Scient. Chérif.*, sér. Bot., No. 12. [171]

NELMES, E. 1951. The Antarctic origin of some British "Carices". In: J. E. Lousley (ed.), *The Study of the Distribution of British Plants*: 108–110. Oxford: B.S.B.I. [109]

NELSSON, E. [125]

NETOLITZKY, F. 1926. Anatomie der Angiospermen-Samen. In: K. Linsbauer (ed.), *Handbuch der Pflanzenanatomie*, **2** (2): 10. Berlin. [165]

NEW, J. K. 1958. A population study of *Spergula arvensis*. I. Two clines and their significance. *Ann. Bot.* (N.S.), **22**: 457–477. [310]

1959. A population study of *Spergula arvensis*. II. Genetics and breeding behaviour. *Ann. Bot.* (N.S.), **23**: 23–33. [164]

NITZELIUS, T. 1963. An account of the use of radiography for the pictorial reproduction and study of plants. *Bot. Not.*, **116**: 238–240. [322]

NORDENSKIÖLD, H. 1937. Intra- and interspecific hybrids of *Phleum pratense* and *P. alpinum*. *Hereditas*, **23**: 304–316. [223]

1945. Cyto-genetic studies in the genus *Phleum*. *Acta Agric. Suecana*, **1**: 1–137. [214]

1949. Synthesis of *Phleum pratense* L. from *P. nodosum* L. *Hereditas*, **35**: 190–202. [225]

1951. Cyto-taxonomical studies in the genus *Luzula*. I. Somatic chromosomes and chromosome numbers. *Hereditas*, **37**: 325–355. [226]

1956. Cyto-taxonomical studies in the genus *Luzula*. II. Hybridization experiments in the *campestris-multiflora* complex. *Hereditas*, **42**: 7–73. [226]

1957. Hybridization experiments in the genus *Luzula*. III. The subgenus *Pterodes*. *Bot. Not.*, **110**: 1–16. [226]

1961. Modes of species differentiation in the genus *Luzula*. In: *Recent Advances in Botany*, **2**: 1469–1473. Toronto Univ. Press. [226]

NORDLINDH, T. 1943. *Studies in the Calenduleae*, I. Lund. [294]

1946. Studies in the *Calenduleae*, II. *Bot. Not.*, **4**: 471–506. [294]

NOWACKI, E. 1960. Systematics of *Genisteae* in the light of chemical analyses. *Genetica Polonica*, **1** (1): 119. [240]

NYGREN, A. 1946. The genesis of some Scandinavian species of *Calamagrostis*. *Hereditas*, **32**: 131–262. [225]

1949. Studies in vivipary in the genus *Deschampsia*. *Hereditas*, **35**: 27–32. [215]

1954. Apomixis in the Angiosperms. *Bot. Rev.*, **20**: 577–649. [384]

NYMAN, C. F. [98]

OLMSTED, C. E. 1944. Photoperiodic responses in twelve geographic strains of side-oats grama. *Bot. Gaz.*, **106**: 46–74. [341]

ONO, M. 1960. Studies on *Heterotropa* and its related genera with special reference to their karyo-morphology and phylogeny. *J. Fac. Sci. Univ. Tokyo*, sect. 3, Bot., **7**: 471–499. [204]

ORTON, G. L. 1955. The role of ontogeny in systematics and evolution. *Evol.*, **9**: 75–83. [159]

OWEN, R. 1848. Report on the archetype and homologies of the vertebrate skeleton. *Rep. 16th Meeting Brit. Assoc. Adv. Sci.*, 169–340. [40]

OWNBEY, M. 1950. Natural hybridisation and amphiploidy in the genus *Tragopogon*. *Amer. J. Bot.*, **37**: 487–499. [450]

PALM, B. T. 1922. The embryosac of *Vittadinia*. *Ann. Jard. Bot. Buitenzorg*, **32**: 88–94. [191]

PALSER, B. F. 1951. Studies of floral morphology in the *Ericales*. I. Organography and vascular anatomy in the *Andromedeae*. *Bot. Gaz.*, **112**: 447–485. [182]

1954. Studies of floral morphology in the *Ericales*. III. Organography and vascular anatomy in several species of the *Arbuteae*. *Phytomorphology*, **4**: 335–354. [182]

1961. Studies of floral morphology in the *Ericales*. V. Organography and vascular anatomy in several United States species of the *Vacciniaceae*. *Bot., Gaz.* **123**: 79–111. [161, 182]

PANDEY, K. K. 1960. Evolution of gametophyte and sporophyte systems of self-incompatibility in Angiosperms. *Evol.*, **14**: 98–115. [372]

PANELATTI, J. 1959. Contribution à l'étude anatomique du genre *Bupleurum* L. au Maroc. *Trav. Inst. Scient. Chérif.*, sér. Bot., No. 15. [171]

PARIS, R. & MOYSE, H. 1957. Examen en electrophorèse sur papier des alcaloides totaux de diverses espèces de pervenches. *C.R. Acad. Sci.*, **245**: 1265–1268. [240]

PARKIN, J. [178]

PAUNERO, E. 1952. Las especies españolas del género *Alopecurus*. *Anal. Jard. Bot. Madrid*, **10**: 301–345. [170]

1953. Las Agrostídeas españolas. *Anal. Inst. Bot. A. J. Cavanilles Madrid*, **11** (1): 319–417. [153]

PAWŁOWSKI, B. [386]

PEARSON, J. 1938. The Tasmanian Bush Opossum: its distribution and colour variations. *Pap. Proc. Roy. Soc. Tasm.* for 1937, 21–29. [318]

PELLEGRIN, F. [170]

PENFOLD, A. R. & MORRISON, F. R. 1927. The occurrence of a number of varieties of *Eucalyptus dives* as determined by chemical analyses of the essential oils. Part I. *J. Proc. Roy. Soc. N.S. Wales*, **61**: 54–67. [251]

PENFOLD, A. R. & WILLIS, J. L. 1961. *The Eucalypts*. London. [161]

PENNEL, F. W. [285, 286]

PÉNZES, A. 1960. Über die Morphologie, Dynamik und zönologische Rolle der Spross-kolonien-bildenden Pflanzen (Polycormone). *Fragm. Flor. Geobot. Krakow*, Ann. VI, pars 4: 501–515. [148 f.n.]

PERCIVAL, M. S. 1961. Types of nectar in the Angiosperms. *New Phytol.*, **60**: 235–281. [248]

PERROT, E. [168]

PERSOON, C. H. 1805–1807. *Synopsis Plantarum, seu Enchiridium Botanicum*. Paris & Tübingen. [98]

——— [287 f.n., 288]

PFEIFFER, L. G. C. [281]

PHILLIPS, E. W. J. 1941. The identification of coniferous woods by their microscopic structure. *J. Linn. Soc. (Bot.)*, **52**: 259–320 [323 f.n.]

1948. *Identification of softwoods by their microscopic structure.* D.S.I.R. Forest Products Research Bulletin No. 22. London. [323 f.n.]

PHILIPSON, W. R. 1937. A revision of the British species of the genus *Agrostis* Linn. *J. Linn. Soc. (Bot.)*, **51**: 73–151. [433]

1961. Relationship and convergence in Angiosperms. *Phytomorphology*, **10**: 367–376. [128, 130]

PICHON, M. [240]

PIJL, L. VAN DER. 1936. Fledermäuse und Blumen. *Flora*, **31**: 1–40. [439]

1960. Ecological aspects of flower evolution. I. Phyletic evolution. *Evol.*, **14**: 403–416. [123, 125, 436, 440]

1961. Ecological aspects of flower evolution. II. Zoophilous flower classes. *Evol.*, **15**: 44–59. [439, 440]

——— [122]

PILGER, R. [345]

PIRIE, N. W. 1955. The principles of microbial classification. Summing up. *J. Gen. Microbiol.*, **12**: 382–386. [234]

PLANTEFOL, L. 1949. *L'Ontogénie de la fleur.* Paris. [178 f.n.]

PLATO. [9, 13]

POJARKOVA, A. I. [163]

PORTER, C. L. 1959. *Taxonomy of Flowering Plants.* San Francisco & London. [57, 265, 268, 271]

POST, G. E. & DINSMORE, J. E. [296]

PRAEGER, R. L. 1951. Hybrids in the Irish flora. *Proc. Roy. Irish Acad.*, **54B**: 1–14. [446]

PRAT, H. 1932. L'épiderme des graminées. Étude anatomique et systématique. *Ann. Sci. Nat.*, Bot. sér. 10, **14**: 119–324. [152]

1960. Vers une classification naturelles des Graminées. *Bull. Soc. Bot. Fr.*, **107**: 32–79. [152, 210]

PRAŽMO, W. 1960. Genetic studies on the genus *Aquilegia* L. I. Crosses between *Aquilegia vulgaris* L. and *Aquilegia ecalcarata* Maxim. *Acta Soc. Bot. Polon.*, **29**: 57–77. [358]

PROCTOR, J. R. & KENDRICK, W. B. 1963. Unequal weighting in numerical taxonomy. *Nature*, **197**: 716–717. [141]

PRYOR, L. D. 1956. Variation in Snow Gum (*Eucalyptus pauciflora* Sieb.). *Abstract Proc. Linn. Soc. N.S. Wales*, No. 665. [414]

PURI, V. 1951. The role of floral anatomy in the solution of morphological problems. *Bot. Rev.*, **17**: 471–553. [179]

1952a. Floral anatomy and inferior ovary. *Phytomorphology*, **2**: 122–129. [179, 182]

1952b. Placentation in Angiosperms. *Bot. Rev.*, **18**: 603–651. [179]

RAMSBOTTOM, J. 1938. Linnaeus and the species concept. *Proc. Linn. Soc. Lond.*, **150**: 192–291. [19]

RANDOLPH, L. F. 1928. Types of supernumerary chromosomes in maize. (Abstract.) *Anat. Rec.*, **41**: 102. [200]

1941. Genetic characteristics of the B chromosomes in maize. *Genetics*, **26**: 608–631. [200]

RANDOLPH, L. F., MITRA, J. & NELSON, I. S. 1961. Cytotaxonomic studies of Louisiana Irises. *Bot. Gaz.*, **123**: 125–133. [483]

RATTER, J. A. 1959. *A cytogenetic study in Spergularia.* (Ph.D. thesis, Liverpool—unpublished.) [483]

RAUNKIAER, C. 1934. *The Life Forms of Plants and statistical Plant Geography.* (transl.) Oxford Univ. Press. [146, 147, 298, 342]

RAVEN, P. H. 1962. The systematics of *Oenothera* subgenus *Chylismia*. *Univ. Calif. Publ. Bot.*, **34**, No. 1. [294, 295]

1963. *Circaea* in the British Isles. *Watsonia*, **5**: 262–272. [475]

RAVEN, P. H. & KYHOS, D. W. 1961. Chromosome numbers in *Compositae*. II. *Helenieae*. *Amer. J. Bot.*, **48**: 842–850. [198]

RAY, J. [14, 15, 17, 20, 22, 26, 78, 95]

RECHINGER, K. H. [296]

REEDER, J. R. 1957. The embryo in grass systematics. *Amer. J. Bot.*, **44**: 756–768. [191]

1959. The grass embryo in systematics. *Proc. 9th Int. Bot. Congr. Montreal*, **2**: 321–322. [191]

REESE, G. 1954. Euploidie, Aneuploidie und B-Chromosomen bei *Caltha vulgaris* L. *Planta*, **44**: 203–268. [201]

1961. Karyotype and Plant Geography. In: *Advances in Botany*, 895–900. Toronto Univ. Press. [213, 231]

REHDER, A. [269]

REICHERT, E. T. 1913. Differentiation and specificity of starches in relation to genera, species, etc. *Carnegie Inst. Washington Publ.* No. 173. [236]

REID, E. M. & CHANDLER, M. E. J. 1933. *The London Clay Flora*. London. [56]

RENDLE, A. B. 1930, 1938. *Classification of Flowering Plants*. Ed. 1. Vol. 1, 1930; vol. 2, 1938. [57, 59, 64]

RENSCH, B. 1929. *Daz Prinzip geographischer Rassenkreise und das Problem der Artbildung*. Berlin. [102, 315]

1947. *Neuere Probleme der Abstammungslehre die transpezifische Evolution*. Stuttgart. [51]

1954. *Neuere Probleme der Abstammungslehre die transpezifische Evolution*. Ed. 2. Stuttgart. [129]

1959. *Evolution above the species level*. (Transl.) London. [32, 424, 428]

——— [356]

RHOADES, M. M. & MCCLINTOCK, B. 1935. The cytogenetics of maize. *Bot. Rev.*, **1**: 292–325. [199]

RIBBANDS, C. R. 1955. The scent perception of the honey bee. *Proc. Roy. Soc. Ser. B*, **143**: 367–379. [435, 436]

RICHENS, R. H. 1955. Studies on *Ulmus*. I. The range of variation of the East Anglian Elms. *Watsonia*, **3**: 138–153. [148]

1958. Studies on *Ulmus*. II. The village Elms of Hertfordshire. *Forestry*, **32**: 138–154. [148]

RICK, C. M. & BARTON, D. W. 1954. Cytological and genetical identification of the primary trisomics of the tomato. *Genetics*, **39**: 640–666. [199]

RICKETT, H. W. 1944. The classification of inflorescences. *Bot. Rev.*, **10**: 187–231. [163]

1955. Materials for a dictionary of botanical terms. III. Inflorescences. *Bull. Torrey Bot. Club*, **82**: 419–445. [163]

RIDGWAY, J. L. [322]

RILEY, H. P. & BRYANT, T. R. 1961. The separation of nine species of the *Iridaceae* by paper chromatography. *Amer. J. Bot.*, **48**: 133–137. [254]

RILEY, R. 1956. The influence of the breeding system on the genecology of *Thlaspi alpestre* L. *New Phytol.*, **55**: 319–330. [368]

1960. The diploidisation of polyploid wheat. *Heredity*, **15**: 407. [207 f.n., 358]

ROBERTS, C. & THORNE, R. S. W. 1960. Biochemical classification of yeasts. *Nature*, **188**: 872–873. [233]

ROCHLEDER, F. [234]

RODRIGUEZ, R. L. 1950. A graphic representation of Bessey's taxonomic system. *Madroño*, **10**: 214–218. [43, 61]

ROGERS, D. J. & TANIMOTO, T. T. 1960. A computer programme for classifying plants. *Science*, **132**: 1115–1118. [1, 133]

RÖHRIG, E. 1957. Über die Schwarzkiefer (*Pinus nigra* Arnold) und ihre Formen. I Teil: Die Natürlichen Standorte. *Silvae Genet.*, **6**: 39–53. [400]

ROLLINS, R. C. 1953. Cytogenetical approaches to the study of genera. *Chronica Botanica*, **14** (3): 133–139. [81, 103]

1954. Interspecific hybridisation and its role in plant evolution. *8me Congr. Int. Bot. Paris Rapp. & Comm.*, sects. 9 & 10: 172–180. [483]

1955. The auriculate-leaves species of *Lesquerella* (*Cruciferae*). *Rhodora*, **57**: 241–264. [294]

1958. The genetic evaluation of a taxonomic character in *Dithyrea* (*Cruciferae*). *Rhodora*, **60**: 145–152. [120]

ROSENHEIM, O. [241 f.n.]
ROSS-CRAIG, S. [322]
ROSSER, E. M. 1955. A new British species of *Senecio*. *Watsonia*, **3**: 228–232. [463]
ROTHMALER, W. 1941. Monographie der Gattung *Petrocoptis* A. Br. *Bot. Jahrb.*, **72**: 117–130. [427]
—— 1944. Systematische Einheiten in der Botanik. *Feddes Rep.*, **54**: 1–22. [92, 99, 100[
—— 1954. Terminologie des subdivisions de l'espèce. *8me Congr. Int. Bot. Paris Rapp. & Comm.*, sect. 4: 67–74. [92, 99]
—— 1955. *Allgemeine Taxonomie und Chorologie der Pflanzen.* Ed. 2, Jena. [99, 316–318]
—— [297]
ROUY, G. & FOUCAUD, F. [98, 100]
ROWLEY, G. D. 1961. The naming of hybrids. *Taxon*, **10**: 211–212. [481]
ROWSON, J. M. 1958. Alkaloids and plant taxonomy. *Proc. Linn. Soc. Lond.*, **169**: 212–216. [238]
ROY, R. P. 1962. Some aspects of genetics of polyploids. *Proceedings of the Summer School of Botany, Darjeeling*: 341-348. New Delhi. [214]
ROYEN, A. VON. [284]
RUBTSOV, N. I. [158]
RUDORF-LAURITZEN, M. 1958. The trisomics of *Antirrhinum majus* L. *Proc. 10th Int. Congr. Genet.*, **2** (Abstracts): 243–244. [199]
RUNEMARK, H. 1961. The species and subspecies concepts in sexual flowering plants. *Bot. Notis.*, **114**: 22–32. [145]
RUTISHAUSER, A. [365]
SACHS, J. VON. 1906. *Geschichte der Botanik vom 16. Jahrhundert bis 1860*. (Transl. H. E. F. Garnsey & I. B. Balfour.) Oxford. [126]
SACHS, L. 1952. Chromosome mosaics in experimental amphiploids in the *Triticinae*. *Heredity*, **6**: 157–170. [194]
ST. JOHN, H. 1958. *A text for the application by the case method of the International Code of Botanical Nomenclature*. *Chronica Botanica*, N.S. Pl. Sci. Book No. 31 [292]
SALISBURY, E. J. 1942. *The Reproductive Capacity of Plants*. London. [124, 165 f.n.]
—— 1961. *Weeds and Aliens*. London. [165 f.n.]
SAMPAIO, G. [100]
SATO, D. 1942. Karyotype alteration and phylogeny in *Liliaceae* and allied families. *Jap. J. Bot.*, **12**: 57–161. [204]
—— 1960. The karyotype analysis in *Zingiberales* with special reference to the protokaryotype and stable karyotype. *Sci. Papers Coll. General Ed. Univ. Tokyo*, **10**: 225–243. [204]
SAVAGE, S. [283, 286]
SAVIDGE, J. P. 1960. The experimental taxonomy of European *Callitriche*. *Proc. Linn. Soc. Lond.*, **178**: 128–130. [330]
SAVILLE, D. B. O. 1952. A study of the species of *Cintactia* on *Carex*, *Kobresia* and *Scirpus* in North America. *Canad. J. Bot.*, **30**: 410–435. [257]
—— 1954. The fungi as aids in the taxonomy of flowering plants. *Science*, **120**: 583–585. [256]
—— 1959. Evolution of *Saxifragaceae* from a mycologist's viewpoint. *Proc. 9th Int. Bot. Congr. Montreal*, **2**: 343. [257]
SAVILE, D. B. O. & CALDER, J. A. 1953. Phylogeny of *Carex* in the light of parasitism by the smut fungi. *Canad. J. Bot.*, **31**: 164–174. [257]
SAX, H. J. 1938. The relation between stomata counts and chromosome numbers. *J. Arn. Arb.*, **19**: 437–441. [154]
SCHMALHAUSEN, I. I. 1960. Evolution and cybernetics. *Evol.*, **14**: 509–524. [336 f.n.]
SCHNEIDER, C. K. [269]
SCHNEIDER, I. 1958. Zytogenetische Untersuchungen an Sippen des Polyploid-Komplexes *Achillea millefolium* L. s. lat. (Zur Phylogenie der Gattung *Achillea*, I). *Öst. Bot. Zeitschr.*, **105**: 111–158. [228]
SCHOTSMANN, H. D. 1961. Contibution à l'étude des *Callitriche* du Canton de Neuchâtel. *Bull. Soc. Neuchât. Sci. Nat.*, **84**: 89–101. [204]
SCHRANK, F. VON P. [306]

SCHRATZ, E. 1960. Die Bedeutung neuere Auffassungen des Artbegriffes für die Pharmakognosie. *Planta Medica*, 8 Jahrg.: 282–296. [237, 252]

SCHULZ, O. E. 1936. *Cruciferae*. In: A. Engler, *Die natürlichen Pflanzenfamilien*, ed. 2, **17B**: 227–658. [154, 161]

——— [166, 210]

SCHWANITZ, F. 1952. Einige kritische Bemerkungen zur Methode der Bestimmung der Polyploidie durch Messung der Pollen- und Spaltöffnungsgrösse. *Züchter*, **22**: 273–275. [154]

SCHWARTEN, L. & RICKETT, H. W. [307]

SCHWARZ, O. 1955. Systematische Monographie der Gattung *Cyclamen*. *Feddes Rep.*, **58**: 234–283. [432]

SCOTT, R. A., BARGHOORN, E. S. & LEOPOLD, E. B. 1960. How old are the Angiosperms? *Amer. J. Sci.*, **258A**: 284–299. [56]

SCOTT-MONCRIEFF, R. 1939. The genetics and biochemistry of flower colour variation. *Ergebn. Enzymforsch.*, **8**: 277–306. [241]

SEALY, R. J. 1958. A Revision of the Genus *Camellia*. London. [71]

SÉBASTIEN, C. 1956. Étude du genre *Phillyrea* Tournefort. *Trav. Inst. Sci. Chérif.*, sér. Bot., No. 6. [171]

SEITHE-VON HOFF, A. 1960. Die Haarformen der Gattung *Rhododendron* L. und die Möglichkeit ihrer taxonomische Verwertung. *Bot. Jahrb.*, **79**: 297–393. [155]

SEWARD. A. C. [56]

SEWERTZOFF, A. N. 1931. *Morphologische Gesetzmassigkeiten der Evolution*. Jena. [35]

SHAN, R. H. & CONSTANCE, L. 1951. The genus *Sanicula* (*Umbelliferae*) in the old world and the new. *Univ. Calif. Publ. Bot.*, **25**. [294]

SHARMA, A. K. & BAL, A. K. 1956. A cytological investigation of some members of the family *Cyperaceae*. *Фyton*, **6**: 7–22. [262]

SHARSMITH, H. 1961. The genus *Hesperolinon* (*Linaceae*). *Univ. Calif. Publ. Bot.*, **32** (4): 235–314. [160, 355, 368, 378, 379]

SHAW, H. K. A. 1950. Post-Darwinian development of taxonomy (botany). In: *Lectures on the Development of Taxonomy*. Linn. Soc. Lond. [57]

SHERFF, E. E. 1940. The delimitation of genera from the conservative point of view. *Bull. Torrey Bot. Club*, **67**: 375–380. [104]

SHERMAN, M. 1946. Karyotype evolution: a cytogenetic study of seven species and six interspecific hybrids of *Crepis*. *Univ. Calif. Publ. Bot.*, **18**: 369–408. [199]

SHULL, G. H. 1929. Point of view of genetics. In: Symposium on the significance of taxonomic units and their natural bases. *Proc. Int. Congr. Plant Sci.* Ithaca, N.Y. 1926: 1578–1586. [460]

SHULL, G. H. [379]

SIMPSON, G. G. 1953a. *The major Features of Evolution*. Columbia Univ. Press. [115, 116, 122, 126, 144]

1953b. The Baldwin effect. *Evol.*, **7**: 110–117. [405]

1961. *Principles of Animal Taxonomy*. New York & London. [6, 8, 9, 11, 12, 34, 41–43, 45, 76, 77, 85, 97, 98, 108 f.n., 127, 129, 412, 414]

——— [10, 11, 32, 44, 54]

SINNOTT, E. W. 1960. *Plant Morphogenesis*. New York, Toronto & London. [336, 338, 339]

SINSKAJA, E. N. 1948. *Dynamics of Species*. (In Russian.) Leningrad. [399]

1950. *Kulturnaya Flora S.S.S.R.* XIII. Vypusk I. Moscow & Leningrad. [458]

1958. Investigations in the composition of ecotypical and varietal populations. A brief survey of some of our works published in Russian. *Report, Scottish Pl. Breeding Station*, **1960**: 31–40. [399]

1960. [The most important wild forage plants of the north Caucasus.] In Russian. *Trud. Priklad. Bot. Genet. Selekc.*, **33**: 149–204. [398, 399]

1961 (ed.). *Flora of Cultivated Plants of the USSR*. XIII. *Perennial Leguminous Plants*, Part I. *Medic, Sweetclover, Fenugreek*. Transl. N. Landau. Israel Program for Scientific Translations. Jerusalem. [461 f.n.]

SIRKS, M. J. 1951. The scope of genetics in problems of evolution. *Report Brisbane Meeting Austral. & New Zeal. Ass. Adv. Sci.*, **28**: 170–177. [354]

SIRKS, M. J. 1952. Variability in the concept of species. *Acta Biotheor.*, ser. A, **10**: 11–22. [20, 89]

SKALIŃSKA, M. 1947. Polyploidy in *Valeriana officinalis* L. in relation to its ecology and distribution. *J. Linn. Soc.* (*Bot.*), **53**: 159–186. [220]

1950. Studies in chromosome numbers of Polish angiosperms. *Acta Soc. Bot. Polon.*, **20**: 45–68. [212]

1951. Studies in cytoecology, geographic distribution and evolution of *Valeriana* L. *Bull. Acad. Polon. Sci. Lett.* B, **1**: 149–175. [220]

SKALIŃSKA, M. BANACH-POGAN, E. & WCISŁO, H. 1957. Further studies in chromosome numbers of Polish angiosperms. *Acta Soc. Bot. Polon.*, **26**: 215–246. [212]

SKALIŃSKA, M. & PIOTROWICZ, M. 1959. Further studies in chromosome numbers of Polish angiosperms (Dicotyledons). *Acta Soc. Bot. Polon.*, **28**: 487–529. [212]

SKALIŃSKA, M. & SOKOŁOWSKA-KULCZYCKA, A. 1961. Further additions to chromosome numbers of Polish angiosperms. *Acta Soc. Bot. Polon.*, **30**: 463–489. [212]

SKOVSTED, Å. 1941. Chromosome numbers in the *Malvaceae* II. *C.R. Trav. Lab. Carlsberg*, sér. physiol., **23**: 195–242. [196]

SLEPYAN, E. J. & KISHKOVSKY, A. N. 1960a. Roentgenographic method of studying leaf morphology. *Bot. Zhurn.*, **45**: 695–703. [322]

1960b. On the use of roentgenography in the study of flower structure. *Bot. Zhurn.*, **45**: 1009–1010. [322]

SMIRNOV, E. S. 1924. Sur l'analyse de la distribution et de la corrélation des caractères dans les unités systématiques. (In Russian.) *C.R. Acad. Sci. U.R.S.S.*, 1924, sér. A: 81–84. [35]

1938. Species construction from a taxonomic point of view. (In Russian.) *Zool. Zhurn.*, **17**: 387–418. [35]

SMITH, D. M. 1961. Variation in the tetraploid sunflowers, *Helianthus decapetalus*, *H. hirsutus* & *H. strumosus*. *Recent Advances in Botany*, 878–881. Toronto Univ. Press. [227–229]

SMITH, H. H. 1943. Effects of genome balance, polyploidy, and single extra chromosomes on size in *Nicotiana*. *Genetics*, **28**: 227–236. [199]

1953. Development of morphologically distinct and genetically isolated populations by interspecific hybridisation and selection. *Proc. 9th Int. Congr. Genet.*, 867–870. [473]

SMITH, W. W. 1933. Some aspects of the bearing of cytology on taxonomy. *Proc. Linn. Soc. Lond.*, **145**: 151–181. [208]

SMITH, W. W. & FLETCHER, H. R. 1947. The genus *Primula*: section *Vernales* Pax. *Trans. Bot. Soc. Edinburgh*, **34**: 402–468. [428, 446]

SMITH, W. W. & FORREST, G. 1928. The sections of the genus *Primula*. *Notes Roy. Bot. Gard. Edinburgh*, **16**: 1–50. [149]

SMITH-WHITE, S. 1948. A survey of chromosome numbers in the *Epacridaceae*. *Proc. Linn. Soc. N.S.W.*, **73**: 37–56. [210]

1954. Chromosome numbers in the *Boronieae* (*Rutaceae*) and their bearing on the evolutionary development of the tribe in the Australian flora. *Austral. J. Bot.*, **2**: 287–303. [207–209, 211]

1955a. Chromosome numbers and pollen types in the *Epacridaceae*. *Austral. J. Bot.*, **3**: 48–67. [210]

1955b. The life history and genetic system of *Leucopogon juniperinus*. *Heredity*, **9**: 79–91. [459]

1959. Chromosome evolution in the Australian flora. *Cold Spring Harb. Symp. Quant. Biol.*, **24**: 273–289. [210, 211]

SNEATH, P. H. A. 1957a. The application of computers to taxonomy. *J. Gen. Microbiol.*, **17**: 201–226. [135]

1957b. Some thoughts on bacterial application. *J. Gen. Microbiol.*, **17**: 184–200. [301]

1958. Some aspects of Adansonian classification and of the taxonomic theory of correlated features. *Ann. Microbiol. Enzymol.*, **8**: 261–268. [94]

1961. In: The mathematical assessment of taxonomic similarity including the use of computers. *Taxon*, **10**: 98. [132]

SNEATH, P. H. A. 1962. The construction of taxonomic groups. In: G. C. Ainsworth & P. H. A. Sneath (ed.), *Microbial Classification*, 289–332. (12th Symp. Soc. Exp. Microbiol.) [111, 133, 134, 136, 137]

SNEATH, P. H. A. & SOKAL, R. R. 1962. Numerical taxonomy. *Nature*, **193**: 855–860. [111, 113, 130, 133, 135, 136]

SNEDECOR, G. W. 1956. *Statistical Methods applied to Experiment in Agriculture and Biology*, ed. 5. Ames, Iowa. [357]

SNOAD, B. 1955. Somatic instability of chromosome number in *Hymenocallis calathinum*. *Heredity*, **9**: 129–134. [194]

SNYDER, L. A. 1950. Morphological variability and hybrid development in *Elymus glaucus*. *Amer. J. Bot.*, **37**: 628–636. [446]

1951. Cytology of inter-strains hybrids and the probable origin of variability in *Elymus glaucus*. *Amer. J. Bot.*, **38**: 195–202. [446]

SOKAL, R. R. 1962. Distance as a measure of taxonomic similarity. *Syst. Zool.*, **10**: 70–79. [135]

SOKAL, R. R. & ROHLF, F. J. 1962. The comparison of dendrograms by objective methods. *Taxon*, **11**: 33–40. [24, 137]

SOLEREDER, H. 1885. *Ueber den systematischen Werth der Holzstruktur bei den Dicotyledonen*. München. [168]

1908. *Systematic Anatomy of the Dicotyledons*. (Transl. L. A. Boodle & F. E. Fritsch.) Oxford. [236]

——— [8]

SOÓ, R. 1948. A new sketch of the phylogenetic of the Angiospermous flowering plants. *Acta Geobot. Hung.*, **6** (2): 114–123. [64]

1961. The present aspect of the evolutionary history of *Telomophyta*. (Some ideas concerning the origin of mosses, a new system of *Gymospermae* and a modern phylogenetical tree of *Angiospermae*.) *Ann. Univ. Sci. Budap.*, sect. Biol., **4**: 167–178. [64]

SOÓ, R. & JAVORKA, S. [268]

SÖRENSEN, N. A. [235]

SÖRENSEN, T. 1953. A revision of the Greenland species of *Puccinellia* Parl. *Medd. om Grønl.*, **136**: 1–180. [152]

1958. Sexual chromosome-aberrants in triploid apomictic *Taraxaca*. *Bot. Tidsskr.*, **54**: 1–22. [384]

SÖRENSEN, T. & GUÐJÓNSSON, G. 1946. Spontaneous chromosome-aberrants in apomictic *Taraxaca*. *Biol. Skr. Dan. Vid. Selsk.*, **4**, No. 2. [200]

SPACH, E. [289]

SPORNE, K. R. 1948a. A note on rapid clearing techniques of wide application. *New Phytol.*, **47**: 290–291. [181]

1948b. Correlation and classification in the Angiosperms. *Proc. Linn. Soc. Lond.*, **160**: 40–47. [38, 39, 40]

1949. A new approach to the problem of the primitive flower. *New Phytol.*, **48**: 259–276. [38, 68, 439]

1954. Statistics and the evolution of Dicotyledons. *Evol.*, **8**: 55–64. [38, 68]

1956. The phylogenetic classification of the Angiosperms. *Biol. Rev.*, **31**: 1–29. [38, 43, 68, 69, 122, 126]

1958. Some aspects of floral vascular systems. *Proc. Linn. Soc. Lond.* **19**: 75–84. [180]

——— [244, 245]

SPRAGUE, T. A. 1936. The art of plant description (rev.). *Nature*, **137**: 637–639. [270 f.n.]

1950. The evolution of botanical taxonomy from Theophrastus to Linneaus. In: *Lectures in the Development of Taxonomy*. Linn. Soc. London. [14]

1952. The British forms of *Valeriana officinalis*. *Watsonia*, **2**: 145–147. [220]

1960. The evolution of the Angiosperms with special reference to the Dicotyledons. *Proc. Cotteswold Nat. Field Club*, **32** (3): 29–42. [57, 58, 60]

STADLER, L. J. 1954. The gene. *Science*, **120**: 811–819. [358]

STANILAND, L. N. [322]

STEARN, W. T. 1938. *Epimedium* and *Vancouveria* (*Berberidaceae*), a monograph. *J. Linn. Soc.* (*Bot.*), **51**: 409–535. [294]

1942. Some lesser known monographs on hardy herbaceous plants. *J. Roy. Hort. Soc.*, **67**: 269–303. [269]

1951. Mapping the distribution of species . In: J. E. Lousley (ed.), *The Study of the Distribution of British Plants*, 48–64. Oxford: Bot. Soc. Brit. Is. [308–313]

1954. The earliest illustrated herbal: Codex Aniciae Julianae. *Graphic* (Zürich), **10**: 322–329. [13]

1956. Keys; shapes of leaves, &c. In: F. J. Chittenden (ed.), *Royal Horticultural Dictionary of Gardening*, supplement: 251–253; 318–322. [117, 301]

1957. An introduction to the "Species Plantarum" and cognate botanical works of Carl Linnaeus. Prefixed to C. Linnaeus, *Species Plantarum*, 1753, Ray Society Facsimile, vol. 1. [20, 285, 286]

1959. The background of Linnaeus' contribution to the nomenclature and methods of systematic biology. *Syst. Zool.*, **8**: 4–22. [16, 17, 19–21, 285, 286]

1960. Notes on Linnaeus "Genera Plantarum". Prefixed to C. Linnaeus, *Genera Plantarum*, 1754, Historiae Naturalis Classica Facsimile, v-xxiv. [15, 16, 285, 286]

1961a. Candolle, the family of de. In: P. Gray (ed.), *Encyclopedia of the Biological Sciences*, 183–185. New York. [26]

1961b. Botanical gardens and botanical literature in the eighteenth century. *Catalogue of Botanical Books in the Collection of Rachel McMasters Miller Hunt*, **2**: xlii-cxl. [5, 16, 19, 23, 283, 286]

——— [27, 285, 286, 317 f.n.]

STEBBINS, G. L. 1938. Cytological characteristics associated with the different growth habits in Dicotyledons. *Amer. J. Bot.*, **25**: 189–198. [213]

1942. The role of isolation in the differentiation of plant species. *Biol. Symposia*, **6**: 217–233. [425]

1947. Types of polyploids: their classification, and significance. *Adv. Genetics*, **1**: 403–429. [213, 214]

1950. *Variation and Evolution in Plants*. Columbia Univ. Press. [32, 38, 121, 124, 157, 197, 203, 204, 213, 214, 216, 222, 223, 377, 383–385, 408, 421, 422, 424, 428, 429, 433, 441–443, 446, 473]

1951. Natural selection and the differentiation of Angiosperm evolution. *Evol.*, **5**: 299–324. [22, 69, 107, 121, 440]

1952. Aridity as an evolutionary stimulus. *Amer. Nat.*, **86**: 33–44. [39, 431]

1954. Les processus de l'évolution aux hautes montagnes. In: P. Chouard (ed.), *Étude Botanique de l'Étage Alpin, particulièrement en France*. Bayeux. [40]

1956. Cytogenetics and evolution of the grass family. *Amer. J. Bot.*, **43**: 890–905. [70, 203, 210, 228]

1957. Self fertilisation and population variability in the higher plants. *Amer. Nat.*, **91**: 337–354. [367, 376, 377, 379]

1958. Longevity, habitat and release of genetic variability in the higher plants. *Cold Spring Harb. Symp. Quant. Biol.*, **23**: 365–378. [377]

1959a. The role of hybridisation in evolution. *Proc. Amer. Phil. Soc.*, **103**: 231–251. [420, 450, 462, 464, 465, 468, 469, 472, 473, 475]

1959b. Genes, chromosomes and evolution. In: W. B. Turrill (ed.), *Vistas in Botany*: 258–290. London. [55, 121, 201, 216, 358, 360, 361, 376, 378, 417, 427, 466]

——— [96, 153, 456]

STEBBINS, G. L. & DALY, K. 1961. Changes in the variation pattern of a hybrid population of *Helianthus* over an eight-year period. *Evol.*, **15**: 60–71. [467]

STEBBINS, G. L. & FERLAN, L. 1956. Population variability, hybridisation and introgression in some species of *Ophrys*. *Evol.*, **10**: 32–46. [434]

STEBBINS, G. L., JENKINS, J. A. & WALTERS, M. S. 1953. Chromosomes and phylogeny in the *Compositae*, tribe *Cichorieae*. *Univ. Calif. Publ. Bot.*, **26**: 401–430. [199, 211]

STEBBINS, G. L. & KHUSH, G. S. 1961. Variation in the organization of the stomatal complex in the leaf epidermis of Monocotyledons and its bearing on their phylogeny. *Amer. J. Bot.*, **48**: 51–59. [49, 154]

STEBBINS, G. L. & TOGBY, H. A. 1944. The cytogenetics of hybrids in *Bromus*. I. Hybrids within the section *Ceratochloa*. *Amer. J. Bot.*, **31**: 1–11. [228]

STEBBINS, G. L. & WALTERS, M. S. 1949. Artificial and natural hybrids in the *Gramineae*, tribe *Hordeae*. III. Hybrids involving *Elymus condensatus* and *E. triticoides*. *Amer. J. Bot.*, **36**: 291–301. [475]

STEBBINS, G. L. & ZOHARY, D. 1959. Cytogenetic and evolutionary studies in the genus *Dactylis*. I. The morphology, distribution and interrelationship of the diploid subspecies. *Univ. Calif. Publ. Bot.*, **31** (1): 1–40. [97, 228]

STEENIS, C. G. G. J. VAN. 1948. *Flora Malesiana*. Ser. 1, **4**, iv–lxix. Djakarta. [144, 345–347, 367]

—— 1954. General principles in the design of Floras. *8me Congr. Int. Bot. Paris Rapp. & Comm.*, sects. 2, 4, 5 & 6: 59–66. [293, 296]

—— 1957. Specific and infraspecific delimitation. *Flora Malesiana*, ser. 1, **5** (3): cixvii–ccxxix. Djakarta. [1, 9, 99, 266, 293, 295, 301, 339, 340, 343, 344, 427] [369, 467]

STERN, F. C. 1956. *Snowdrops and Snowflakes: a Study of the Genera Galanthus and Leucojum*. London. [149, 432] [313]

STERN, K. R. 1961. Revision of *Dicentra* (*Fumariaceae*). *Brittonia*, **13**: 1–57. [34, 294]

STEYERMARK, J. A. 1947. Notes on drying plants. *Rhodora*, **49**: 220–227. [265]

STIX, E. 1960. Pollenmorphologische Untersuchungen an Compositen. *Grana Palynologica*, **22**: 41–114. [185]

STRAUB, J. [376]

STRAW, R. M. 1955. Hybridisation, homogamy and sympatric speciation. *Evol.*, **9**: 441–444. [474]

—— 1956a. Adaptive morphology of the *Pentstemon* flower. *Phytomorphology*, **6**: 112–119. [123]

—— 1956b. Floral isolation in *Pentstemon*. *Amer. Nat.*, **90**: 47–53. [474]

SUMMERHAYES, V. S. 1951. *Wild Orchids of Britain*. London [50]

SVENSON, H. K. 1945. On the descriptive method of Linnaeus. *Rhodora*, **47**: 273–302, 363–388. [282, 284–286]

SVOBODA, B. 1953. *Lesni dřeviny*, vol. 1. Praha. [400]

SWAMY, B. G. L. & BAILEY, I. W. 1949. The morphology and relationships of *Cercidiphyllum*. *J. Arn. Arb.*, **30**: 187–210. [168]

SWANSON, L. P. 1958. *Cytology and Cytogenetics*. London. [198, 202, 213, 214, 421]

SYLVESTER-BRADLEY, P. C. (ed.). 1956. *The Species Concept in Palaeontology*. Systematics Assoc. Publ. No. 2. London. [57]

TÄCKHOLM, V. & DRAR, M. [297 f.n.]

TAKHTAJAN, A. L. 1958. *Origins of Angiospermous Plants*. Amer. Inst. Biol. Sci. (Transl. from Russian edition of 1954.) [38, 64, 175]

—— 1959a. *Essays on the evolutionary Morphology of Plants*. Amer. Inst. Biol. Sci. (Transl. from Russian edition of 1954.) [157, 158, 343, 358] [64]

—— 1959b. *Die Evolution der Angiospermen*. Jena. [35, 38, 63, 108]

TARNAVASCHI, I. T. 1948. Die Chromosomenzahlen der Anthophyten-Flora von Rumänien mit einem Ausblick auf das Polyploidie-Problem. *Bull. Grad. Bot. Cluj.*, **28**, suppl. 1: 1–130. [195]

TATEOKA, T. 1962. Starch grains of endosperm in grass systematics. *Bot. Mag. Tokyo*, **75**: 377–383. [236]

TEPFER, S. S. 1953. Floral anatomy and ontogeny in *Aquilegia formosa* var. *truncata* and *Ranunculus repens*. *Univ. Calif. Publ. Bot.*, **25**: 513–648. [179, 180]

TERENTJEV, P. V. 1931. Biometrische Untersuchungen über die morphologischen Merkmale von *Rana ridibunda* Pall. *Biometrika*, **23**: 23–51. [35]

—— 1959. The method of correlation pleiades. *Leningrad State Univ. Vestnik*, **9**: 127–141. (In Russian.) [35]

TERRELL, E. E. 1963. Symbols and terms for morphological intergradation and hybridisation. *Taxon*, **12**: 105–108. [484]

TÉTÉNYI, P. 1958. Proposition à propos de la nomenclature des races chimiques. *Taxon*, **7**: 40–41. [252]

1960. Das Problem der Nomenclatur chemischer Taxa. *Taxon*, **9**: 241–242. [252]

1962. Subspecific chemical taxa of plants. *Bot. Zhurn.*, **47**: 1731–1741. [252]

THEOPHRASTUS, E. [12, 13, 17]

THERMAN, E. 1956. Cytotaxonomy of the tribe *Polygonatae*. *Amer. J. Bot.*, **49**: 134–142. [203]

THOMAS, P. T. 1936. Genotypic control of chromosome size. *Nature*, **138**: 402. [202]

THOMSON, J. MCLEAN. [178]

TIEGHEM, P. VAN. [191]

TIPPO, O. 1938. Comparative anatomy and morphology of the *Moraceae* and their presumed allies. *Bot. Gaz.*, **100**: 1–99. [176]

TISCHLER, G. 1950. *Die Chromosomenzahlen der Gefässpflanzen Mitteleuropas*. The Hague. [195]

TOGBY, H. A. 1943. A cytological study of *Crepis fuliginosa, C. neglecta*, and their F_1 hybrid, and its bearing on the mechanism of phylogenetic reduction in chromosome number. *J. Genetics.*, **45**: 67–111. [199]

TOME, G. A. & JOHNSON, I. J. 1945. Self and cross fertility relationships in *Lotus corniculatus* L. and *L. tenuis* Waldst. & Kit. *J. Amer. Soc. Agron.*, **37**: 1011–1023. [222]

TOMLINSON, P. B. 1960. Seedling leaves in Palms and their morphological significance. *J. Arn. Arb.*, **41**: 414–428. [149, 159]

1961. Morphological and anatomical characteristics of the *Marantaceae*. *J. Linn. Soc.* (*Bot.*), **58**: 55–78. [170]

1962. The leaf base in Palms. Its morphology and mechanical biology. *J. Arn. Arb.*, **43**: 23–50. [145, 151]

TOURNEFORT, J. P. DE. [14, 15, 23]

TOWERS, G. H. N. & GIBBS, R. D. 1953. Lignin chemistry and the taxonomy of higher plants. *Nature*, **172**: 25–26. [183, 247]

TOWNSEND, C. C. 1962. Some notes on *Galeopsis ladanum* L. and *G. angustifolia* Ehrh. ex Hoffm. *Watsonia*, **5**: 143–149. [155]

TROLL, W. [125]

TRONCHET, A. 1930. Recherches sur les types d'organisation le plus répandus de la plantule des dicotylédones. Leurs principales modifications, leurs rapports. *Archiv. Bot.*, **4**: 1–252. [160 f.n.]

TUCKER, J. & SAUER, J. 1958. Aberrant *Amaranthus* populations of the Sacramento-San Joaquin Delta, California. *Madroño*, **14**: 252–261. [474]

TURESSON, G. 1922a. The species and variety as ecological units. *Hereditas*, **3**: 100–113. [54, 388]

1922b. The genotypical response of the plant species to the habitat. *Hereditas*, **3**: 211–350. [54, 388, 389, 404, 449]

1925. The plant species in relation to habitat and climate. *Hereditas*, **6**: 147–236. [388]

1929. Zur Natur und Begrenzung der Arteneinheiten. *Hereditas*, **12**: 323–334. [385, 388]

1931. The geographical distribution of the alpine ecotype of some Eurasiatic plants. *Hereditas*, **15**: 329–346. [388]

1943. Variation in the apomictic microspecies of *Alchemilla vulgaris* L. *Bot. Notiser*, **1943**: 413–427. [366, 385]

1956. Variation in the apomictic microspecies of *Alchemilla vulgaris* L. II. *Bot. Notiser*, **109**: 400–404. [366, 384]

——— [387, 391, 392, 394, 398, 406, 410, 412, 414, 461]

TURRILL, W. B. 1934. The correlation of morphological variation with distribution in some species of *Ajuga*. *New Phytol.*, **33**: 218–230. [447]

1938a. The expansion of taxonomy. *Biol. Rev.*, **13**: 342–373. [4, 82]

1938b. Statistical studies in two populations of *Fraxinus*. *New Phytol.*, **37**: 160–172. [449]

TURRILL, W. B. 1942. Taxonomy and phylogeny. *Bot. Rev.*, **8**: 247–270, 473–532, 655–707. [13, 29, 57, 68]

1946. The ecotype concept. A consideration with appreciation and criticism especially of recent trends. *New Phytol.*, **45**: 34–43. [410]

1948. *British Plant Life*. London. [345]

1950. Modern trends in the classification of plants. *Advancement of Science*, **26**: 1-16. [69]

TUTIN, T. G. 1955. Species problems in plants with reduced floral structure. In: J. E. Lousley (ed.), *Species Studies in the British Flora*: 21–26. Oxford: Bot. Soc. Brit. Is. [145, 152]

1956. Generic criteria in flowering plants. *Watsonia*, 3 : 317–323. [79, 166]

1957. A contribution to the experimental taxonomy of *Poa annua* L. *Watsonia*, **4**: 1–10. [223, 450]

1963. Scale effects and other subjective influences in taxonomy. *Bot. Notiser*, **116**: 122–126. [84]

U, N. 1935. Genome analysis in *Brassica* with special reference to the experimental formation of *B. napus* and peculiar mode of fertilization. *Jap. J. Bot.*, **7**: 389–452. [223]

UPHOF, J. C. T. 1938. Cleistogamic flowers. *Bot. Rev.*, **4**: 21–49. [367]

1962. Plant hairs. In: W. Zimmermann & P. G. Ozenda (ed.), *Handbuch der Pflanzenanatomie* (ed. 2), **4** (5): 1–206. [154]

VAARAMA, A. 1949. Spindle abnormalities and variation in chromosome number in *Ribes nigrum*. *Hereditas*, **35**: 136–162. [194]

VAARTAJA, O. 1954. Photoperiodic ecotypes of trees. *Canad. J. Bot.*, **32**: 392–399. [341]

VALENTINE, D. H. 1949. The units of experimental taxonomy. *Acta Biotheor.*, **9**: 75–88. [216, 390, 420]

1953. Evolutionary aspects of species differences in *Primula*. *Symp. Soc. Exp. Biol.*, **7**: 145–159. [428]

1960. The treatment of apomictic groups in *Flora Europaea*. *Feddes Rep.*, **63**: 114–127. [382, 383, 384]

1961. Evolution in the genus *Primula*. In: P. J. Wanstall (ed.), *A Darwin Centenary*: 71–87. London: Bot. Soc. Brit. Is. [443, 445, 452]

1962. Sections and genera in the *Primulaceae*. *Taxon*, **11** (3): 71–74. [22] [386]

———

VALENTINE, D. H. & LÖVE, Á. 1958. Taxonomic and biosystematic categories. *Brittonia*, **10**: 153–166. [3, 93–96, 99, 101, 381, 382, 385, 390, 445]

VAN CAMPO, M. 1959. Importance des caractères de l'endexine des grains de pollen en systématique et en phylogénie. *Proc. 9th Int. Bot. Congr. Montreal*, **2**: 410–411. [186]

VASSILZENKO, I. T. 1961. On the relationship of ontogenesis and phylogenesis in higher plants. *Bot. Zhurn.*, **46**: 1734–1739. [158]

VAVILOV, N. I. 1922. The law of homologous series in variation. *J. Genetics*, **12**: 47–89. [82]

VERMEULEN, P. 1959. The different structure of the rostellum in *Ophrydeae* and *Neottieae*. *Acta Bot. Neerl.*, **8**: 338–355. [161]

VEUILLET, J. M. 1959. Contribution à l'étude morphologique et anatomique du genre *Elaeoselinum* Koch au Maroc. *Trav. Inst. Sci. Chérif.*, sér. Bot., No. 18. [171]

VICIOSO, C. 1953. *Genisteas españolas, I.* Publ. Inst. For. Inv. Exper. Madrid, núm. 67. [171]

VIDAKOVIĆ, M. 1957. Oblici crnog bora u Jugoslaviji na temelju anatomije iglica. Forms of *Pinus nigra* in Yugoslavia on the basis of anatomical structure of the needles. *Glasnik za Šumske Pokuse*, **13**: 111–248. [169]

VINDT, J. 1960. Monographie des Euphorbiacées du Maroc. Deuxième partie (Anatomie). *Trav. Inst. Sci. Chérif.*, sér. Bot., No. 19. [171]

VOSS, E. G. 1952. The history of keys and phylogenetic trees in systematic biology. *J. Sci. Lab. Denison Univ.*, **43**: 1–25. [43]

WADDINGTON, C. H. 1953. Genetic assimilation of an acquired character. *Evol.*, **7**: 118–126. [405]

1957. *The Strategy of the Genes*. London. [335, 336, 405, 406, 427]

WADDINGTON, C. H. 1960. *The Ethical Animal.* London. [90]

1962. *New Patterns in Genetics and Development.* Columbia Univ. Press. [90, 336 f.n., 406]

WAGENITZ, G. 1955. Pollenmorphologie und Systematik in der Gattung *Centaurea* L. s.l. *Flora,* **142**: 213. [237]

WALKER, E. H. [268, 322 f.n.]

WALKER, T. G. 1958. Hybridisation in some species of *Pteris* L. *Evol.,* **12**: 82–92. [483]

WALKER, S. 1962. Further studies in the genus *Dryopteris:* the origin of *D. clintoniana, D. celsa* and related taxa. *Amer. J. Bot.,* **49**: 497–503. [483]

WALTERS, S. M. 1949. *Alchemilla vulgaris* L. agg. in Britain. *Watsonia,* **1**: 6–18. [148, 365, 366]

1951. The study of plant distribution. In J. E. Lousley (ed.), *The Study of the Distribution of British Plants*: 12-23. Oxford: Bot. Soc. Brit. Is. [426]

1953. *Eleocharis mamillata* Lindb. fil. and allied species. *Ber. Schweiz. Bot. Ges.,* **63**: 271–286. [152]

WALTERS, S. M. 1961. The shaping of Angiosperm taxonomy. *New Phytol.,* **60**: 74–84. [21, 22, 24, 84, 107, 426]

1962. Generic and specific concepts in the European Flora. *Preslia,* **34**: 207–226. [418, 419]

—— [163, 386, 450, 452]

WALTERS, S. M. & PERRING, F. [309]

WANSCHER, J. H. 1948. The interrelation of the different terms used in plant systematics. *K. VetHojsk. Aarsskr.,* **1948**: 1–28. [460]

WARBURG, E. F. 1938. Taxonomy and relationship in the *Geraniales* in the light of their cytology. *New Phytol.,* **37**: 130–159, 189–209. [207, 210, 460]

1949. Cytology and critical groups. In: A. J. Wilmott (ed.), *British Flowering Plants and modern systematic Methods:* 11–16. London: Bot. Soc. Brit. Is. [196]

WARD, G. H. 1953. *Artemisia* section *Seriphidium* in North America, a cytotaxonomic study. *Contr. Dudley Herb.,* **4**: 155–205. [228]

1959. Adaptation and modification in a polyploid complex. *Proc. 9th Int. Bot. Congr. Montreal,* **2**: 422. [228]

WARDLAW, C. W. 1952. *Phylogeny and Morphogenesis.* London. [338]

WAREING, P. F. [340]

WARMING, J. E. B. [58]

WATSON, L. 1962. The taxonomic significance of stomatal distribution and morphology in *Epacridaceae. New Phytol.,* **61**: 36–40. [153]

WATSON, P. J. & CLAUSEN, J. 1961. Phenotypic responses to contrasting environments in the genus *Poa. Report, Scottish Plant Breeding Station,* **1961**: 64–78. [407]

WATSON, W. C. R. 1958. *Handbook of the Rubi of Great Britain and Ireland.* Cambridge Univ. Press. [384]

WEATHERBY, C. P. 1942. Subspecies. *Rhodora,* **44**: 151–167. [100]

WEBB, D. A. [271 f.n.]

WEEVERS, T. 1943. The relation between taxonomy and chemistry of plants. *Blumea,* **5**: 412–422. [235]

WEIDITZ, H. [13]

WENDELBO, P. 1961a. Studies in the *Primulaceae,* I. A monograph of the genus *Dionysia. Årbok Univ. Bergen, 1961, Mat.-Naturv. serie,* No. 3. [146, 149]

1961b. Studies in *Primulaceae.* III. On the genera related to *Primula* with special reference to their pollen morphology. *Årbok Univ. Bergen, 1961, Mat.-Naturv. serie,* No. 19. [187]

WENT, F. W. 1957. *The experimental Control of Plant Growth, with special Reference to the Earhart Plant Research Laboratory at the California Institute of Technology.* Waltham, Mass. [340]

WERNHAM, H. F. 1912. Floral evolution: with particular reference to the sympetalous Dicotyledons; IX. Summary and conclusion. *New Phytol.,* **11**: 373–397. [121]

WETTSTEIN, R. VON. 1898. *Grundzüge der geographisch-morphologischen Methode der Pflanzensystematik.* Jena. [99, 315]

WETTSTEIN R. VON. 1896. Die europäischen Arten der Gattung *Gentiana* aus der Sektion *Endotricha* Froel. und ihr entwicklungsgeschichtlicher Zusammenanhang. *Denkschr. Math.-Nat. Cl. Akad. Wiss.*, **64**: 1–74. [315]

1930–1935. *Handbuch der systematischen Botanik*, Ed. 4. Leipzig & Wien. [315]

——— [58, 64, 92]

WHITAKER, T. W. 1944. The inheritance of certain characters in a cross of two American species of *Lactuca*. *Bull. Torrey Bot. Club*, **71**: 347–355. [431]

WHITE, F. 1962. Geographic variation and speciation in Africa with particular reference to *Diospyros*. *Systematics Assoc. Publ.*, No. 4: 71–103. London. [310, 311]

WHITEHEAD, F. H. 1951. Ecology of the altipiano of Monte Maiella, Italy. *J. Ecol.*, **39**: 330–355. [344]

1954. An example of taxonomic discrimination by biometric method. *New Phytol.*, **53**: 496–510. [138]

1955. Taxonomic studies in the genus *Cerastium* I. *C. atrovirens*, *C. pumilum*, *C. semidecandrum*. *Watsonia*, **3**: 213–227. [138]

1962. Experimental studies of the effect of wind on plant growth and anatomy. II. *Helianthus annuus*. *New Phytol.*, **61**: 59–62. [344]

WHITEHOUSE, H. L. K. 1950. Multiple-allelomorph incompatibility of pollen and style in the evolution of the Angiosperms. *Ann. Bot.* (N.S.), **14**: 199–216. [372]

WHITMORE, T. 1962a. Studies in systematic bark morphology. I. Bark morphology in *Dipterocarpaceae*. II. General features of bark contruction in *Dipterocarpaceae*. *New Phytol.*, **61**: 191–220. [147 f.n.]

1962b. Studies in systematic bark morphology. III. Bark taxonomy in *Diptero-carpaceae*. *Gardens' Bull. Singapore*, **19**: 321–371. [147 f.n.]

WILKIE, J. S. 1962. Some reasons for the rediscovery of Mendel's work in the first years of the present century. *Brit. J. Hist. Sci.*, **1**: 5–17. [31]

WILKINS, D. A. 1959. Sampling for genecology. *Report, Scottish Plant Breeding Station*, **1959**: 92–96. [406, 407]

1960a. The measurement and genetical analysis of lead tolerance in *Festuca ovina*. *Report, Scottish Plant Breeding Station*, **1960**: 85–98. [398]

1960b. Recognizing adaptive variants. *Proc. Linn. Soc. Lond.*, **171**: 122–126. [409]

WILLDENOW, C. L. [281]

WILLIAMS, F. N. [116]

WILLIAMS, W. T. & LAMBERT, J. M. 1961. Multivariate methods in taxonomy. *Taxon*, **10**: 205–211. [140]

WILLIS, J. C. 1948. *A Dictionary of the Flowering Plants and Ferns*, Ed. 6. Cambridge. • [59, 270]

——— [22, 315]

WILLKOMM, H. M. & LANGE, J. M. [303]

WILMOTT, A. J. 1949. Infraspecific categories of variation. In: A. J. Wilmott (ed.), *British Flowering Plants and modern systematic Methods*: 28–45. London: Bot. Soc. Brit. Is. [49, 405, 415]

1950a. Systematic botany from Linnaeus to Darwin. In: *Lectures on the Development of Taxonomy*. Linn. Soc. London. [126]

1950b. A new method for the identification and study of critical groups. *Proc. Linn. Soc. Lond.*, **162**: 83–98. [329]

WILSON, C. L. 1953. The telome theory. *Bot. Rev.*, **19**: 417–437. [180]

WILSON, J. Y. 1959. Vegetative reproduction in the bluebell, *Endymion nonscriptus* (L.) Garcke. *New Phytol.*, **58**: 155–163. [364]

WIMMER, F. E. 1957. *Campanulaceae-Lobelioideae*. *Pflanzenreich*, **276f** (IV. 106). [294]

WINGE, O. 1940. Taxonomic and evolutionary studies in *Erophila* based on cytogenetic investigations. *C.R. Trav. Lab. Carlsberg*, sér. physiol., **24**: 41–74. [378]

WIT, H. C. D. DE. [284]

WITTWER, S. H. & HELLYER, I. G. 1954. Chemical induction of male sterility in Cucurbits. *Science*, **120**: 893–894. [376]

WODEHOUSE, R. P. 1935. *Pollen Grains*. New York. [183]

1936a. Evolution of pollen grains. *Bot. Rev.*, **2**: 67–84. [183]

WODEHOUSE, R. P. 1936b. Pollen grains in the identification and classification of plants, viii. *Amer. J. Bot.*, **23**: 535. [183]

WOLF, E. & ZERRAHN-WOLF, G. 1935. The effect of light intensity, area and flicker-frequency in the visual reactions of the honey bee. *J. Gen. Physiol.*, **18**: 853–863. [435]

WOODGER, J. H. 1952. *Biology and Language*. Cambridge Univ. Press. [76]

WOODSON, R. E. 1941. The North American *Asclepiadaceae*, I. Perspective of the genera. *Ann. Missouri Bot. Gard.*, **28**: 193–248. [323, 433]

1947. Some dynamics of leaf variation in *Asclepias tuberosa*. *Ann. Missouri Bot. Gard.*, **34**: 353–432. [277, 318]

WRIGHT, C. W. 1950. Palaeontologic classification. *J. Palaeontol.*, **24**: 746–748. [454]

WRIGHT, J. W. 1962. *A Bibliography of Forest Genetics and Forest Tree Improvement, 1958–59*. U.S. Department of Agriculture, Miscellaneous Publ., No. 906. [483 f.n.]

WRIGHT, S. 1943. Isolation by distance. *Genetics*, **28**: 114–138. [361]

——— [426]

WYCHERLEY, P. R. 1953. Proliferation of spikelets in British grasses. *Watsonia*, **3**: 41. [383]

YAKOVLEV, M. S. 1950. [The structure of endosperm and embryo in cereals as a systematic feature.] *Trud. Bot. Inst. Akad. Nauk Armyansk SSR.*, **1**: 121–218. [191]

YAKOVLEV, M. S. & YOFFE, M. D. 1959. On a new type of embryogenesis in *Paeonia* L. *Proc. 9th Int. Bot. Congr. Montreal*, **2**: 437–438. [191]

YAMPOLSKY, C. & YAMPOLSKY, H. 1922. Distribution of sex forms in the phanerogamic flora. *Bibl. Genet.*, **3**: 1–62. [370]

ZAHUR, M. Z. 1959. Comparative study of secondary phloem of 423 species of woody dicotyledons belonging to 85 families. *Cornell Univ. Agr. Stat. Mem.*, **358**: 1–160. [175]

ZÉSIGER, F. 1961. Recherches cytotaxinomiques sur les Joubarbes (genres *Sempervivum* L. et *Jovibarba* Opiz.). *Ber. Scweiz Bot. Ges.*, **71**: 113–117.

ZIMMERMAN, W. 1930. *Die Phylogenie der Pflanzen*. Jena.

1959. *Die Phylogenie der Pflanzen*. Stuttgart.

ZOHARY, D. & FELDMAN, M. 1962. Hybridisation between amphidiploids and the evolution of polyploids in the wheat (*Aegilops-Triticum*) group. *Evol.*, **16**: 44–61. [463 f.n., 465]

ZOHARY, M. 1950. Evolutionary trends in the fruiting head of *Compositae*. *Evol.*, **4**: 103–109. [124]

1956. The genus *Tamarix* in Israel. *Trop. Woods*, No. 104: 24–60. [161]

SUPPLEMENTARY BIBLIOGRAPHY

Since the completion of this work in 1963 a large amount of relevant literature has been published. We have listed below a small selection of the more important books, reviews or general papers that have come to our notice, in an attempt to bring our bibliography up to date as regards the major developments in taxonomy. In addition a few major works which were omitted through ignorance or neglect are given.

ALSTON, R. E. & TURNER, B. L. 1963. *Biochemical Systematics.* New Jersey.

[ASLIB]. 1962. *Classification: an interdisciplinary problem. Aslib Proc.*, **14**: 222-262.

BENNET, E. 1964. Historical perspectives in genecology. *Record Scottish Pl. Breed. Station*, **1964**: 49-115.

CONSTANCE, L. 1964. Systematic Botany—an unending synthesis. *Taxon*, **13**: 257-273.

DAVIS, P. H. & CULLEN, J. 1965. *The identification of Flowering Plant families.* Edinburgh & London.

DU PRAW, E. J. 1965. Non-Linnaean taxonomy and the systematics of honeybees. *Syst. Zool.*, **14**: 1-24.

EHRLICH, P. R. 1964. Some axioms of taxonomy. *Syst. Zool.*, **13**: 109-123.

ERDTMAN, H. 1963. Some aspects of chemotaxonomy. *Pure and Applied Chemistry*, **6**: 679-708.

FOSBERG, F. R. & SACHET, MARIE-HÉLÈNE. 1965. *Manual for Tropical Herbaria. Regn. Veg.*, **39**. Utrecht.

GRANT, V. 1963. *The Origin of Adaptations.* New York & London.

1964. The biological composition of a taxonomic species in *Gilia. Adv. Gen.*, **12**: 281-328.

GUINOCHET, M. 1955. *Logique et dynamique du peuplement végétal.* Paris.

HEGNAUER, R. 1962. *Chemotaxonomie der Pflanzen, eine Übersicht über die Verbreitung und die systematische Bedeutung der Pflanzenstoffe.* Band **1**. *Thallophyten, Bryophyten, Pteridophyten und Gymnospermen.* Basel & Stuttgart.

1963. *Chemotaxonomie der Pflanzen . . .* Band **2**. *Monocotyledoneae.* Basel & Stuttgart.

1965. *Chemotaxonomie der Pflanzen . . .* Band **3**. *Dicotyledoneae: Acanthaceae-Cyrillaceae.* Basel & Stuttgart.

HENNIG, W. 1950. *Grundzuge einer Theorie der phylogenetischen Systematik.* Berlin.

HESLOP-HARRISON, J. 1964. Forty years of genecology. *Adv. Ecol. Res.*, **2**: 159-247.

HEYWOOD, V. H. & MCNEILL, J. 1964. (eds.). Phenetic and Phylogenetic Classification. *Systematics Assoc. Publ.*, No. 6. London.

HUTCHISON, J. 1964. *The Genera of Flowering Plants* (*Angiospermae*). *Dicotyledones*, vol. **1**. Oxford.

LEONE, C. A. 1964 (ed.). *Taxonomic Biochemistry and Serology.* New York.

LEPPIK, E. E. 1963. Fossil evidence of floral evolution. *Lloydia*, **26**: 91-115.

MAYR, E. 1963. *Animal Species and Evolution.* London.

MELCHIOR, H. 1964. A. ENGLER'S *Syllabus der Pflanzenfamilien.* Band II. *Angiospermen.* Berlin-Nikolassee.

MICHENER, C. D. 1963. Some future developments in taxonomy. *Syst. Zool.*, **12**: 151-172.

REMANE, A. 1956. *Die Grundlagen des natürlichen Systems, der vergleichenden Anatomie und der Phylogenetik. Theoretische Morphologie und Systematik.* I. ed. 2. Leipzig.

SIMPSON, G. G., ROE, A. & LEWONTIN, R. C. 1960. *Quantitative Zoology*, ed. 2. New York.

SOKAL, R. R. 1962. Typology and empiricism in taxonomy. *J. theoret. Biol.*, 3: 230-267.

SOKAL, R. R. & SNEATH, P. H. A. 1963. *Principles of Numerical Taxonomy*. San Francisco & London.

SOÓ, R. 1963. *Fejlodéstörténeti Növényrendszertan*. Budapest.

STEARN, W. T. 1965. *Botanical Latin*. London.

SWAIN, T. 1963. (ed.). *Chemical Plant Taxonomy*. London & New York.

THORNE, R. F. 1963. Some problems and guiding principles of angiosperm phylogeny. *Amer. Nat*, **97**: 287-305.

TRAUB, H. P. 1964. *Lineagics*. Limited 1st. ed. The American Plant Life Society, La Jolla, California.

TROLL, W. 1965. Die Infloreszenzen, Typologie und Stellung im Aufblau des Vegetationskörpers. Band **1**. Jena & Stuttgart.

TURRILL, W. B. 1964 (ed.). *Recent Researches in plant taxonomy*. *Vistas in Botany*, **4**.

TUTIN, T. G., HEYWOOD, V. H., BURGES, N. A., VALENTINE, D. H., WALTERS, S. M. & WEBB, D. A. 1964. *Flora Europaea* vol. **1**. Cambridge Univ. Press.

ZIMMERMAN, W. 1963. Gibt es ausser dem phylogenetischen System "natürliche" Systeme der Organismen ? *Biol. Zentralbl.*, **82**: 525-568.

1965. *Die Telomtheorie*. Stuttgart.

INDEX OF SUBJECTS AND ORGANISMS